D1621320

Applied Mathematics
and Modeling for
Chemical Engineers

Wiley Series In Chemical Engineering

Bird, Armstrong, and Hassager: *Dynamics of Polymeric Liquids*, Vol. I *Fluid Mechanics*

Bird, Hassager, Armstrong, and Curtiss: *Dynamics of Polymeric Liquids*, Vol. II *Kinetic Theory*

Bird, Stewart, and Lightfoot: *Transport Phenomena*

Brownell and Young: *Process Equipment Design*: *Vessel Design*

Felder and Rousseau: *Elementary Principles of Chemical Processes*, 2nd Edition

Franks: *Modeling and Simulation in Chemical Engineering*

Froment and Bischoff: *Chemical Reactor Analysis and Design*

Henley and Seader: *Equilibrium—Stage Separation Operations in Chemical Engineering*

Hill: *An Introduction to Chemical Engineering Kinetics and Reactor Design*

Jawad and Farr: *Structural Analysis and Design of Process Equipment*

Levenspiel: *Chemical Reaction Engineering*, 2nd Edition

Reklaitis: *Introduction to Material and Energy Balances*

Rice and Do: *Applied Mathematics and Modeling for Chemical Engineers*

Sandler: *Chemical and Engineering Thermodynamics*

Smith and Corripio: *Principles and Practice of Automatic Process Control*

Smith and Missen: *Chemical Reaction Equilibrium Analysis*

Ulrich: *A Guide to Chemical Engineering Process Design and Economics*

Welty, Wicks, and Wilson: *Fundamentals of Momentum, Heat, and Mass Transfer*, 3rd Edition

Applied Mathematics and Modeling for Chemical Engineers

Richard G. Rice
Louisiana State University

Duong D. Do
University of Queensland
St. Lucia, Queensland, Australia

John Wiley & Sons, Inc.
New York • Chichester • Brisbane • Toronto • Singapore

Acquisitions Editor	Cliff Robichaud
Marketing Manager	Susan J. Elbe
Senior Production Editor	Savoula Amanatidis
Designer	Pedro A. Noa
Cover Designer	Ben Arrington
Manufacturing Manager	Lori Bulwin
Illustration Coordinator	Eugene P. Aiello

This book was typeset in Times Roman by Science Typographers, Inc. and printed and bound by Hamilton Printing Company. The cover was printed by Phoenix Color Corp.

Library of Congress Cataloging-in-Publication Data

Rice, Richard G.
   Applied mathematics and modeling for chemical engineers / Richard
G. Rice. Duong D. Do.

      p.   cm.—(Wiley series in chemical engineering)
   Includes bibliographical references and index.
   ISBN 0-471-30377-1
      1. Differential equations.   2. Chemical processes—Mathematical
models.   3. Chemical engineering—Mathematics.   I. Duong, D. Do.
II. Title.   III. Series.
QA371.R37   1994
660'.2842'015118—dc20   94-5245
                                                                                          CIP

Printed in the United States of America.

10 9 8 7 6 5 4 3 2

To Judy, Todd, Andrea, and William,
for making it all worthwhile.
RGR

To An and Binh, for making
my life full.
DDD

Preface

The revolution created in 1960 by the publication and widespread adoption of the textbook *Transport Phenomena* by Bird et al. ushered in a new era for chemical engineering. This book has nurtured several generations on the importance of problem formulation by elementary differential balances. Modeling (or idealization) of processes has now become standard operating procedure, but, unfortunately, the sophistication of the modeling exercise has not been matched by textbooks on the solution of such models in quantitative mathematical terms. Moreover, the widespread availability of computer software packages has weakened the generational skills in classical analysis.

The purpose of this book is to attempt to bridge the gap between classical analysis and modern applications. Thus, emphasis is directed in Chapter 1 to the proper representation of a physicochemical situation into correct mathematical language. It is important to recognize that if a problem is incorrectly posed in the first instance, then any solution will do. The thought process of "idealizing," or approximating an actual situation, is now commonly called "modeling." Such models of natural and man-made processes can only be fully accepted if they fit the reality of experiment. We try to give emphasis to this well-known truth by selecting literature examples, which sustain experimental verification.

Following the model building stage, we introduce classical methods in Chapters 2 and 3 for solving ordinary differential equations (ODE), adding new material in Chapter 6 on approximate solution methods, which include perturbation techniques and elementary numerical solutions. This seems altogether appropriate, since most models are approximate in the first instance. Finally, because of the propensity of staged processing in chemical engineering, we introduce analytical methods to deal with important classes of finite-difference equations in Chapter 5.

In Chapters 7 to 12 we deal with numerical solution methods, and partial differential equations (PDE) are presented. Classical techniques, such as combination of variables and separation of variables, are covered in detail. This is followed by Chapter 11 on PDE transform methods, culminating in the generalized Sturm–Liouville transform. This allows sets of PDEs to be solved as handily as algebraic sets. Approximate and numerical methods close out the treatment of PDEs in Chapter 12.

This book is designed for teaching. It meets the needs of a modern undergraduate curriculum, but it can also be used for first year graduate students. The homework problems are ranked by numerical subscript or an asterisk. Thus, subscript 1 denotes mainly computational problems, whereas subscripts 2 and 3 require more synthesis and analysis. Problems with an asterisk are the most difficult and are suited for graduate students. Chapters 1 through 6 comprise a suitable package for a one-semester, junior level course (3 credit hours). Chapters 7 to 12 can be taught as a one-semester course for advanced senior or graduate level students.

Academics find increasingly less time to write textbooks, owing to demands on the research front. RGR is most grateful for the generous support from the faculty of the Technical University of Denmark (Lynbgy), notably Aa. Fredenslund and K. Ostergaard, for their efforts in making sabbatical leave there in 1991 so successful, and extends a special note of thanks to M. Michelsen for his thoughtful reviews of the manuscript and for critical discussions on the subject matter. He also acknowledges the influence of colleagues at all the universities where he took residence for short and lengthy periods including: University of Calgary, Canada; University of Queensland, Australia; University of Missouri, Columbia; University of Wisconsin, Madison; and of course Louisiana State University, Baton Rouge.

Richard G. Rice
Louisiana State University
September 1994

Duong D. Do
University of Queensland
September 1994

Contents

Chapter 1   Formulation of Physicochemical Problems   **3**

    1.1   Introduction   3
    1.2   Illustration of the Formulation Process (Cooling of Fluids)   4
    1.3   Combining Rate and Equilibrium Concepts (Packed Bed Adsorber)   10
    1.4   Boundary Conditions and Sign Conventions   13
    1.5   Summary of the Model Building Process   16
    1.6   Model Hierarchy and Its Importance in Analysis   17
    1.7   References   28
    1.8   Problems   28

**Chapter 2   Solution Techniques for Models Yielding Ordinary
Differential Equations (ODE)**   **37**

    2.1   Geometric Basis and Functionality   37
    2.2   Classification of ODE   39
    2.3   First Order Equations   39
        2.3.1   Exact Solutions   41
        2.3.2   Equations Composed of Homogeneous Functions   43
        2.3.3   Bernoulli's Equation   45
        2.3.4   Riccati's Equation   45
        2.3.5   Linear Coefficients   49
        2.3.6   First Order Equations of Second Degree   50
    2.4   Solution Methods for Second Order Nonlinear Equations   51
        2.4.1   Derivative Substitution Method   52
        2.4.2   Homogeneous Function Method   58
    2.5   Linear Equations of Higher Order   61
        2.5.1   Second Order Unforced Equations:
                Complementary Solutions   63
        2.5.2   Particular Solution Methods for Forced Equations   72
        2.5.3   Summary of Particular Solution Methods   88
    2.6   Coupled Simultaneous ODE   89
    2.7   Summary of Solution Methods for ODE   96
    2.8   References   97
    2.9   Problems   97

Chapter 3   Series Solution Methods and Special Functions   104

   3.1   Introduction to Series Methods   104
   3.2   Properties of Infinite Series   106
   3.3   Method of Frobenius   108
          3.3.1   Indicial Equation and Recurrence Relation   109
   3.4   Summary of the Frobenius Method   126
   3.5   Special Functions   127
          3.5.1   Bessel's Equation   128
          3.5.2   Modified Bessel's Equation   130
          3.5.3   Generalized Bessel Equation   131
          3.5.4   Properties of Bessel Functions   135
          3.5.5   Differential, Integral and Recurrence Relations   137
   3.6   References   141
   3.7   Problems   142

Chapter 4   Integral Functions   148

   4.1   Introduction   148
   4.2   The Error Function   148
          4.2.1   Properties of Error Function   149
   4.3   The Gamma and Beta Functions   150
          4.3.1   The Gamma Function   150
          4.3.2   The Beta Function   152
   4.4   The Elliptic Integrals   152
   4.5   The Exponential and Trigonometric Integrals   156
   4.6   References   158
   4.7   Problems   158

Chapter 5   Staged-Process Models: The Calculus of Finite Differences   164

   5.1   Introduction   164
          5.1.1   Modeling Multiple Stages   165
   5.2   Solution Methods for Linear Finite Difference Equations   166
          5.2.1   Complementary Solutions   167
   5.3   Particular Solution Methods   172
          5.3.1   Method of Undetermined Coefficients   172
          5.3.2   Inverse Operator Method   174
   5.4   Nonlinear Equations (Riccati Equation)   176
   5.5   References   179
   5.6   Problems   179

Chapter 6   Approximate Solution Methods for ODE: Perturbation Methods   184

   6.1   Perturbation Methods   184
          6.1.1   Introduction   184
   6.2   The Basic Concepts   189
          6.2.1   Gauge Functions   189
          6.2.2   Order Symbols   190
          6.2.3   Asymptotic Expansions and Sequences   191
          6.2.4   Sources of Nonuniformity   193

6.3   The Method of Matched Asymptotic Expansion   195
       6.3.1   Matched Asymptotic Expansions for Coupled Equations   202
6.4   References   207
6.5   Problems   208

Chapter 7   Numerical Solution Methods (Initial Value Problems)   225

7.1   Introduction   225
7.2   Type of Method   230
7.3   Stability   232
7.4   Stiffness   243
7.5   Interpolation and Quadrature   246
7.6   Explicit Integration Methods   249
7.7   Implicit Integration Methods   252
7.8   Predictor-Corrector Methods and Runge–Kutta Methods   253
       7.8.1   Predictor-Corrector Methods   253
       7.8.2   Runge–Kutta Methods   254
7.9   Extrapolation   258
7.10   Step Size Control   258
7.11   Higher Order Integration Methods   260
7.12   References   260
7.13   Problems   261

**Chapter 8   Approximate Methods for Boundary Value Problems:
               Weighted Residuals   268**

8.1   The Method of Weighted Residuals   268
       8.1.1   Variations on a Theme of Weighted Residuals   271
8.2   Jacobi Polynomials   285
       8.2.1   Rodrigues Formula   285
       8.2.2   Orthogonality Conditions   286
8.3   Lagrange Interpolation Polynomials   289
8.4   Orthogonal Collocation Method   290
       8.4.1   Differentiation of a Lagrange Interpolation Polynomial 291
       8.4.2   Gauss–Jacobi Quadrature   293
       8.4.3   Radau and Lobatto Quadrature   295
8.5   Linear Boundary Value Problem—
       Dirichlet Boundary Condition   296
8.6   Linear Boundary Value Problem—
       Robin Boundary Condition   301
8.7   Nonlinear Boundary Value Problem—
       Dirichlet Boundary Condition   304
8.8   One-Point Collocation   309
8.9   Summary of Collocation Methods   311
8.10   Concluding Remarks   313
8.11   References   313
8.12   Problems   314

Chapter 9   Introduction to Complex Variables and Laplace Transforms   **331**

9.1   Introduction   331
9.2   Elements of Complex Variables   332
9.3   Elementary Functions of Complex Variables   334
9.4   Multivalued Functions   335
9.5   Continuity Properties for Complex Variables: Analyticity   337
    9.5.1   Exploiting Singularities   341
9.6   Integration: Cauchy's Theorem   341
9.7   Cauchy's Theory of Residues   345
    9.7.1   Practical Evaluation of Residues   347
    9.7.2   Residues at Multiple Poles   349
9.8   Inversion of Laplace Transforms by Contour Integration   350
    9.8.1   Summary of Inversion Theorem for Pole Singularities   353
9.9   Laplace Transformations: Building Blocks   354
    9.9.1   Taking the Transform   354
    9.9.2   Transforms of Derivatives and Integrals   357
    9.9.3   The Shifting Theorem   360
    9.9.4   Transform of Distribution Functions   361
9.10   Practical Inversion Methods   363
    9.10.1   Partial Fractions   363
    9.10.2   Convolution Theorem   366
9.11   Applications of Laplace Transforms for Solutions of ODE   368
9.12   Inversion Theory for Multivalued Functions:
    The Second Bromwich Path   378
    9.12.1   Inversion when Poles and Branch Points Exist   382
9.13   Numerical Inversion Techniques   383
    9.13.1   The Zakian Method   383
    9.13.2   The Fourier Series Approximation   388
9.14   References   390
9.15   Problems   390

Chapter 10   Solution Techniques for Models Producing PDEs   **397**

10.1   Introduction   397
    10.1.1   Classification and Characteristics of Linear Equations   402
10.2   Particular Solutions for PDEs   405
    10.2.1   Boundary and Initial Conditions   406
10.3   Combination of Variables Method   409
10.4   Separation of Variables Method   420
    10.4.1   Coated Wall Reactor   421
10.5   Orthogonal Functions and Sturm–Liouville Conditions   426
    10.5.1   The Sturm–Liouville Equation   426
10.6   Inhomogeneous Equations   434
10.7   Applications of Laplace Transforms for Solutions of PDEs   443
10.8   References   454
10.9   Problems   455

Chapter 11   Transform Methods for Linear PDEs   **486**

    11.1   Introduction   486
    11.2   Transforms in Finite Domain: Sturm–Liouville Transforms   487
            11.2.1 Development of Integral Transform Pairs   487
            11.2.2 The Eigenvalue Problem and the Orthogonality
                     Condition   494
            11.2.3 Inhomogeneous Boundary Conditions   504
            11.2.4 Inhomogeneous Equations   511
            11.2.5 Time-Dependent Boundary Conditions   513
            11.2.6 Elliptic Partial Differential Equations   516
    11.3   Generalized Sturm–Liouville Integral Transform   521
            11.3.1 Introduction   521
            11.3.2 The Batch Adsorber Problem   521
    11.4   References   537
    11.5   Problems   538

Chapter 12   Approximate and Numerical Solution Methods for PDEs   **546**

    12.1   Polynomial Approximation   546
    12.2   Singular Perturbation   562
    12.3   Finite Difference   572
            12.3.1 Notations   573
            12.3.2 Essence of the Method   574
            12.3.3 Tridiagonal Matrix and the Thomas Algorithm   576
            12.3.4 Linear Parabolic Partial Differential Equations   578
            12.3.5 Nonlinear Parabolic Partial Differential Equations 586
            12.3.6 Elliptic Equations   588
    12.4   Orthogonal Collocation for Solving PDEs   592
            12.4.1 Elliptic PDE   592
            12.4.2 Parabolic PDE: Example 1   598
            12.4.3 Coupled Parabolic PDE: Example 2   600
    12.5   Orthogonal Collocation on Finite Elements   603
    12.6   References   615
    12.7   Problems   616

Appendix A: Review of Methods for Nonlinear Algebraic Equations   **630**

    A.1   The Bisection Algorithm   630
    A.2   The Successive Substitution Method   632
    A.3   The Newton–Raphson Method   635
    A.4   Rate of Convergence   639
    A.5   Multiplicity   641
    A.6   Accelerating Convergence   642
    A.7   References   643

Appendix B: Vectors and Matrices   **644**

    B.1   Matrix Definition   644
    B.2   Types of Matrices   646
    B.3   Matrix Algebra   647

B.4   Useful Row Operations   649
B.5   Direct Elimination Methods   651
       B.5.1   Basic Procedure   651
       B.5.2   Augmented Matrix   652
       B.5.3   Pivoting   654
       B.5.4   Scaling   655
       B.5.5   Gauss Elimination   656
       B.5.6   Gauss–Jordan Elimination   656
       B.5.7   LU Decomposition   658
B.6   Iterative Methods   659
       B.6.1   Jacobi Method   659
       B.6.2   Gauss–Seidel Iteration Method   660
       B.6.3   Successive Overrelaxation Method   660
B.7   Eigenproblems   660
B.8   Coupled Linear Differential Equations   661
B.9   References   662

Appendix C: Derivation of the Fourier–Mellin Inversion Theorem   663

Appendix D: Table of Laplace Transforms   671

Appendix E: Numerical Integration   676

E.1   Basic Idea of Numerical Integration   676
E.2   Newton Forward Difference Polynomial   677
E.3   Basic Integration Procedure   678
       E.3.1   Trapezoid Rule   678
       E.3.2   Simpson's Rule   680
E.4   Error Control and Extrapolation   682
E.5   Gaussian Quadrature   683
E.6   Radau Quadrature   687
E.7   Lobatto Quadrature   690
E.8   Concluding Remarks   693
E.9   References   693

Nomenclature   694

Postface   698

Index   701

PART ONE

Myself when young did eagerly frequent
Doctor and Saint, and heard great argument
About it and about: but evermore
Came out by the same door as in I went.

Rubáiyát of Omar Khayyám, XXX.

Chapter 1

Formulation of Physicochemical Problems

1.1 INTRODUCTION

Modern science and engineering requires high levels of qualitative logic before the act of precise problem formulation can occur. Thus, much is known about a physicochemical problem beforehand, derived from experience or experiment (i.e., empiricism). Most often, a theory evolves only after detailed observation of an event. Thus, the first step in problem formulation is necessarily qualitative (fuzzy logic). This first step usually involves drawing a picture of the system to be studied.

The second step is the bringing together of all applicable physical and chemical information, conservation laws, and rate expressions. At this point, the engineer must make a series of critical decisions about the conversion of mental images to symbols, and at the same time, how detailed the model of a system must be. Here, one must classify the real purposes of the modeling effort. Is the model to be used only for explaining trends in the operation of an existing piece of equipment? Is the model to be used for predictive or design purposes? Do we want steady-state or transient response? The scope and depth of these early decisions will determine the ultimate complexity of the final mathematical description.

The third step requires the setting down of finite or differential volume elements, followed by writing the conservation laws. In the limit, as the differential elements shrink, then differential equations arise naturally. Next, the problem of boundary conditions must be addressed, and this aspect must be treated with considerable circumspection.

When the problem is fully posed in quantitative terms, an appropriate mathematical solution method is sought out, which finally relates dependent (responding) variables to one or more independent (changing) variables. The final result may be an elementary mathematical formula, or a numerical solution portrayed as an array of numbers.

1.2 ILLUSTRATION OF THE FORMULATION PROCESS (COOLING OF FLUIDS)

We illustrate the principles outlined above and the hierarchy of model building by way of a concrete example: the cooling of a fluid flowing in a circular pipe. We start with the simplest possible model, adding complexity as the demands for precision increase. Often, the simple model will suffice for rough, qualitative purposes. However, certain economic constraints weigh heavily against overdesign, so predictions and designs based on the model may need to be more precise. This section also illustrates the "need to know" principle, which acts as a catalyst to stimulate the garnering together of mathematical techniques. The problem posed in this section will appear repeatedly throughout the book, as more sophisticated techniques are applied to its complete solution.

Model 1—Plug Flow

As suggested in the beginning, we first formulate a mental picture and then draw a sketch of the system. We bring together our thoughts for a simple plug flow model in Fig. 1.1*a*. One of the key assumptions here is *plug flow*, which means that the fluid velocity profile is plug shaped, in other words uniform at all radial positions. This almost always implies turbulent fluid flow conditions, so that fluid elements are well-mixed in the radial direction, hence the fluid temperature is fairly uniform in a plane normal to the flow field (i.e., the radial direction).

If the tube is not too long or the temperature difference is not too severe, then the physical properties of the fluid will not change much, so our second

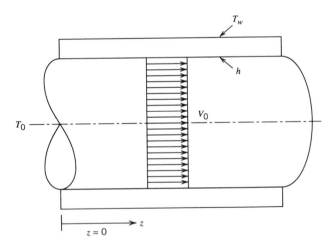

Figure 1.1*a* Sketch of plug flow model formulation.

step is to express this and other assumptions as a list:

1. A steady-state solution is desired.
2. The physical properties (ρ, density; C_p, specific heat; k, thermal conductivity, etc.) of the fluid remain constant.
3. The wall temperature is constant and uniform (i.e., does not change in the z or r direction) at a value T_w.
4. The inlet temperature is constant and uniform (does not vary in r direction) at a value T_0, where $T_0 > T_w$.
5. The velocity profile is plug shaped or flat, hence it is uniform with respect to z or r.
6. The fluid is well-mixed (highly turbulent), so the temperature is uniform in the radial direction.
7. Thermal conduction of heat along the axis is small relative to convection.

The third step is to sketch, and act upon, a differential volume element of the system (in this case, the flowing fluid) to be modeled. We illustrate this elemental volume in Fig. 1.1b, which is sometimes called the "control volume."

We act upon this elemental volume, which spans the whole of the tube cross section, by writing the general conservation law

$$\text{Rate in} - \text{Rate out} + \text{Rate of Generation} = \text{Rate of Accumulation} \quad (1.1)$$

Since steady state is stipulated, the accumulation of heat is zero. Moreover, there are no chemical, nuclear, or electrical sources specified within the volume element, so heat generation is absent. The only way heat can be exchanged is through the perimeter of the element by way of the temperature difference between wall and fluid. The incremental rate of heat removal can be expressed as a positive quantity using Newton's law of cooling, that is,

$$\Delta Q = (2\pi R \Delta z)h\left[\overline{T}(z) - T_w\right] \quad (1.2)$$

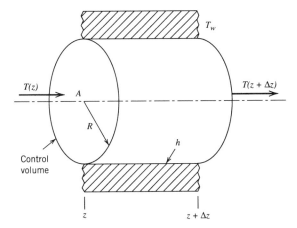

Figure 1.1b Elemental or control volume for plug flow model.

As a convention, we shall express all such rate laws as positive quantities, invoking positive or negative signs as required when such expressions are introduced into the conservation law (Eq. 1.1). The contact area in this simple model is simply the perimeter of the element times its length.

The constant heat transfer coefficient is denoted by h. We have placed a bar over T to represent the average between $T(z)$ and $T(z + \Delta z)$

$$\overline{T}(z) \simeq \frac{T(z) + T(z + \Delta z)}{2} \tag{1.3}$$

In the limit, as $\Delta z \to 0$, we see

$$\lim_{\Delta z \to 0} \overline{T}(z) \to T(z) \tag{1.4}$$

Now, along the axis, heat can enter and leave the element only by convection (flow), so we can write the elemental form of Eq. 1.1 as

$$\underbrace{v_0 A \rho C_p T(z)}_{\text{Rate heat flow in}} - \underbrace{v_0 A \rho C_p T(z + \Delta z)}_{\text{Rate heat flow out}} - \underbrace{(2\pi R \Delta z) h(\overline{T} - T_w)}_{\text{Rate heat loss through wall}} = 0 \tag{1.5}$$

The first two terms are simply mass flow rate times local enthalpy, where the reference temperature for enthalpy is taken as zero. Had we used $C_p(T - T_{ref})$ for enthalpy, the term T_{ref} would be cancelled in the elemental balance. The final step is to invoke the fundamental lemma of calculus, which defines the act of differentiation

$$\lim_{\Delta z \to 0} \frac{T(z + \Delta z) - T(z)}{\Delta z} \to \frac{dT}{dz} \tag{1.6}$$

We rearrange the conservation law into the form required for taking limits, and then divide by Δz

$$-v_0 A \rho C_p \frac{T(z + \Delta z) - T(z)}{\Delta z} - (2\pi R h)(\overline{T} - T_w) = 0 \tag{1.7}$$

Taking limits, one at a time, then yields the sought-after differential equation

$$v_0 A \rho C_p \frac{dT}{dz} + 2\pi R h[T(z) - T_w] = 0 \tag{1.8}$$

where we have cancelled the negative signs.

Before solving this equation, it is good practice to group parameters into a single term (lumping parameters). For such elementary problems, it is convenient to lump parameters with the lowest order term as follows:

$$\frac{dT(z)}{dz} + \lambda[T(z) - T_w] = 0 \tag{1.9}$$

where

$$\lambda = 2\pi Rh / (v_0 A \rho C_p)$$

It is clear that λ must take units of reciprocal length.

As it stands, the above equation is classified as a linear, inhomogeneous equation of first order, which in general must be solved using the so-called Integrating-Factor method, as we discuss later in Chapter 2 (Section 2.3).

Nonetheless, a little common sense will allow us to obtain a final solution without any new techniques. To do this, we remind ourselves that T_w is everywhere constant, and that differentiation of a constant is always zero, so we can write

$$\frac{d(T(z) - T_w)}{dz} = \frac{dT(z)}{dz} \tag{1.10}$$

This suggests we define a new dependent variable, namely

$$\theta = T(z) - T_w \tag{1.11}$$

hence Eq. 1.9 now reads simply

$$\frac{d\theta(z)}{dz} + \lambda\theta(z) = 0 \tag{1.12}$$

This can be integrated directly by separation of variables, so we rearrange to get

$$\frac{d\theta}{\theta} + \lambda dz = 0 \tag{1.13}$$

Integrating term by term yields

$$\ln\theta + \lambda z = \ln K \tag{1.14}$$

where $\ln K$ is any (arbitrary) constant of integration. Using logarithm properties, we can solve directly for θ

$$\theta = K \exp(-\lambda z) \tag{1.15}$$

It now becomes clear why we selected the form $\ln K$ as the arbitrary constant in Eq. 1.14.

All that remains is to find a suitable value for K. To do this, we recall the boundary condition denoted as T_0 in Fig. 1.1a, which in mathematical terms has the meaning

$$T(0) = T_0; \quad \text{or} \quad \theta(0) = T(0) - T_w = T_0 - T_w \tag{1.16}$$

Thus, when $z = 0$, $\theta(0)$ must take a value $T_0 - T_w$, so K must also take this value.

Our final result for computational purposes is

$$\frac{T(z) - T_w}{T_0 - T_w} = \exp\left(\frac{-2\pi R h z}{v_0 A \rho C_p}\right) \tag{1.17}$$

We note that all arguments of mathematical functions must be dimensionless, so the above result yields a dimensionless temperature

$$\frac{T(z) - T_w}{T_0 - T_w} = \psi \tag{1.18}$$

and a dimensionless length scale

$$\frac{2\pi R h z}{v_0 A \rho C_p} = \zeta \tag{1.19}$$

Thus, a problem with six parameters, two external conditions (T_0, T_w) and one each dependent and independent variable has been reduced to only two elementary (dimensionless) variables, connected as follows

$$\psi = \exp(-\zeta) \tag{1.20}$$

Model II—Parabolic Velocity

In the development of Model I (plug flow), we took careful note that the assumptions used in this first model building exercise implied "turbulent flow" conditions, such a state being defined by the magnitude of the Reynolds number $(v_0\, d/\nu)$, which must always exceed 2100 for this model to be applicable. For slower flows, the velocity is no longer plug shaped, and in fact when $Re < 2100$, the shape is parabolic

$$v_z = 2v_0\left[1 - (r/R)^2\right] \tag{1.21}$$

where v_0 now denotes the average velocity, and v_z denotes the locally varying value (Bird et al. 1960). Under such conditions, our earlier assumptions must be carefully reassessed; specifically, we will need to modify items 5, 6, and 7 in the previous list:

5. The z-directed velocity profile is parabolic shaped and depends on the position r.
6. The fluid is not well mixed in the radial direction, so account must be taken of radial heat conduction.
7. Because convection is smaller, axial heat conduction may also be important.

These new physical characteristics cause us to redraw the elemental volume as shown in Fig. 1.1c. The control volume now takes the shape of a ring of thickness Δr and length Δz. Heat now crosses two surfaces, the annular area

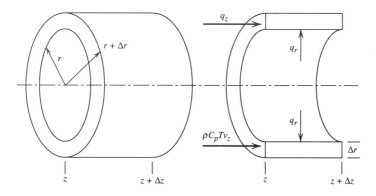

Figure 1.1c Control volume for Model II.

normal to fluid flow, and the area along the perimeter of the ring. We shall need to designate additional (vector) quantities to represent heat flux (rate per unit normal area) by molecular conduction:

$$q_r(r, z) = \text{molecular heat flux in radial direction} \qquad (1.22)$$

$$q_z(r, z) = \text{molecular heat flux in axial direction} \qquad (1.23)$$

The net rate of heat gain (or loss) by conduction is simply the flux times the appropriate area normal to the flux direction. The conservation law (Eq. 1.1) can now be written for the element shown in Fig. 1.1c.

$$v_z(2\pi r \Delta r)\rho C_p T(z, r) - v_z(2\pi r \Delta r)\rho C_p T(z + \Delta z, r)$$

$$+ (2\pi r \Delta r q_z)|_z - (2\pi r \Delta r q_z)|_{z + \Delta z} + (2\pi r \Delta z q_r)|_r - (2\pi r \Delta z q_r)|_{r + \Delta r} = 0$$

$$(1.24)$$

The new notation is necessary, since we must deal with products of terms, either or both of which may be changing.

We rearrange this to a form appropriate for the fundamental lemma of calculus. However, since two position coordinates are now allowed to change, we must define the process of partial differentiation, for example,

$$\lim_{\Delta z \to 0} \frac{T(z + \Delta z, r) - T(z, r)}{\Delta z} = \left(\frac{\partial T}{\partial z}\right)_r \qquad (1.25)$$

which of course implies holding r constant as denoted by subscript (we shall delete this notation henceforth). Thus, we divide Eq. 1.24 by $2\pi \Delta z \Delta r$ and rearrange to get

$$-v_z \rho C_p r \frac{T(z + \Delta z, r) - T(z, r)}{\Delta z}$$

$$-\frac{[rq_z]|_{z+\Delta z} - [rq_z]|_z}{\Delta z} - \frac{[rq_r]|_{r+\Delta r} - [rq_r]|_r}{\Delta r} = 0 \qquad (1.26)$$

Taking limits, one at a time, then yields

$$-v_z \rho C_p r \frac{\partial T}{\partial z} - \frac{\partial(rq_z)}{\partial z} - \frac{\partial(rq_r)}{\partial r} = 0 \qquad (1.27)$$

The derivative with respect to (wrt) z implies holding r constant, so r can be placed outside this term; thus, dividing by r and rearranging shows

$$-\frac{\partial q_z}{\partial z} - \frac{\partial}{r \partial r}(rq_r) = v_z \rho C_p \frac{\partial T}{\partial z} \qquad (1.28)$$

At this point, the equation is insoluble since we have one equation and three unknowns (T, q_z, q_r). We need to know some additional rate law to connect fluxes q to temperature T. Therefore, it is now necessary to introduce the famous Fourier's law of heat conduction, the vector form of which states that heat flux is proportional to the gradient in temperature

$$q = -k \nabla T \qquad (1.29)$$

and the two components of interest here are

$$q_r = -k \frac{\partial T}{\partial r}; \qquad q_z = -k \frac{\partial T}{\partial z} \qquad (1.30)$$

Inserting these two new equations into Eq. 1.28 along with the definition of v_z, yields finally a single equation, with one unknown $T(r,z)$

$$k \frac{\partial^2 T}{\partial z^2} + k \frac{1}{r} \frac{\partial}{\partial r}\left(r \frac{\partial T}{\partial r}\right) = 2v_0 \rho C_p \left[1 - \left(\frac{r}{R}\right)^2\right] \frac{\partial T}{\partial z} \qquad (1.31)$$

The complexity of Model II has now exceeded our poor powers of solution, since we have much we need to know before attempting such second-order partial differential equations. We shall return to this problem occasionally as we learn new methods to effect a solution, and as new approximations become evident.

1.3   COMBINING RATE AND EQUILIBRIUM CONCEPTS (PACKED BED ADSORBER)

The occurrence of a rate process and a thermodynamic equilibrium state is common in chemical engineering models. Thus, certain parts of a whole system may respond so quickly that, for practical purposes, local equilibrium may be assumed. Such an assumption is an integral (but often unstated) part of the qualitative modeling exercise.

To illustrate the combination of rate and equilibrium principles, we next consider a widely used separation method, which is inherently unsteady: *packed bed adsorption*. We imagine a packed bed of finely granulated (porous) solid (e.g., charcoal) contacting a binary mixture, one component of which selectively adsorbs (physisorption) onto and within the solid material. The physical process of adsorption is so fast relative to other slow steps (diffusion within the solid particle), that in and near the solid particles, local equilibrium exists

$$q = KC^* \tag{1.32}$$

where q denotes the average composition of the solid phase, expressed as moles solute adsorbed per unit volume solid particle, and C^* denotes the solute composition (moles solute per unit volume fluid), which would exist at equilibrium. We suppose that a single film mass transport coefficient controls the transfer rate between flowing and immobile (solid) phase.

It is also possible to use the same model even when intraparticle diffusion is important (Rice 1982) by simply replacing the film coefficient with an "effective" coefficient. Thus, the model we derive can be made to have wide generality.

We illustrate a sketch of the physical system in Fig. 1.2. It is clear in the sketch that we shall again use the plug flow concept, so the fluid velocity profile is flat. If the stream to be processed is dilute in the adsorbable species (adsorbate), then heat effects are usually ignorable, so isothermal conditions will be taken. Finally, if the particles of solid are small, the axial diffusion effects, which are Fickian-like, can be ignored and the main mode of transport in the mobile fluid phase is by convection.

Interphase transport from the flowing fluid to immobile particles obeys a rate law, which is based on departure from the thermodynamic equilibrium state. Because the total interfacial area is not known precisely, it is common practice

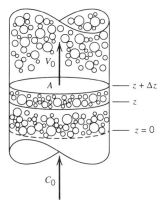

Figure 1.2 Packed bed adsorber.

to define a volumetric transfer coefficient, which is the product $k_c a$ where a is the total interfacial area per unit volume of packed column. The incremental rate expression (moles/time) is then obtained by multiplying the volumetric transfer coefficient $(k_c a)$ by the composition linear driving force and this times the incremental volume of the column $(A \Delta z)$

$$\Delta R = k_c a (C - C^*) \cdot A \, \Delta z \tag{1.33}$$

We apply the conservation law (Eq. 1.1) to the adsorbable solute contained in both phases, as follows

$$v_0 A C(z, t) - v_0 A C(z + \Delta z, t) = \varepsilon A \, \Delta z \frac{\partial C}{\partial t} + (1 - \varepsilon) A \, \Delta z \frac{\partial q}{\partial t} \tag{1.34}$$

where v_0 denotes superficial fluid velocity (velocity that would exist in an empty tube), ε denotes the fraction void (open) volume, hence $(1 - \varepsilon)$ denotes the fractional volume taken up by the solid phase. Thus, ε is volume fraction between particles and is often called interstitial void volume; it is the volume fraction through which fluid is convected. The rate of accumulation has two possible sinks: accumulation in the fluid phase (C) and in the solid phase (q).

By dividing through by $A \, \Delta z$, taking limits as before, we deduce that the overall balance for solute obeys

$$-v_0 \frac{\partial C}{\partial z} = \varepsilon \frac{\partial C}{\partial t} + (1 - \varepsilon) \frac{\partial q}{\partial t} \tag{1.35}$$

Similarly, we may make a solute balance on the immobile phase alone, using the rate law, Eq. 1.33, noting adsorption *removes* material from the flowing phase and *adds* it to the solid phase. Now, since the solid phase loses no material and generates none (assuming chemical reaction is absent), then the solid phase balance is

$$A(1 - \varepsilon) \, \Delta z \frac{\partial q}{\partial t} = k_c a (C - C^*) A \, \Delta z \tag{1.36}$$

which simply states: Rate of accumulation equals rate of transfer to the solid. Dividing out the elementary volume, $A \, \Delta z$, yields

$$(1 - \varepsilon) \frac{\partial q}{\partial t} = k_c a (C - C^*) \tag{1.37}$$

We note that as equilibrium is approached (as $C \to C^*$)

$$\frac{\partial q}{\partial t} \to 0$$

Such conditions correspond to "saturation," hence no further molar exchange occurs. When this happens to the whole bed, the bed must be "regenerated," for example by passing a hot, inert fluid through the bed, thereby desorbing solute.

The model of the system is now composed of Eqs. 1.35, 1.37, and 1.32: There are three equations and three unknowns (C, C^*, q).

To make the system model more compact, we attempt to eliminate q, since $q = KC^*$; hence we have

$$v_0 \frac{\partial C}{\partial z} + \varepsilon \frac{\partial C}{\partial t} + (1 - \varepsilon) K \frac{\partial C^*}{\partial t} = 0 \qquad (1.38)$$

$$(1 - \varepsilon) K \frac{\partial C^*}{\partial t} = k_c a (C - C^*) \qquad (1.39)$$

The solution to this set of partial differential equations (PDEs) can be effected by suitable transform methods (e.g., the Laplace transform) for certain types of boundary and initial conditions (BC and IC). For the adsorption step, these are

$$q(z, 0) = 0 \text{ (initially clean solid)} \qquad (1.40)$$

$$C(0, t) = C_0 \text{ (constant composition at bed entrance)} \qquad (1.41)$$

The condition on q implies (cf. Eq. 1.32)

$$C^*(z, 0) = 0 \qquad (1.42)$$

Finally, if the bed was indeed initially clean, as stated above, then it must also be true

$$C(z, 0) = 0 \text{ (initially clean interstitial fluid)} \qquad (1.43)$$

We thus have three independent conditions (note, we could use either Eq. 1.40 or Eq. 1.42, since they are linearly dependent) corresponding to three derivatives:

$$\frac{\partial C^*}{\partial t}, \ \frac{\partial C}{\partial t}, \ \frac{\partial C}{\partial z}$$

As we demonstrate later, in Chapter 10, linear systems of equations can only be solved exactly when there exists one BC or IC for each order of a derivative. The above system is now properly posed, and will be solved as an example in Chapter 10 using Laplace transform.

1.4 BOUNDARY CONDITIONS AND SIGN CONVENTIONS

As we have seen in the previous sections, when time is the independent variable, the boundary condition is usually an initial condition, meaning we must specialize the state of the dependent variable at some time t_0 (usually $t_0 = 0$). For the steady state, we have seen that integrations of the applicable equations always produce arbitrary constants of integration. These integration constants must be evaluated, using stipulated boundary conditions to complete the model's solution.

For the physicochemical problems occurring in chemical engineering, most boundary or initial conditions are (or can be made to be) of the homogeneous type; *a condition or equation is taken to be homogeneous if, for example, it is satisfied by* y(x), *and is also satisfied by* λy(x), *where* λ *is an arbitrary constant.* The three classical types for such homogeneous boundary conditions at a point, say x_0, are the following:

$$\text{(i)} \quad y(x) = 0 \qquad @ \qquad x = x_0$$

$$\text{(ii)} \quad \frac{dy}{dx} = 0 \qquad @ \qquad x = x_0$$

$$\text{(iii)} \quad \beta y + \frac{dy}{dx} = 0 \qquad @ \qquad x = x_0$$

Most often, the boundary values for a derived model are not homogeneous, but can be made to be so. For example, Model II in Section 1.2 portrays cooling of a flowing fluid in a tube. Something must be said about the fluid temperature at the solid wall boundary, which was specified to take a constant value T_w. This means all along the tube length, we can require

$$T(r, z) = T_w \quad @ \quad r = R, \quad \text{for all } z$$

As it stands, this does not match the condition for homogeneity. However, if we define a new variable θ

$$\theta = T(r, z) - T_w \tag{1.44}$$

then it is clear that the wall condition will become homogeneous, of type (i)

$$\theta(r, z) = 0 \quad @ \quad r = R, \quad \text{for all } z \tag{1.45}$$

When redefining variables in this way, one must be sure that the original defining equation is unchanged. Thus, since the derivative of a constant (T_w) is always zero, then Eq. 1.31 for the new dependent variable θ is easily seen to be unchanged

$$k\frac{\partial^2\theta}{\partial z^2} + k\left(\frac{\partial^2\theta}{\partial r^2} + \frac{1}{r}\frac{\partial\theta}{\partial r}\right) = 2v_0\rho C_p\left[1 - (r/R)^2\right]\frac{\partial\theta}{\partial z} \tag{1.46}$$

It often occurs that the heat (or mass) flux at a boundary is controlled by a heat (or mass) transfer coefficient, so for a circular tube the conduction flux is proportional to a temperature difference

$$q_r = -k\frac{\partial T}{\partial r} = U(T - T_c) \quad @ \quad r = R, \text{ for all } z; \quad T_c = \text{constant} \tag{1.47}$$

Care must be taken to ensure that sign conventions are obeyed. In our cooling problem (Model II, Section 1.2), it is clear that

$$q_r > 0, \frac{\partial T}{\partial r} \leq 0$$

so that $U(T - T_c)$ must be positive, which it is, since the coolant temperature $T_c < T(R, z)$.

This boundary condition also does not identify exactly with the type (iii) homogeneous condition given earlier. However, if we redefine the independent variable to be $\theta = T - T_c$, then we have

$$\left(\frac{U}{k}\right)\theta + \frac{\partial \theta}{\partial r} = 0 \quad @ \quad r = R, \quad \text{for all } z \tag{1.48}$$

which is identical in form with the type (iii) homogeneous boundary condition when we note the equivalence: $\theta = y$, $U/k = \beta$, $r = x$, and $R = x_0$. It is also easy to see that the original convective-diffusion Eq. 1.31 is unchanged when we replace T with θ. This is a useful property of linear equations.

Finally, we consider the type (ii) homogeneous boundary condition in physical terms. For the pipe flow problem, if we had stipulated that the tube wall was well insulated, then the heat flux at the wall is nil, so

$$q_r = -k\frac{\partial T}{\partial r} = 0 \quad @ \quad r = R, \quad \text{for all } z \tag{1.49}$$

This condition is of the homogeneous type (ii) without further modification.

Thus, we see that models for a fluid flowing in a circular pipe can sustain any one of the three possible homogeneous boundary conditions.

Sign conventions can be troublesome to students, especially when they encounter type (iii) boundary conditions. It is always wise to double-check to ensure that the sign of the left-hand side is the same as that of the right-hand side. Otherwise, negative transport coefficients will be produced, which is thermodynamically impossible. To guard against such inadvertent errors, it is useful to produce a sketch showing the qualitative shape of the expected profiles.

In Fig. 1.3 we sketch the expected shape of temperature profile for a fluid being cooled in a pipe. The slope of temperature profile is such that $\partial T/\partial r \leq 0$. If we exclude the centerline ($r = 0$), where exactly $\partial T/\partial r = 0$ (the symmetry condition), then always $\partial T/\partial r < 0$. Now, since fluxes (which are vector quantities) are always positive when they move in the positive direction of the coordinate system, then it is clear why the negative sign appears in Fourier's law

$$q_r = -k\frac{\partial T}{\partial r} \tag{1.50}$$

Thus, since $\partial T/\partial r < 0$, then the product $-k\partial T/\partial r > 0$, so that flux $q_r > 0$. This convention thus ensures that heat moves down a temperature gradient, so transfer is always from hot to cold regions. For a heated tube, flux is always in the anti-r direction, hence it must be a negative quantity. Similar arguments

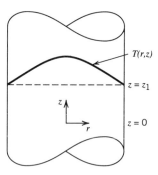

Figure 1.3 Expected temperature profile for cooling fluids in a pipe at an arbitrary position z_1.

hold for mass transfer where Fick's law is applicable, so that the radial component of flux in cylindrical coordinates would be

$$J_r = -D\frac{\partial C}{\partial r} \tag{1.51}$$

1.5 SUMMARY OF THE MODEL BUILDING PROCESS

These introductory examples are meant to illustrate the essential qualitative nature of the early part of the model building stage, which is followed by more precise quantitative detail as the final image of the desired model is made clearer. It is a property of the human condition that minds change as new information becomes available. Experience is an important factor in model formulation, and there have been recent attempts to simulate the thinking of experienced engineers through a format called *Expert Systems*. The following step-by-step procedure may be useful for beginners.

1. Draw a sketch of the system to be modeled and label/define the various geometric, physical and chemical quantities.
2. Carefully select the important dependent (response) variables.
3. Select the possible independent variables (e.g., z, t), changes in which must necessarily affect the dependent variables.
4. List the parameters (physical constants, physical size, and shape) that are expected to be important; also note the possibility of nonconstant parameters [e.g., viscosity changing with temperature, $\mu(T)$].
5. Draw a sketch of the expected behavior of the dependent variable(s), such as the "expected" temperature profile we used for illustrative purposes in Fig. 1.3.
6. Establish a "control volume" for a differential or finite element (e.g., CSTR) of the system to be modeled; sketch the element and indicate all inflow–outflow paths.

7. Write the "conservation law" for the volume element: Express flux and reaction rate terms using general symbols, which are taken as positive quantities, so that signs are introduced only as terms are inserted according to the rules of the conservation law, Eq. 1.1.
8. After rearrangement to the proper differential format, invoke the fundamental lemma of calculus to produce a differential equation.
9. Introduce specific forms of flux (e.g., $J_r = -D\partial C/\partial r$) and rate ($R_A = kC_A$); note, the opposite of generation is depletion, so when a species is depleted, then its loss rate must be entered with the appropriate sign in the conservation law (i.e., replace "+ generation" with "− depletion" in Eq. 1.1).
10. Write out all possibilities for boundary values of the dependent variables; the choice among these will be made in conjunction with the solution method selected for the defining (differential) equation.
11. Search out solution methods, and consider possible approximations for: (i) the defining equation, (ii) the boundary conditions, and (iii) an acceptable final solution.

It is clear that the modeling and solution effort should go hand in hand, tempered of course by available experimental and operational evidence. A model that contains unknown and unmeasurable parameters is of no real value.

1.6 MODEL HIERARCHY AND ITS IMPORTANCE IN ANALYSIS

As pointed out in Section 1.1 regarding the real purposes of the modeling effort, the scope and depth of these decisions will determine the complexity of the mathematical description of a process. If we take this scope and depth as the barometer for generating models, we will obtain a hierarchy of models where the lowest level may be regarded as a black box and the highest is where all possible transport processes known to man in addition to all other concepts (such as thermodynamics) are taken into account. Models, therefore, do not appear in isolation, but rather they belong to a family where the hierarchy is dictated by the number of rules (transport principles, thermodynamics). It is this family that provides engineers with capabilities to predict and understand the phenomena around us. The example of cooling of a fluid flowing in a tube (Models I and II) in Section 1.2 illustrated two members of this family. As the level of sophistication increased, the mathematical complexity increased. If one is interested in exactly how heat is conducted through the metal casing and is disposed of to the atmosphere, then the complexity of the problem must be increased by writing down a heat balance relation for the metal casing (taking it to be constant at a value T_w is, of course, a model, albeit the simplest one). Further, if one is interested in how the heat is transported near the entrance section, one must write down heat balance equations before the start of the tube, in addition to the Eq. 1.31 for the active, cooling part of the tube. Furthermore, the nature of the boundary conditions must be carefully scrutinized before and after the entrance zone in order to properly describe the boundary conditions.

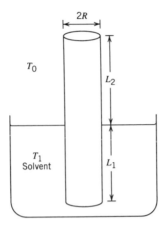

Figure 1.4 Schematic diagram of a heat removal from a solvent bath.

To further demonstrate the concept of model hierarchy and its importance in analysis, let us consider a problem of heat removal from a bath of hot solvent by immersing steel rods into the bath and allowing the heat to dissipate from the hot solvent bath through the rod and thence to the atmosphere (Fig. 1.4).

For this elementary problem, it is wise to start with the simplest model first to get some feel about the system response.

Level 1

In this level, let us assume that:

(a) The rod temperature is uniform, that is, from the bath to the atmosphere.
(b) Ignore heat transfer at the two flat ends of the rod.
(c) Overall heat transfer coefficients are known and constant.
(d) No solvent evaporates from the solvent air interface.

The many assumptions listed above are necessary to simplify the analysis (i.e., to make the model tractable).

Let T_0 and T_1 be the atmosphere and solvent temperatures, respectively. The steady-state heat balance (i.e., no accumulation of heat by the rod) shows a balance between heat collected in the bath and that dissipated by the upper part of the rod to atmosphere

$$h_L(2\pi R L_1)(T_1 - T) = h_G(2\pi R L_2)(T - T_0) \qquad (1.52)$$

where T is the temperature of the rod, and L_1 and L_2 are lengths of rod exposed to solvent and to atmosphere, respectively. Obviously, the volume elements are finite (not differential), being composed of the volume above the liquid of length L_2 and the volume below of length L_1.

Solving for T from Eq. 1.52 yields

$$T = \frac{(T_0 + \alpha T_1)}{(1 + \alpha)} \tag{1.53}$$

where

$$\alpha = \frac{h_L L_1}{h_G L_2} \tag{1.54}$$

Equation 1.53 gives us a very quick estimate of the rod temperature and how it varies with exposure length. For example, if α is much greater than unity (i.e., long L_1 section and high liquid heat transfer coefficient compared to gas coefficient), the rod temperature is then very near T_1. Taking the rod temperature to be represented by Eq. 1.53, the rate of heat transfer is readily calculated from Eq. 1.52 by replacing T:

$$Q = \frac{h_L 2\pi R L_1}{(1 + \alpha)}(T_1 + T_0) = \frac{h_L L_1}{\left(1 + \dfrac{h_L L_1}{h_G L_2}\right)} 2\pi R(T_1 - T_0) \tag{1.55a}$$

$$Q = \frac{1}{\left(\dfrac{1}{h_L L_1} + \dfrac{1}{h_G L_2}\right)} 2\pi R(T_1 - T_0) \tag{1.55b}$$

When $\alpha = h_L L_1 / h_G L_2$ is very large, the rate of heat transfer becomes simply

$$Q \cong 2\pi R h_G L_2(T_1 - T_0) \tag{1.55c}$$

Thus, the heat transfer is controlled by the segment of the rod exposed to the atmosphere. It is interesting to note that when the heat transfer coefficient contacting the solvent is very high (i.e., $\alpha \gg 1$), it does not really matter how much of the rod is immersed in the solvent.

Thus for a given temperature difference and a constant rod diameter, the rate of heat transfer can be enhanced by either increasing the exposure length L_2 or by increasing the heat transfer rate by stirring the solvent. However, these conclusions are tied to the assumption of constant rod temperature, which becomes tenuous as atmospheric exposure is increased.

To account for effects of temperature gradients in the rod, we must move to the next level in the model hierarchy, which is to say that a differential volume must be considered.

Level 2

Let us relax part of the assumption (i) of the first model by assuming only that the rod temperature below the solvent liquid surface is uniform at a value T_1. This is a reasonable proposition, since the liquid has a much higher thermal

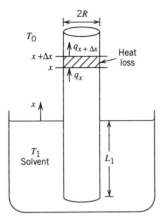

Figure 1.5 Shell element and
the system coordinate.

conductivity than air. The remaining three assumptions of the level 1 model are
retained.

Next, choose an upward pointing coordinate x with the origin at the solvent-air
surface. Figure 1.5 shows the coordinate system and the elementary control
volume.

Applying a heat balance around a thin shell segment with thickness Δx gives

$$\pi R^2 q(x) - \pi R^2 q(x + \Delta x) - 2\pi R \Delta x h_G (T - T_0) = 0 \qquad (1.56)$$

where the first and second terms represent heat conducted into and out of the
element and the last term represents heat loss to atmosphere. We have decided,
by writing this, that temperature gradients are likely to exist in the part of the
rod exposed to air, but are unlikely to exist in the submerged part.

Dividing Eq. 1.56 by $\pi R^2 \Delta x$ and taking the limit as $\Delta x \to 0$ yields the
following first order differential equation for the heat flux, q:

$$\frac{dq}{dx} + \frac{2}{R} h_G (T - T_0) = 0 \qquad (1.57)$$

Assuming the rod is homogeneous, i.e., the thermal conductivity is uniform, the
heat flux along the axis is related to the temperature according to Fourier's law
of heat conduction (Eq. 1.29). Substitution of Eq. 1.29 into Eq. 1.57 yields

$$k \frac{d^2 T}{dx^2} = \frac{2 h_G}{R} (T - T_0) \qquad (1.58)$$

Equation 1.58 is a second order ordinary differential equation, and to solve this,
two conditions must be imposed. One condition was stipulated earlier:

$$x = 0, \qquad T = T_1 \qquad (1.59a)$$

The second condition (heat flux) can also be specified at $x = 0$ or at the other end of the rod, i.e., $x = L_2$. Heat flux is the sought-after quantity, so it cannot be specified a priori. One must then provide a condition at $x = L_2$. At the end of the rod, one can assume Newton's law of cooling prevails, but since the rod length is usually longer than the diameter, most of the heat loss occurs at the rod's lateral surface, and the flux from the top surface is small, so write approximately:

$$x = L_2; \qquad \frac{dT}{dx} \simeq 0 \qquad\qquad (\textbf{1.59}b)$$

Equation 1.58 is subjected to the two boundary conditions (Eq. 1.59) to yield the solution

$$T = T_0 + (T_1 - T_0)\frac{\cosh\left[m(L_2 - x)\right]}{\cosh\left(mL_2\right)} \qquad\qquad (\textbf{1.60})$$

where

$$m = \sqrt{\frac{2h_G}{Rk}} \qquad\qquad (\textbf{1.61})$$

We will discuss the method of solution of such second order equations in Chapter 2.

Once we know the temperature distribution of the rod above the solvent-air interface, then the rate of heat loss can be calculated either of two ways. In the first, we know that the heat flow through area πR^2 at $x = 0$ must be equal to the heat released into the atmosphere, that is,

$$Q = -\pi R^2 k \frac{\partial T}{\partial x}\bigg|_{x=0} \qquad\qquad (\textbf{1.62})$$

Applying Eq. 1.60 to Eq. 1.62 gives

$$Q = 2\pi R h_G L_2 \eta (T_1 - T_0) \qquad\qquad (\textbf{1.63}a)$$

where

$$\eta = \frac{\tanh\left(mL_2\right)}{mL_2}$$

This dimensionless group (called *effectiveness factor*) represents the ratio of actual heat loss to the (maximum) loss rate when gradients are absent.

The following figure (Fig. 1.6) shows the log–log plot of η versus the dimensionless group mL_2. We note that the effectiveness factor approaches unity when mL_2 is much less than unity and it behaves like $1/mL_2$ as mL_2 is very large.

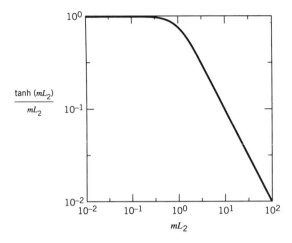

Figure 1.6 A plot of the effectiveness factor versus mL_2.

In the limit for small mL_2, we can write

$$\eta = \frac{\tanh(mL_2)}{mL_2} \approx 1 \qquad (1.63b)$$

which is the most effective heat transfer condition. This is physically achieved when

(a) The rod thermal conductivity is large.
(b) The segment exposed to atmosphere (L_2) is short.

For such a case, we can write the elementary result

$$Q = 2\pi R h_G L_2 (T_1 - T_0) \qquad (1.64)$$

which is identical to the first model (Eq. 1.55c). Thus, we have learned that the first model is valid only when $mL_2 \ll 1$. Another way of calculating the heat transfer rate is carrying the integration of local heat transfer rate along the rod

$$Q = \int_0^{L_2} dq = \int_0^{L_2} h_G (T - T_0)(2\pi R dx) \qquad (1.65)$$

where T is given in Eq. 1.60 and the differential transfer area is $2\pi R dx$. Substituting T of Eq. 1.60 into Eq. 1.65 yields the same solution for Q as given in Eq. 1.63a.

Levels 1 and 2 solutions have one assumption in common: The rod temperature below the solvent surface was taken to be uniform. The validity of this modelling assumption will not be known until we move up one more level in the model hierarchy.

Level 3

In this level of modelling, we relax the assumption (i) of the first level by allowing for temperature gradients in the rod for segments above and below the solvent-air interface.

Let the temperature below the solvent-air interface be T^I and that above the interface be T^{II}. Carrying out the one-dimensional heat balances for the two segments of the rod, we obtain

$$\frac{d^2 T^I}{dx^2} = \frac{2h_L}{Rk}(T^I - T_1) \tag{1.66}$$

and

$$\frac{d^2 T^{II}}{dx^2} = \frac{2h_G}{Rk}(T^{II} - T_0) \tag{1.67}$$

We shall still maintain the condition of zero flux at the flat ends of the rod. This means of course that

$$x = -L_1; \qquad \frac{dT^I}{dx} = 0 \tag{1.68}$$

$$x = L_2; \qquad \frac{dT^{II}}{dx} = 0 \tag{1.69}$$

Equations 1.68 and 1.69 provide two of the four necessary boundary conditions. The other two arise from the continuity of temperature and flux at the $x = 0$ position, that is,

$$x = 0; \qquad T^I = T^{II} \tag{1.70a}$$

$$x = 0; \qquad \frac{dT^I}{dx} = \frac{dT^{II}}{dx} \tag{1.70b}$$

Solutions of Eqs. 1.66 and 1.67 subject to conditions 1.68 and 1.69 are easily solved by methods illustrated in the next chapter (Example 2.25)

$$T^I = T_1 + A \cosh\left[n(x + L_1)\right] \tag{1.71}$$

and

$$T^{II} = T_0 + B \cosh\left[m(L_2 - x)\right] \tag{1.72}$$

where m is defined in Eq. 1.61 and

$$n = \sqrt{\frac{2h_L}{Rk}} \tag{1.73}$$

The constants of integration, A and B, can be found by substituting Eqs. 1.71

and 1.72 into the continuity conditions (Eqs. 1.70a,b) to finally get

$$B = \frac{(T_1 - T_0)}{\left[\cosh(mL_2) + \dfrac{m}{n}\dfrac{\sinh(mL_2)}{\sinh(nL_1)}\cosh(nL_1)\right]} \tag{1.74}$$

$$A = -\frac{(T_1 - T_0)}{\left[\cosh(nL_1) + \dfrac{n}{m}\dfrac{\sinh(nL_1)}{\sinh(mL_2)}\cosh(mL_2)\right]} \tag{1.75}$$

The rate of heat transfer can be obtained by using either of the two ways mentioned earlier, that is, using flux at $x = 0$, or by integrating around the lateral surface. In either case we obtain

$$Q = -\pi R^2 k \frac{dT^1(0)}{dx} \tag{1.76}$$

$$Q = 2\pi R h_G L_2 \eta \frac{(T_1 - T_0)}{\left[1 - \dfrac{m \tanh(mL_2)}{n \tanh(nL_1)}\right]} \tag{1.77}$$

where the effectiveness factor η is defined in Eq. 1.63.

You may note the difference between the solution obtained by the level 2 model and that obtained in the third level. Because of the allowance for temperature gradients (which represents the rod's resistance to heat flow) in the segment underneath the solvent surface, the rate of heat transfer calculated at this new level is less than that calculated by the level two model where the rod temperature was taken to be uniform at T_1 below the liquid surface.

This implies from Eq. 1.77 that the heat resistance in the submerged region is negligible compared to that above the surface only when

$$\frac{m \tanh(mL_2)}{n \tanh(nL_1)} \ll 1 \tag{1.78}$$

When the criterion (1.78) is satisfied, the rate of heat transfer given by Model 2 is valid. This is controlled mainly by the ratio $m/n = (h_G/h_L)^{1/2}$, which is always less than unity.

What we have seen in this exercise is simply that higher levels of modeling yield more information about the system and hence provide needed criteria to validate the model one level lower. In our example, the level 3 model provides the criterion (1.78) to indicate when the resistance to heat flow underneath the solvent bath can be ignored compared to that above the surface, and the level 2 model provides the criterion (1.63b) to indicate when there is negligible conduction-resistance in the steel rod.

The next level of modelling is by now obvious: At what point and under what conditions do radial gradients become significant? This moves the modelling exercise into the domain of partial differential equations.

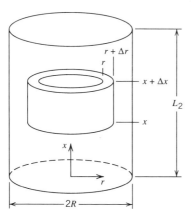

Figure 1.7 Schematic diagram of shell for heat balance.

Level 4

Let us investigate the fourth level of model where we include radial heat conduction. This is important if the rod diameter is large relative to length. Let us assume in this model that there is no resistance to heat flow underneath the solvent interface, so as before, take temperature $T = T_1$ when $x \le 0$. This then leaves only the portion above the solvent surface to study.

Setting up the annular shell shown in Fig. 1.7 and carrying a heat balance in the radial and axial directions, we obtain the following heat conduction equation:

$$(2\pi r\Delta x q_r)|_r - (2\pi r\Delta x q_r)|_{r+\Delta r} + (2\pi r\Delta r q_x)|_x - (2\pi r\Delta r q_x)|_{x+\Delta x} = 0$$

Dividing this equation by $2\pi\Delta r\Delta x$ and taking limits, we obtain

$$-\frac{\partial}{\partial r}(rq_r) - r\frac{\partial q_x}{\partial x} = 0$$

Next, insert the two forms of Fourier's laws

$$q_r = -k\frac{\partial T}{\partial r}; \qquad q_x = -k\frac{\partial T}{\partial x}$$

and get finally,

$$k\left[\frac{1}{r}\frac{\partial}{\partial r}\left(r\frac{\partial T}{\partial r}\right) + \frac{\partial^2 T}{\partial x^2}\right] = 0 \qquad (1.79)$$

Here we have assumed that the conductivity of the steel rod is isotropic and constant, that is, the thermal conductivity k is uniform in both x and r directions, and does not change with temperature.

Equation 1.79 is an elliptic partial differential equation. The physical boundary conditions to give a suitable solution are the following:

$$r = 0, \qquad \frac{\partial T}{\partial r} = 0 \tag{1.80a}$$

$$r = R; \qquad -k\frac{\partial T}{\partial r} = h_G(T - T_0) \tag{1.80b}$$

$$x = 0; \qquad T = T_1 \tag{1.80c}$$

$$x = L_2; \qquad \frac{\partial T}{\partial x} = 0 \tag{1.80d}$$

Equation 1.80a implies symmetry at the center of the rod, whereas at the curved outer surface of the rod the usual Newton cooling condition is applicable (Eq. 1.80b). Equation 1.80d states that there is no heat flow across the flat end of the rod. This is tantamount to saying that either the flat end is insulated or the flat end area is so small compared to the curved surface of the rod that heat loss there is negligible. Solutions for various boundary conditions can be found in Carslaw and Jaeger (1959).

When dealing with simple equations (as in the previous three models), the dimensional equations are solved without recourse to the process of nondimensionalisation. Now, we must deal with partial differential equations, and to simplify the notation during the analysis and also to deduce the proper dimensionless parameters, it is necessary to reduce the equations to nondimensional form. To achieve this, we introduce the following nondimensional variables and parameters:

$$u = \frac{T - T_0}{T_1 - T_0}, \; \xi = \frac{r}{R}, \; \zeta = \frac{x}{L_2} \tag{1.81a}$$

$$\Delta = \left(\frac{R}{L_2}\right), \quad Bi = \frac{h_G R}{k} \quad \text{(Biot number)} \tag{1.81b}$$

where it is clear that only two dimensionless parameters arise: Δ and Bi. The dimensionless heat transfer coefficient $(h_G R/k)$, called the *Biot number*, represents the ratio of convective film transfer to conduction in the metal rod.

The nondimensional relations now become

$$\frac{1}{\xi}\frac{\partial}{\partial \xi}\left(\xi\frac{\partial u}{\partial \xi}\right) + \Delta^2\frac{\partial^2 u}{\partial \zeta^2} = 0 \tag{1.82}$$

$$\xi = 0; \qquad \frac{\partial u}{\partial \xi} = 0 \tag{1.83a}$$

$$\xi = 1; \qquad \frac{\partial u}{\partial \xi} = -Bi\, u \tag{1.83b}$$

$$\zeta = 0; \qquad u = 1 \tag{1.83c}$$

$$\zeta = 1; \qquad \frac{\partial u}{\partial \zeta} = 0 \tag{1.83d}$$

It is clear that these independent variables (ξ and ζ) are defined relative to the maximum possible lengths for the r and x variables, R and L_2, respectively. However, the way u (nondimensional temperature) is defined is certainly not unique. One could easily define u as follows

$$u = \frac{T}{T_0} \quad \text{or} \quad u = \frac{T}{T_1} \quad \text{or} \quad u = \frac{T - T_0}{T_0} \quad \text{or} \quad u = \frac{T - T_0}{T_1} \quad (1.84)$$

and so on. There are good reasons for the selection made here, as we discuss in Chapters 10 and 11. The solution of Eq. 1.82 subject to boundary conditions (1.83) is given in Chapter 11 and its expression is given here only to help illustrate model hierarchy. The solution u is

$$u = \frac{T - T_0}{T_1 - T_0} = \sum_{n=1}^{\infty} \frac{\langle 1, K_n \rangle}{\langle K_n, K_n \rangle} K_n(\xi) \frac{\cosh\left[\frac{\beta_n}{\Delta}(1 - \zeta)\right]}{\cosh\left[\frac{\beta_n}{\Delta}\right]} \quad (1.85)$$

where the functions are defined as

$$K_n(\xi) = J_0(\beta_n \xi) \quad (1.86a)$$

and the many characteristic values (eigenvalues) are obtained by trial-and-error from

$$\beta_n J_1(\beta_n) = Bi J_0(\beta_n) \quad (1.86b)$$

The other functional groups are defined as

$$\langle 1, K_n \rangle = \frac{J_1(\beta_n)}{\beta_n} \quad (1.86c)$$

$$\langle K_n, K_n \rangle = \frac{J_1^2(\beta_n)}{2}\left[1 + \left(\frac{\beta_n}{Bi}\right)^2\right] \quad (1.86d)$$

where $J_0(\beta)$ and $J_1(\beta)$ are tabulated relations called Bessel functions, which are discussed at length in Chapter 3. The rate of heat transfer can be calculated using the heat flux entering at position $x = 0$, but we must also account for radial variation of temperature so that the elemental area is $2\pi r\,dr$; thus integrating over the whole base gives

$$Q = \int_0^R \left[-k\frac{\partial T}{\partial x}\right]_{x=0} 2\pi r\,dr \quad (1.87)$$

Putting this in nondimensional form, we have

$$Q = \frac{2\pi R^2 k}{L_2}(T_1 - T_0)\int_0^1 \left(-\frac{\partial u(0)}{\partial \zeta}\right)\xi\,d\xi \quad (1.88)$$

Inserting dimensionless temperature from Eq. 1.85, we obtain the following somewhat complicated result for heat transfer rate:

$$Q = \frac{2\pi R^2 k(T_1 - T_0)}{L_2 \Delta} \sum_{n=1}^{\infty} \frac{\beta_n \langle 1, K_n \rangle^2}{\langle K_n, K_n \rangle} \tan h\left(\frac{\beta_n}{\Delta}\right) \qquad (1.89)$$

This illustrates how complexity grows quickly as simplifications are relaxed.

For small $Bi \ll 1$, it is not difficult to show from the transcendental equation (1.86b) that the smallest eigenvalue is

$$\beta_1 \cong (2Bi)^{1/2} \qquad (1.90)$$

Substituting this into Eq. 1.89, we will obtain Eq. 1.63a. Thus, the fourth model shows that the radial heat conduction inside the rod is unimportant when

$$Bi \ll 1 \qquad (1.91)$$

In summary, we have illustrated how proper model hierarchy sets limits on the lower levels. In particular, one can derive criteria (like Eq. 1.91) to show when the simpler models are valid. Some solutions for the simpler models can be found in Walas (1991).

The obvious question arises: When is a model of a process good enough? This is not a trivial question, and it can only be answered fully when the detailed economics of design and practicality are taken into account. Here, we have simply illustrated the hierarchy of one simple process, and how to find the limits of validity of each more complicated model in the hierarchy. In the final analysis, the user must decide when tractability is more important than precision.

1.7 REFERENCES

1. Bird, R. B., W. E. Stewart, and E.N. Lightfoot. *Transport Phenomena*. John Wiley & Sons, Inc., New York (1960).
2. Carslaw, H. S., and J. C. Jaeger, *Conduction of Heat in Solids*, 2nd ed., Oxford University Press, New York (1959).
3. Rice, R.G., "Approximate Solutions for Batch, Packed Tube and Radial Flow Adsorbers—Comparison with Experiment," *Chem. Eng. Sci.* 37, 83–97 (1982).
4. Walas, S.M., *Modeling with Differential Equations in Chemical Engineering*, Butterworth-Heinemann, Boston (1991).

1.8 PROBLEMS

1.1₁. Length Required for Cooling Coil

A cooling coil made of copper tube is immersed in a regulated constant temperature bath held at a temperature of 20° C. The liquid flowing through the tube enters at 22° C, and the coil must be sufficiently long to ensure the exit liquid sustains a temperature of 20.5° C. The bath is so well stirred that heat transfer resistance at the tube–bath interface is minimal, and the copper wall resistance can also be ignored. Thus, the tube wall temperature can be taken equal to the bath temperature. Use Eq. 1.17 to estimate the required tube length (L) under the following conditions for the flowing liquid:

C_p = 1 kcal/kg $\cdot°$ C
R = 0.01 m
v_0 = 1 m/s
ρ = 10^3 kg/m^3
μ = 0.001 kg/m $\cdot$ s
k = 1.43 $\cdot$ 10^{-4} kcal/(s $\cdot$ m $\cdot$ 3° K)

Since the Reynolds number is in the turbulent range, use the correlation of Sieder and Tate (Bird et al. 1960) to calculate h

$$Nu = 0.026 \, Re^{0.8} \, Pr^{1/3}$$

where

$$Nu = \frac{hD}{k} \quad (D = 2R), \quad \text{Nusselt number}$$

$$Pr = \frac{C_p \mu}{k}, \quad \text{Prandtl number}$$

$$Re = \frac{Dv_0\rho}{\mu}, \quad \text{Reynolds number}$$

Answer: $(L/D) = 353.5$

1.2$_2$. Cooling of Fluids in Tube Flow: Locally Varying h

Apply the conditions of Model I (plug flow) and rederive the expression for temperature change when the heat transfer coefficient is not constant but varies according to the law:

$$h = \gamma/\sqrt{z}$$

Answer: $\psi = \exp\left(-2\beta\sqrt{z}\right), \quad \beta = (2\pi R\gamma)/(v_0 A\rho C_p)$

1.3$_2$. Dissolution of Benzoic Acid

Initially pure water is passed through a tube constructed of solid benzoic acid. Since benzoic acid is slightly soluble in water (denote solubility as C^* moles acid/cm^3 solution), the inner walls of the tube will dissolve very slowly. By weighing the dried tube before and after exposure, it is possible to calculate the rate of mass transfer.

(a) Take a quasisteady state material balance for plug velocity profiles and show that the ODE obtained is

$$-v_0\frac{dC}{dx} + k_C\left(\frac{4}{D}\right)(C^* - C) = 0$$

where D denotes the inner tube diameter (taken as approximately invariant), v_0 is liquid velocity, and k_C is the (constant) mass transfer coefficient.

(b) Define $\theta = (C - C^*)$ and show that the solution to part (a) is

$$\theta = K \exp\left(-\frac{4}{D}\frac{k_C}{v_0}x\right)$$

(c) If pure water enters the tube, evaluate K and obtain the final result

$$\frac{C(x)}{C^*} = 1 - \exp\left(-\frac{4}{D}\frac{k_C}{v_0}x\right)$$

(d) If the tube is allowed to dissolve for a fixed time Δt, show that the weight change can be calculated from

$$\Delta W = M_B C^* \Delta t v_0 \left(\frac{\pi}{4}D^2\right)\left[1 - \exp\left(-\frac{4}{D}\frac{k_C}{v_0}L\right)\right]$$

where L is tube length, and M_B is molecular weight acid.

(e) Rearrange the result in part (d) to solve directly for k_C, under condition when $4k_C L/Dv_0 < 1$, and show

$$k_C \approx \frac{\Delta W}{[M_B C^* \Delta t \pi DL]}$$

(f) Discuss the assumptions implied in the above analysis and deduce a method of estimating the maximum possible experimental error in calculating k_C; note, experimental quantities subject to significant errors are: ΔW, Δt, and D.

1.4$_1$. Lumped Thermal Model for Thermocouple

We wish to estimate the dynamics of a cold thermocouple probe suddenly placed in a hot flowing fluid stream for the purpose of temperature measurement. The probe consists of two dissimilar metal wires joined by soldering at the tip, and the wires are then encased in a metal sheath and the tip is finally coated with a bead of plastic to protect it from corrosion. Take the mass of the soldered tip plus plastic bead to be m, with specific heat C_p. Denote the transfer coefficient as h.

(a) If the effects of thermal conductivity can be ignored, show that the temperature response of the probe is described by

$$mC_p\frac{dT}{dt} = hA(T_f - T)$$

where A denotes the exposed area of probe tip, and $T(t)$ is its temperature.

(b) Lump the explicit parameters to form the system time constant, and for constant T_f, define a new variable $\theta = (T_f - T)$ and show that the compact form results

$$\tau \frac{d\theta}{dt} = -\theta$$

where the system time constant is defined as

$$\tau = \frac{mC_p}{hA} \quad (\text{sec})$$

(c) Integrate the expression in (b), using the initial condition $T(0) = T_0$ and show that

$$\frac{T - T_f}{T_0 - T_f} = \exp\left(-\frac{t}{\tau}\right)$$

(d) Rearrange the expression in (c) to obtain

$$\frac{T_0 - T}{T_0 - T_f} = 1 - \exp\left(-\frac{t}{\tau}\right)$$

and thus show the temperature excess is 63% of the final steady-state value after a time equivalent to one time constant has elapsed. This also represents a quick and easy way to deduce system time constants based on an elementary experiment.

1.5$_3$. Distributed Thermal Model for Thermocouple

If the plastic bead covering the tip of the thermocouple described in Problem 1.4 is quite large, and since plastic usually sustains a very low value of thermal conductivity, then the simple lumped model solution becomes quite inaccurate. To improve the model, we need to account for thermal conductivity in the (assumed) spherical shape of the plastic bead.

(a) Assuming the bead is a perfect sphere, contacted everywhere by external fluid of temperature T_f, perform a shell balance on an element of volume $4\pi r^2 \Delta r$ and show that

$$\rho C_p \frac{\partial T}{\partial t} = k \frac{1}{r^2} \frac{\partial}{\partial r}\left(r^2 \frac{\partial T}{\partial r}\right)$$

(b) Perform an elemental heat balance at the surface of the sphere and deduce

$$-k \frac{\partial T}{\partial r} = h(T - T_f) \quad \text{at} \quad r = R$$

where R is the radius of the plastic sphere.

1.6$_2$. Modeling of Piston with Retaining Spring

The schematic figure shows a piston fitted snugly into a cylinder. The piston is caused to move by increasing or decreasing pressure P. As air is admitted by way of valve V1, the increased pressure drives the piston to

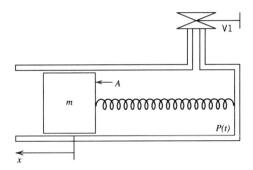

the left, while the attached spring exerts a force to restrain the piston. At the same time, a highly viscous lubricant sealant at the juncture of piston and cylinder exerts a resisting force to damp the piston movement; the forces can be represented by

$$F_\mu = \alpha\mu \frac{dx}{dt}; \qquad \mu = \text{lubricant viscosity}$$

$$F_x = Kx; \qquad K = \text{spring constant}$$

(a) Perform a force balance on the piston and show that

$$m\frac{d^2x}{dt^2} + \alpha\mu\frac{dx}{dt} + Kx = AP(t)$$

(b) Arrange this equation to obtain the standard form of the damped inertial equation:

$$\tau^2\frac{d^2x}{dt^2} + 2\zeta\tau\frac{dx}{dt} + x = f(t)$$

and hence, deduce an expression for damping coefficient ζ and time constant τ. Such equations are used to model pneumatic valves, shock absorbers, manometers, and so on.

Answer: $\tau = \sqrt{m/K}, \qquad \zeta = \alpha\mu/2\sqrt{Km}$

1.7₃. Mass Transfer in Bubble Column

Bubble columns are used for liquid aeration and gas–liquid reactions. Thus, finely suspended bubbles produce large interfacial areas for effective mass transfer, where the contact area per unit volume of emulsion is calculated from the expression $a = 6\varepsilon/d_B$, where ε is the volume fraction of injected gas. While simple to design and construct, bubble columns sustain rather large eddy dispersion coefficients, and this must be accounted for in the modeling process. For cocurrent operation, liquid of superficial velocity u_{0L} is injected in parallel with gas superficial velocity u_{0G}. The liquid velocity profile can be taken as plug shaped, and the gas voidage can be treated as uniform throughout the column. We wish to model a column used to aerate water, such that liquid enters with a composition C_0. Axial dispersion can be modeled using a Fickian-like relationship

$$J = -D_e \frac{dC}{dx} \qquad \left(\frac{\text{moles}}{\text{liquid area} - \text{time}}\right)$$

while the solubility of dissolved oxygen is denoted as C^*. We shall denote distance from the bottom of the column as x.

(a) Derive the steady-state oxygen mole balance for an incremental volume of $A\Delta x$ (A being the column cross-sectional area) and show that the liquid phase balance is

$$(1 - \varepsilon) D_e \frac{d^2 C}{dx^2} - u_{0L} \frac{dC}{dx} + k_c a (C^* - C) = 0$$

(b) Lump parameters by defining a new dimensionless length as

$$z = \frac{k_c a x}{u_{0L}}$$

and define the excess concentration as $y = (C - C^*)$, and so obtain the elementary, second order, ordinary differential equation

$$\alpha \frac{d^2 y}{dz^2} - \frac{dy}{dz} - y = 0$$

where

$$\alpha = \frac{(1 - \varepsilon) D_e k_c a}{u_{0L}^2} \qquad \text{(dimensionless)}$$

Note, the usual conditions in practice are such that $\alpha \leq 1$.

(c) Perform a material balance at the entrance to the column as follows. Far upstream, the transport of solute is mainly by convection: $Au_{0L}C_0$. At the column entrance ($x = 0$), two modes of transport are present, hence, show that one of the boundary conditions should be:

$$u_{0L} C_0 = -(1 - \varepsilon) D_e \frac{dC}{dx}\bigg|_{x=0} + u_{0L} C(0)$$

The second necessary boundary condition is usually taken to be

$$\frac{dC}{dx} = 0 \qquad \text{at} \qquad x = L$$

where L denotes the position of the liquid exit. What is the physical meaning of this condition? The two boundary conditions are often referred as the Danckwerts type, in honor of P. V. Danckwerts.

1.8₃. Dissolution and Reaction of Gas in Liquids

Oxygen dissolves into and reacts irreversibly with aqueous sodium sulphite solutions. If the gas solubility is denoted as C_A^* at the liquid–gas interface, derive the elementary differential equation to describe the steady-state composition profiles of oxygen in the liquid phase when the rate of oxygen reaction is represented by $R_A = kC_A^n$ and the local oxygen diffusion flux is described by $J_A = -D_A \, dC_A/dz$, where D_A is diffusivity and z is distance from the interface into the liquid.

$$\text{Answer: } D_A \frac{d^2 C_A}{dz^2} - kC_A^n = 0$$

1.9₃. Modeling of a Catalytic Chemical Reactor

Your task as a design engineer in a chemical company is to model a fixed bed reactor packed with the company proprietary catalyst of spherical shape. The catalyst is specific for the removal of a toxic gas at very low concentration in air, and the information provided from the catalytic division is that the reaction is first order with respect to the toxic gas concentration. The reaction rate has units of moles of toxic gas removed per mass of catalyst per time.

The reaction is new and the rate constant is nonstandard, that is, its value does not fall into the range of values known to your group of design engineers. Your first attempt, therefore, is to model the reactor in the simplest possible way so that you can develop some intuition about the system before any further modeling attempts are made to describe it exactly.

(a) For simplicity, assume that there is no appreciable diffusion inside the catalyst and that diffusion along the axial direction is negligible. An isothermal condition is also assumed (this assumption is known to be invalid when the reaction is very fast and the heat of reaction is high). The coordinate z is measured from entrance of the packed bed. Perform the mass balance around a thin shell at the position z with the shell thickness of Δz and show that in the limit of the shell thickness approaching zero, the following relation is obtained

$$u_0 \frac{dC}{dz} = -(1 - \varepsilon)\rho_p(kC)$$

where u_0 is the superficial velocity, C is the toxic gas concentration, ε is the bed porosity, ρ_p is the catalyst density, and (kC) is the chemical reaction rate per unit catalyst mass.

(b) Show that this lumped parameter model has the solution

$$\ln\left(\frac{C}{C_0}\right) = -\frac{(1 - \varepsilon)\rho_p k}{u_0} z$$

where C_0 denotes the entrance condition.

(c) The solution given in part (b) yields the distribution of the toxic gas concentration along the length of the reactor. Note the exponential decline of the concentration. Show that the toxic concentration at the exit, which is required to calculate the conversion, is

$$C_L = C_0 \exp\left[-\frac{(1 - \varepsilon)\rho_p kL}{u_0} \right]$$

(d) In reactor design, the information normally provided is throughput and mass of catalyst. So if you multiply the denominator and numerator of the bracketed term in the last part by the cross-sectional area, A, show that the exit concentration is

$$C_L = C_0 \exp\left[-\frac{Wk}{F} \right]$$

where W is the mass of catalyst, and F is the volumetric gas flow rate.

(e) The dimensionless argument Wk/F is a key design parameter. To achieve a 95% conversion, where the conversion is defined as

$$X = \frac{C_0 - C_L}{C_0}$$

show that the nondimensional group Wk/F must be equal to 3. This means that if the throughput is provided and the rate constant is known the mass of catalyst required is simply calculated as

$$W = 3\frac{F}{k}$$

(f) The elementary Model 1 is of the lumped parameter type, and its validity is questionable because of a number of assumptions posed. To check its validity, you wish to relax one of the assumptions and move one level up the hierarchy ladder. Suppose you relax the axial diffusion assumption, and hence show that the mass balance, when diffusion is important, becomes

$$D\varepsilon\frac{d^2C}{dz^2} - u_0\frac{dC}{dz} - (1 - \varepsilon)\rho_p kC = 0$$

Since this is a second order ordinary differential equation, two boundary conditions must be required. The two possible conditions after Danckwerts are

$$z = 0; \qquad u_0 C_0 = u_0 C|_{z=0} - D\varepsilon\frac{dC}{dz}\Big|_{z=0}$$

$$z = L; \qquad \frac{dC}{dz} = 0$$

(g) Define the following nondimensional variables and parameters

$$y = \frac{C}{C_0}; \quad x = \frac{z}{L}; \quad Pe = \frac{u_0 L}{D\varepsilon}; \quad N = \frac{Wk}{F}$$

and show that the resulting modeling equations are

$$\frac{1}{Pe}\frac{d^2 y}{dx^2} - \frac{dy}{dx} - Ny = 0$$

$$x = 0; \quad 1 = y|_{x=0} - \frac{1}{Pe}\frac{dy}{dx}\bigg|_{x=0}$$

$$x = 1; \quad \frac{dy}{dx} = 0$$

Compare this model (hereafter called Model 2) with Model 1, and show that the axial diffusion may be ignored when $Pe \gg 1$ (this can be accomplished several ways: by decreasing porosity ε or by reducing D, or by increasing velocity or length).

(h) To study the effect of the mass transfer inside the catalyst particle, we need to remove the assumption of no diffusion resistance inside the particle. This means that the mass balance within the particle must be linked with the external composition. To investigate this effect, we shall ignore the axial diffusion (which is usually small for packing made up of finely granulated solid) and the external film resistance surrounding the particle.

Set up a thin spherical shell (control volume) inside the particle, and show that the mass balance equation is

$$D_e \frac{1}{r^2}\frac{\partial}{\partial r}\left(r^2 \frac{\partial C_p}{\partial r}\right) - \rho_p k C_p = 0$$

where C_p is the toxic gas concentration within the particle, and D_e is the effective diffusivity and is defined as Fickian-like:

$$J_p = -D_e \frac{\partial C_p}{\partial r} \quad \left\{\frac{\text{moles transported by diffusion}}{\text{cross-sectional area-time}}\right\}$$

and suitable boundary conditions for negligible film resistance and particle symmetry are

$$C_p(R) = C; \quad \partial C_p/\partial r = 0 \quad \text{at} \quad r = 0$$

where R denotes particle radius.

(i) Next, set up the mass balance around the thin element spanning the whole column cross section (as in Model 1), but this time the control volume will exclude the catalyst volume. This means that material is lost to the various sinks made up by the particles. Show that the mass balance equation on this new control volume is

$$-u_0 \frac{dC}{dz} = (1 - \varepsilon)\frac{3}{R}D_e \frac{\partial C_p}{\partial r}\bigg|_{r=R}$$

Chapter 2

Solution Techniques for Models Yielding Ordinary Differential Equations (ODE)

2.1 GEOMETRIC BASIS AND FUNCTIONALITY

In Chapter 1, we took the point of view that elementary differential balances gave rise to certain differential equations, and for first order systems, we could represent these in general by

$$\frac{dy}{dx} = f(x, y) \tag{2.1}$$

where $f(x,y)$ is some arbitrary, but known, function $((x^2 + y^2)^{1/2}$, xy^2, etc.). Our strategy was to find suitable solutions to the ODE so that we could compute y for any given x, such as

$$y = g(x, c) \tag{2.2}$$

where g(x,c) is some elementary function ($cx^{1/2}$, $c\sin x$, $\sinh(c + x)^{1/2}$, etc.), and c is some arbitrary constant of integration, to be found from a boundary, or initial condition. The most elementary first step in this strategy is to separate variables, so that direct integration can take place; that is, $\int g(x)dx = \int f(y)dy$. All of the techniques to follow aim for this final goal.

However, the early mathematicians were also concerned with the *inverse problem*: Given the geometric curve $y = g(x)$, what is the underlying differential equation defining the curve? Thus, many of the classical mathematical methods take a purely geometric point of view, with obvious relevance to observations in astronomy.

Suppose we have a family of curves; represented by the function

$$f(x, y; \lambda) = 0 \qquad (2.3)$$

where λ is an arbitrary parameter. Note, each distinct value of λ will generate a separate curve in the y-x coordinate system, so we then speak of a *family of curves*.

To determine the defining ODE, we write the total differential of the family using the chain rule:

$$df = 0 = \frac{\partial f}{\partial x} dx + \frac{\partial f}{\partial y} dy \qquad (2.4)$$

If we eliminate λ from this using Eq. 2.3, then the sought after ODE results.

EXAMPLE 2.1

Suppose a family of curves can be represented by

$$x^2 + y^2 - \lambda x = 0 \qquad (2.5)$$

Deduce the defining differential equation.
Write the total differential, taking

$$f(x, y) = x^2 + y^2 - \lambda x = 0 \qquad (2.6)$$

hence,

$$\frac{df}{dx} = \frac{\partial f}{\partial x} + \frac{\partial f}{\partial y}\frac{dy}{dx} = 2x - \lambda + 2y\frac{dy}{dx} = 0 \qquad (2.7)$$

Eliminate λ using Eq. 2.6 to obtain finally,

$$2yx\frac{dy}{dx} + x^2 - y^2 = 0 \qquad (2.8)$$

Quite often, taking a geometric point of view will assist in resolving a solution. Observation of the total differential may also lead to solutions. Thus, consider the relation

$$\frac{dy}{dx} - \left(\frac{y}{x}\right) = 0 \qquad (2.9)$$

which is in fact the exact differential

$$d\left(\frac{y}{x}\right) = 0 \qquad (2.10)$$

Hence, direct integration yields

$$y = cx \tag{2.11}$$

In the following section, we shall exploit this property of *exactness* in a formal way to effect solutions to certain classes of problems.

2.2   CLASSIFICATION OF ODE

The broad classification of equations with which engineers must contend are

(a) Linear equations.
(b) Nonlinear equations.

Much is known about linear equations, and in principle all such equations can be solved by well-known methods. On the other hand, there exists no general solution method for nonlinear equations. However, a few special types are amenable to solution, as we show presently.

We begin this chapter by studying first order ODE, classifying techniques based on certain properties. We next extend these techniques to second and higher order equations. Finally, we present detailed treatment and very general methods for classes of linear equations, about which much is known.

2.3   FIRST ORDER EQUATIONS

The most commonly occurring first order equation in engineering analysis is the linear first order equation (the **I**-factor equation)

$$\frac{dy}{dx} + \alpha(x)y = f(x) \tag{2.12}$$

which is sometimes called the first order equation with forcing function, $f(x)$ being the forcing function. We saw an example of this equation in Chapter 1, Eq. 1.9

$$\frac{dT}{dz} + \lambda T(z) = \lambda T_w(z) \tag{2.13}$$

where we have denoted locally varying wall temperature $T_w(z)$, rather than the special case $T_w = $ constant. If the wall temperature was fixed and constant, we showed earlier that a common sense change of variables allowed direct integration after separation of variables. However, such is not the case if the wall temperature varies with position z.

For the general case, given in functional form as Eq. 2.12, we shall allow the coefficient α and the forcing function f to vary with respect to the independent variable x. It is clear that separation of variables is not possible in the present state.

To start the solution of Eq. 2.12, we put forth the proposition that there exists an elementary, separable solution

$$\frac{d}{dx}[\mathbf{I}(x)y] = \mathbf{I}(x)f(x) \tag{2.14}$$

If such a form existed, then variables are separated and the solution is straightforward by direct integration

$$y = \frac{1}{\mathbf{I}(x)} \int \mathbf{I}(x)f(x) \, dx + \frac{c}{\mathbf{I}(x)} \tag{2.15}$$

where c is an arbitrary constant of integration.

To prove such a solution exists, we must specify the function $\mathbf{I}(x)$. To do this, we rearrange Eq. 2.14 into exactly the same form as Eq. 2.12

$$\frac{dy}{dx} + \frac{1}{\mathbf{I}(x)} \frac{d\mathbf{I}(x)}{dx} y = f(x) \tag{2.16}$$

In order for these two equations to be identical, we must require

$$\frac{1}{\mathbf{I}(x)} \frac{d\mathbf{I}(x)}{dx} = \alpha(x) \tag{2.17}$$

or, within a constant of integration

$$\mathbf{I}(x) = \exp\left(\int \alpha(x) \, dx\right) \tag{2.18}$$

where $\mathbf{I}(x)$ is called the *integrating factor*. This is the necessary and sufficient condition for the general solution, Eq. 2.15, to exist.

EXAMPLE 2.2

To solve the differential equation

$$\frac{dy}{dx} - \frac{2}{x} y = x \tag{2.19}$$

we first determine the I-factor

$$\mathbf{I} = \exp\left(\int \left(-\frac{2}{x}\right) dx\right) = \exp(-2\ln x) = \frac{1}{x^2} \tag{2.20}$$

Hence,

$$y = x^2 \int \frac{1}{x^2} x \, dx + cx^2 \tag{2.21}$$

so finally,

$$y = x^2 \ln x + cx^2 \tag{2.22}$$

Two other types of equations occur so frequently they deserve special treatment; namely,

(a) Equations possessing *exact* solutions,
(b) Equations composed of homogeneous functions.

2.3.1 Exact Solutions

As mentioned in Section 2.1, occasionally a solution exists which is an exact differential

$$d\varphi(x, y) = 0 \tag{2.23}$$

but the functional relation φ is not obvious to the untrained eye. If such a function exists, then it must also be true, according to the chain rule, that

$$d\varphi = \frac{\partial \varphi}{\partial x} dx + \frac{\partial \varphi}{\partial y} dy = 0 \tag{2.24}$$

But how do we use this information to find y as a function x? Actually, the key to uncovering the existence of an exact solution resides in the well-known property of continuous functions, which stipulates

$$\frac{\partial}{\partial x} \left(\frac{\partial \varphi}{\partial y} \right) = \frac{\partial}{\partial y} \left(\frac{\partial \varphi}{\partial x} \right) \tag{2.25}$$

Thus, suppose there exists an equation of the form

$$M(x, y) \, dx + N(x, y) \, dy = 0 \tag{2.26}$$

By comparing this with Eq. 2.24, we conclude it must be true

$$M(x, y) = \frac{\partial \varphi}{\partial x}, \qquad N(x, y) = \frac{\partial \varphi}{\partial y} \tag{2.27}$$

if φ is to exist as a possible solution. By invoking the continuity condition, Eq. 2.25, then the necessary and sufficient condition for φ to exist is

$$\frac{\partial N}{\partial x} = \frac{\partial M}{\partial y} \tag{2.28}$$

EXAMPLE 2.3

Solve the equation

$$(2xy^2 + 2)\, dx + (2x^2y + 4y)\, dy = 0 \qquad (2.29)$$

This equation is exact, since

$$\frac{\partial M}{\partial y} = 4xy, \qquad \frac{\partial N}{\partial x} = 4xy \qquad (2.30)$$

The unknown function $\varphi(x,y)$ must therefore be described by the relations

$$\frac{\partial \varphi}{\partial x} = 2xy^2 + 2 \qquad (2.31)$$

$$\frac{\partial \varphi}{\partial y} = 2x^2y + 4y \qquad (2.32)$$

Note, the partial of φ with respect to (wrt) x implies holding y constant, while the partial wrt y implies holding x constant. We shall use this important information to help find the connection between y and x.

First, we integrate Eq. 2.31 with respect to x (holding y constant)

$$\varphi = x^2y^2 + 2x + f(y) \qquad (2.33)$$

Note, since y was held constant, we must add an arbitrary function, $f(y)$, rather than an arbitrary constant to be perfectly general in the analysis. Next, we insert Eq. 2.33 into Eq. 2.32

$$2x^2y + \frac{df}{dy} = 2x^2y + 4y \qquad (2.34)$$

so we see that integration yields

$$f(y) = 2y^2 + C_2 \qquad (2.35)$$

and finally, adding this to Eq. 2.33 yields

$$\varphi = x^2y^2 + 2x + 2y^2 + C_2 \qquad (2.36)$$

This is not the most useful form for our result. Now, since Eq. 2.23 integrates to yield $\varphi = C_1$, then φ also equals to some arbitrary constant. Combining C_1 and C_2 into another arbitrary constant yields the sought after connection

$$x^2y^2 + 2x + 2y^2 = K = C_1 - C_2 \qquad (2.37)$$

We could write this as $y = f(x)$ as follows

$$y^2 = \frac{(K - 2x)}{(x^2 + 2)} \qquad (2.38)$$

Note further that two possible branches exist

$$y = \pm \sqrt{\frac{(K - 2x)}{(x^2 + 2)}} \qquad (2.39)$$

We see the arbitrary constant is implicit, the expected case for nonlinear equations.

2.3.2   Equations Composed of Homogeneous Functions

A function $f(x,y)$ is said to be homogeneous of degree n if there exists a constant n such that for every parameter λ

$$f(\lambda x, \lambda y) = \lambda^n f(x, y) \qquad (2.40)$$

Thus, the functions $x^2 + xy$ and $\tan(x/y)$ are both homogeneous, the first of degree 2 and the second of degree 0. However, the function $(y^2 + x)$ is *not* homogeneous.

The first order equation

$$P(x, y)\, dx + Q(x, y)\, dy = 0 \qquad (2.41)$$

is said to be homogeneous if P and Q are *both* homogeneous of the same degree n, for some constant n (including zero).

This implies that first order equations composed of homogeneous functions can always be arranged in the form

$$\frac{dy}{dx} = f\left(\frac{y}{x}\right) \qquad (2.42)$$

This form is easy to remember, since the dimensional ratio y/x appears throughout. The occurrence of this ratio suggests the substitution $y/x = v(x)$.

EXAMPLE 2.4

The nonlinear equation

$$y^2 + x^2 \frac{dy}{dx} = xy \frac{dy}{dx} \qquad (2.43)$$

can be rearranged to the form

$$\frac{dy}{dx} = \frac{\left(\frac{y}{x}\right)^2}{\left(\frac{y}{x} - 1\right)}$$

(2.44)

which is clearly homogeneous. Replacing $y/x = v(x)$, shows

$$\frac{dy}{dx} = x\frac{dv}{dx} + v$$

(2.45)

so the equation for $v(x)$ is

$$x\frac{dv}{dx} + v = \frac{v^2}{v-1}$$

(2.46)

or

$$x\frac{dv}{dx} = \frac{v}{v-1}$$

(2.47)

Separation of variables yields

$$\frac{(v-1)}{v}\,dv = \frac{dx}{x}$$

(2.48)

Integrating term by term produces

$$v - \ln v = \ln x + \ln K$$

(2.49)

where $\ln K$ is an arbitrary constant of integration. Using the properties of logarithms shows

$$Kx = \frac{\exp(v)}{v} = \frac{\exp\left(\frac{y}{x}\right)}{\left(\frac{y}{x}\right)}$$

(2.50)

This is an *implicit* relation between y and x, which is typical of nonlinear solutions.

We have reviewed standard methods of solutions, including equations that are exact, those that contain homogeneous functions (of the same degree), and the frequently occurring **I**-factor equation. All of these methods include, in the final analysis, making the dependent and independent variables separable, so that direct integration completes the solution process.

However, some equations are not amenable to standard methods and require considerable ingenuity to effect a solution. Often, an elementary change of variables reduces a nonlinear equation to a linear form. We illustrate some of these uncommon types with examples in the next sections.

2.3.3   Bernoulli's Equation

The Bernoulli equation

$$\frac{dy}{dx} + P(x)y = Q(x)y^n; \qquad n \neq 1 \tag{2.51}$$

is similar to the first order forced equation (**I**-factor) discussed earlier, except for the nonlinear term on the right-hand side, y^n.

If we divide y^n throughout, a substitution may become evident

$$y^{-n}\frac{dy}{dx} + P(x)y^{-n+1} = Q(x) \tag{2.52}$$

Inspecting the first term suggests incorporating the coefficient into the differential, so that

$$y^{-n}\frac{dy}{dx} = \frac{1}{1-n}\frac{d(y^{1-n})}{dx} \tag{2.53}$$

This immediately reveals the substitution

$$v = y^{1-n} \tag{2.54}$$

hence, the original equation is now linear in v

$$\left(\frac{1}{1-n}\right)\frac{dv}{dx} + P(x)v = Q(x) \tag{2.55}$$

which is easily solved using the **I**-factor method.

2.3.4   Riccati's Equation

A nonlinear equation, which arises in both continuous and staged (i.e., finite difference) processes, is Riccati's equation

$$\frac{dy}{dx} = P(x)y^2 + Q(x)y + R(x) \tag{2.56}$$

A frequently occurring special form is the case when $P(x) = -1$

$$\frac{dy}{dx} + y^2 = Q(x)y + R(x) \tag{2.57}$$

A change of variables given by

$$y = \frac{1}{u}\frac{du}{dx} \tag{2.58}$$

yields the derivative

$$\frac{dy}{dx} = \frac{1}{u} \cdot \frac{d^2u}{dx^2} - \frac{1}{u^2} \cdot \left(\frac{du}{dx}\right)^2 \tag{2.59}$$

Inserting this into Eq. 2.57 eliminates the nonlinear term

$$\frac{d^2u}{dx^2} - Q(x)\frac{du}{dx} - R(x)u = 0 \tag{2.60}$$

which is a linear second order equation with nonconstant coefficients. This may be solved in general by the Frobenius series method, as will be shown in Chapter 3.

EXAMPLE 2.5

A constant-volume batch reactor undergoes the series reaction sequence

$$A \xrightarrow{k_1} B \xrightarrow{k_2} C$$

The initial concentration of A is denoted by C_{A0}, whereas B and C are initially nil. The reaction rates per unit reactor volume are described by

$$R_A = k_1 C_A^n, \qquad R_B = k_1 C_A^n - k_2 C_B^m$$

Find the solutions of the differential equations describing $C_B(t)$ for the following cases

(a) $n = 1, \qquad m = 2$
(b) $n = 2, \qquad m = 1$
(c) $n = 1, \qquad m = 1$

CASE (a) $n = 1, \qquad m = 2$

The material balances are written as

$$\frac{dC_A}{dt} = -k_1 C_A$$

$$\frac{dC_B}{dt} = k_1 C_A - k_2 C_B^2$$

The solution for C_A is straightforward

$$C_A = C_{A0} \exp(-k_1 t) \tag{2.61}$$

Hence, the expression for C_B is nonlinear with exponential forcing

$$\frac{dC_B}{dt} = k_1 C_{A0} \exp(-k_1 t) - k_2 C_B^2 \qquad (2.62)$$

If we scale time by replacing $\theta = k_2 t$, the above expression becomes identical to the special form of the Riccati equation. Thus, on comparing with Eq. 2.57, we take

$$Q(\theta) = 0, \qquad R(\theta) = \frac{k_1}{k_2} C_{A0} \exp\left(-\frac{k_1}{k_2}\theta\right)$$

to see that

$$\frac{dC_B}{d\theta} + C_B^2 = R(\theta) \qquad (2.63)$$

If we make the Riccati transformation

$$C_B = \frac{1}{u}\frac{du}{d\theta}$$

we finally obtain

$$\frac{d^2 u(\theta)}{d\theta^2} - \frac{k_1}{k_2} C_{A0} \exp\left(-\frac{k_1}{k_2}\theta\right) u(\theta) = 0 \qquad (2.64)$$

We have thus transformed a nonlinear first order equation to a solvable, linear second order equation.

CASE (b) n = 2,   m = 1

The simultaneous equations for this case are

$$\frac{dC_A}{dt} = -k_1 C_A^2 \quad \therefore \quad C_A = \left[\frac{C_{A0}}{1 + k_1 C_{A0} t}\right]$$

$$\frac{dC_B}{dt} = k_1 C_A^2 - k_2 C_B$$

Inserting $C_A(t)$ yields the classic inhomogeneous (**I**-factor) equation discussed in Section 2.3:

$$\frac{dC_B}{dt} + k_2 C_B = k_1 \left[\frac{C_{A0}}{1 + k_1 C_{A0} t}\right]^2 \qquad (2.65)$$

The integrating factor is $\mathbf{I} = \exp(k_2 t)$; hence, the solution is

$$C_B(t) = k_1 \exp(-k_2 t) \int \exp(k_2 t) \left[\frac{C_{A0}}{1 + k_1 C_{A0} t} \right]^2 dt + C \exp(-k_2 t) \quad (2.66)$$

where C is the constant of integration. The integral is tabulated in the form

$$\int \frac{\exp(ax)}{x^2} dx$$

so we next substitute

$$\tau = 1 + k_1 C_{A0} t, \quad dt = \frac{d\tau}{(k_1 C_{A0})}, \quad a = \frac{k_2}{k_1 C_{A0}}$$

hence, we obtain

$$C_B = C_{A0} \exp(-k_2 t) \exp\left(-\frac{k_2}{k_1 C_{A0}}\right) \int \frac{\exp(a\tau)}{\tau^2} d\tau + C \exp(-k_2 t) \quad (2.67)$$

Performing the integration yields finally,

$$C_B = C \exp(-k_2 t) + C_{A0} \exp\left[-\left(k_2 t + \frac{k_2}{k_1 C_{A0}} \right) \right]$$

$$\left[-\frac{\exp(a\tau)}{\tau} + a\left(\ln \tau + \frac{a\tau}{(1)(1!)} + \frac{(a\tau)^2}{(2)(2!)} + \frac{(a\tau)^3}{(3)(3!)} + \cdots \right) \right] \quad (2.68)$$

Now, since $\tau = 1$ when $t = 0$, the arbitrary constant C becomes, since $C_B(0) = 0$,

$$C = C_{A0} \exp\left(-\frac{k_2}{k_1 C_{A0}} \right) \left[\exp(a) - a\left(\frac{a}{(1)(1!)} + \frac{a^2}{(2)(2!)} + \frac{a^3}{(3)(3!)} + \cdots \right) \right]$$

$$(2.69)$$

CASE (c) $n = 1$, $m = 1$

The linear case is described by

$$\frac{dC_A}{dt} = -k_1 C_A \quad \therefore \quad C_A = C_{A0} \exp(-k_1 t) \quad (2.70)$$

$$\frac{dC_B}{dt} = k_1 C_A - k_2 C_B \quad (2.71)$$

This also yields the $\mathbf{I}$-factor equation, if the time variation is desired. Often, the relationship between C_A and C_B is desired, so we can use a different approach by dividing the two equations to find

$$\frac{dC_B}{dC_A} = -1 + \frac{k_2}{k_1}\left(\frac{C_B}{C_A}\right) \tag{2.72}$$

This takes the homogeneous form, so according to Section 2.3.2, let

$$\frac{C_B}{C_A} = V, \qquad \frac{dC_B}{dC_A} = C_A\frac{dV}{dC_A} + V \tag{2.73}$$

hence,

$$C_A\frac{dV}{dC_A} = -1 + \frac{k_2}{k_1}(V) - V \tag{2.74}$$

so we obtain

$$\frac{dV}{\left[-1 + \left(\frac{k_2}{k_1} - 1\right)V\right]} = \frac{dC_A}{C_A} \tag{2.75}$$

Integrating, noting $V = 0$ when $C_A = C_{A0}$ yields finally

$$\frac{C_B}{C_A} = \frac{k_1}{k_2 - k_1}\left[1 - \left(\frac{C_A}{C_{A0}}\right)^{(k_2/k_1)-1}\right] \tag{2.76}$$

2.3.5 Linear Coefficients

Equations of first order with linear coefficients,

$$(ax + by + c)\,dx - (\alpha x + \beta y + \gamma)\,dy = 0$$

can be reduced to the classic homogeneous functional form by suitable change of variables. Putting the above into the usual form to test for homogeneous functions

$$\frac{dy}{dx} = \frac{ax + by + c}{\alpha x + \beta y + \gamma} = \frac{a + b\left(\frac{y}{x}\right) + \frac{c}{x}}{\alpha + \beta\left(\frac{y}{x}\right) + \frac{\gamma}{x}} \tag{2.77}$$

we see that the last terms in the numerator and denominator prevent the right-hand side from becoming the proper form, that is $f(y/x)$. This suggests a dual change of variables with two adjustable parameters, ε and δ

$$y = u + \varepsilon, \qquad dy = du \tag{2.78}$$

$$x = v + \delta, \qquad dx = dv \tag{2.79}$$

hence,

$$\frac{du}{dv} = \frac{a(v + \delta) + b(u + \varepsilon) + c}{\alpha(v + \delta) + \beta(u + \varepsilon) + \gamma} \tag{2.80}$$

We have two degrees of freedom to eliminate two constants, so we eliminate ε and γ by selecting

$$a\delta + b\varepsilon + c = 0 \tag{2.81}$$

$$\alpha\delta + \beta\varepsilon + \gamma = 0 \tag{2.82}$$

Solving these for ε, δ yields

$$\varepsilon = \frac{\alpha c - \gamma a}{\beta a - \alpha b} \tag{2.83}$$

$$\delta = -\frac{c}{a} - \frac{b}{a} \cdot \frac{\alpha c - \gamma a}{\beta a - \alpha b} \tag{2.84}$$

An inconsistency arises if $\beta a = \alpha b$, which is obvious. It is clear that the final result is now of the homogeneous type

$$\frac{du}{dv} = \frac{a + b\left(\dfrac{u}{v}\right)}{\alpha + \beta\left(\dfrac{u}{v}\right)} \tag{2.85}$$

and this can be solved directly by substituting $z = u/v$.

2.3.6   First Order Equations of Second Degree

The order of a differential equation corresponds to the highest derivative, whereas the degree is associated with the power to which the highest derivative is raised.

A nonlinear equation, which is first order and second degree, is

$$\left(\frac{dy}{dx}\right)^2 - 2\frac{dy}{dx} + y = x - 1 \tag{2.86}$$

This requires a different approach, as nonlinear systems often do. The first step is to replace $p = dy/dx$ and solve the remaining quadratic equation for p

$$p = \frac{dy}{dx} = 1 \pm \sqrt{x - y} \tag{2.87}$$

We observe two branches are possible, depending on selection of sign. The appearance of the linear difference $(x - y)$ suggests replacing $u = x - y$, so that we have

$$\frac{du}{dx} = 1 - \frac{dy}{dx} \tag{2.88}$$

We now have the separable equation

$$\frac{du}{dx} = \pm \sqrt{u} \tag{2.89}$$

Integration yields the general solution

$$2\sqrt{u} = \pm x + c \tag{2.90}$$

Replacing $u = x - y$ shows finally

$$4y = 4x - (c \pm x)^2 \tag{2.91}$$

Again, we observe that the arbitrary constant of integration is implicit, which is quite usual for nonlinear systems.

We reinspect the original equation

$$\frac{dy}{dx} = 1 \pm \sqrt{x - y} \tag{2.92}$$

and observe that a solution $y = x$ also satisfies this equation. This solution cannot be obtained by specializing the arbitrary constant c, and is thus called a *singular solution* (Hildebrand 1965). This unusual circumstance can only occur in the solution of nonlinear equations. The singular solution sometimes describes an "envelope" of the family of solutions, but is not in general a curve belonging to the family of curves (since it cannot be obtained by specializing the arbitrary constant c).

2.4   SOLUTION METHODS FOR SECOND ORDER NONLINEAR EQUATIONS

As stated earlier, much is known about linear equations of higher order, but no general technique is available to solve the nonlinear equations that arise frequently in natural and man-made systems. When analysis fails to uncover the analytical solution, the last recourse is to undertake numerical solution methods, as introduced in Chapters 7 and 8.

We begin this section by illustrating the types of nonlinear problems that can be resolved using standard methods. Some important nonlinear second order equations are

$$\frac{d^2y}{dx^2} + \frac{2}{x}\frac{dy}{dx} + y^\alpha = 0 \qquad \text{(Lane–Emden equation)} \qquad \textbf{(2.93)}$$

$$\frac{d^2\psi}{dt^2} + \omega^2 \sin\psi = 0 \qquad \text{(Nonlinear Pendulum equation)} \qquad \textbf{(2.94)}$$

$$\frac{d^2y}{dx^2} + ay + by^3 = 0 \qquad \text{(Duffing equation)} \qquad \textbf{(2.95)}$$

$$\frac{d^2y}{dx^2} + a(y^2 - 1)\frac{dy}{dx} + y = 0 \qquad \text{(Van der Pol equation)} \qquad \textbf{(2.96)}$$

The general strategy for attacking nonlinear equations is to reduce them to linear form. Often, inspection of the equation suggests the proper approach. The two most widely used strategies are as follows.

1. Derivative substitution method: replace $p = dy/dx$ if either y is not explicit or x is not explicit.
2. Homogeneous function method: replace $v = y/x$ if the equation can be put into the homogeneous format

$$x\frac{d^2y}{dx^2} = f\left(\frac{dy}{dx}, \frac{y}{x}\right)$$

Note, as before, the dimensional ratio y/x appears throughout, which is a good indicator that the technique may work.

2.4.1 Derivative Substitution Method

Two classes of problems arise, which can be categorized by inspection: Either y or x does not appear alone (is not explicit). In both cases, we start with the substitution $p = dy/dx$.

EXAMPLE 2.6

The nonlinear Pendulum problem is a case where x is not explicit

$$\frac{d^2y}{dx^2} + \omega^2 \sin(y) = 0 \qquad \textbf{(2.97)}$$

Make the substitution

$$p = \frac{dy}{dx} \qquad \textbf{(2.98)}$$

therefore,

$$\frac{dp}{dx} + \omega^2 \sin(y) = 0 \tag{2.99}$$

Since x is not explicit, assume $p(y)$, so that

$$\frac{dp(y)}{dx} = \frac{dp}{dy}\frac{dy}{dx} = \frac{dp}{dy}p \tag{2.100}$$

Substituting above yields the separable form

$$p\frac{dp}{dy} + \omega^2 \sin(y) = 0 \tag{2.101}$$

Integrating yields

$$p^2 = 2\omega^2 \cos(y) + C_1 \tag{2.102}$$

Two branches are possible on taking square roots

$$p = \frac{dy}{dx} = \pm\sqrt{2\omega^2 \cos(y) + C_1} \tag{2.103}$$

So finally, the integral equation results

$$\int \frac{dy}{\sqrt{2\omega^2 \cos(y) + C_1}} = \pm x + C_2 \tag{2.104}$$

As we show in Chapter 4, the above result can be re-expressed as an elliptic integral of the first kind for certain realizable boundary conditions.

EXAMPLE 2.7

Soluble gas (A) reactant dissolves into the flat interface of a deep body of liquid reagent with which it reacts through the nonlinear (irreversible) rate law,

$$R_A = k_n C_A^n \quad \text{(mole/volume} \cdot \text{time)} \tag{2.105}$$

Taking the coordinate system z pointing down from the interface at $z = 0$, we can write the steady-state material balance for flux through an elemental slice Δz thick with area normal to flux designated as A_n

$$A_n J_z(z) - A_n J_z(z + \Delta z) - k_n C_A^n(A_n \Delta z) = 0 \tag{2.106}$$

Taking limits, cancelling A_n yields

$$-\frac{dJ_z}{dz} - k_n C_A^n = 0 \tag{2.107}$$

Introducing Fick's law of diffusion

$$J_z = -D\frac{\partial C_A}{\partial z} \tag{2.108}$$

yields finally,

$$D\frac{d^2 C_A}{dz^2} - k_n C_A^n = 0 \tag{2.109}$$

We note that z (independent variable) is not explicit, so we again write

$$p = \frac{dC_A}{dz} \tag{2.110}$$

and taking $p(C_A)$, then

$$\frac{d^2 C_A}{dz^2} = \frac{dp(C_A)}{dz} = \frac{dp}{dC_A}\frac{dC_A}{dz} = \frac{dp}{dC_A}p \tag{2.111}$$

Inserting the second derivative yields the separable form

$$p\frac{dp}{dC_A} - \left(\frac{k_n}{D}\right)C_A^n = 0 \tag{2.112}$$

Integrating yields

$$p^2 = 2\left(\frac{k_n}{D}\right)\left(\frac{C_A^{n+1}}{n+1}\right) + C_1 \tag{2.113}$$

Two branches again appear

$$p = \frac{dC_A}{dz} = \pm\sqrt{\frac{2k_n}{D(n+1)}C_A^{n+1} + C_1} \tag{2.114}$$

At this point, we sketch a curve indicating the expected behavior of $C_A(z)$, as shown in Fig. 2.1.

This shows we expect C_A to diminish rapidly (by reaction depletion) as we penetrate deeper into the liquid. Moreover, the slope of the curve shown is everywhere negative! Thus, we must select the negative root; hence,

$$\frac{dC_A}{dz} = -\sqrt{\frac{2k_n}{D(n+1)}C_A^{n+1} + C_1} \tag{2.115}$$

Since the reaction rate is irreversible, eventually all of the species A will be consumed, so that $C_A \to 0$ as $z \to \infty$. Now, as $C_A \to 0$, we expect the flux also to diminish to zero, so that $J_z \to 0$, as $C_A \to 0$. This suggests that we should

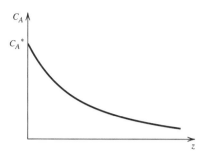

Figure 2.1 Expected behavior of $C_A(z)$.

take $C_1 = 0$, for an unbounded liquid depth; hence,

$$\frac{dC_A}{\sqrt{C_A^{n+1}}} = -\sqrt{\frac{2k_n}{D(n+1)}} \, dz \tag{2.116}$$

and this integral is

$$C_A^{(1-n)/2} = -\left(\frac{1-n}{2}\right)\left(\frac{2k_n}{D(n+1)}\right)^{1/2} z + C_2 \tag{2.117}$$

At the interface ($z = 0$), we denote the gas solubility (Henry's law) as C_A^*, so C_2 is evaluated as

$$C_2 = (C_A^*)^{(1-n)/2} \tag{2.118}$$

The final form for computational purposes is explicit in z, implicit in C_A

$$C_A^{*(1-n)/2} - C_A^{(1-n)/2} = \left(\frac{2k_n}{D(n+1)}\right)^{1/2} z \tag{2.119}$$

However, engineers usually want to know the mass transfer rate at the interface (where $C_A = C_A^*$), so that

$$W_A = A_n\left(-D\frac{dC_A}{dz}\right)_{z=0} = A_n\sqrt{\frac{2k_n D}{n+1} C_A^{*(n+1)}} \tag{2.120}$$

For linear reaction rates ($n = 1$), we get the familiar result

$$W_A = A_n C_A^* \sqrt{k_1 D} \tag{2.121}$$

This expression is found useful for determining interfacial area in bubble column reactors (sulphite oxidation). Thus, by measuring transfer rate W_A (moles/sec) for known values of C_A^*, k_1, and D, then area A_n can be calculated. The method is not restricted to linear kinetics, in light of Eq. 2.120.

EXAMPLE 2.8

Find the solution using the initial condition $y = 1, x = 0$

$$\frac{d^2y}{dx^2} = \frac{1.25}{y}\left(\frac{dy}{dx}\right)^2 - A_1 y + \frac{A_2}{y} - \frac{A_3}{y^3} \tag{2.122}$$

This dimensionless equation has recently appeared in the literature (Prince and Blanch 1990) to describe the physics of bubble coalescence. Here, y represents dimensionless film thickness (joining two touching bubbles) and x represents dimensionless time. The solution (credited to R.G. Rice) is a classic example on applications of the p-substitution method, so replace $p = dy/dx$, and for the second derivative, assume $p(y)$ since x is not explicit

$$\frac{d^2y}{dx^2} = \frac{dp}{dx} = \frac{dp(y)}{dy}\frac{dy}{dx} = \frac{dp}{dy}p = \frac{1}{2}\frac{d}{dy}(p^2) \tag{2.123}$$

Inserting this and multiplying by 2 produces the **I**-factor first order equation

$$\frac{d(p^2)}{dy} - \frac{2.5}{y}p^2 = 2\left[-A_1 y + \frac{A_2}{y} - \frac{A_3}{y^3}\right] \tag{2.124}$$

The integrating factor is

$$\mathbf{I}(y) = \exp\int\left(-\frac{2.5}{y}\right)dy = \frac{1}{y^{2.5}} \tag{2.125}$$

so the solution is

$$p^2 = y^{2.5}\int\frac{2}{y^{2.5}}\left(-A_1 y + \frac{A_2}{y} - \frac{A_3}{y^3}\right)dy + K_0 y^{2.5} \tag{2.126}$$

This integral then yields the dimensionless rate of film thinning between two contacting bubbles, which is p

$$p = \frac{dy}{dx} = \pm\left[4A_1 y^2 - \frac{2}{2.5}A_2 + \frac{2}{4.5}\frac{A_3}{y^2} + K_0 y^{2.5}\right]^{1/2} \tag{2.127}$$

Since film thinning occurs, $dy/dx < 0$, hence we select the negative branch. The final integral is straightforward, if we take $K_0 = 0$,

$$\frac{2cy^2 + b}{\sqrt{4ac - b^2}} = \sinh\left[\sqrt{c(\beta - 2x)}\right] \tag{2.128}$$

where

$$a = \frac{2}{4.5}A_3, \quad b = -\frac{2}{2.5}A_2, \quad c = 4A_1 \tag{2.129}$$

and the arbitrary second constant of integration is obtained from $y = 1$, $x = 0$,

$$\beta = \frac{1}{\sqrt{c}} \sinh^{-1}\left[\frac{2c + b}{\sqrt{4ac - b^2}}\right] \tag{2.130}$$

A transition in behavior occurs when $4ac = b^2$, which was used by these authors to deduce conditions conducive to coalescence. This transition point was independent of the arbitrary constant K_0, so setting this to zero had no effect on the final conclusion. This is often the case with dynamic systems, wherein the character of the solution form is most important.

EXAMPLE 2.9

Find the solution to the linear equation

$$\frac{d^2y}{dx^2} + 2x\frac{dy}{dx} = x \tag{2.131}$$

The p-substitution method can also be used to good effect on linear equations with nonconstant coefficients, such as the above. First, replace $p = dy/dx$ to get

$$\frac{dp}{dx} + 2xp = x \tag{2.132}$$

This is the familiar **I**-factor linear equation, so let

$$\mathbf{I} = \exp\int 2x\,dx = \exp(x^2) \tag{2.133}$$

hence, the solution for p is

$$p = \exp(-x^2)\int x\exp(+x^2)\,dx + A\exp(-x^2) \tag{2.134}$$

Noting that $xdx = \frac{1}{2}dx^2$ yields

$$p = \frac{1}{2} + A\exp(-x^2) \tag{2.135}$$

Integrating again produces

$$y = \frac{1}{2}x + A\int \exp(-x^2)\,dx + B \tag{2.136}$$

We could replace the indefinite integral with a definite one, since this would only change the already arbitrary constant

$$y = \frac{1}{2}x + A \int_0^x \exp(-\alpha^2)\, d\alpha + B \qquad (2.137)$$

This integral is similar to a tabulated function called the *error function*, which will be discussed in Chapter 4

$$\text{erf}(x) = \frac{2}{\sqrt{\pi}} \int_0^x \exp(-\alpha^2)\, d\alpha \qquad (2.138)$$

Using this, we can now write our final solution in terms of known functions and two arbitrary constants

$$y = \frac{1}{2}x + C\,\text{erf}(x) + B \qquad (2.139)$$

EXAMPLE 2.10

Solve the nonlinear second order equation

$$\frac{d^2 y}{dx^2} + \left(\frac{dy}{dx}\right)^2 - x = 0 \qquad (2.140)$$

This nonlinear equation can be put into a familiar form, again by replacing $p = dy/dx$

$$\frac{dp}{dx} + p^2 - x = 0 \qquad (2.141)$$

This is exactly like the special form of Ricatti's equation with $Q = 0, R = x$, so let

$$p = \frac{1}{z}\frac{dz}{dx} \qquad (2.142)$$

giving the linear equation

$$\frac{d^2 z}{dx^2} - xz = 0 \qquad (2.143)$$

This is the well-known Airy equation, which will be discussed in Chapter 3. Its solution is composed of Airy functions, which are also tabulated.

2.4.2   Homogeneous Function Method

In a manner similar to the first order case, we attempt to rearrange certain equations into the homogeneous format, which carries the dimensional ratio y/x,

$$x\frac{d^2 y}{dx^2} = f\left(\frac{dy}{dx}, \frac{y}{x}\right) \qquad (2.144)$$

If this can be done, then a possible solution may evolve by replacing $v = y/x$.

Often, a certain class of linear equations also obey the homogeneous property, for example, the Euler equation (or Equidimensional equation),

$$x^2 \frac{d^2 y}{dx^2} + Ax \frac{dy}{dx} + By = 0; \qquad A, B \text{ constant} \qquad (2.145)$$

Note that units of x cancel in the first two terms. This linear equation with nonconstant coefficients can be reduced to a constant coefficient linear equation by the simple change of variables

$$x = e^t \qquad \text{or} \qquad t = \ln(x) \qquad (2.146)$$

Changing variables starting with the first derivative

$$\frac{dy(x)}{dx} = \frac{dy(t)}{dt} \frac{dt}{dx} = \frac{dy(t)}{dt} \frac{1}{x} \qquad (2.147)$$

$$\frac{d^2 y}{dx^2} = \frac{d}{dx} \left[\frac{dy(t)}{dt} \frac{1}{x} \right] = \frac{d}{dt} \left[\frac{dy}{dt} e^{-t} \right] \frac{dt}{dx} \qquad (2.148)$$

$$\frac{d^2 y}{dx^2} = \frac{d}{dt} \left[\frac{dy}{dt} e^{-t} \right] \frac{1}{x} \qquad (2.149)$$

$$\frac{d^2 y}{dx^2} = \left[\frac{d^2 y}{dt^2} e^{-t} - \frac{dy}{dt} e^{-t} \right] \frac{1}{x} \qquad (2.150)$$

$$\frac{d^2 y}{dx^2} = \left[\frac{d^2 y}{dt^2} - \frac{dy}{dt} \right] \frac{1}{x^2} \qquad (2.151)$$

Inserting these into the defining equation causes cancellation of x

$$\frac{d^2 y}{dt^2} + (A - 1) \frac{dy}{dt} + By = 0 \qquad (2.152)$$

The method of solving such linear constant coefficient equations will be treated in the next section.

EXAMPLE 2.11

Consider the nonlinear homogeneous equation

$$x \frac{d^2 y}{dx^2} + \left(\frac{dy}{dx} \right)^2 - \left(\frac{y}{x} \right)^2 = 0 \qquad (2.153)$$

Under conditions when the boundary conditions are $dy/dx = 1$, $y = 0$ at $x = 1$, find a suitable solution. Replace $y/x = v(x)$ so that

$$x\left[x\frac{d^2v}{dx^2} + 2\frac{dv}{dx}\right] + \left[x\frac{dv}{dx} + v\right]^2 - v^2 = 0 \tag{2.154}$$

hence,

$$x^2\frac{d^2v}{dx^2} + 2x\frac{dv}{dx} + x^2\left(\frac{dv}{dx}\right)^2 + 2xv\frac{dv}{dx} = 0 \tag{2.155}$$

This has the Euler-Equidimensional form, so let $x = e^t$

$$\left[\frac{d^2v}{dt^2} - \frac{dv}{dt}\right] + 2\frac{dv}{dt} + \left(\frac{dv}{dt}\right)^2 + 2v\left(\frac{dv}{dt}\right) = 0 \tag{2.156}$$

Now, since the independent variable (t) is missing, write $p = dv/dt$

$$\frac{d^2v}{dt^2} = \frac{dp}{dt} = \frac{dp}{dv}\frac{dv}{dt} = p\frac{dp}{dv} \tag{2.157}$$

so that

$$p\frac{dp}{dv} + p + p^2 + 2vp = 0 \tag{2.158}$$

which can be factored to yield two possible solutions

$$p\left[\frac{dp}{dv} + p + (1 + 2v)\right] = 0 \tag{2.159}$$

This can be satisfied by $p = 0$, or

$$\frac{dp}{dv} + p = -(1 + 2v) \tag{2.160}$$

This latter result is the **I**-factor equation, which yields for $\mathbf{I} = \exp(v)$

$$p = 1 - 2v + c\exp(-v) \tag{2.161}$$

We pause to evaluate c noting

$$p = \frac{dv}{dt} = x\frac{dv}{dx} = \frac{dy}{dx} - \frac{y}{x} \tag{2.162}$$

hence, at $x = 1$, then $p = 1$ and $v = y/x = 0$, so $c = 0$.
  Integrating again

$$\frac{dv}{1 - 2v} = dt \tag{2.163}$$

yields

$$\frac{K}{\sqrt{1 - 2v}} = e^t \tag{2.164}$$

Replacing $v = y/x$ and $x = e^t$, and since $y = 0$ at $x = 1$, then $K = 1$ so that squaring yields

$$y = \frac{(x^2 - 1)}{2x} = \frac{x}{2} - \frac{1}{2x} \tag{2.165}$$

The *singular solution*, $p = 0$, which is $dv/dt = 0$, so that y/x = constant is a solution. This solution cannot satisfy the two boundary conditions.

2.5 LINEAR EQUATIONS OF HIGHER ORDER

The most general linear differential equation of nth order can be written in the standard form

$$\frac{d^n y}{dx^n} + a_{n-1}(x)\frac{d^{n-1}y}{dx^{n-1}} + \cdots + a_1(x)\frac{dy}{dx} + a_0(x)y = f(x) \tag{2.166}$$

where engineers denote $f(x)$ as the forcing function. We recall in Section 1.4 the definition of *homogeneous type* equations: "a condition (e.g., boundary condition) or equation is taken to be homogeneous if it is satisfied by $y(x)$ and is also satisfied by $cy(x)$, where c is an arbitrary constant." Thus, the above equation is not homogeneous. In fact, mathematicians call it the nth order inhomogeneous equation, because of the appearance of $f(x)$. If $f(x) = 0$, then the above equation is homogeneous. The first part of this section deals with the unforced, or homogeneous nth order equation

$$\frac{d^n y}{dx^n} + a_{n-1}(x)\frac{d^{n-1}y}{dx^{n-1}} + \cdots + a_1(x)\frac{dy}{dx} + a_0(x)y = 0 \tag{2.167}$$

If we denote P as the linear differential operator

$$P = \frac{d^n}{dx^n} + a_{n-1}(x)\frac{d^{n-1}}{dx^{n-1}} + \cdots + a_1(x)\frac{d}{dx} + a_0(x) \tag{2.168}$$

then we can abbreviate the lengthy representation of the nth order equation as

$$P[y] = 0 \tag{2.169}$$

It is clear that

$$P[y] = P[cy] = cP[y] = 0 \tag{2.170}$$

so that the equation is indeed homogeneous.

The most general solution to Eq. 2.167 is called the *homogeneous* or *complementary* solution. The notation complementary comes about when $f(x)$ is different from zero. Thus, when the forcing function $f(x)$ is present, it produces an additional solution, which is particular to the specific form taken by $f(x)$. Hence, solutions arising because of the presence of finite $f(x)$ are called *particular* solutions. These solutions are *complemented* by solutions obtained when $f(x) = 0$.

We first focus our efforts in solving the unforced or homogeneous equation, and then concentrate on dealing with solutions arising when forcing is applied through $f(x)$ (i.e., the particular solution).

It is clear in the homogeneous Eq. 2.167 that if all coefficients $a_0, \ldots a_{n-1}(x)$ were zero, then we could solve the final equation by n successive integrations of

$$\frac{d^n y}{dx^n} = 0 \tag{2.171}$$

which produces the expression

$$y = C_1 + C_2 x + C_3 x^2 + \cdots + C_n x^{n-1} \tag{2.172}$$

containing n arbitrary constants of integration.

As a matter of fact, within any defined interval (say, $0 \le x \le L$) wherein the coefficients $a_0(x), \ldots a_{n-1}(x)$ are continuous, then there exists a continuous solution to the homogeneous equation containing *exactly* n independent, arbitrary constants.

Moreover, because the homogeneous equation is linear, it is easily seen that any linear combination of individual solutions is also a solution, provided that each individual solution is *linearly independent* of the others. We define linearly independent to mean: an individual solution *cannot* be obtained from another solution by multiplying it by any arbitrary constant. For example, the solution $y_1 = c_1 exp(x)$ is linearly independent of $y_2 = C_2 exp(-x)$, since we cannot multiply the latter by any constant to obtain the former. However, the solution $y_3 = 4x^2$ is not linearly independent of $y_4 = 2x^2$, since it is obvious that y_3 can be obtained by multiplying y_4 by 2.

Thus, if n linearly independent solutions $(y_1, y_2, \ldots y_n)$ to the associated homogeneous equation:

$$P[y(x)] = 0 \tag{2.173}$$

can be found, then the sum (theorem of superposition)

$$y = c_1 y_1(x) + c_2 y_2(x) + \cdots + c_n y_n(x) = \sum_{k=1}^{n} c_k y_k(x) \tag{2.174}$$

is the *general solution* to the linear, homogeneous, unforced, nth order equation. When we must also deal with the case $f(x) \ne 0$, we shall call the above solution the general, complementary solution and denote it as $y_c(x)$. Thus, it is now clear that if we could find the integral of

$$P[y_p] = f(x) \tag{2.175}$$

where y_p is the *particular solution*, then the complete solution, by superposition, is

$$y = y_p(x) + y_c(x) = y_p(x) + \sum_{k=1}^{n} c_k y_k(x) \tag{2.176}$$

It should now be clear that we have satisfied the original forced equation, since

$$Py = P(y_p + y_c) = Py_p + Py_c = f(x) \tag{2.177}$$

since by definition

$$Py_c = 0 \tag{2.178}$$

$$Py_p = f(x) \tag{1.279}$$

Thus, it is now clear that the process of solving an ordinary linear differential equation is composed of two parts. The first part is to find the n linearly independent solutions to the unforced (homogeneous) equation, denoting this as the complementary solution, $y_c(x)$. The second part is to find the particular solution arising from the forcing function $f(x)$. Finally, we must insure that the *particular solution* is *linearly independent* of each of the solutions comprising the complementary solution; if this were not true, then a particular integral could reproduce one of the complementary solutions, and no new information is added to the final result. We also note in passing that the arbitrary constants of integration are found (via boundary conditions) using the *complete solution* $(y_c + y_p)$, not just the complementary part.

2.5.1   Second Order Unforced Equations: Complementary Solutions

The second order linear equation is of great importance and arises frequently in engineering. We shall reserve treatment of the case of nonconstant coefficients; that is,

$$\frac{d^2 y}{dx^2} + a_1(x)\frac{dy}{dx} + a_0(x)y = 0 \tag{2.180}$$

to Chapter 3, where the general Frobenius series method is introduced. In this section, we shall treat the case of constant coefficients, so that a_0, a_1 = constants. The method described below is directly applicable to nth order systems provided *all coefficients* are again constant.

Thus, for constant coefficients, we shall assume there exists complementary solutions of the form

$$y_c = A \exp(rx); \quad A, r = \text{constant} \tag{2.181}$$

where r represents a characteristic root (or eigenvalue) of the equation and A

is the integration constant (arbitrary). It is of course necessary that such a proposed solution *satisfies* the defining equation, so it must be true that

$$\frac{d^2}{dx^2}[A\exp(rx)] + a_1\frac{d}{dx}[A\exp(rx)] + a_0[A\exp(rx)] = 0 \quad (2.182)$$

Performing the indicated operations yields

$$A[r^2 + a_1r + a_0]\exp(rx) = 0 \qquad (2.183)$$

There are three ways to satisfy this equation, two of them are trivial (remember, zero is always a solution to homogeneous equations, a trivial result). We thereby deduce that the root(s) must be satisfied by

$$r^2 + a_1r + a_0 = 0 \qquad (2.184)$$

if our proposed solution is to have any nontrivial existence. This quadratic equation sustains two roots, given by

$$r_{1,2} = \frac{-a_1 \pm \sqrt{a_1^2 - 4a_0}}{2} \qquad (2.185)$$

which we denote for convenience as r_1 and r_2. Since two possible roots exist, then the theorem of superposition suggests that two linearly independent solutions exist for the complementary solution

$$y_c = A\exp(r_1x) + B\exp(r_2x) \qquad (2.186)$$

It is easy to see that each of these, taken one at a time, satisfies the original equation, so that the general solution is the sum of the two. But are the solutions linearly independent? To answer this, we need to know the nature of the two roots. Are they real or complex? Are they unequal or equal? We consider these possibilities by considering a series of examples.

EXAMPLE 2.12

Find the complementary solutions for the second order equation

$$\frac{d^2y}{dx^2} + 5\frac{dy}{dx} + 4y = 0 \qquad (2.187)$$

Inspection shows that the characteristic equation can be obtained by replacing

dy/dx with r and d^2y/dx^2 with r^2, so that

$$r^2 + 5r + 4 = 0 \qquad (2.188)$$

$$r_{1,2} = \frac{-5 \pm \sqrt{5^2 - 4 \cdot 4}}{2} = \frac{-5 \pm 3}{2} = -1, -4 \qquad (2.189)$$

Thus, the solution is

$$y_c = A \exp(-x) + B \exp(-4x) \qquad (2.190)$$

It is clear that the roots are real and distinct, so the two solutions are linearly independent.

EXAMPLE 2.13

Solve the second order equation with boundary conditions

$$\frac{d^2y}{dx^2} + 4\frac{dy}{dx} + 4y = 0 \qquad (2.191)$$

where $y(0) = 0$ and $dy(0)/dx = 1$.
   The characteristic equation is

$$r^2 + 4r + 4 = 0 \qquad (2.192)$$

so that

$$r = \frac{-4 \pm \sqrt{(4)^2 - 4 \cdot 4}}{2} = -2 \qquad (2.193)$$

which shows that only one root results (i.e., a double root); hence, we might conclude that the general solution is

$$y_1 = A_0 \exp(-2x) \qquad (2.194)$$

Clearly, a single arbitrary constant cannot satisfy the two stated boundary conditions, so it should be obvious that one solution, along with its arbitrary constant, is missing. As stipulated earlier, an nth order equation must yield n arbitrary constants, and n linearly independent solutions. For the present case, $n = 2$, so that we need to find an *additional* linearly independent solution.
   To find the second solution, we *use the definition of linear independence* to propose a new solution, so that we write

$$y_2 = v(x)\exp(-2x) \qquad (2.195)$$

Now, if $v(x)$ is *not* a simple constant, then the second solution will be linearly independent of $y_1 = A_0 \exp(-2x)$. Thus, we have used the first solution (and the definition of linear independence) to construct the second one. Inserting y_2

into the defining equation shows after some algebra

$$\frac{d^2v}{dx^2} = 0 \tag{2.196}$$

so that

$$v = Bx + C \tag{2.197}$$

hence,

$$y_2 = (Bx + C)\exp(-2x) \tag{2.198}$$

The arbitrary constant C can be combined with A_0 and call it A, hence our two linearly independent solutions yield the complementary solution

$$y_c = A \exp(-2x) + Bx \exp(-2x) \tag{2.199}$$

This analysis is in fact a general result for any second order equation when *equal roots occur*; that is,

$$y_c = A \exp(rx) + Bx \exp(rx) \tag{2.200}$$

since the second solution was generated from $y = v(x)\exp(rx)$, and it is easy to show in general this always leads to $d^2v/dx^2 = 0$.

Applying the boundary conditions to Eq. 2.200 shows

$$y_c(0) = 0 = A(1) + B(0)(1)$$

hence $A = 0$. To find B, differentiate

$$\frac{dy_c(0)}{dx} = 1 = B(1) + B(0)$$

therefore, $B = 1$; hence, the complementary solution satisfying the stipulated boundary conditions is

$$y_c = x \exp(-2x) \tag{2.201}$$

EXAMPLE 2.14

Solve the second order equation

$$\frac{d^2y}{dx^2} + y = 0 \tag{2.202}$$

We immediately see difficulties, since the characteristic equation is

$$r^2 + 1 = 0 \tag{2.203}$$

so complex roots occur

$$r_{1,2} = \pm\sqrt{-1} = \pm i \tag{2.204}$$

This defines the complex variable i (a subject dealt with in Chapter 9), but we could still proceed and write the solution

$$y_c = A \exp(+ix) + B \exp(-ix) \tag{2.205}$$

This form is not particularly valuable for computation purposes, but it can be put into more useful form by introducing the Euler formula

$$e^{ix} = \cos(x) + i \sin(x) \tag{2.206}$$

which allows representation in terms of well-known, transcendental functions. Thus, the complex function e^{ix} can be represented as the linear sum of a real part plus a complex part. This allows us to write

$$y_c = A[\cos(x) + i \sin(x)] + B[\cos(x) - i \sin(x)] \tag{2.207}$$

or

$$y_c = (A + B)\cos(x) + (A - B)i \sin(x) \tag{2.208}$$

Now, since A and B are certainly arbitrary, hence in general $(A + B)$ is different from $(A - B)i$, then we can define these groups of constants as new constants, so

$$y_c = D \cos(x) + E \sin(x) \tag{2.209}$$

which is the computationally acceptable general result. The Euler formula will be discussed in Chapter 9. Suffice to say, it arises naturally from the power series expansion of the exponential function.

EXAMPLE 2.15

The differential equation

$$\frac{d^2 y}{dx^2} - 2\frac{dy}{dx} + 2y = 0 \tag{2.210}$$

has properties similar to the last example. The characteristic equation is

$$r^2 - 2r + 2 = 0 \tag{2.211}$$

hence,

$$r = \frac{2 \pm \sqrt{2^2 - 4 \cdot 2}}{2} = 1 \pm i \tag{2.212}$$

so the solutions are

$$y_c = \exp(x)\left[A \exp(+ix) + B \exp(-ix)\right] \qquad (2.213)$$

Introducing the Euler formula as before shows

$$y_c = \exp(x)\left[C \cos(x) + D \sin(x)\right] \qquad (2.214)$$

EXAMPLE 2.16

Find the relation to predict the composition profile in a packed tube reactor undergoing isothermal linear kinetics with axial diffusion. The packed tube, heterogeneous catalytic reactor is used to convert species B by way of the reaction

$$B \rightarrow \text{products}; \qquad R_B = kC_B\left(\frac{\text{moles}}{\text{time-volume bed}}\right)$$

into products under (assumed) isothermal conditions. Diffusion along the axis is controlled by Fickian-like expression so that, in parallel with transport by convection due to superficial velocity v_0, there is also a diffusion-like flux represented by a Fickian relation

$$J_E = -D_E \frac{\partial C_B}{\partial z} \qquad \left(\frac{\text{mole}}{\text{area-time}}\right) \qquad (2.215)$$

acting along the longitudinal axis (z-coordinate). For linear kinetics, the conservation law of species B for an element Δz long of the tube with cross-sectional area A, is

$$v_0 AC_B|_z - y_0 AC_B|_{z+\Delta z} + AJ_E|_z - AJ_E|_{z+\Delta z} - R_B(A\Delta z) = 0 \quad (2.216)$$

Dividing by the element volume $A\Delta z$ and taking limits produces the transport equation

$$-v_0 \frac{dC_B}{dz} - \frac{dJ_E}{dz} - R_B = 0 \qquad (2.217)$$

Introducing the flux vector, Eq. 2.215, and the rate expression R_B then produces a constant coefficient, second order linear differential equation

$$D_E \frac{d^2 C_B}{dz^2} - v_0 \frac{dC_B}{dz} - kC_B = 0 \qquad (2.218)$$

If we divide through by D_E, this has a form identical with Eq. 2.180, where $a_1 = -v_0/D_E$ and $a_0 = -k/D_E$. The characteristic equation for a solution of the form $C_B = A \exp(rz)$ is

$$r^2 - \left(\frac{v_0}{D_E}\right) r - \left(\frac{k}{D_E}\right) = 0 \tag{2.219}$$

so the roots are

$$r_{1,2} = \frac{1}{2}\left(\frac{v_0}{D_E}\right) \pm \sqrt{\frac{1}{4}\left(\frac{v_0}{D_E}\right)^2 + \left(\frac{k}{D_E}\right)} \tag{2.220}$$

hence, the general complementary solution is

$$C_B = A \exp(r_1 z) + B \exp(r_2 z) \tag{2.221}$$

where we denote r_1 as possessing the positive argument and r_2 as the negative. It is often found convenient (e.g., in applying boundary conditions) to express this solution using hyperbolic functions; thus, we note

$$\cosh(x) = \frac{\exp(x) + \exp(-x)}{2} \tag{2.222}$$

and

$$\sinh(x) = \frac{\exp(x) - \exp(-x)}{2} \tag{2.223}$$

Now, we could write our solution in symbolic form

$$C_B = \exp(\alpha z)[A \exp(\beta z) + B \exp(-\beta z)] \tag{2.224}$$

where

$$\alpha = \frac{v_0}{2D_E}, \qquad \beta = \sqrt{\frac{1}{4}\left(\frac{v_0}{D_E}\right)^2 + \left(\frac{k}{D_E}\right)} \tag{2.225}$$

Now, add and subtract terms within the brackets and rearrange to see

$$C_B = \exp(\alpha z)[(A - B)\sinh(\beta z) + (A + B)\cosh(\beta z)] \tag{2.226}$$

Moreover, since the new arbitrary constant groups are not generally equal, we redefine constants

$$C_B = \exp(\alpha z)[F \sinh(\beta z) + E \cosh(\beta z)] \tag{2.227}$$

This is a convenient form for computation and also for evaluating constants by way of boundary conditions, since

$$\frac{d \sinh(x)}{dx} = \cosh(x) \tag{2.228}$$

$$\frac{d \cosh(x)}{dx} = \sinh(x) \tag{2.229}$$

For packed beds of finite length L, the most widely used boundary values are the famous *Danckwerts Conditions*:

$$v_0 C_0 = -D_E \frac{dC_A(z)}{dz} + v_0 C_A(z); \qquad z = 0 \tag{2.230}$$

$$\frac{dC_A(z)}{dz} = 0; \qquad z = L \tag{2.231}$$

The first simply states the convective flux of fresh solution far upstream of known composition C_0 exactly equals the combined convective and diffusion flux at the bed entrance. This accounts for so-called backmixing at the bed entrance. The second condition states that diffusion flux at the exit tends to zero. Both these boundary conditions require differentiation of the general solution, hence we justify the use of the hyperbolic function form (i.e., it is impossible to make a sign error on differentiating hyperbolic functions!).

Applying the above *Boundary Conditions* to the general solution in Eq. 2.227 leads to the algebraic equations needed to find integration constants F and E in terms of known parameters; the first condition yields

$$
\begin{aligned}
C_0 v_0 = \; & -D_E \beta [F \cosh(0) + E \sinh(0)] \exp(0) \\
& - D_E \alpha [F \sinh(0) + E \cosh(0)] \exp(0) \\
& + v_0 [F \sinh(0) + E \cosh(0)] \exp(0)
\end{aligned}
\tag{2.232}
$$

Noting that $\sinh(0) = 0$ and $\cosh(0) = 1$, we find

$$C_0 v_0 = E(v_0 - D_E \alpha) + F(-D_E \beta) \tag{2.233}$$

The second condition gives

$$0 = (\alpha F + \beta E)\sinh(\beta L) + (\alpha E + \beta F)\cosh(\beta L) \tag{2.234}$$

Solving for F and E yields

$$E = \frac{C_0[\alpha \tanh(\beta L) + \beta]}{\left(\dfrac{k}{v_0}\right)\tanh(\beta L) + \alpha \tanh(\beta L) + \beta} \tag{2.235}$$

$$F = \frac{-C_0[\alpha + \beta \tanh(\beta L)]}{\left(\dfrac{k}{v_0}\right)\tanh(\beta L) + \alpha \tanh(\beta L) + \beta} \tag{2.236}$$

where we have used $\beta^2 - \alpha^2 = k/D_E$. Recognizing that both α and β have units of reciprocal length, we can write the final solution in dimensionless form in terms of two dimensionless numbers. Thus, defining dimensionless composition to be $\psi = C_B/C_0$ and dimensionless length to be $\xi = z/L$, we have

$$\psi = \frac{\exp(\alpha L \xi)[e(\alpha, \beta)\cosh(\beta L \xi) - f(\alpha, \beta)\sinh(\beta L \xi)]}{g(\alpha, \beta)} \quad (\textbf{2.237})$$

where

$$e(\alpha, \beta) = \alpha L \tanh(\beta L) + \beta L \quad (\textbf{2.238})$$

$$f(\alpha, \beta) = \beta L \tanh(\beta L) + \alpha L \quad (\textbf{2.239})$$

$$g(\alpha, \beta) = \left(\frac{kL}{v_0} + \alpha L\right)\tanh(\beta L) + \beta L \quad (\textbf{2.240})$$

There are actually only two dimensionless groups hidden in the above maze; these are

$$\alpha L = \frac{1}{2}\left(\frac{Lv_0}{D_E}\right); \qquad \frac{Lv_0}{D_E} = Pe \quad \text{(Peclet number)} \quad (\textbf{2.241})$$

$$\beta L = \sqrt{\frac{1}{4}\left(\frac{Lv_0}{D_E}\right)^2 + \frac{L^2 k}{D_E}} \; ; \qquad \frac{L^2 k}{D_E} = Ha \quad \text{(Hatta number)} \quad (\textbf{2.242})$$

From these two (Pe and Ha), we can relate the remaining group

$$\frac{kL}{v_0} = \frac{Ha}{Pe} \quad (\textbf{2.243})$$

From Eq. 2.237, we see at the entrance

$$\psi(0) = \frac{e(\alpha, \beta)}{g(\alpha, \beta)} \quad (\textbf{2.244})$$

This example illustrates the complexity of even elementary second order equations when practical boundary conditions are applied. Fig. 2.2 illustrates the reactor composition profiles predicted for plug and dispersive flow models. Under quite realizable operating conditions, the effect of backmixing (diffusion) is readily seen. Diffusion tends to reduce the effective number of stages for a packed column. Its effect can be reduced by using smaller particle sizes, since Klinkenberg and Sjenitzer (1956) have shown that the effective diffusion coefficient (D_E) varies as $v_0 d_p$, where d_p is particle size. This also implies that Peclet number is practically independent of velocity ($Pe \sim L/d_p$).

Under conditions where effective diffusion is small, so that $D_E \sim 0$, then we would have solved the plug flow model

$$v_0 \frac{dC_B}{dz} + kC_B = 0 \quad (\textbf{2.245})$$

Using the entrance condition $C_B(0) = C_0$, so that separating variables yields

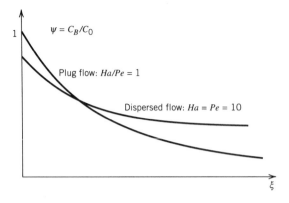

Figure 2.2 Composition profiles in catalytic reactor.

finally

$$\psi = \frac{C_B}{C_0} = \exp\left(-\frac{kL}{v_0}\xi\right) \tag{2.246}$$

To compare the two solutions illustrated in Fig. 2.2, we recall that $kL/v_0 = Ha/Pe$.

2.5.2 Particular Solution Methods for Forced Equations

We have seen at the beginning of the chapter that forced equations give rise to solutions, which are particular to the form taken by the forcing function $f(x)$. We consider the case of constant coefficients as follows:

$$\frac{d^2y}{dx^2} + a_1\frac{dy}{dx} + a_0y = f(x) \tag{2.247}$$

where again we note the general solution is comprised of two parts, viz

$$y = y_c(x) + y_p(x) \tag{2.248}$$

We discussed methods of finding $y_c(x)$ in the previous section, and now we discuss general methods of finding the particular integral, $y_p(x)$.

There are three widely used methods to find $y_p(x)$; the first two are applicable only to the case of constant coefficients

1. *Method of Undetermined Coefficients* This is a rather evolutionary technique, which builds on the functional form taken by $f(x)$.
2. *Method of Inverse Operators* This method builds on the property that integration as an operation is the inverse of differentiation.
3. *Method of Variation Parameters* This method is the most general approach and can be applied even when coefficients are nonconstant; it is based on the principles of *linear independence* and *superposition*, and exploits these necessary properties to construct a particular integral.

1. Method of Undetermined Coefficients

This widely used technique is somewhat intuitive, and is also easily implemented. It has certain disadvantages, since it is not completely fail-safe in the hands of a novice. The first step in finding y_p is to produce a collection of functions obtained by differentiating $f(x)$. Each of these generated functions are multiplied by an undetermined coefficient and the sum of these plus the original function are then used as a "trial expression" for y_p. The unknown coefficients are determined by inserting the trial solution into the defining equation. Thus, for a second order equation, two differentiations are needed. However, for an nth order equation, n differentiations are necessary (a serious disadvantage).

EXAMPLE 2.17

Find the complementary and particular solutions for the linear equation

$$\frac{d^2y}{dx^2} - y = x^2 \qquad (2.249)$$

and evaluate arbitrary constants using $y(0) = 1$, $dy(0)/dx = 0$.

The first step is to find the complementary solutions. We shall need to know these to insure that our particular solution is in fact linearly independent of each complementary solution. The characteristic equation for the unforced equation is

$$r^2 - 1 = 0; \qquad r_{1,2} = \pm 1 \qquad (2.250)$$

so the roots are real and distinct; hence,

$$y_c = A \exp(x) + B \exp(-x) \qquad (2.251)$$

To construct the particular solution, we note that repeated differentiation of $f(x) = x^2$ yields x and 1, so that we propose the linear combinations

$$y_p = ax^2 + bx + c \qquad (2.252)$$

The undetermined coefficients (a, b, c) are to be determined by inserting our proposed solution into the left-hand side of the defining equation; thus,

$$2a - (ax^2 + bx + c) = x^2 + (0)(x) + (0)(1) \qquad (2.253)$$

Note, we have written $f(x)$ as a descending series with the last two coefficients of magnitude zero. This will help in deducing the values for a, b, c. We next equate all multipliers of x^2 on left- and right-hand sides, all multipliers of x, and all multipliers of unity. These deductive operations produce

$$
\begin{aligned}
x^2 &: -a = 1 && \therefore a = -1 \\
x &: -b = 0 && \therefore b = 0 \\
1 &: 2a - c = 0 && \therefore c = 2a = -2
\end{aligned}
$$

We solve for the coefficients in sequence and see that

$$y_p = -x^2 - 2 \tag{2.254}$$

The complete solution is then

$$y = A \exp(x) + B \exp(-x) - (x^2 + 2) \tag{2.255}$$

It is clear that all solutions are linearly independent. Finally, we apply boundary conditions $y(0) = 1$ and $dy(0)/dx = 0$ to see

$$
\begin{aligned}
1 &= A + B - 2 \\
0 &= A - B
\end{aligned}
$$

This shows

$$A = B = \frac{3}{2}$$

EXAMPLE 2.18

Find the linearly independent particular solutions for

$$\frac{d^2y}{dx^2} - y = \exp(x) \tag{2.256}$$

The complementary solution is the same as the last problem. Repeated differentiation of the exponential function reproduces the exponential function. We are keenly aware that a trial solution $y_p = a \exp(x)$ is *not* linearly independent of one of the complementary solutions. We respond to this difficulty by invoking the definition of linear independence

$$y_p = v(x)\exp(x) \tag{2.257}$$

Clearly, if $v(x)$ is not a constant, then this particular solution will be linearly independent of the complementary function $\exp(x)$. Inserting the proposed $y_p(x)$ in the left-hand side of the defining equation yields

$$\frac{d^2v}{dx^2} + 2\frac{dv}{dx} = 1 \tag{2.258}$$

To find $v(x)$, we replace $dv/dx = p$

$$\frac{dp}{dx} + 2p = 1 \tag{2.259}$$

This is the **I**-factor equation with solution

$$p = \frac{dv}{dx} = \frac{1}{2} + ce^{-2x} \tag{2.260}$$

Integrating again shows

$$v = \frac{1}{2}x - \frac{c}{2}e^{-2x} + D \tag{2.261}$$

This suggests a particular solution

$$y_p = \left(\frac{1}{2}x\right)\exp(x) \tag{2.262}$$

since the other two terms yield contributions that are not linearly independent (they could be combined with the complementary parts).

The complete solution is

$$y = A\exp(x) + B\exp(-x) + \frac{1}{2}x\exp(x) \tag{2.263}$$

and all three solutions are linearly independent.

Another way to construct the particular integrals under circumstances when the forcing function duplicates one of the complementary solutions is to write

$$y_p = ax\exp(x) \tag{2.264}$$

Inserting this into the defining equation shows $a = 1/2$ as before. In fact, if an identity is not produced (i.e., a is indeterminate), then the next higher power is used, $ax^2\exp(x)$, and so on, until the coefficient is found. We demonstrate this in Example 2.19.

EXAMPLE 2.19

Find the complementary and particular solution for

$$\frac{d^2y}{dx^2} - 8\frac{dy}{dx} + 16y = 6xe^{4x} \tag{2.265}$$

The characteristic equation is

$$r^2 - 8r + 16 = (r - 4)^2 \tag{2.266}$$

Thus, we have repeated roots

$$r_{1,2} = 4 \tag{2.267}$$

As we learned earlier in Example 2.13, the second complementary solution is obtained by multiplying the first by x, so that

$$y_c = Ae^{4x} + Bxe^{4x} \tag{2.268}$$

However, the forcing function has the same form as xe^{4x}, so our first trial for the y function is

$$y_p = ax^2e^{4x} \tag{2.269}$$

which is linearly independent of both parts of the complementary solution. Differentiating twice yields

$$y_p' = 2axe^{4x} + 4ax^2e^{4x}$$
$$y_p'' = 2ae^{4x} + 8axe^{4x} + 8axe^{4x} + 16ax^2e^{4x} \tag{2.270}$$

inserting these relations into the defining equation yields

$$\left[2a + 16ax + 16ax^2\right]e^{4x} - \left[16ax + 32ax^2\right]e^{4x}$$
$$+ 16[ax^2]e^{4x} = 6xe^{4x} \tag{2.271}$$

Cancelling terms shows the null result

$$2ae^{4x} = 6xe^{4x} \tag{2.272}$$

hence, a is indeterminate. Next, try the higher power

$$y_p = ax^3e^{4x} \tag{2.273}$$
$$y_p' = 3ax^2e^{4x} + 4ax^3e^{4x} \tag{2.274}$$
$$y_p'' = 6axe^{4x} + 12ax^2e^{4x} + 12ax^2e^{4x} + 16ax^3e^{4x} \tag{2.275}$$

Inserting these yields

$$\left[6ax + 24ax^2 + 16ax^3\right]e^{4x} - \left[24ax^2 + 32ax^3\right]e^{4x}$$
$$+ \left[16ax^3\right]e^{4x} = 6xe^{4x} \tag{2.276}$$

Cancelling terms, what remains identifies the undetermined coefficient

$$6axe^{4x} = 6xe^{4x} \tag{2.277}$$

hence, $a = 1$.

The complete solution can now be written

$$y = (A + Bx + x^3)e^{4x} \qquad (2.278)$$

We see some serious disadvantages with this technique, especially the large amount of algebraic manipulation (which produces human errors) required for only moderately complex problems. The above particular solution could have been worked out in two lines of calculation, without trial and error, using the *Inverse Operator* technique, as we show next.

2. *Method of Inverse Operators*

This method builds on the Heaviside differential operator, defined as

$$Dy = \frac{dy}{dx} \qquad (2.279)$$

where D is the elementary operation d/dx. It follows certain algebraic laws, and must always *precede* a function to be operated upon; thus it is clear that repeated differentiation can be represented by

$$D(Dy) = D^2y = \frac{d^2y}{dx^2} \qquad (2.280)$$

$$D(D^2y) = D^3y = \frac{d^3y}{dx^3} \qquad (2.281)$$

$$D^ny = \frac{d^ny}{dx^a} \qquad (2.282)$$

Because the operator D is a linear operator, it can be summed and factored

$$\frac{d^2y}{dx^2} - 8\frac{dy}{dx} + 16y = D^2y - 8Dy + 16y = 0 \qquad (2.283)$$

The operators can be collected together as a larger operator

$$(D^2 - 8D + 16)y = 0 \qquad (2.284)$$

This also can be factored, again maintaining order of operations

$$(D - 4)^2y = 0 \qquad (2.285)$$

In manipulating the Heaviside operator D, the laws of algebraic operation must be followed. These basic laws are as follows.

(a) The Distributive Law. For algebraic quantities A, B, C, this law requires

$$A(B + C) = AB + AC \qquad (2.286)$$

We used this above law when we wrote

$$(D^2 - 8D + 16)y = D^2y - 8Dy + 16y \qquad (2.287)$$

The operator D is in general distributive.

(b) The Commutative Law. This law sets rules for the order of operation

$$AB = BA \tag{2.288}$$

which does not generally apply to the Heaviside operator, since obviously

$$\mathrm{D}y \neq y\mathrm{D} \tag{2.289}$$

However, operators do commute with themselves, since

$$(\mathrm{D} + 4)(\mathrm{D} + 2) = (\mathrm{D} + 2)(\mathrm{D} + 4) \tag{2.290}$$

(c) The Associative Law. This law sets rules for sequence of operation

$$A(BC) = (AB)C \tag{2.291}$$

and does not in general apply to D, since sequence for differentiation must be preserved. However, it is true that

$$\mathrm{D}(\mathrm{D}y) = (\mathrm{DD})y \tag{2.292}$$

but that

$$\mathrm{D}(xy) \neq (\mathrm{D}x)y \tag{2.293}$$

since we know that $\mathrm{D}(xy) = (\mathrm{D}x)y + x\mathrm{D}y$.

To use the operators (in an inverse fashion) we have only two rules which must be remembered. We will lead up to these rules gradually by considering, first, the operation on the most prevalent function, the exponential $\exp(rx)$. We have seen in the last section that all complementary solutions have origins in the exponential function.

Operation on exponential. It is clear that differentiation of $\exp(rx)$ yields

$$\mathrm{D}(e^{rx}) = re^{rx} \tag{2.294}$$

and repeated differentiation gives

$$\mathrm{D}^2(e^{rx}) = r^2 e^{rx} \tag{2.295}$$

$$\mathrm{D}^n(e^{rx}) = r^n e^{rx} \tag{2.296}$$

and a sum of operators, forming a polynomial such as $P(D)$

$$P(\mathrm{D})(e^{rx}) = P(r)e^{rx} \tag{2.297}$$

This forms the basis for Rule 1. We have alluded to this earlier in Section 2.5.1 in discussing the characteristic equation for Example 2.12

$$(\mathrm{D}^2 + 5\mathrm{D} + 4)e^{rx} = (r^2 + 5r + 4)e^{rx} \tag{2.298}$$

Operation on products with exponentials. The second building block to make operators useful for finding particular integrals is the operation on a general function $f(x)$

$$D(f(x)e^{rx}) = e^{rx}Df + fD(e^{rx}) = e^{rx}(D + r)f(x) \qquad (2.299)$$

Repeated differentiation can be shown to yield

$$D^2(fe^{rx}) = e^{rx}(D + r)^2 f(x) \qquad (2.300)$$

$$D^n(fe^{rx}) = e^{rx}(D + r)^n f(x) \qquad (2.301)$$

and for any polynomial of D, say $P(D)$

$$P(D)(f(x)e^{rx}) = e^{rx}P(D + r)f(x) \qquad (2.302)$$

We learn in this sequence of examples that operation on the product $f(x)e^{rx}$ with D simply requires shifting the exponential to the front and operating on $f(x)$ with $(D + r)$. This forms the basis for Rule 2.

The inverse operator. Modern calculus often teaches that integration as an operation is the inverse of differentiation. To see this, write

$$\frac{d}{dx} \int f(x)\, dx = D \int f(x)\, dx = f(x) \qquad (2.303)$$

which implies

$$\int f(x)\, dx = D^{-1}f(x) \qquad (2.304)$$

Thus, the operation $D^{-1}f(x)$ implies integration with respect to x, whereas $Df(x)$ denotes differentiation with respect to x. This "integrator," D^{-1}, can be treated like any other algebraic quantity, provided the rules of algebra, mentioned earlier, are obeyed.

We have already seen that polynomials of operator D obey two important rules:

1. Rule 1: $P(D)e^{rx} = P(r)e^{rx}$
2. Rule 2: $P(D)(f(x)e^{rx}) = e^{rx}P(D + r)f(x)$

We show next that these rules are also obeyed by inverse operators.

EXAMPLE 2.20

Find the particular solution for

$$\frac{dy}{dx} - 2y = e^x \qquad (2.305)$$

Write this in operator notation, noting that if $f(x)$ appears on the right-hand side, we are obviously seeking a particular solution

$$(D - 2) y_p = e^x \tag{2.306}$$

hence keeping the order of operation in mind

$$y_p = \frac{1}{D - 2} e^x \tag{2.307}$$

Clearly, any polynomial in the denominator can be expanded into an ascending series by synthetic division; in the present case, we can use the binomial theorem written generally as

$$(1 + f)^p = 1 + pf + \frac{p(p - 1)}{(1)(2)} f^2 + \frac{p(p - 1)(p - 2)}{(1)(2)(3)} f^3 + \cdots + \tag{2.308}$$

To put our polynomial operator in this form, write

$$\frac{1}{D - 2} = \frac{1}{-2\left(1 - \dfrac{D}{2}\right)} \tag{2.309}$$

so that we see the equivalence $f = -D/2$, $p = -1$; hence,

$$\frac{1}{-2\left(1 - \dfrac{D}{2}\right)} = -\frac{1}{2}\left[1 + \left(\frac{1}{2}D\right) + \left(\frac{1}{2}D\right)^2 + \left(\frac{1}{2}D\right)^3 + \cdots + \right] \tag{2.310}$$

hence operating on $\exp(x)$ using Rule 1

$$y_p = \frac{1}{D - 2} e^x = -\frac{1}{2}\left[1 + \left(\frac{1}{2}D\right) + \left(\frac{1}{2}D\right)^2 + \cdots + \right] e^x \tag{2.311}$$

yields

$$y_p = e^x\left[1 + \left(\frac{1}{2}\right) + \left(\frac{1}{2}\right)^2 + \left(\frac{1}{2}\right)^3 + \cdots \right]\left(-\frac{1}{2}\right) \tag{2.312}$$

But the series of terms is a geometrical progression and the sum to infinity is equal to 2, so we have finally

$$y_p = -e^x \tag{2.313}$$

and the general solution is, since $y_c = A \exp(2x)$

$$y(x) = A \exp(2x) - \exp(x) \tag{2.314}$$

This example simply illustrates that an inverse operator can always be expanded in series as a polynomial and so our previous Rule 1 is also applicable to inverse operators.

RULE 1: INVERSE OPERATORS

We see in general that polynomials in the denominator can be operated upon by Rule 1

$$\frac{1}{P(\mathrm{D})}e^{rx} = \frac{1}{P(r)}e^{rx} \qquad (2.315)$$

Thus, we could have applied this rule directly to the previous example without series expansion; since $r = 1$, we have

$$y_p = (\mathrm{D} - 2)^{-1}e^x = -e^x \qquad (2.316)$$

which is quite easy and efficient to use.

Occasionally, when applying Rule 1 to find a particular integral y_p , we encounter the circumstance $P(r) = 0$. This is an important fail-safe feature of the inverse operator method, since it tells the analyst that the requirements of linear independence have failed. The case when $P(r) = 0$ arises when the forcing function $\mathrm{f}(x)$ is of the exact form as one of the complementary solutions.

RULE 2: INVERSE OPERATORS

The difficulty above can always be overcome by a clever use of Rule 2. If $P(r) = 0$ then obviously $P(\mathrm{D})$ contains a root equal to r; that is, if we could factor $P(\mathrm{D})$ then

$$\frac{1}{P(\mathrm{D})} = \frac{1}{(\mathrm{D} - r)} \cdot \frac{1}{g(\mathrm{D})} \qquad (2.317)$$

For n repeated roots, this would be written

$$\frac{1}{P(\mathrm{D})} = \frac{1}{(\mathrm{D} - r)^n} \cdot \frac{1}{g(\mathrm{D})} \qquad (2.318)$$

Now, since $g(\mathrm{D})$ contains no roots r, then Rule 1 can be used. However, we must modify operation of $1/(\mathrm{D} - r)^n$ when it operates on $\exp(rx)$. Thus, we plan to operate on $\exp(rx)$ in precise sequence. Consider Rule 2 for polynomials in the denominator

$$\frac{1}{P(\mathrm{D})}[f(x)e^{rx}] = e^{rx}\frac{1}{P(\mathrm{D} + r)}f(x) \qquad (2.319)$$

and suppose $f(x) = 1$, then if $P(\mathrm{D}) = (\mathrm{D} - r)^n$, we have

$$\frac{1}{(\mathrm{D} - r)^n}[(1)e^{rx}] = e^{rx}\frac{1}{\mathrm{D}^n}1 \qquad (2.320)$$

This suggests n repeated integrations of unity

$$\frac{1}{D^n} 1 = \iiint \cdots \int_n 1 \, dx = \frac{x^n}{n!} \tag{2.321}$$

Now, reconsider the general problem for a forcing function $\exp(rx)$

$$\frac{1}{P(D)} \exp(rx) = \frac{1}{(D-r)^n g(D)} \exp(rx) \tag{2.322}$$

First, operate on $\exp(rx)$ using Rule 1 as $g(D)^{-1} \exp(rx)$, then shift $\exp(rx)$ to get

$$\frac{1}{(D-r)^n} \exp(rx) \frac{1}{g(r)}$$

Next, operate on $\exp(rx)$ using Rule 2, taking $f(x) = 1$; hence (since $g(r)$ is finite),

$$\exp(rx) \frac{1}{D^n} \frac{1}{g(r)} = \frac{\exp(rx)}{g(r)} \iiint_n \cdots \int dx = \frac{e^{rx}}{g(r)} \frac{x^n}{n!} \tag{2.323}$$

We finally conclude, when roots of the complementary solutions appear as the argument in an exponential forcing function, we will arrive at $P(r) = 0$, implying loss of linear independence. By factoring out such roots, and applying Rule 2, a particular solution can always be obtained.

EXAMPLE 2.21

Find the particular solution for

$$\frac{d^2 y}{dx^2} - 4\frac{dy}{dx} + 4y = xe^{2x} \tag{2.324}$$

Applying the operator D and factoring

$$(D^2 - 4D + 4) y_p = (D - 2)^2 y_p = xe^{2x} \tag{2.325}$$

and solve for y_p

$$y_p = \frac{1}{(D-2)^2} xe^{2x} \tag{2.326}$$

If we apply Rule 1, we see $P(2 - 2) = 0$. So, apply Rule 2, noting that

$f(x) = x$, hence replacing $(D - 2)$ with $(D + 2 - 2)$

$$y_p = e^{2x} \frac{1}{(D + 2 - 2)^2} x \tag{2.327}$$

$$y_p = e^{2x} \frac{1}{D^2} x = \frac{x^3 e^{2x}}{6} \tag{2.328}$$

As we saw earlier, for repeated roots, the general complementary solution is $(A + Bx) \exp(2x)$, so that the complete solution is

$$y = (A + Bx)e^{2x} + \frac{1}{6} x^3 e^{2x} \tag{2.329}$$

The reader can clearly see the speed and efficiency of this method compared to the tedious treatment required by the Method of Undetermined Coefficients, as done in Example 2.19.

Inverse operators on trigonometric functions. We have treated periodic functions such as $\sin(x)$, $\cos(x)$ in the study of complementary solution (Eq. 2.206), and found the Euler formula useful

$$e^{ix} = \cos(x) + i \sin(x) \tag{2.330}$$

Thus, we say the Real part of e^{ix} is $\cos(x)$ and the Imaginary part is $\sin(x)$

$$Re(e^{ix}) = \cos(x) \tag{2.331}$$

$$Im(e^{ix}) = \sin(x) \tag{2.332}$$

Thus, if $\cos(x)$ appears as a forcing function $f(x)$, then to use the inverse operators acting on exponential functions, we would write, for example,

$$\frac{d^2 y}{dx^2} - y = \cos(x) = Re(e^{ix}) \tag{2.333}$$

Now, in solving for the particular integral, it is also implied that we must extract only the Real Part of the final solution; thus, using Rule 1 with $r = i$, we get

$$y_p = \text{Real} \left[\frac{1}{(D^2 - 1)} e^{ix} \right] = \text{Real} \left[-\frac{1}{2} e^{ix} \right] \tag{2.334}$$

since $i^2 = -1$. Thus, we have finally,

$$y_p = -\frac{1}{2} \cos(x) \tag{2.335}$$

We can verify this using the method of Undetermined Coefficients. Repeated differentiation yields only two functions, so that

$$y_p = a \cos (x) + b \sin (x)$$

Inserting this into the defining equation

$$[-a \cos (x) - b \sin (x)] - [a \cos (x) + b \sin (x)] = \cos (x) \quad \textbf{(2.336)}$$

Therefore, we conclude: $-2a = 1, b = 0$ so that

$$y_p = -\frac{1}{2} \cos (x) \quad \textbf{(2.337)}$$

as required, but this method requires much more algebra. Had the forcing function been $\sin (x)$, then we would have extracted the Imaginary Part of the answer. We illustrate this next.

EXAMPLE 2.22

Find the particular solution for

$$\frac{d^2 y}{dx^2} + y = \sin (x) \quad \textbf{(2.338)}$$

Inserting operators and solving for y gives

$$y_p = Im \left[\frac{1}{D^2 + 1} e^{ix} \right] = Im \left[\frac{1}{(D - i)(D + i)} e^{ix} \right] \quad \textbf{(2.339)}$$

We use Rule 1 first on the nonzero factor $(D + i)$, then operate on $\exp (ix)$ with $(D - i)$ using Rule 2; so the first step is

$$y_p = Im \left[\frac{1}{D - i} e^{ix} \frac{1}{2i} \right] \quad \textbf{(2.340)}$$

Now, apply Rule 2, noting that $f(x) = 1$

$$y_p = Im \left[\left(\frac{e^{ix}}{2i} \right) \frac{1}{D} \cdot 1 \right] \quad \textbf{(2.341)}$$

Here, we see that $D^{-1}1 = x$, and thus, the imaginary part is

$$Im \left(\frac{e^{ix}}{2i} \right) = Im \left(\frac{ie^{ix}}{-2} \right) = -\frac{1}{2} \cos (x) \quad \textbf{(2.342)}$$

Hence, we finally obtain

$$y_p = -\frac{1}{2}x \cos(x) \tag{2.343}$$

Had we used the Method of Undetermined Coefficients, it would have been necessary to make the first guess (to insure linear independence from the complementary solutions which are $\sin(x)$ and $\cos(x)$)

$$y_p = ax \sin(x) + bx \cos(x) \tag{2.344}$$

which would lead to a lengthy and tedious analysis, as the reader can verify.

In general, the Inverse Operator method is not recommended for product functions such as $x \sin(x)$, etc., because of difficulty in expanding operators in series to operate on polynomial functions (i.e., $a + bx + cx^2$, etc). In such cases, the Method of Variation of Parameters, which follows, may be used to good effect.

3. Method of Variation of Parameters

As mentioned at the beginning of this section, this method can be applied even when coefficients are nonconstant, so that we treat the general case

$$\frac{d^2y}{dx^2} + a_1(x)\frac{dy}{dx} + a_0(x)y = f(x) \tag{2.345}$$

At the outset, it is assumed that the two linearly independent complementary solutions are known

$$y_c(x) = Au(x) + Bv(x) \tag{2.346}$$

The Variation of Parameters method is based on the premise that the particular solutions are linearly independent of $u(x)$ and $v(x)$. We start by proposing

$$y_p(x) = F_u(x)u(x) + F_v(x)v(x) \tag{2.347}$$

where obviously F_u and F_v are not constant. It is clear that if we insert this proposed solution into the defining equation, we shall obtain one equation, but we have two unknowns: F_u and F_v. Thus, we must propose one additional equation, as we show next, to have a solvable system. Performing the required differentiation shows using prime to denote differentiation

$$\frac{dy_p}{dx} = (uF_u' + vF_v') + (u'F_u + v'F_v) \tag{2.348}$$

It is clear that a second differentiation will introduce second derivatives of the unknown functions F_u, F_v. To avoid this complication, we take as our second proposed equation

$$uF_u' + vF_v' = 0 \qquad (2.349)$$

This is the most convenient choice, as the reader can verify. We next find y_p''

$$\frac{d^2 y_p}{dx^2} = (F_u u'' + F_v v'') + (F_u' u' + F_v' v'). \qquad (2.350)$$

Inserting dy_p/dx and $d^2 y_p/dx^2$ into the defining equation we obtain, after rearrangement

$$F_u[u'' + a_1(x)u' + a_0(x)u] + F_v[v'' + a_1(x)v' + a_0(x)v] \qquad (2.351)$$
$$+ F_u' u' + F_v' v' = f(x)$$

It is obvious that the bracketed terms vanish, because they satisfy the homogeneous equation [when $f(x) = 0$] since they are complementary solutions. The remaining equation has two unknowns,

$$u'F_u' + v'F_v' = f(x) \qquad (2.352)$$

This coupled with our second proposition

$$uF_u' + vF_v' = 0 \qquad (2.353)$$

forms a system of two equations with two unknowns. Solving these by defining $p = F_u'$ and $q = F_v'$ shows; first, from Eq. 2.353

$$p = -\frac{v}{u}q \qquad (2.354)$$

Inserting this into Eq. 2.352 gives

$$u'\left(-\frac{v}{u}q\right) + v'q = f(x) \qquad (2.355)$$

hence

$$q = \frac{dF_v}{dx} = \frac{-uf(x)}{u'v - v'u} \qquad (2.356)$$

and this allows p to be obtained as

$$p = \frac{dF_u}{dx} = \frac{vf(x)}{u'v - v'u} \qquad (2.357)$$

These are now separable, so that within an arbitrary constant:

$$F_u(x) = \int \frac{vf(x)}{u'v - v'u}\, dx \tag{2.358}$$

$$F_v(x) = \int \frac{-uf(x)}{u'v - v'u}\, dx \tag{2.359}$$

These integrations, then, produce the particular solutions, worth repeating as

$$y_p = u(x)F_u(x) + v(x)F_v(x) \tag{2.360}$$

The denominators in Eqs. 2.358 and 2.359 represent the negative of the so-called Wronskian determinant

$$W(u,v) = \begin{vmatrix} u & v \\ u' & v' \end{vmatrix} = uv' - u'v$$

which is nonzero if u and v are indeed linearly independent. For the second order systems considered here, linear independence can be deduced by inspection. For higher order systems, the application of the Wronskian is the most direct way to inspect linear independence (Hildebrand 1962).

EXAMPLE 2.23

The second order equation with nonconstant coefficients

$$4x\frac{d^2y}{dx^2} + 6\frac{dy}{dx} + y = f(x) \tag{2.361}$$

has complementary solutions (when $f(x) = 0$) obtainable by the Frobenius series method (Chapter 3)

$$y_c(x) = A\frac{\sin(\sqrt{x})}{\sqrt{x}} + B\frac{\cos(\sqrt{x})}{\sqrt{x}}. \tag{2.362}$$

Find the particular solution when $f(x) = 1/x^{3/2}$.
   Here, we take the complementary functions to be

$$u = \frac{\sin(\sqrt{x})}{\sqrt{x}}; \qquad v = \frac{\cos(\sqrt{x})}{\sqrt{x}}$$

We first compute the denominator for the integrals in Eqs. 2.358 and 2.359

$$u'v - v'u = \frac{1}{2}\frac{1}{x^{3/2}}\left(\cos^2(\sqrt{x}) + \sin^2(\sqrt{x})\right) = \frac{1}{2}\frac{1}{x^{3/2}} \tag{2.363}$$

Inserting this into the same integrals yields:

$$F_u = \int 2 \frac{\cos\left(\sqrt{x}\right)}{\sqrt{x}} x^{3/2} \frac{1}{x^{3/2}} \, dx = 4 \sin\left(\sqrt{x}\right) \tag{2.364}$$

$$F_v = -\int 2 \frac{\sin\left(\sqrt{x}\right)}{\sqrt{x}} x^{3/2} \frac{1}{x^{3/2}} \, dx = 4 \cos\left(\sqrt{x}\right) \tag{2.365}$$

so that we finally have the particular solution

$$y_p = 4 \frac{\sin^2\left(\sqrt{x}\right)}{\sqrt{x}} + 4 \frac{\cos^2\left(\sqrt{x}\right)}{\sqrt{x}} = \frac{4}{\sqrt{x}} \tag{2.366}$$

which is linearly independent of the complementary solutions.

2.5.3 Summary of Particular Solution Methods

We have illustrated three possible methods to find particular solutions. Each has certain advantages and disadvantages, which are summarized as follows.

1. Method of Undetermined Coefficients

This technique has advantages for elementary polynomial forcing functions (e.g., $2x^2 + 1$, $5x^3 + 3$, etc.), and it is easy to apply and use. However, it becomes quite tedious to use on trigonometric forcing functions, and it is not fail-safe in the sense that some experience is necessary in constructing the trial function. Also, it does not apply to equations with nonconstant coefficients.

2. Method of Inverse Operators

This method is the quickest and safest to use with exponential or trigonometric forcing functions. Its main disadvantage is the necessary amount of new material a student must learn to apply it effectively. Although it can be used on elementary polynomial forcing functions (by expanding the inverse operators into ascending polynomial form), it is quite tedious to apply for such conditions. Also, it cannot be used on equations with nonconstant coefficients.

3. Method of Variation of Parameters

This procedure is the most general method, since it can be applied to equations with variable coefficients. Although it is fail-safe, it often leads to intractable integrals to find F_v and F_u. It is the method of choice when treating forced problems in transport phenomena, since both cylindrical and spherical coordinate systems always lead to equations with variable coefficients.

2.6 COUPLED SIMULTANEOUS ODE

In principle, any set of n linear first order coupled equations is equivalent to the nth order inhomogeneous equation given earlier as

$$\frac{d^n y}{dt^n} + a_{n-1} \frac{d^{n-1} y}{dt^{n-1}} + \cdots + a_1 \frac{dy}{dt} + a_0 y = f(t) \tag{2.367}$$

To see this, we redefine variables as follows:

$$x_1 = y, \quad x_2 = \frac{dx_1}{dt}, \quad x_3 = \frac{dx_2}{dt}, \quad x_n = \frac{dx_{n-1}}{dt} \tag{2.368}$$

These definitions turn the nth order equation into the coupled set of first order equations

$$\frac{dx_1}{dt} = 0 \cdot x_1 + 1 \cdot x_2 + 0 \cdot x_3 + 0 \cdot x_4 + \cdots + 0 \cdot x_n$$

$$\frac{dx_2}{dt} = 0 \cdot x_1 + 0 \cdot x_2 + 1 \cdot x_3 + 0 \cdot x_4 + \cdots + 0 \cdot x_n$$

$$\vdots \tag{2.369}$$

$$\frac{dx_{n-1}}{dt} = 0 \cdot x_1 + 0 \cdot x_2 + 0 \cdot x_3 + 0 \cdot x_4 + \cdots + 1 \cdot x_n$$

$$\frac{dx_n}{dt} = -a_0 x_1 - a_1 x_2 - a_2 x_3 - \cdots - a_{n-1} x_n + f(t)$$

In vector form, these can be abbreviated as

$$\frac{d\mathbf{x}}{dt} = \mathbf{A} \cdot \mathbf{x} + \mathbf{f} \tag{2.370}$$

where the vectors are

$$\frac{d\mathbf{x}}{dt} = \left[\frac{dx_1}{dt}, \frac{dx_2}{dt}, \ldots, \frac{dx_n}{dt} \right]^{\mathsf{T}} \tag{2.371}$$

$$\mathbf{x} = [x_1, x_2, \ldots, x_n]^{\mathsf{T}}$$

$$\mathbf{f} = [0, 0, 0, \ldots, f(t)]^{\mathsf{T}}$$

and the matrix of coefficients is

$$\mathbf{A} = \begin{bmatrix} 0 & 1 & 0 & 0 & \cdots & 0 \\ 0 & 0 & 1 & 0 & \cdots & 0 \\ & \cdot & \cdot & \cdot & \cdots & \cdot \\ & \cdot & \cdot & \cdot & \cdots & \cdot \\ -a_0 & -a_1 & -a_2 & \cdot & \cdots & -a_{n-1} \end{bmatrix}. \tag{2.372}$$

Treatment of elementary matrix methods is reviewed in Appendix A at the end of this book. Students interested in advanced material on the subject should consult the excellent text by Amundson (1966). Suffice to say that these specialized techniques must ultimately solve the same required characteristic equation as taught here, namely,

$$r^n + a_{n-1}r^{n-1} + \cdots + a_1 r + a_0 = 0 \qquad (2.373)$$

so that the n eigenvalues r must be found as before.

However, useful methods exist that treat simultaneous equations without resorting to formalized methods of multilinear algebra. We shall discuss two of these methods because of their utility and frequent occurrence in practical problems:

1. Elimination of Independent Variables.
2. Elimination of Dependent Variables.

These common-sense methods often escape the notice of an analyst, because the structure and complexity of a problem may be so intimidating. We illustrate the above principles with a few examples as follows.

EXAMPLE 2.24

Finely dispersed catalyst particles in a bed are used to promote the irreversible nth order gas-phase reaction in a batch, adiabatic reactor of constant volume

$$A \rightarrow \text{Products}; \qquad R_A = k_n(T)C_A^n$$

where C_A denotes concentration of A and $k_n(T)$ is the temperature dependent rate constant, which obeys the Arrhenius expression

$$k_n = \alpha \exp\left(-\frac{E}{RT}\right) \qquad (2.374)$$

The product of gas volume and exothermic heat is given by λ, and the heat capacity of the gas is much smaller than the solid catalyst. Find the maximum temperature sustained by the insulated bed if there is no volume change in reacting A to products for a mass m of solid particles.

The simultaneous heat and mole balances can be written as

$$\frac{dC_A}{dt} = -C_A^n \alpha \exp\left(-\frac{E}{RT}\right) \qquad (2.375)$$

$$mC_p\frac{dT}{dt} = +\lambda C_A^n \alpha \exp\left(-\frac{E}{RT}\right) \qquad (2.376)$$

This is a rather intimidating set of highly nonlinear equations, if a time-domain solution is sought. However, we need only to find the relationship between C_A

and T, so divide the equations (to eliminate time) and see

$$mC_p \frac{dT}{dC_A} = -\lambda \qquad (2.377)$$

which is separable, so that we get

$$T = -\left(\frac{\lambda}{mC_p}\right)C_A + K \qquad (2.378)$$

where K is a constant of integration. If we take the initial conditions to be $T(0) = T_0$ and $C_A(0) = C_0$, then we have

$$K = T_0 + \frac{\lambda}{mC_p}C_0 \qquad (2.379)$$

and the general result is

$$(T - T_0) = \frac{\lambda}{mC_p}(C_0 - C_A) \qquad (2.380)$$

Clearly, the maximum temperature occurs after all of the reactant A is devoured, so setting $C_A = 0$ yields

$$T_{\max} = T_0 + \frac{\lambda}{mC_p}C_0 \qquad (2.381)$$

To find the transient equation $T(t)$ describing temperature between two extremes $(T_0 \rightarrow T_{\max})$ we can use Eq. 2.380 to eliminate $C_A(t)$ in Eq. 2.376, so that we need only solve the single nonlinear equation

$$mC_p \frac{dT}{dt} = \alpha\lambda\left[C_0 - \frac{mC_p}{\lambda}(T - T_0)\right]^n \exp\left(-\frac{E}{RT}\right) \qquad (2.382)$$

Approximations for this are possible for modest temperature rise by expanding the exponential function in series and retaining the low order terms, leading to a linearized expression.

EXAMPLE 2.25

The double pipe, cocurrent heat exchanger is used to cool a distillate product using cold water circulating through the jacket as illustrated in Fig. 2.3. The overall heat transfer coefficient is taken to be U and the mass flow of distillate and water is W_i and W_0, respectively. Under turbulent flow conditions, the fluid temperatures are taken to be uniform across individual flow cross sections. Find the relationship to predict how steady-state temperature changes with axial position, and from this, deduce an expression to compute the average ΔT

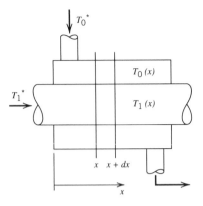

Figure 2.3 Double-pipe heat exchanger.

between streams. Ignore axial conduction effects and use constant physical properties. Assume the inner pipe of diameter d is quite thin.

We first apply the conservation law to each fluid in turn, as follows:

$$W_i C_{pi} T_i(x) - W_i C_{pi} T_i(x + \Delta x) - U(\pi \, d\Delta x)(\overline{T}_i - \overline{T}_0) = 0 \quad \textbf{(2.383)}$$

$$W_0 C_{p0} T_0(x) - W_0 C_{p0} T_0(x + \Delta x) + U(\pi \, d\Delta x)(\overline{T}_i - \overline{T}_0) = 0 \quad \textbf{(2.384)}$$

where as noted in Chapter 1, the overbar represents the average between positions x and $x + \Delta x$, and in the limit: $\lim_{\Delta x \to 0} \overline{T} \to T$. Dividing by Δx and taking limits as before yields the set of coupled equations:

$$W_i C_{pi} \frac{dT_i}{dx} + U\pi \, d(T_i - T_0) = 0 \quad \textbf{(2.385)}$$

$$W_0 C_{p0} \frac{dT_0}{dx} - U\pi \, d(T_i - T_0) = 0 \quad \textbf{(2.386)}$$

where $T_i > T_0$. Combining parameters as taught in Chapter 1, we rearrange to get

$$\frac{dT_i}{dx} + \lambda_i (T_i - T_0) = 0 \quad \textbf{(2.387)}$$

$$\frac{dT_0}{dx} - \lambda_0 (T_i - T_0) = 0 \quad \textbf{(2.388)}$$

where

$$\lambda_i = \frac{U\pi \, d}{W_i C_{pi}} \; ; \qquad \lambda_0 = \frac{U\pi \, d}{W_0 C_{p0}}$$

The solutions to these equations are conveniently obtained using the Heaviside operator $D = d/dx$. Thus, rewrite using operators to see

$$(D + \lambda_i)T_i = \lambda_i T_0(x) \tag{2.389}$$

$$(D + \lambda_0)T_0 = \lambda_0 T_i(x) \tag{2.390}$$

The equations have identical structures, and this suggests the following procedure. Apply the second operator $(D + \lambda_0)$ to the first equation and see

$$(D + \lambda_0)(D + \lambda_i)T_i = \lambda_i(D + \lambda_0)T_0 \tag{2.391}$$

But the defining Eq. 2.390 shows that $(D + \lambda_0) = \lambda_0 T_i$; hence, we can decouple the equations to get a single equation for T_i

$$(D + \lambda_0)(D + \lambda_i)T_i = \lambda_i \lambda_0 T_i \tag{2.392}$$

This can be simplified further owing to cancellation of terms

$$D^2 T_i + (\lambda_0 + \lambda_i)DT_i = 0 \tag{2.393}$$

This homogeneous equation has characteristic roots

$$r^2 + (\lambda_0 + \lambda_i)r = 0 \tag{2.394}$$

hence,

$$r_1 = 0; \qquad r_2 = -(\lambda_0 + \lambda_i)$$

so the complementary solution is

$$T_i(x) = B_i + C_i \exp\left[-(\lambda_0 + \lambda_i)x\right] \tag{2.395}$$

Performing the same operation on the equation for T_0, that is, applying the operation $(D + \lambda_i)$ and then using $(D + \lambda_i)T_i = \lambda_i T_0$ yields an identical equation, as might be expected

$$(D + \lambda_i)(D + \lambda_0)T_0 = \lambda_0 \lambda_i T_0 \tag{2.396}$$

The solution is the same as for $T_i(x)$, except in general the arbitrary constants are different

$$T_0(x) = B_0 + C_0 \exp\left[-(\lambda_0 + \lambda_i)x\right] \tag{2.397}$$

At first glance it may appear that we have four arbitrary constants. However, we can show that B_i, C_i are linearly connected to B_0, C_0 as follows. Inserting the two solutions into either of the original heat balances shows, for example using

Eq. 2.387

$$-C_i(\lambda_0 + \lambda_i)\exp\left[-(\lambda_0 + \lambda_i)x\right]$$
$$+ \lambda_i\left[B_i - B_0 + (C_i - C_0)\exp\left[-(\lambda_0 + \lambda_i)x\right]\right] = 0 \tag{2.398}$$

This can only be satisfied if we stipulate the conditions

$$B_i = B_0 \tag{2.399}$$

$$(\lambda_0 + \lambda_i)C_i = \lambda_i(C_i - C_0) \tag{2.400}$$

which reduces to

$$C_i = -\frac{\lambda_i}{\lambda_0}C_0 = -\frac{W_0 C_{p0}}{W_i C_{pi}}C_0 \tag{2.401}$$

Thus, there exists only two independent constants of integration, and the two equations for temperature distribution are

$$T_i(x) = B_0 - \frac{\lambda_i}{\lambda_0}C_0 \exp\left[-(\lambda_0 + \lambda_i)x\right] \tag{2.402}$$

$$T_0(x) = B_0 + C_0 \exp\left[-(\lambda_0 + \lambda_i)x\right] \tag{2.403}$$

We can find the two constants B_0, C_0 using boundary conditions

$$T_i(0) = T_i^*; \qquad T_0(0) = T_0^* \tag{2.404}$$

where T_i^* and T_0^* denote the inlet temperatures of hot distillate and cool water, respectively, hence $T_i^* > T_0^*$. Solving for B_0, C_0 yields:

$$B_0 = T_0^* + \Delta T^*\left[\frac{\lambda_i}{\lambda_0} + 1\right]^{-1} \tag{2.405}$$

$$C_0 = -\Delta T^*\left[\frac{\lambda_i}{\lambda_0} + 1\right]^{-1} \tag{2.406}$$

where

$$\Delta T^* = T_i^* - T_0^*$$

Inserting these and rearranging to a more suitable form gives finally the dimensionless results, using $N = \lambda_i/\lambda_0$:

$$\psi_i = \frac{T_i(x) - T_0^*}{T_i^* - T_0^*} = \frac{1}{N + 1}\left\{1 + N\exp\left[-(\lambda_0 + \lambda_i)x\right]\right\} \tag{2.407}$$

$$\psi_0 = \frac{T_0(x) - T_0^*}{T_i^* - T_0^*} = \frac{1}{N + 1}\left\{1 - \exp\left[-(\lambda_0 + \lambda_i)x\right]\right\} \tag{2.408}$$

As x increases, it is easy to see that $T_i(x)$ decreases, as required, and also $T_0(x)$ increases as expected. For convenience, we have denoted dimensionless temperatures as ψ_i and ψ_0, respectively, and these groups appear naturally in the course of analysis. The right-hand sides are also dimensionless, since

$$N = \frac{\lambda_i}{\lambda_0} = \frac{W_0 C_{p0}}{W_i C_{pi}}$$

and of course the argument of the exponential is also dimensionless. This could be used to define a dimensionless axial coordinate as we did in Chapter 1

$$\zeta = (\lambda_0 + \lambda_i)x = \frac{U(\pi d)x(W_i C_{pi} + W_0 C_{p0})}{W_i W_0 C_{pi} C_{p0}} \qquad (2.409)$$

We now write the final solutions in very compact form:

$$\psi_i(\zeta) = \frac{1}{N+1}[1 + N\exp(-\zeta)] \qquad (2.410)$$

$$\psi_0(\zeta) = \frac{1}{N+1}[1 - \exp(-\zeta)] \qquad (2.411)$$

Thus, we see

$$\psi_i(0) = 1 \qquad \text{and} \qquad \psi_0(0) = 0$$

as required. To find the average $\Delta T = (T_i - T_0)_{\text{avg}}$, we could perform the operation

$$(T_i - T_0)_{\text{avg}} = \frac{1}{L}\int_0^L [T_i(x) - T_0(x)]\,dx \qquad (2.412)$$

where L denotes the distance to the exit. It is less tedious to perform this operation in dimensionless form, noting that we can define the equivalent dimensionless temperature difference as

$$\left(\frac{\Delta T}{\Delta T^*}\right)_{\text{avg}} = (\psi_i - \psi_0)_{\text{avg}} = \frac{1}{\zeta_L}\int_0^{\zeta_L}[\psi_i(\zeta) - \psi_0(\zeta)]\,d\zeta \qquad (2.413)$$

where

$$\zeta_L = (\lambda_i + \lambda_0)L = (N+1)\lambda_0 L$$

This is easily integrated, since

$$\psi_i - \psi_0 = \exp(-\zeta) \qquad (2.414)$$

Integrating this, we find

$$\left(\frac{\Delta T}{\Delta T^*}\right)_{\text{avg}} = \frac{1}{\zeta_L}[1 - \exp(-\zeta_L)] \qquad (2.415)$$

We could eliminate the exponential term by noting

$$\Psi_i(L) - \psi_0(L) = \exp(-\zeta_L) = \frac{T_i(L) - T_0(L)}{\Delta T^*} \qquad (2.416)$$

so now we have

$$\left(\frac{\Delta T}{\Delta T^*}\right)_{avg} = \frac{1}{\zeta_L}\left(1 - \frac{T_i(L) - T_0(L)}{\Delta T^*}\right) \qquad (2.417)$$

Moreover, taking logarithms of $[\psi_i(L) - \psi_0(L)]$ yields

$$\ln[\psi_i(L) - \psi_0(L)] = -\zeta_L = \ln\left(\frac{T_i(L) - T_0(L)}{\Delta T^*}\right) \qquad (2.418)$$

Inserting this and multiplying through by ΔT^*, noting the definition

$$\Delta T^* = T_i^* - T_0^* = T_i(0) - T_0(0) \qquad (2.419)$$

we thus obtain the expected result, which defines the Log-Mean ΔT

$$\Delta T_{avg} = \frac{[T_i(L) - T_0(L)] - [T_i(0) - T_0(0)]}{\ln\left(\dfrac{T_i(L) - T_0(L)}{T_i(0) - T_0(0)}\right)} \qquad (2.420)$$

How would this average ΔT change if we had included effects arising from axial conduction?

2.7 SUMMARY OF SOLUTION METHODS FOR ODE

We started this chapter by delineating the two fundamental types of equations, either nonlinear or linear. We then introduced the few techniques suitable for nonlinear equations, noting the possibility of so-called singular solutions when they arose. We also pointed out that nonlinear equations describing model systems usually lead to the appearance of "implicit" arbitrary constants of integration, which means they appear within the mathematical arguments, rather than as simple multipliers as in linear equations. The effect of this implicit constant often shows up in startup of dynamic systems. Thus, if the final steady state depends on the way a system is started up, one must be suspicious that the system sustains nonlinear dynamics. No such problem arises in linear models, as we showed in several extensive examples. We emphasized that no general technique exists for nonlinear systems of equations.

The last and major parts of this chapter dealt with linear equations, mainly because such equations are always solvable by general methods. We noted that forced equations contain two sets of solutions: the particular solutions [related directly to the type of forcing function $f(x)$] and the complementary solution [the solution obtainable when $f(x) = 0$], so that in all cases: $y(x) = y_c(x) + y_p(x)$. We emphasize again that the arbitrary constants are found (in conjunction with *B.C.* and *I.C.*) using the complete solution $y(x)$. We illustrated three methods to find the particular integral: Undetermined Coefficients, Inverse Operators, and Variation of Parameters. Only the last of these was applicable to linear equations with nonconstant coefficients.

Linear homogeneous equations containing nonconstant coefficients were not treated, except for the elementary Euler-Equidimensional equation, which was reduced to a constant coefficient situation by letting $x = \exp(t)$. In the next chapter, we deal extensively with the nonconstant coefficient case, starting with the premise that all continuous solutions are in fact representable by an infinite series of terms, for example: $\exp(x) = 1 + x + x^2/2! + x^3/3! + \cdots$. This leads to a formalized procedure, called the Method of Frobenius, to find all the linearly independent solutions of homogeneous equations, even if coefficients are nonconstant.

2.8 REFERENCES

1. Amundson, N. R. *Mathematical Methods in Chemical Engineering, Matrices and Their Application.* Prentice Hall, Englewood Cliffs, New Jersey (1966).
2. Hildebrand, F. B. *Advanced Calculus for Applications.* Prentice Hall, Englewood Cliffs, New Jersey, pp. 5, 34 (1962).
3. Klinkenberg, A., and F. Sjenitzer, "Holding Time Distributions of the Gaussian Type," *Chem. Eng. Sci.* 5, 258 (1956).
4. Prince, M. J., and H. W. Blanch, "Transition Electrolyte Concentrations for Bubble Coalescence," *AIChE J.* 36, 1425–29 (1990).
5. Rice, R. G., "Transpiration Effects in Solids Dissolution," *Chem. Eng. Sci.* 10, 1465 (1982).
6. Rice. R. G., and M. A. Littlefield, "Dispersion Coefficients for Ideal Bubbly Flow in Truly Vertical Bubble Column," *Chem. Eng. Sci.* 42, 2045 (1987).

2.9 PROBLEMS

2.1$_2$. A tall, cylindrical tank is being filled, from an initially empty state, by a constant inflow of q liters/sec of liquid. The flat tank bottom has corroded and sustains a leak through a small hole of area A_0. If the cross-sectional area of the tank is denoted by A, and time-varying height of liquid is $h(t)$, then:

(a) find the dynamic relationship describing tank height, if the volumetric leak rate obeys Torricelli's law, $q_0 = A_0\sqrt{2gh(t)}$ (g is gravitational acceleration).

(b) determine the relationship to predict the final steady-state liquid height in the tank.

(c) define $x = \sqrt{h}$, separate variables and deduce the implicit solution for h:

$$t = \left(\frac{qA}{A_0^2 g}\right)\ln\left[\frac{q}{q - A_0\sqrt{2gh}}\right] - 2\left(\frac{A}{A_0}\right)\sqrt{\frac{h}{2g}}$$

(d) sketch the curve for h versus t, and compare with the case for a nonleaking tank.

2.2$_2$. Two vertical, cylindrical tanks, each 10 m high, are installed side-by-side in a tank farm, their bottoms at the same level. The tanks are connected at their bottoms by a horizontal pipe 2 meters long, with pipe inside diameter 0.03 m. The first tank (1) is full of oil and the second tank (2) is empty. Moreover, tank 1 has a cross-sectional area twice that of tank 2. The first tank also has another outlet (to atmosphere) at the bottom, composed of a short horizontal pipe 2 m long, 0.03-m diameter. Both of the valves for the horizontal pipes are opened simultaneously. What is the maximum oil level in tank 2? Assume laminar flow in the horizontal pipes, and neglect kinetic, entrance–exit losses.

Answer: 4.07 m

2.3$_2$. The consecutive, second order, irreversible reactions are carried out in a batch reactor

$$A + S \xrightarrow{k_1} X$$

$$X + S \xrightarrow{k_2} Y$$

One mole of A and two moles of S are initially added. Find the mole fraction X remaining in solution after half the A is consumed; take $k_2/k_1 = 2$.

Answer: $y_x = 1/9$

2.4$_2$. Solve the following first order equations:

(a) $\rho \dfrac{dT}{d\rho} - \dfrac{4}{\rho}\dfrac{d\rho}{dT} = 0;$ $T = T_0$ when $\rho = \rho_0$

(b) $(x^2 y + x)\,dy + (xy^2 - y)\,dx = 0$

(c) $\dfrac{dy}{dx} + \dfrac{y}{x} = \sin(ax)$

(d) $\dfrac{dy}{dx} - \dfrac{2}{x}y = y^3$

2.5$_3$. Solve the following second order equations:

(a) $x\dfrac{d^2 y}{dx^2} + \left(\dfrac{dy}{dx}\right)^2 - \left(\dfrac{y}{x}\right)^2 = 0;$ $y(1) = 2;$ $y'(1) = -1$

(b) $x^2\dfrac{d^2 y}{dx^2} + x\dfrac{dy}{dx} = \ln(x);$ $y(1) = 1;$ $y'(1) = 0$

(c) $y\dfrac{d^2 y}{dx^2} + \left(\dfrac{dy}{dx}\right)^2 = \dfrac{dy}{dx};$ $y(0) = 1;$ $y'(0) = 2$

2.6*. The reversible set of reactions represented by

$$A \underset{k_2}{\overset{k_1}{\rightleftharpoons}} B \underset{k_4}{\overset{k_3}{\rightleftharpoons}} C$$

is carried out in a batch reactor under conditions of constant volume and temperature. Only one mole of A is present initially, and any time t the moles are N_A, N_B, N_C. The net rate of disappearance of A is given by

$$\frac{dN_A}{dt} = -k_1 N_A + k_2 N_B$$

and for B, it is

$$\frac{dN_B}{dt} = -(k_2 + k_3) N_B + k_1 N_A + k_4 N_C$$

and for all times, the stoichiometry must be obeyed

$$N_A + N_B + N_C = 1$$

(a) Show that the behavior of $N_A(t)$ is described by the second order ODE

$$\frac{d^2 N_A}{dt^2} + (k_1 + k_2 + k_3 + k_4)\frac{dN_A}{dt} + (k_1 k_3 + k_2 k_4 + k_1 k_4) N_A = k_2 k_4$$

(b) One initial condition for the second order equation in part (a) is $N_A(0) = 1$; what is the second necessary initial condition?
(c) Find the complete solution for $N_A(t)$, using the conditions in part (b) to evaluate the arbitrary constants of integration.

2.7$_3$. Solid, stubby, cylindrical metal rods (length-to-diameter ratio = 3) are used as heat promoters on the exterior of a hot surface with surface temperature of 700° C. The ambient air flowing around the rod-promoters has a temperature of 30° C. The metal conductivity (k) takes a value of 0.247 cal/(sec · cm ·° K). The heat transfer coefficient (h) around the surface of the promoter is constant at 3.6 Kcal/(m² · hr ·° C).

(a) Analyze a single rod of 4-mm diameter and show that the steady-state differential balance yields the following differential equation

$$\frac{d^2 T}{dx^2} - \left(\frac{2h}{Rk}\right)(T - T_A) = 0; \qquad 2R = \text{diameter}$$

for the case when metal temperature changes mainly in the x direction (x is directed outward from the hot surface, and rod radius is R).

(b) Find the characteristic roots for the ODE in part (a). What are the physical units of these roots?

(c) Find the solution of part (a) using the conditions:

$$T = T_H, \qquad x = 0 \qquad \text{(hot surface)}$$

$$-k\frac{dT}{dx} = h(T - T_A), \qquad x = L \qquad \text{(exposed flat tip)}$$

and show the temperature profile is represented by

$$\frac{T - T_A}{T_H - T_A} = \cosh\left[2\left(\frac{x}{L}\right)\left(\frac{L}{D}\right)\sqrt{Bi}\right]$$

$$- \frac{2\tanh\left(2\frac{L}{D}\sqrt{Bi}\right) + \sqrt{Bi}}{2 + \sqrt{Bi}\,\tanh\left(2\frac{L}{D}\sqrt{Bi}\right)}\,\sinh\left[2\frac{x}{L}\frac{L}{D}\sqrt{Bi}\right]$$

where $Bi = hD/k$ (Biot number, dimensionless; the ratio of film to metal transfer rates).

(d) Use the definitions of total heat flow and find the effectiveness factor for the present promoter

$$\eta = \frac{Q}{Q_{max}}$$

and show that the general expression for η is

$$\eta = \left(\frac{1}{2}\right)\frac{1}{\sqrt{Bi}\left(\frac{L}{D} + \frac{1}{4}\right)}\left[\frac{\tanh\left(2\frac{L}{D}\sqrt{Bi}\right) + \frac{1}{2}\sqrt{Bi}}{1 + \frac{1}{2}\sqrt{Bi}\,\tanh\left(2\frac{L}{D}\sqrt{Bi}\right)}\right]$$

(e) For small arguments, $2(L/D)\sqrt{Bi} \ll 1$, show that the effectiveness factor becomes approximately

$$\eta \approx \frac{1}{\left(1 + \frac{L}{D}\cdot Bi\right)}$$

[Hint: look at the series expansion for $\tanh(u)$]

(f) Compute η for the present promoter.

2.8$_2$. Find the complementary and particular solutions and thereby write the general solutions for the following:

(a) $$\frac{d^2y}{dx^2} + y = x\sin(x)$$

(b) $$\frac{d^2y}{dx^2} - 2\frac{dy}{dx} + y = xe^x$$

(c) $$x^2\frac{d^2y}{dx^2} + x\frac{dy}{dx} - y = x$$

2.9*. When gas is injected into a column of water, a liquid circulation pattern develops. Thus, upflow at a rate Q_u (m^3/s) rises in the central core and downflow occurs at a rate Q_d in the annulus. If liquid of composition C_0 is also injected at the column base at a rate Q_0, with outflow at the same rate, then $Q_u = Q_d + Q_0$ (if density is constant).

(a) The injected gas contains a soluble component (with solubility C^* moles/m^3) so that mass transfer occurs by way of a constant volumetric mass transfer coefficient denoted as $k_c a$. There is also an exchange of solute between upflowing and downflowing liquid at a rate per unit height equal to $K_E(C_u - C_d)$. If the flow areas for upflow and downflow areas are equal (A), perform a material balance and show that

$$Q_u \frac{dC_u}{dz} = k_c a A (C^* - C_u) - K_E(C_u - C_d)$$

$$- Q_d \frac{dC_d}{dz} = k_c a A (C^* - C_d) + K_E(C_u - C_d)$$

where z is distance from column base.

(b) Define new variables to simplify matters as

$$\theta_u = C_u - C^*$$

$$\theta_d = C_d - C^*$$

$$\zeta = z(k_c a A + K_E)/Q_0 \qquad \text{(dimensionless distance)}$$

$$q_u = Q_u/Q_0 \qquad \text{(dimensionless upflow)}$$

$$q_d = Q_d/Q_0 \qquad \text{(dimensionless downflow)}$$

and show that the coupled relations are

$$q_u \frac{d\theta_u}{d\zeta} + \theta_u = \alpha \theta_d$$

$$- q_d \frac{d\theta_d}{d\zeta} + \theta_d = \alpha \theta_u$$

where $\alpha = K_E/(K_E + k_c a A)$

(c) Use the operator method to find solutions for $\theta_u(\zeta)$ and $\theta_d(\zeta)$.

(d) Show that the resulting four arbitrary constants are not independent and then write the solutions in terms of only two unknown integration constants.

(e) Apply the saturation condition

$$\theta_d, \theta_u \to 0 \text{ as } \zeta \to \infty$$

and a material balance at the entrance

$$Q_d C_d(0) + Q_0 C_0 = Q_u C_u(0)$$

to evaluate the remaining arbitrary constants of integration and thereby obtain relations to predict composition profiles along the axis.

(f) Deduce asymptotic solutions for the case when $Q_d \to 0$ and when $K_E \to \infty (\alpha \to 1.0)$; this corresponds to the plug-flow, nonrecirculating result.

2.10*. When an insoluble bubble rises in a deep pool of liquid, its volume increases according to the ideal gas law. However, when a soluble bubble rises from deep submersion, there is a competing action of dissolution that tends to reduce size. Under practical conditions, it has been proved (Rice 1982) that the mass transfer coefficient (k_c) for spherical particles (or bubbles) in free-fall (or free-rise) is substantially constant. Thus, for sparingly soluble bubbles released from rest, the following material balance is applicable

$$\frac{d\left(C \cdot \frac{4}{3}\pi R^3\right)}{dt} = -k_c \cdot C^* \cdot 4\pi R^2(t)$$

where $C = P/R_g T$ is the (ideal) molar density of gas, C^* is molar solubility of gas in liquid, and $R(t)$ is the changing bubble radius. The pressure at a distance z from the top liquid surface is $P = P_A + \rho_L g z$ and the rise velocity is assumed to be quasisteady and follows the intermediate law according to Rice and Littlefield (1987) to give a linear relation between speed and size

$$\frac{dz}{dt} = U = \left(\frac{2g}{15\nu^{1/2}}\right)^{2/3} \cdot 2R(t) = \beta \cdot R(t)$$

where g is gravitational acceleration and ν is liquid kinematic viscosity.

(a) Show that a change of variables allows the material balance to be written as

$$R\frac{dR}{dP} + \frac{1}{3}\frac{R^2}{P} = -\frac{\lambda}{P}$$

where

$$\lambda = \frac{k_c R_g T C^*}{\rho_L g \beta}$$

(b) Solve the equation in part (a) subject to the initial condition $R(0) = R_0$, $P(0) = P_0 = P_A + \rho g z_0$ and prove that

$$\frac{P}{P_0} = \left(\frac{R_0^2 + 3\lambda}{R^2 + 3\lambda}\right)^{3/2}$$

This expression could be used to find the distance or time required to cause a soluble bubble to completely disappear ($R \to 0$).

2.11₂. Considerable care must be exercised in applying the inverse operator method to forcing functions composed of products. Consider the equation

$$\frac{d^2y}{dx^2} - \frac{dy}{dx} = xe^x$$

for which we wish to find the particular solution.

(a) Apply the inverse operator and show

$$y_p = \frac{1}{D(D-1)} xe^x$$

then apply Rule 2 to the bracketed expression:

$$y_p = \frac{1}{D}\left[\frac{1}{D-1} xe^x\right]$$

(b) Complete the indicated operations to show

$$y_p = \left(\frac{x^2}{2} - x + 1\right)e^x$$

(c) Write the general solution in terms of only linearly independent solutions and two arbitrary constants.

Chapter 3

Series Solution Methods and Special Functions

3.1 INTRODUCTION TO SERIES METHODS

In Chapter 2 you learned that all linear differential equations with constant coefficients of the type, for example,

$$\frac{d^2y}{dx^2} - 2\frac{dy}{dx} + y = 0 \tag{3.1}$$

sustained complementary solutions of the form $\exp(rx)$, where the characteristic roots in this case satisfy $r^2 - 2r + 1 = 0$. It may not be obvious, but these solutions are in fact Power Series representations, since the exponential function can be represented by an infinite series

$$\exp(rx) = 1 + rx + \frac{(rx)^2}{2!} + \frac{(rx)^3}{3!} + \cdots \tag{3.2}$$

This series representation can be written in compact form

$$\exp(rx) = \sum_{n=0}^{\infty} \frac{(rx)^n}{n!} \tag{3.3}$$

where it is clear that $0! = 1$. As a matter of fact, we could have attempted a solution of Eq. 3.1 using the representation

$$y = \sum_{n=0}^{\infty} a_n x^n \tag{3.4}$$

We know in advance that $a_n = r^n/n!$, but nonetheless such a procedure may be quite useful for cases when an analytical solution is not available, such as when nonconstant coefficients arise [i.e., $a_1(x), a_0(x)$].

Let us proceed to solve Eq. 3.1 by the series representation, Eq. 3.4. Assume that the series coefficients a_n are unknown and we must find them. First, differentiate the series, Eq. 3.4, and insert the several derivatives into the defining Eq. 3.1 to see if useful generalizations can be made.

$$y' = \sum_{n=0}^{\infty} a_n(n) x^{n-1} \tag{3.5}$$

$$y'' = \sum_{n=0}^{\infty} a_n(n)(n-1) x^{n-2} \tag{3.6}$$

Inserting these into Eq. 3.1 yields the necessary condition

$$\sum_{n=0}^{\infty} a_n(n)(n-1) x^{n-2} - 2 \sum_{n=0}^{\infty} a_n(n) x^{n-1} + \sum_{n=0}^{\infty} a_n x^n = 0 \tag{3.7}$$

We can see that the first two terms of the first series contribute nothing, so we can increment the indices n upward twice with no loss of information and replace it with

$$\sum_{n=0}^{\infty} a_{n+2}(n+2)(n+1) x^n$$

The second series has zero for its first term, so we increment upward once and replace it with

$$\sum_{n=0}^{\infty} a_{n+1}(n+1) x^n$$

Now, since all series groups have in common x^n, then write the general result

$$\sum_{n=0}^{\infty} \left[a_{n+2}(n+2)(n+1) - 2(n+1)a_{n+1} + a_n \right] x^n = 0 \tag{3.8}$$

The only nontrivial generalization we can make is

$$a_{n+2}(n+2)(n+1) - 2a_{n+1}(n+1) + a_n = 0 \tag{3.9}$$

This so-called recurrence relationship is a finite-difference equation, which can be treated by the methods of Chapter 5. It has a structure similar to the Euler-Equidimensional ODE ($x^2 y'' - 2xy' + y = 0$) and it is easy to show the solution is (a recurrence relation)

$$a_n = a_0 \frac{r^n}{n!} \tag{3.10}$$

where r satisfies $r^2 - 2r + 1 = 0$, which has a double root at $r = 1$ so that we finally obtain the first solution

$$y(x) = a_0 \sum x^n/n! = a_0 \exp(x) \tag{3.11}$$

as expected at the outset. It is always possible to find such recurrence relations to relate the series coefficients, but they are not always straightforward, as we show later by examples.

We learned in Chapter 2, when only a single root obtains for a second order equation, then the second linearly independent solution was obtained by multiplying the first solution by x, so that the complete complementary solution to Eq. 3.1 is

$$y = a_0 \sum_{n=0}^{\infty} \frac{x^n}{n!} + b_0 x \sum_{n=0}^{\infty} \frac{x^n}{n!} \tag{3.12}$$

or explicitly

$$y = a_0 \exp(x) + b_0 x \exp(x) \tag{3.13}$$

We could have foreshadowed such problems (i.e., repeated or nondistinct roots) by proposing the more general form for series representation

$$y = \sum a_n x^{n+c} \tag{3.14}$$

where c is a variable index (it would obviously take values 0 and 1 in the previous example). This is the form used in the method of Frobenius to be discussed later in the chapter. At this point, it should be clear that this method will allow the two linearly independent solutions to be obtained in one pass, provided the series representation is indeed convergent.

It is the purpose of this chapter to show how linear equations with variable coefficients can be solved using a series representation of the type shown by Eq. 3.14. This powerful technique has great utility in solving transport problems, since these problems always give rise to variable coefficients owing to the cylindrical or spherical coordinate systems used in practical systems of interest.

3.2 PROPERTIES OF INFINITE SERIES

An expansion of the type

$$a_0 + a_1(x - x_0) + \cdots + a_n(x - x_0)^n + \cdots = \sum_{n=0}^{\infty} a_n(x - x_0)^n \tag{3.15}$$

is called a Power Series around the point x_0. If such a series is to converge (be convergent), it must approach a finite value as n tends to infinity. We wish to know those values of x that insure convergence of the series. Presently, we shall take x, x_0 and the coefficients a_n to be real variables. Complex power series can be treated by the method taught in Chapter 9. We assert that only convergent series solutions are of value for solving differential equations.

To determine the values of x, which lead to convergent series, we can apply the ratio test (Boas 1983), which states that if the absolute value of the ratio of the $(n + 1)^{st}$ term to nth term approaches a limit ε as $n \to \infty$, then the series itself converges when $\varepsilon < 1$ and diverges when $\varepsilon > 1$. The test fails if $\varepsilon = 1$. In the case of the Power Series, Eq. 3.15, we see

$$\varepsilon = \lim_{n \to \infty} \left| \frac{a_{n+1}}{a_n} \right| |x - x_0| = \frac{1}{R} |x - x_0| \qquad (3.16)$$

where

$$R = \lim_{n \to \infty} \left| \frac{a_n}{a_{n+1}} \right| \qquad (3.17)$$

if the limit indeed exists. Thus, it is seen that for convergence, $\epsilon < 1$; therefore,

$$|x - x_0| < R \qquad (3.18)$$

which is the condition on the values of x to insure convergence. Thus, if we can find a value R, then the *range of convergence* for the series is given by

$$(x_0 - R) < x < (x_0 + R) \qquad (3.19)$$

The distance R is called the *radius of convergence*.

Now, within the interval of convergence, the original Power Series can be treated like any other continuously differentiable function. Such series formed by differentiation or integration are thus guaranteed to be convergent. Consider a series we have come to know very well

$$\exp(x) = 1 + x + \frac{x^2}{2!} + \cdots + \frac{x^n}{n!} + \frac{x^{n+1}}{(n+1)!} + \cdots \qquad (3.20)$$

so that we see the radius of convergence is (note the expansion is around the point $x_0 = 0$)

$$R = \lim_{n \to \infty} \left| \frac{a_n}{a_{n+1}} \right| = \lim_{n \to \infty} \left| \frac{(n+1)!}{n!} \right| = \lim_{n \to \infty} (n + 1) = \infty \qquad (3.21)$$

It is thus clear for any finite x that

$$\varepsilon = \frac{1}{R} |x - x_0| < 1.0 \qquad (3.22)$$

since $R \to \infty$. Obviously, this well-known series is convergent in the region

$$-\infty < x < \infty \qquad (3.23)$$

Consider the Binomial series discussed earlier in Section 2.5.2

$$(1 + x)^p = 1 + px + \frac{p(p-1)}{(1)(2)}x^2 + \frac{p(p-1)(p-2)}{(1)(2)(3)}x^3 + \cdots \quad (3.24)$$

The nth term can be seen to be

$$a_n = \frac{p(p-1)(p-2)\cdots(p-n+1)}{n!} \quad (3.25)$$

so that the radius of convergence is

$$R = \lim_{n\to\infty}\left|\frac{a_n}{a_{n+1}}\right| = \lim_{n\to\infty}\left|\frac{n+1}{p-n}\right| \quad (3.26)$$

Dividing numerator and denominator by n shows

$$R = \lim_{n\to\infty}\left|\frac{1 + 1/n}{\dfrac{p}{n} - 1}\right| \to 1 \quad (3.27)$$

Therefore, to insure

$$\varepsilon = |x|/R < 1 \quad (3.28)$$

then $|x| < 1$, so the Binomial series around the point $x_0 = 0$ is convergent provided $|x| < 1$.

3.3 METHOD OF FROBENIUS

In solving Transport Phenomena problems, cylindrical and spherical coordinate systems always give rise to equations of the form

$$x^2\frac{d^2y}{dx^2} + xP(x)\frac{dy}{dx} + Q(x)y = 0 \quad (3.29)$$

which contain variable coefficients. On comparing with the general second order, variable coefficient case, Eq. 2.167, we see that $a_1(x) = P(x)/x$ and $a_0(x) = Q(x)/x^2$.

We shall assume that the functions $Q(x)$ and $P(x)$ are convergent around the point $x_0 = 0$ with radius of convergence R

$$P(x) = P_0 + P_1x + P_2x^2 + \cdots \quad (3.30)$$

$$Q(x) = Q_0 + Q_1x + Q_2x^2 + \cdots \quad (3.31)$$

Under these conditions, the equation can be solved by the series Method of Frobenius. Such series will also be convergent for $|x| < R$.

Thus, for the general case, we start with the expansion

$$y = \sum_{n=0}^{\infty} a_n x^{n+c} \tag{3.32}$$

where c is a variable index (to be determined) and $a_0 \neq 0$ (since, as we saw earlier in Eq. 3.11, a_0 is in fact an arbitrary constant of integration).

The first stage of the analysis is to find suitable values for c through the *indicial relation*. The second stage is to find the relations for a_n from the *recurrence relation*. This second stage has many twists and turns and can best be learned by example. We consider these relations in the next section.

3.3.1 Indicial Equation and Recurrence Relation

As stated earlier, the first stage is to find the values for c, through an indicial equation. This is obtained by inspecting the coefficients of the *lowest powers* in the respective series expansions. Consider Eq. 3.29 with $P(x)$, $Q(x)$ given by Eqs. 3.30 and 3.31; first, perform the differentiations

$$\frac{dy}{dx} = \sum_{n=0}^{\infty} a_n(n+c)x^{n+c-1} = y' \tag{3.33}$$

$$\frac{d^2y}{dx^2} = \sum_{n=0}^{\infty} a_n(n+c-1)(n+c)x^{n+c-2} = y'' \tag{3.34}$$

Insert these into the defining Eq. 3.29

$$\sum_{n=0}^{\infty} a_n(n+c-1)(n+c)x^{n+c} + (P_0 + P_1 x + \cdots)$$
$$\sum_{n=0}^{\infty} a_n(n+c)x^{n+c} + (Q_0 + Q_1 x + \cdots)\sum_{n=0}^{\infty} a_n x^{n+c} = 0 \tag{3.35}$$

To satisfy this equation, all coefficients of x^c, x^{c+1}, x^{c+n}, and so on, must be identically zero. Taking the lowest coefficient, x^c, we see

$$a_0[(c-1)c + P_0 \cdot c + Q_0]x^c = 0 \tag{3.36}$$

This can be satisfied three possible ways, only one of which is nontrivial

$$c(c-1) + P_0 c + Q_0 = 0 \tag{3.37}$$

This quadratic equation is called the *indicial relationship*. Rearrangement shows it yields two possible values of c

$$c^2 + c(P_0 - 1) + Q_0 = 0 \tag{3.38}$$

$$c_{1,2} = \frac{(1 - P_0) \pm \sqrt{(P_0 - 1)^2 - 4Q_0}}{2} \tag{3.39}$$

The remainder of the solution depends on the character of the values c_1, c_2. If they are distinct (not equal) and do not differ by an integer, the remaining analysis is quite straightforward. However, if they are equal or differ by an integer, special techniques are required.

We shall treat these special cases by way of example. The special cases can be categorized as:

Case I The values of c are distinct and do not differ by an integer.
Case II The values of c are equal.
Case III The values of c are distinct but differ by an integer; two situations arise in this category, denoted as case IIIa or IIIb.

EXAMPLE 3.1 *DISTINCT ROOTS (NOT DIFFERING BY INTEGER, CASE I)*

Consider the second order equation with variable coefficients

$$4x\frac{d^2y}{dx^2} + 6\frac{dy}{dx} - y = 0 \tag{3.40}$$

Comparison with Eq. 3.29 shows that (on multiplying through by x and dividing by 4)

$$P_0 = 6/4, \; P_1 = P_2 = P_n = 0, \text{ and}$$

$$Q_0 = 0, \; Q_1 = -1/4, \; Q_2 = Q_3 = Q_n = 0$$

First, write the differentiations of

$$y = \sum_{n=0}^{\infty} a_n x^{n+c}$$

as

$$y' = \sum_{n=0}^{\infty} a_n(n+c)x^{n+c-1} \tag{3.41}$$

$$y'' = \sum_{n=0}^{\infty} a_n(n+c)(n+c-1)x^{n+c-2} \tag{3.42}$$

Insert these into Eq. 3.40 to yield

$$4\sum_{n=0}^{\infty} a_n(n+c)(n+c-1)x^{(n+c-1)}$$

$$+6\sum_{n=0}^{\infty} a_n(n+c)x^{n+c-1} - \sum_{n=0}^{\infty} a_n x^{n+c} = 0 \tag{3.43}$$

Remove the lowest coefficient (x^{c-1}) and form the *indicial equation*

$$a_0[4c(c-1) + 6c]x^{c-1} = 0 \tag{3.44}$$

Now, since $a_0 \neq 0$, then we must have

$$[4c^2 + 2c] = 0; \therefore c_2 = 0, c_1 = -\frac{1}{2} \tag{3.45}$$

Now, since the first two summations have had their first terms removed, then summing should begin with $n = 1$, so that we should write

$$4 \sum_{n=1}^{\infty} a_n(n+c)(n+c-1)x^{n+c-1} + 6 \sum_{n=1}^{\infty} a_n(n+c)x^{n+c-1}$$

$$- \sum_{n=0}^{\infty} a_n x^{n+c} = 0$$

Another way to represent these sums is to increment n upward (i.e., replace n with $n + 1$), and this allows the sums to start from $n = 0$. It also yields the same power on x (i.e., x^{n+c}) in each summation so that they can all be combined under one summation

$$\sum_{n=0}^{\infty} [4a_{n+1}(n+1+c)(n+c) + 6a_{n+1}(n+1+c) - a_n]x^{n+c} = 0 \tag{3.46}$$

The *recurrence relation* is obviously the bracketed terms set to zero. In most practical problems, uncovering the recurrence relation is not so straightforward, and the coefficients $a_0 = f(a_1)$, $a_1 = f(a_2)$, and so on, must be found one at a time, as we show later in example Ex. 3.2. However, in the present case, we have the general result for any value c

$$a_{n+1} = \frac{a_n}{(4n + 4c + 6)(n + 1 + c)} \tag{3.47}$$

We next consider the two cases $c = 0$ and $c = -1/2$; first, when $c = 0$, we see

$$a_{n+1} = \frac{a_n}{(4n + 6)(n + 1)} = \frac{a_n}{(2n + 3)(2n + 2)} \tag{3.48}$$

We note this yields, when $n = 0$

$$a_1 = \frac{a_0}{(3)(2)}$$

To find a general relation for the $(n + 1)$th or nth coefficient, in terms of a_0 (which we treat as an arbitrary constant of integration, so that it is always the lead term), write the series of products

$$\frac{a_{n+1}}{a_0} = \left(\frac{a_{n+1}}{a_n}\right) \cdot \left(\frac{a_n}{a_{n-1}}\right) \cdots \cdots \left(\frac{a_1}{a_0}\right) \tag{3.49}$$

There are $(n + 1)$ of these products; inserting the recurrence relation, Eq. 3.48, shows

$$\frac{a_{n+1}}{a_0} = \frac{1}{(2n + 3)(2n + 2)} \cdot \frac{1}{(2n + 1)(2n)} \cdots \frac{1}{(3)(2)} \tag{3.50}$$

It is clear this defines a factorial

$$\frac{a_{n+1}}{a_0} = \frac{1}{(2n + 3)!} \tag{3.51}$$

To find a_n, increment downward (replace n with $n - 1$)

$$\frac{a_n}{a_0} = \frac{1}{(2n + 1)!} \tag{3.52}$$

which yields the sought-after general coefficient a_n in terms of the lead coefficient a_0, so that we have the first solution

$$y_1 = a_0 \sum_{n=0}^{\infty} \frac{1}{(2n + 1)!} x^n = a_0\left(1 + \frac{x}{3!} + \frac{x^2}{5!} + \cdots\right) \tag{3.53}$$

This series of terms may be recognizable, if it is written in terms of $\sqrt{x}$

$$y_1 = \frac{a_0}{\sqrt{x}}\left(\sqrt{x} + \frac{(\sqrt{x})^3}{3!} + \frac{(\sqrt{x})^5}{5!} + \cdots\right) \tag{3.54}$$

This is clearly the hyperbolic sine series, so that

$$y_1 = a_0 \sinh(\sqrt{x})/\sqrt{x} \tag{3.55}$$

For the second linearly independent solution, take $c = -1/2$, to see the recurrence relation, from Eq. 3.47

$$b_{n+1} = \frac{b_n}{(4n + 4)\dfrac{(2n + 1)}{2}} = \frac{b_n}{(2n + 2)(2n + 1)} \tag{3.56}$$

Here, we use b_n (instead of a_n) to distinguish the second from the first solution.

As before, we first note that $b_1 = b_0/(2)(1)$. Next, form the $(n + 1)$ products

$$\frac{b_{n+1}}{b_0} = \left(\frac{b_{n+1}}{b_n}\right)\left(\frac{b_n}{b_{n-1}}\right) \cdots \left(\frac{b_1}{b_0}\right) \qquad (3.57)$$

Inserting the recurrence relation, suitably incremented, shows

$$\frac{b_{n+1}}{b_0} = \frac{1}{(2n + 2)(2n + 1)} \cdot \frac{1}{(2n)(2n - 1)} \cdots \frac{1}{(2)(1)} \qquad (3.58)$$

This again produces a factorial, so

$$\frac{b_{n+1}}{b_0} = \frac{1}{(2n + 2)!} \qquad (3.59)$$

Incrementing downward then yields the required result

$$\frac{b_n}{b_0} = \frac{1}{(2n)!} \qquad (3.60)$$

so that we have the second solution

$$y_2 = \frac{b_0}{\sqrt{x}} \sum_{n=0}^{\infty} \frac{1}{(2n)!} x^n = \frac{b_0}{\sqrt{x}} \left(1 + \frac{x}{2!} + \frac{x^2}{4!} + \cdots \right) \qquad (3.61)$$

The series terms may be recognized as the hyperbolic cosine operating on the argument $\sqrt{x}$

$$y_2 = b_0 \cosh(\sqrt{x})/\sqrt{x} \qquad (3.62)$$

which is linearly independent of the first solution. It is unusual, and unexpected, that such series solutions are expressible as elementary, tabulated functions. Had the coefficient multiplying y in Eq. 3.40 been positive, then it is easy to show by Frobenius that the two linearly independent solutions would be: $\sin \sqrt{x}/\sqrt{x}$ and $\cos \sqrt{x}/\sqrt{x}$.

EXAMPLE 3.2   *DISTINCT VALUES OF C (DIFFERING BY AN INTEGER, CASE IIIB).*

We applied the Ricatti transformation to a nonlinear equation in Example 2.10 and arrived at the linear Airy equation

$$\frac{d^2z}{dx^2} - xz = 0 \qquad (3.63)$$

This equation arises frequently in physical science, and it also illustrates some of the pitfalls in undertaking a Frobenius analysis. Unlike the first example, we shall see that, most often, series coefficients must be deduced one at a time. As before, we propose a solution of the form

$$z = \sum_{n=0}^{\infty} a_n x^{n+c} \tag{3.64}$$

Differentiating this and inserting z'' and z into the defining equation we find

$$\sum_{n=0}^{\infty} a_n(n+c)(n+c-1)x^{n+c-2} - \sum_{n=0}^{\infty} a_n x^{n+c+1} = 0 \tag{3.65}$$

If we remove the first three terms of the z'' series, and then increment upward three times (replace n with $n+3$), we get

$$\sum_{n=0}^{\infty} [a_{n+3}(n+3+c)(n+2+c) - a_n]x^{n+1+c}$$

$$\tag{3.66}$$

$$+\underline{a_0c(c-1)x^{c-2}} + a_1(c+1)cx^{c-1} + a_2(c+2)(c+1)x^c = 0$$

The underlined term is the lowest power and produces the indicial equation

$$c(c-1) = 0; \therefore c_2 = 1, \quad c_1 = 0 \tag{3.67}$$

Now, the remaining terms must also be identically zero, and there are several ways to accomplish this task. For example, the solution corresponding to $c = c_1 = 0$ will automatically cause the term $a_1(c+1)cx^{c-1} = 0$, so that nothing can be said about the value of a_1, hence it is indeterminate; we then treat a_1 as an arbitrary constant of integration, just as we did for a_0. Finally, it is clear that the only way to remove the term $a_2(c+2)(c+1)x^c$, regardless of the value of c, is to set $a_2 = 0$. This now leaves us with the general recurrence relation

$$a_{n+3} = \frac{a_n}{(n+3+c)(n+2+c)} \tag{3.68}$$

Consider the solution corresponding to $c = c_1 = 0$ (the smallest root of c)

$$a_{n+3} = \frac{a_n}{(n+3)(n+2)} \tag{3.69}$$

We repeat that both a_0 and a_1 are arbitrary constants for this case; moreover, we also required earlier that for all cases $a_2 = 0$. Thus, we can write the coefficients for this solution one at a time using Eq. 3.69, first in terms of a_0

$$a_3 = \frac{a_0}{3 \cdot 2}; \quad a_6 = \frac{a_3}{6 \cdot 5} = \frac{a_0}{6 \cdot 5 \cdot 3 \cdot 2}, \text{ etc.} \tag{3.70}$$

and also in terms of a_1

$$a_4 = \frac{a_1}{4 \cdot 3}; a_7 = \frac{a_4}{7 \cdot 6} = \frac{a_1}{7 \cdot 6 \cdot 4 \cdot 3}, \text{etc.} \tag{3.71}$$

All terms linearly connected to a_2 must be zero (i.e., $a_5 = a_2/(5 \cdot 4)$, since $a_2 = 0$).

We can now write the two linearly independent solutions containing two arbitrary constants a_0, a_1 as

$$z = a_0\left(1 + \frac{x^3}{3!} + \frac{1 \cdot 4}{6!}x^6 + \cdots\right) + a_1\left(x + \frac{1 \cdot 2}{4!}x^4 + \frac{2 \cdot 5}{7!}x^7 + \cdots\right)$$

$$\tag{3.72}$$

or

$$z = a_0 f(x) + a_1 g(x) \tag{3.73}$$

This special circumstance for Case III will be called Case IIIb; it is distinguished by the fact that one of the coefficients a_j becomes *indeterminate* when using the *smallest index value c*, and $j = c_2 - c_1$. Thus, we see $a_j = a_1$ in the present case, but this could have been foreshadowed by inspecting $(c_2 - c_1) = j$. It is easy to show that the solution for $c = 1$ reproduces one of the above solutions. Thus, the lowest index c produces two linearly independent solutions at once.

Forms of the two series solutions generated above are widely tabulated and are called Airy functions (Abramowitz and Stegun 1965); however, they are represented in slightly different form

$$z = K_1 Ai(x) + K_2 Bi(x) \tag{3.74}$$

where in terms of our previous functions $f(x)$ and $g(x)$

$$Ai(x) = B_1 f(x) - B_2 g(x) \tag{3.75}$$

and

$$Bi(x) = \sqrt{3}\left(B_1 f(x) + B_2 g(x)\right) \tag{3.76}$$

and the constants are defined as

$$B_1 = Ai(0) = Bi(0)/\sqrt{3} = 3^{-2/3}/\Gamma(2/3) \tag{3.77}$$

$$B_2 = Ai'(0) = Bi'(0)/\sqrt{3} = 3^{-1/3}/\Gamma(1/3) \tag{3.78}$$

The tabulated Gamma functions, $\Gamma(x)$, will be discussed in Chapter 4.

EXAMPLE 3.3 *EQUAL VALUES OF C (CASE II)*

Porous, cylindrical-shaped pellets are used as catalyst for the reaction $A \xrightarrow{k}$ products, in a packed bed. We wish to model the steady-state diffusion-reaction processes within the particle. When the pellet length to diameter ratio $(L/2R) > 3$, flux to the particle end caps can be ignored. Assume the surface composition is C_{As} and that reaction kinetics is controlled by a linear rate expression $R_A = kC_A$ (mole/volume · time), and diffusive flux obeys $J_A = -D_A \partial C_A / \partial r$.

Taking an annular shell of the type depicted in Fig. 1.1c, the steady-state conservation law is, ignoring axial diffusion,

$$(2\pi r L J_A)|_r - (2\pi r L J_A)|_{r+\Delta r} - (2\pi r \Delta r L) R_A = 0 \tag{3.79}$$

Dividing by $(2\pi L \Delta r)$, and taking limits, yields

$$-\frac{d}{dr}(r J_A) - r R_A = 0 \tag{3.80}$$

Introducing the flux and rate expressions gives

$$D_A \frac{1}{r} \frac{d}{dr}\left(r\frac{dC_A}{dr}\right) - kC_A = 0 \tag{3.81}$$

Differentiating, and then rearranging, shows

$$r^2 \frac{d^2 C_A}{dr^2} + r\frac{dC_A}{dr} - r^2\left(\frac{k}{D_A}\right)C_A = 0 \tag{3.82}$$

Defining the new variables

$$y = C_A/C_{As}, \qquad x = r\sqrt{\frac{k}{D_A}} \tag{3.83}$$

yields the variable coefficient, linear ODE

$$x^2 \frac{d^2 y}{dx^2} + x\frac{dy}{dx} - x^2 y = 0 \tag{3.84}$$

We propose to solve this by the Frobenius method, so let

$$y = \sum_{n=0}^{\infty} a_n x^{n+c},$$

hence insert the derivatives to get

$$\sum_{n=0}^{\infty} a_n(n+c)(n+c-1)x^{n+c} + \sum_{n=0}^{\infty} a_n(n+c)x^{n+c} - \sum_{n=0}^{\infty} a_n x^{n+c+2} = 0$$

(3.85)

Removing the coefficients of the lowest power (x^c) and setting to zero gives the *indicial equation*

$$a_0[c(c-1)+c]x^c = 0$$

(3.86)

hence, we see $c^2 = 0$, so $c_1 = c_2 = 0$. Since we have removed a term from the first two series, we must rezero the summations by incrementing each upward by 1 (replace n with $n+1$) to get

$$\sum_{n=0}^{\infty} a_{n+1}(n+1+c)(n+c)x^{n+c+1} + \sum_{n=0}^{\infty} a_{n+1}(n+1+c)x^{n+1+c}$$

$$- \sum_{n=0}^{\infty} a_n x^{n+c+2} = 0$$

(3.87)

It is not yet possible to get a recurrence relation from this, since the first two series have lower powers than the last one. This situation suggests removing a term from each of the first two, and rezeroing again to get

$$\sum_{n=0}^{\infty} [a_{n+2}(n+2+c)(n+1+c) + a_{n+2}(n+2+c) - a_n]x^{n+c+2}$$

(3.88)

$$+ a_1[(c+1)(c)+(c+1)]x^{c+1} = 0$$

It is clear that the coefficient of x^{c+1} does not disappear when $c = 0$, so that we must set $a_1 = 0$. As we shall see, this implies that all odd coefficients become zero. We can now write the *recurrence relation* in general to find

$$a_{n+2}[n+2+c]^2 = a_n$$

(3.89)

or for the case $c = 0$

$$a_{n+2} = \frac{a_n}{(n+2)^2}$$

(3.90)

The general case, Eq. 3.89, shows that if $a_1 = 0$, then $a_3 = a_5 = a_7 = \cdots = 0$, so all odd coefficients are identically zero. Thus, we know at this point that both linearly independent solutions are composed of even functions. This suggests replacing n with $2k$, so that the original form now reads

$$y(x) = \sum_{k=0}^{\infty} a_{2k} x^{2k+c}$$

(3.91)

and the *general recurrence relation* is

$$a_{2k+2} = a_{2k}/[2k + 2 + c]^2 \qquad (3.92)$$

To find the first solution, set $c = 0$; hence,

$$a_{2k+2} = a_{2k}/[2k + 2]^2 = \frac{a_{2k}}{(2^2)(k + 1)^2} \qquad (3.93)$$

We also see that $a_2 = a_0/(2^2)$, so that we write the series of products to attempt to relate a_{2k+2} to a_0

$$\frac{a_{2k+2}}{a_0} = \frac{a_{2k+2}}{a_{2k}} \frac{a_{2k}}{a_{2k-2}} \cdots \frac{a_2}{a_0} \qquad (3.94)$$

hence

$$\frac{a_{2k+2}}{a_0} = \frac{1}{(2k+2)^2} \cdot \frac{1}{(2k)^2} \cdots \frac{1}{2^2} \qquad (3.95)$$

To find a_{2k}, replace k with $(k - 1)$, so incrementing downward gives

$$\frac{a_{2k}}{a_0} = \frac{1}{(2k)^2} \cdot \frac{1}{(2k-2)^2} \cdots \frac{1}{2^2} \qquad (3.96)$$

There are exactly k terms in the series of products (one term if $k = 1$, two terms if $k = 2$, etc). The factor 2^2 can be removed from each term, so that the sequence becomes more recognizable

$$\frac{a_{2k}}{a_0} = \frac{1}{2^2(k)^2} \cdot \frac{1}{2^2(k-1)^2} \cdots \frac{1}{2^2 \cdot 1} \qquad (3.97)$$

Thus, 2^2 is repeated k times, and the factorial sequence $k \cdot (k - 1) \cdot (k - 2)\ldots 1$ is also repeated twice, so that we can write the compact form

$$\frac{a_{2k}}{a_0} = \frac{1}{(2^2)^k} \frac{1}{(k!)^2} \qquad (3.98)$$

We can now write the first solution, for $c = 0$

$$y_1(x) = a_0 \sum_{k=0}^{\infty} \left(\frac{x}{2}\right)^{2k} \frac{1}{(k!)^2} = a_0 I_0(x) \qquad (3.99)$$

The series of terms denoted as $I_0(x)$ is the zero order, modified Bessel function of the first kind, to be discussed later in section 3.5.1.

To find the second linearly independent solution, we could form the product as taught in Chapter 2

$$y_2(x) = v(x)I_0(x) \tag{3.100}$$

which will be linearly independent of $I_0(x)$ if $v(x) \neq$ constant. The reader can see this approach will be tedious.

Another approach is to treat the index c like a continuous variable. We denote a function $u(x, c)$, which satisfies the general recurrence relation, Eq. 3.89, but not the indicial relation, so that

$$x^2 \frac{d^2u(x, c)}{dx^2} + x\frac{du(x, c)}{dx} - x^2u(x, c) + b_0c^2x^c = 0 \tag{3.101}$$

where

$$u(x, c) = b_0 \sum_{k=0}^{\infty} \frac{1}{(2k + c)^2} \cdots \frac{1}{(c + 4)^2} \cdot \frac{1}{(c + 2)^2} x^{2k+c} \tag{3.102}$$

We have replaced a_n with b_n to distinguish between the first and second solutions. Note, in passing, that if $c \to 0$, then $u(x, 0)$ satisfies the original equation and is identical to the first solution $y_1(x)$. Now, if we differentiate $u(x, c)$ partially with respect to c, we obtain

$$x^2 \frac{d^2}{dx^2}\left(\frac{\partial u}{\partial c}\right) + x\frac{d}{dx}\left(\frac{\partial u}{\partial c}\right) - x^2\frac{\partial u}{\partial c} + 2b_0cx^c + b_0c^2x^c \ln x = 0 \tag{3.103}$$

We see in the limit as $c \to 0$, the two residual terms disappear, and we conclude that

$$\partial u/\partial c|_{c=0}$$

is another solution, which we can write as

$$y_2(x) = \frac{\partial u(x, c)}{\partial c}\bigg|_{c=0} \tag{3.104}$$

Now, it remains only to develop a systematic way of differentiating $u(x, c)$. Inspection of Eq. 3.102 shows that $u(x, c)$ is composed of a product of functions of c

$$g(c) = f_1(c) \cdot f_2(c) \cdot f_3(c) \cdots f_k(c)f_{k+1}(c) \tag{3.105}$$

where starting from the lowest:

$$f_1(c) = \frac{1}{(c + 2)^2}$$

$$f_2(c) = \frac{1}{(c + 4)^2}$$

$$f_3(c) = \frac{1}{(c + 6)^2}$$

$$\vdots$$

$$f_k(c) = \frac{1}{(c + 2k)^2}$$

$$f_{k+1}(c) = x^{2k+c} = \exp[(2k + c)\ln x]$$

We can use the properties of logarithms to conveniently differentiate these product functions, since

$$\ln g = \ln f_1 + \ln f_2 + \cdots \ln f_k + (2k + c)\ln x \qquad (3.106)$$

and, since

$$\frac{dg}{dc} = g\frac{d(\ln g)}{dc} \qquad (3.107)$$

then we can differentiate one at a time to see

$$\frac{dg}{dc} = f_1 \cdot f_2 \cdot f_3 \cdots f_k, f_{k+1}\left[\frac{f_1'}{f_1} + \frac{f_2'}{f_2} + \cdots + \frac{f_k'}{f_k} + \ln x\right] \qquad (3.108)$$

since

$$\frac{d\ln f_1}{dc} = \frac{1}{f_1}\frac{df_1}{dc}, \ \frac{d\ln f_2}{dc} = \frac{1}{f_2}\frac{df_2}{dc}, \text{etc.}$$

The tabulation of these derivatives are the following:

$$\frac{f_1'}{f_1} = \frac{-2}{(c + 2)} \xrightarrow{c=0} \frac{-2}{2}$$

$$\frac{f_2'}{f_2} = \frac{-2}{(c + 4)} \xrightarrow{c=0} \frac{-2}{4}$$

$$\vdots \qquad\qquad (3.109)$$

$$\frac{f_k'}{f_k} = \frac{-2}{(c + 2k)} \xrightarrow{c=0} \frac{-2}{(2k)}$$

$$\frac{f_{k+1}'}{f_{k+1}} = \ln x$$

Inserting these into

$$\frac{\partial u}{\partial c}\bigg|_{c=0}$$

gives

$$y_2(x) = \frac{\partial u}{\partial c}\bigg|_{c=0} = b_0 \sum_{k=0}^{\infty} \frac{1}{2^2} \cdot \frac{1}{4^2} \cdots \frac{x^{2k}}{(2k)^2}\left[\frac{-2}{2} + \frac{-2}{4} + \cdots + \frac{-2}{(2k)} + \ln x\right]$$

(3.110)

Combining the k products of 2^2 with x^{2k} gives

$$y_2(x) = b_0 \sum_{k=0}^{\infty} \frac{1}{(k!)^2}\left(\frac{x}{2}\right)^{2k}\ln x - b_0 \sum_{k=0}^{\infty} \frac{1}{(k!)^2}\left[1 + \frac{1}{2} + \frac{1}{3} + \cdots + \frac{1}{k}\right]\left(\frac{x}{2}\right)^{2k}$$

(3.111)

The summing function with k terms

$$\varphi(k) = 1 + \frac{1}{2} + \frac{1}{3} + \cdots + \frac{1}{k}$$

(3.112)

takes values

$$\varphi(0) = 0, \varphi(1) = 1, \varphi(2) = 1 + \frac{1}{2}, \varphi(3) = 1 + \frac{1}{2} + \frac{1}{3}, \text{etc.}$$ (3.113)

The solutions to this equation are tabulated and are called zero order, modified Bessel functions. We discuss them in the next section. Thus, the solution to Eq. 3.84 could have been written directly as

$$y(x) = a_0 I_0(x) + b_0 K_0(x)$$

(3.114)

where $I_0(x)$ and $K_0(x)$ denote the modified Bessel functions of the first and second kind. As with the Airy functions, the tabulated form of the second function is slightly different (within an arbitrary constant) from the series solutions worked out here, as can be seen by comparison with the terms given in the next section. It is also clear that the second solution tends to $-\infty$ as x tends to zero because of the appearance of $\ln(x)$. Thus, for the present catalyst pellet problem, we would set $b_0 = 0$ to insure admissibility (continuity), so our final result is

$$y(x) = \frac{C_A}{C_{As}} = a_0 I_0\left(r\sqrt{\frac{k}{D_A}}\right)$$

(3.115)

Now, it can be seen from our first solution, $y_1(x)$ in Eq. 3.99, that $I_0(0) = 1$, so that the present solution is finite at the centerline. At the outer edge of the pellet, where $r = R$, $C_A(R) = C_{As}$ so evaluating a_0 gives a value

$1/I_0[R\sqrt{(k/D_A)}]$, hence our final result for computational purposes is

$$\frac{C_A}{C_{As}} = \frac{I_0\left(r\sqrt{\dfrac{k}{D_A}}\right)}{I_0\left(R\sqrt{\dfrac{k}{D_A}}\right)} \tag{3.116}$$

We can use this to compute catalyst effectiveness factor, as we demonstrate in the next section.

EXAMPLE 3.4   *DISTINCT VALUES OF C DIFFERING BY AN INTEGER (CASE IIIA)*

The following equation is satisfied by the confluent hypergeometric function of Kummer

$$x\frac{d^2y}{dx^2} + (\alpha - x)\frac{dy}{dx} - \beta y = 0 \tag{3.117}$$

Use the method of Frobenius to find the two linearly independent solutions when $\alpha = \beta = 2$. Introducing

$$y = \sum_{n=0}^{\infty} a_n x^{n+c}$$

and its derivatives yields

$$\sum_{n=0}^{\infty} \left[a_n(n + c)(n + c - 1) + 2a_n(n + c)\right]x^{n+c-1}$$

$$- \sum_{n=0}^{\infty} \left[a_n(n + c) + 2a_n\right]x^{n+c} = 0 \tag{3.118}$$

Removing the lowest power x^{c-1} gives the indicial equation

$$a_0[c(c - 1) + 2c]x^{c-1} = 0 \tag{3.119}$$

so that $c = 0, -1$, which corresponds to Case III, since $c_2 - c_1 = 1$; thus be wary of a_1. Since we removed one term from the first series, rezero this summation by incrementing n upward. This immediately gives the recurrence relation, since all terms multiply x^{n+c}

$$\sum_{n=0}^{\infty} \left[a_{n+1}(n + 1 + c)(n + c + 2) - a_n(n + c + 2)\right]x^{n+c} = 0 \tag{3.120}$$

Thus, the general recurrence relation is

$$a_{n+1}[(n + 1 + c)(n + c + 2)] = a_n[n + c + 2] \tag{3.121}$$

In Example (3.2), we foreshadowed a problem would arise in Case III for the term a_j, where $c_2 - c_1 = j$. Since we are alert to this problem, we shall give Eq. 3.121 a closer inspection, by writing the relation for a_1 (since $j = 1$ in the present case)

$$a_1(c + 1)(c + 2) = a_0(c + 2) \tag{3.122}$$

We see immediately that, when $c = c_1 = -1$, then $a_1 \to \infty$, which is inadmissible. However, we can still proceed with the case $c = 0$, which gives $a_1 = a_0$, and

$$a_{n+1} = \frac{a_n}{n + 1} \tag{3.123}$$

or

$$a_n = \frac{a_{n-1}}{n} \tag{3.124}$$

We can write the products to connect a_n to a_0

$$\frac{a_n}{a_0} = \frac{a_n}{a_{n-1}} \frac{a_{n-1}}{a_{n-2}} \cdots \frac{a_1}{a_0} \tag{3.125}$$

hence,

$$\frac{a_n}{a_0} = \frac{1}{n} \cdot \frac{1}{n - 1} \cdots \frac{1}{1} = \frac{1}{n!} \tag{3.126}$$

This allows the first solution to be written

$$y_1(x) = a_0 \sum_{n=0}^{\infty} \frac{x_n}{n!} = a_0 \exp(x) \tag{3.127}$$

Now, the case when $c = -1$ has a discontinuity, or singularity. We can also relate, in general, the coefficients a_n to a_0 as before, using Eqs. 3.121 and 3.122

$$\frac{a_n}{a_0} = \frac{1}{(n + c)} \cdot \frac{1}{n - 1 + c} \cdots \frac{1}{(c + 2)} \cdot \frac{1}{(c + 1)} \tag{3.128}$$

From this, we can write the general solution $u(x, c)$ (which does not yet satisfy the indicial solution), using b_n to distinguish from the first solution

$$u(x, c) = b_0 \sum_{n=0}^{\infty} \frac{1}{(c + 1)} \frac{1}{(c + 2)} \cdots \frac{1}{(n + c)} \cdot x^{n+c} \tag{3.129}$$

As it stands, this function satisfies the relation

$$x\frac{d^2 u(x,c)}{dx^2} + (2-x)\frac{du(x,c)}{dx} - 2u(x,c) + b_0 c(c+1)x^{c-1} = 0 \quad (3.130)$$

since we have stipulated that the indicial equation is not yet satisfied. The singularity in Eq. 3.129 can be removed by defining the new function $(c+1)u(x,c)$, so that

$$x\frac{d^2(c+1)u}{dx^2} + (2-x)\frac{d(c+1)u}{dx} - 2(c+1)u + b_0 c(c+1)^2 x^{c-1} = 0$$

$$(3.131)$$

It is now clear that the new function $(c+1)u(x,c)$ no longer blows up when $c = -1$, and it also satisfies the original defining equation. But is it linearly independent of the first solution? To check this, we use Eq. 3.129 to see

$$(c+1)u(x,c)|_{c=-1} = b_0 \sum_{n=0}^{\infty} \frac{1}{(n-1)!}x^{n-1} \quad (3.132)$$

Now, by definition $(-1)! \rightarrow \infty$ (as we show in relation to the Gamma function in Chapter 4), so that the first term in the new series contributes nothing. This suggests incrementing upward (replace n with $n+1$) to get

$$(c+1)u(x,c)|_{c=-1} = b_0 \sum_{n=0}^{\infty} \frac{1}{n!}x^n \quad (3.133)$$

But this is the *same* as the first solution (within an arbitrary constant) and is *not* therefore linearly independent! Our elementary requirements for the second solution are simple: It must satisfy the original ODE, and it must be linearly independent of the first solution.

However, if we used the same procedure as in Example 3.3 (Case II) and partially differentiate Eq. 3.131, we see the appealing result

$$x\frac{d^2}{dx^2}\frac{\partial}{\partial c}(c+1)u + (2-x)\frac{d}{dx}\frac{\partial}{\partial c}(c+1)u$$

$$- 2\frac{\partial}{\partial c}(c+1)u + b_0(c+1)^2 x^{c-1} \quad (3.134)$$

$$+ 2b_0 c(c+1)x^{c-1} + b_0 c(c+1)^2 x^{c-1} \ln x = 0$$

Taking limits as $c \rightarrow -1$ thereby removes the residual terms and we see, indeed, that a new solution is generated

$$y_2 = \frac{\partial}{\partial c}(c+1)u(x,c)|_{c=-1} \quad (3.135)$$

so that we must perform the operations

$$y_2 = b_0 \sum_{n=0}^{\infty} \frac{\partial}{\partial c} \left[\frac{1}{c+2} \cdot \frac{1}{c+3} \cdots \frac{1}{c+n} \cdot x^{n+c} \right]_{c=-1} \quad \textbf{(3.136)}$$

Since we have removed the $(c+1)$ term, there are n functions of c in the product

$$g(c) = \frac{1}{c+2} \cdot \frac{1}{c+3} \cdots \frac{1}{c+n} \cdot x^{n+c} = f_1 \cdot f_2 \cdots f_n \quad \textbf{(3.137)}$$

The differentiation of this product is done as before

$$\frac{dg}{dc} = g \left[\frac{f_1'}{f_1} + \frac{f_2'}{f_2} + \cdots + \frac{f_{n-1}'}{f_{n-1}} + \ln x \right] \quad \textbf{(3.138)}$$

where

$$\frac{f_1'}{f_1} = -\frac{1}{c+2}, \quad \frac{f_2'}{f_2} = -\frac{1}{c+3}, \quad \frac{f_{n-1}'}{f_{n-1}} = -\frac{1}{c+n}, \quad \frac{f_n'}{f_n} = \ln x$$

Inserting these and taking limits as $c \to -1$

$$y_2(x) = b_0 \sum_{n=0}^{\infty} \frac{x^{n-1}}{(n-1)!} \left[-1 - \frac{1}{2} - \frac{1}{3} \cdots + \frac{-1}{n-1} + \ln x \right] \quad \textbf{(3.139)}$$

Earlier, we indicated $(-1)! \to \infty$, so that the first term contributes zero, hence replace n with $n+1$ to get finally,

$$y_2(x) = b_0 \ln x \sum_{n=0}^{\infty} \frac{x^n}{n!} - b_0 \sum_{n=0}^{\infty} \frac{x^n}{n!} \varphi(n) \quad \textbf{(3.140)}$$

where

$$\varphi(n) = 1 + \frac{1}{2} + \frac{1}{3} + \cdots + \frac{1}{n}$$

and as before

$$\varphi(0) = 0, \ \varphi(1) = 1, \ \varphi(2) = 1 + \frac{1}{2}, \text{ etc.}$$

Again, the first term is zero, so that we could increment the second series upward and replace it with

$$-b_0 \sum_{n=0}^{\infty} \frac{x^{n+1}}{(n+1)!} \varphi(n+1)$$

We note, as in Example 3.3 (Case II), that $\ln x$ occurs naturally in the second solution, and this will always occur for Case IIIa and Case II. This simply means that at $x = 0$, these second solutions will tend to infinity, and for conservative physicochemical systems (the usual case in chemical engineering), such solutions must be excluded, for example, by stipulating $b_0 = 0$. If $x = 0$ is not in the domain of the solution, then the solution must be retained, unless other boundary conditions justify elimination. In the usual case, the condition of symmetry at the centerline (or finiteness) will eliminate the solution containing $\ln x$.

3.4 SUMMARY OF THE FROBENIUS METHOD

The examples selected and the discussion of convergence properties focused on solutions around the point $x_0 = 0$. This is the usual case for chemical engineering transport processes, since cylindrical or spherical geometry usually includes the origin as part of the solution domain. There are exceptions, of course, for example, annular flow in a cylindrical pipe. For the exceptional case, series solutions can always be undertaken for finite x_0 expressed as powers of $(x - x_0)$, which simply implies a shift of the origin.

Thus, we have shown that any second (or higher) order, variable coefficient equation of the type

$$x^2 \frac{d^2 y}{dx^2} + xP(x)\frac{dy}{dx} + Q(x)y = 0 \qquad (3.141)$$

can be solved by writing

$$y = \sum_{n=0}^{\infty} a_n x^{n+c} \qquad (a_0 \neq 0) \qquad (3.142)$$

provided $P(x)$ and $Q(x)$ can be expanded in a convergent series of (nonnegative) powers of x for all $|x| < R$, and under such conditions the solutions are guaranteed to be convergent in $|x| < R$. The following procedure was consistently obeyed:

1. The *indicial* equation was found by inspecting the coefficients of the lowest power of x.
2. The *recurrence* relation was found by progressively removing terms and rezeroing summations until all coefficients multiplied a common power on x.
3. Three general cases would arise on inspecting the character of the roots $c = c_1$, $c = c_2$, where $c_2 - c_1 = j$:
   (a) **Case I** If the roots are distinct (not equal) and do not differ by an integer, then the two linearly independent solutions are obtained by adding the solutions containing c_1 and c_2, respectively.
   (b) **Case II** If the roots are equal ($c_1 = c_2$), one solution is obtained directly using c_1 in Eq. 3.142 and the second is obtained from

$$y_2 = \left[\frac{\partial u(x,c)}{\partial c} \right]_{c=c_1}$$

where

$$u(x, c) = \sum_{n=0}^{\infty} b_n x^{n+c}$$

which does not in general satisfy the indicial equation.

(c) **Case III** If the roots c_1 and c_2 are distinct, but differ by an integer such that $c_2 - c_1 = j$, then the coefficient a_j will sustain one of the two behavior patterns: a_j tends to ∞ (a discontinuity) or a_j becomes indeterminate. These subsets are treated as follows.

Case IIIa When $a_j \rightarrow \infty$, the singularity causing the bad behavior (the smallest root, c_1) is removed, and the second solution is generated using

$$y_2 = \frac{\partial}{\partial c}(c - c_1)u(x, c)\big|_{c = c_1}$$

where $u(x, c)$ is defined as in Case II.

Case IIIb When a_j is indeterminate, then the complete solution is obtained using the smallest root c_1, taking a_0 and a_j as the two arbitrary constants of integration.

3.5   SPECIAL FUNCTIONS

Several variable coefficient ODE arise so frequently that they are given names and their solutions are widely tabulated in mathematical handbooks. One of these special functions occupies a prominent place in science and engineering, and it is the solution of Bessel's equation

$$x^2 \frac{d^2 y}{dx^2} + x \frac{dy}{dx} + (x^2 - p^2)y = 0 \qquad (3.143)$$

or equivalently

$$x \frac{d}{dx}\left(x \frac{dy}{dx}\right) + (x^2 - p^2)y = 0$$

The solutions to this equation are called Bessel functions of order p; that is, $J_p(x)$, $J_{-p}(x)$. Because of its importance in chemical engineering, we shall reserve a complete section to the study of Bessel's equation.

We mentioned earlier in Example 3.4 the confluent hypergeometric function of Kummer, which satisfies the equation

$$x \frac{d^2 y}{dx^2} + (c - x)\frac{dy}{dx} - ay = 0 \qquad (3.144)$$

and which has tabulated solutions denoted as $M(a, c; x)$. In fact, if $c = 1$ and $a = -n$ (where n is a positive integer or zero), then one solution is the *Laguerre* polynomial, $y = L_n(x)$.

An equation of some importance in numerical solution methods (Chapter 8) is the Jacobi equation

$$x(1 - x)\frac{d^2y}{dx^2} + [a - (1 + b)x]\frac{dy}{dx} + n(b + n)y = 0 \qquad (3.145)$$

which is satisfied by the nth order Jacobi polynomial, $y = J_n^{(a,b)}(x)$. This solution is called regular, since it is well behaved at $x = 0$, where $J_n^{(a,b)}(0) = 1$.

The *Legendre* polynomial is a special case of the Jacobi polynomial which satisfies (see Homework Problem 3.7)

$$(1 - x^2)\frac{d^2y}{dx^2} - 2x\frac{dy}{dx} + n(n + 1)y = 0 \qquad (3.146)$$

The nth Legendre polynomial is denoted as $P_n(x)$.

The *Chebyshev* polynomial satisfies

$$(1 - x^2)\frac{d^2y}{dx^2} - x\frac{dy}{dx} + n^2y = 0 \qquad (3.147)$$

The regular form we denote as $y = T_n(x)$, and n is a positive integer or zero.

3.5.1   Bessel's Equation

The method of Frobenius can be applied to Eq. 3.143 (Bessel's equation) to yield two linearly independent solutions, which are widely tabulated as

$$y(x) = AJ_p(x) + BJ_{-p}(x) \qquad (3.148)$$

for real values of x. It was clear from Example 3.3 that different forms arise depending on whether $2p$ is integer (or zero), or simply a real (positive) number.

Thus, if $2p$ is not integer (or zero), then the required solutions are expressed as

$$J_p(x) = \sum_{n=0}^{\infty} \frac{(-1)^n \left(\frac{1}{2}x\right)^{2n+p}}{n!\Gamma(n + p + 1)} \qquad (3.149)$$

where $\Gamma(n + p + 1)$ is the tabulated Gamma function (to be discussed in Chapter 4). The second solution is obtained by simply replacing p with $-p$ in Eq. 3.149.

Now when p is zero, we can express the Gamma function as a factorial, $\Gamma(n + 1) = n!$, so the zero order Bessel function of the first kind is represented by

$$J_0(x) = \sum_{n=0}^{\infty} \frac{(-1)^n \left(\frac{1}{2}x\right)^{2n}}{n! \cdot n!} = 1 - \frac{\left(\frac{1}{2}x\right)^2}{(1!)^2} + \frac{\left(\frac{1}{2}x\right)^4}{(2!)^2} - \cdots \qquad (3.150)$$

The reader can see, had we applied the method of Frobenius, the second solution would take the form

$$y_2 = b_0 \left[J_0(x) \cdot \ln x - \sum_{m=0}^{\infty} \frac{(-1)^{m+1} \left(\frac{1}{2}x\right)^{2m+2}}{[(m+1)!]^2} \left(1 + \frac{1}{2} + \cdots + \frac{1}{m+1}\right) \right]$$

$$(3.151)$$

The bracketed function is called the Neumann form of the second solution. However, the most widely used and tabulated function is the Weber form, obtained by adding $(\gamma - \ln 2)J_0(x)$ to the above, and multiplying the whole by $2/\pi$. This is the standard form tabulated, and is given the notation

$$Y_0(x) = \frac{2}{\pi} \left[\ln\left(\frac{1}{2}x\right) + \gamma \right] J_0(x)$$

$$- \frac{2}{\pi} \sum_{m=0}^{\infty} \frac{(-1)^{m+1} \left(\frac{1}{2}x\right)^{2m+2}}{[(m+1)!]^2} \varphi(m+1) \qquad (3.152)$$

where as before,

$$\varphi(m+1) = 1 + \frac{1}{2} + \cdots + \frac{1}{m+1}$$

The Euler constant γ is defined as

$$\gamma = \lim_{m \to \infty} \left(1 + \frac{1}{2} + \frac{1}{3} + \cdots + \frac{1}{m} - \ln(m)\right) = 0.5772 \qquad (3.153)$$

Thus, the general solution when $p = 0$ is represented using symbols as

$$y(x) = AJ_0(x) + BY_0(x) \qquad (3.154)$$

It is obvious that $\lim_{x \to 0} Y_0(x) = -\infty$, so that for conservative systems, which include $x = 0$ in the domain of the physical system, obviously one must require $B = 0$ to attain finiteness at the centerline (symmetry).

When p is integer we again use the Weber form for the second solution and write (after replacing p with integer k)

$$y(x) = AJ_k(x) + BY_k(x) \qquad (3.155)$$

Thus, $J_k(x)$ has the same representation as given in Eq. 3.149, except we replace $\Gamma(n + k + 1) = (n + k)!$.

3.5.2 Modified Bessel's Equation

As we saw in Example 3.3, another form of Bessel's equation arises when a negative coefficient occurs in the last term, that is,

$$x^2\frac{d^2y}{dx^2} + x\frac{dy}{dx} - (x^2 + p^2)y = 0 \tag{3.156}$$

This can be obtained directly from Eq. 3.143 by replacing x with ix (since $i^2 = -1$). The solution when p is not integer or zero yields

$$y = AJ_p(ix) + BJ_{-p}(ix) \tag{3.157}$$

and if p is integer or zero, write

$$y = AJ_k(ix) + BY_k(ix) \tag{3.158}$$

However, because of the complex arguments, we introduce the modified Bessel function, which contains real arguments, so if p is not integer (or zero), write

$$y = AI_p(x) + BI_{-p}(x) \tag{3.159}$$

or, if p is integer k (or zero), write

$$y = AI_k(x) + BK_k(x) \tag{3.160}$$

The modified Bessel functions can be computed from the general result for any p

$$I_p(x) = \sum_{n=0}^{\infty} \frac{\left(\frac{1}{2}x\right)^{2n+p}}{n!\Gamma(n + p + 1)} \tag{3.161}$$

If p is integer k, then replace $\Gamma(n + k + 1) = (n + k)!$ The second solution for integer k is after Weber

$$K_k(x) = (-1)^{k+1}\left[\ln\left(\frac{1}{2}x\right) + \gamma\right]I_k(x)$$

$$+ \frac{1}{2}\sum_{m=0}^{k-1} \frac{(-1)^m(k - m - 1)!}{m!}\left(\frac{1}{2}x\right)^{2m-k} \tag{3.162}$$

$$+ \frac{1}{2}\sum_{m=0}^{\infty} \frac{(-1)^k\left(\frac{1}{2}x\right)^{2m+k}}{m!(m + k)!}[\varphi(m + k) + \varphi(m)]$$

Plots of $J_0(x)$, $J_1(x)$, $Y_0(x)$, $I_0(x)$, and $K_0(x)$ are shown in Fig. 3.1.

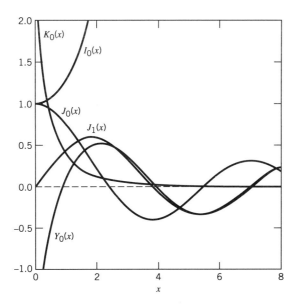

Figure 3.1 Plots of $J_0(x)$, $J_1(x)$, $Y_0(x)$, $I_0(x)$, and $K_0(x)$.

3.5.3 Generalized Bessel Equation

Very often, Bessel's equation can be obtained by an elementary change of variables (either dependent, independent, or both variables). For the general case, we can write

$$x^2 \frac{d^2 y}{dx^2} + x(a + 2bx^r) \frac{dy}{dx} \tag{3.163}$$
$$+ \left[c + dx^{2s} - b(1 - a - r)x^r + b^2 x^{2r} \right] y = 0$$

Representing $Z_{\pm p}(x)$ as one of the Bessel functions, then a general solution can be written

$$y = x^{\left(\frac{1-a}{2} \right)} e^{-\left(\frac{bx^r}{r} \right)} \left[A Z_p \left(\frac{\sqrt{|d|}}{s} x^s \right) + B Z_{-p} \left(\frac{\sqrt{|d|}}{s} x^s \right) \right] \tag{3.164}$$

where

$$p = \frac{1}{s} \sqrt{ \left(\frac{1-a}{2} \right)^2 - c } \tag{3.165}$$

The types of Bessel functions that arise depend on the character of $(d)^{1/2}/s$

and the values of p

1. If $\sqrt{d}/s$ *real* and p is *not* integer (or zero), then Z_p denotes J_p and Z_{-p} denotes J_{-p}.
2. If $\sqrt{d}/s$ is *real* and p is zero or integer k, then Z_p denotes J_k and Z_{-p} denotes Y_k.
3. If $\sqrt{d}/s$ is *imaginary* and p is *not* zero or integer, then Z_p denotes I_p and Z_{-p} denotes I_{-p}.
4. If $\sqrt{d}/s$ is *imaginary* and p is zero or integer k, then Z_p denotes I_k and Z_{-p} denotes K_k.

EXAMPLE 3.5

Pin-promoters of the type shown in Fig. 3.2a and 3.2b are used in heat exchangers to enhance heat transfer by promoting local wall turbulence and by extending heat transfer area. Find an expression to compute the temperature profile, assuming temperature varies mainly in the x direction. The plate temperature T_b, fluid temperature T_a, and heat transfer coefficient h are constant.

We first denote the coordinate x as starting at the pin tip, for geometric simplicity (i.e., similar triangles). Heat is conducted along the pin axis and is lost through the perimeter of incremental area $A_s = 2\pi y \,\Delta s$, so that writing the steady-state conservation law gives, where at any plane through the pin, the cross-sectional area is $A = \pi y^2$

$$(q_x A)\big|_x - (q_x A)\big|_{x+\Delta x} - h 2\pi y \Delta s(T - T_a) = 0 \qquad (3.166)$$

Now, similar triangles show $y/x = b/H$ and the incremental length $\Delta s = \Delta x/\cos\beta$, so that replacing y and dividing by Δx yields

$$-\frac{d(q_x A)}{dx} - \frac{2\pi x b h}{H \cos\beta}(T - T_a) = 0 \qquad (3.167)$$

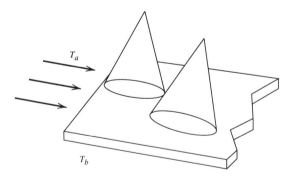

Figure 3.2a Pin-promoters attached to heat exchange surface.

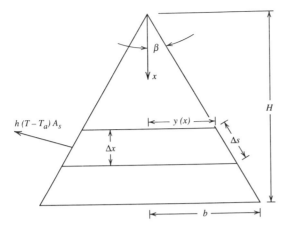

Figure 3.2b Geometry of a single pin.

Replacing

$$q_x = -k\frac{dT}{dx} \quad \text{and} \quad A = \pi x^2 \left(\frac{b}{H}\right)^2$$

gives

$$\frac{d}{dx}\left(x^2\frac{dT}{dx}\right) - x\left[\frac{2h}{k\left(\dfrac{b}{H}\right)\cos\beta}\right](T - T_a) = 0 \qquad \textbf{(3.168)}$$

The equation can be made homogeneous by replacing $(T - T_a) = \theta$, and we shall denote the group of terms as

$$\lambda = \frac{2h}{\left[k\left(\dfrac{b}{H}\right)\cos\beta\right]}$$

so differentiating yields

$$x^2\frac{d^2\theta}{dx^2} + 2x\frac{d\theta}{dx} - x\lambda\theta = 0 \qquad \textbf{(3.169)}$$

We now introduce a dimensionless length scale, $z = \lambda x$ to get finally

$$z^2\frac{d^2\theta}{dz^2} + 2z\frac{d\theta}{dz} - z\theta = 0 \qquad \textbf{(3.170)}$$

Comparing this with the Generalized Bessel relation, Eq. 3.163, indicates we should let

$$b = 0, a = 2, c = 0, d = -1, s = 1/2$$

and moreover, Eq. 3.165 suggests

$$p = 1 \quad \text{and} \quad \frac{1}{s}\sqrt{d} = 2i \text{ (imaginary)} \tag{3.171}$$

Comparing with Eq. 3.164, and item (4) in the list following, we arrive at

$$\theta = \frac{1}{\sqrt{z}}\left[A_0 I_1(2\sqrt{z}) + B_0 K_1(2\sqrt{z}) \right] \tag{3.172}$$

We first check for finiteness at $z = 0$ (i.e., $x = 0$), using the expansion in Eq. 3.161

$$\lim_{z \to 0} \frac{1}{\sqrt{z}} I_1(2\sqrt{z}) = \lim_{z \to 0} \left[\frac{1}{\sqrt{z}} \frac{\sqrt{z}}{1} + \frac{1}{\sqrt{z}} \frac{(\sqrt{z})^3}{1!2!} + \cdots \right] = 1 \tag{3.173}$$

so we see that this solution is admissible at $z = 0$. It is easy to show for small arguments that $K_n(x) \simeq 2^{n-1}(n-1)!x^{-n}$, hence approximately $K_1(2\sqrt{z}) \simeq 0.5/\sqrt{z}$, so that in the limit

$$\lim_{z \to 0} z^{-1/2} K_1(2\sqrt{z}) \to \infty$$

Thus, this function is inadmissible in the domain of $z = 0$, so take $B_0 = 0$. This same conclusion would have been reached following a Frobenius analysis, since the series solution would obviously contain $\ln(x)$ as taught in the previous sections.

Whereas boundary conditions were not explicitly stated, it is clear in Fig. 3.2a that $T(H) = T_b$ or $\theta(\lambda H) = T_b - T_a$; hence, the arbitrary constant A_0 can be evaluated

$$A_0 = \sqrt{\lambda H}\,(T_b - T_a)/I_1(2\sqrt{\lambda H}) \tag{3.174}$$

In terms of x, the final solution is

$$\frac{T(x) - T_a}{T_b - T_a} = \sqrt{\frac{H}{x}} \frac{I_1(2\sqrt{\lambda x})}{I_1(2\sqrt{\lambda H})} \tag{3.175}$$

The dimensionless temperature $(T(x) - T_a)/(T_b - T_a)$ arises naturally and can be computed directly from expansions for I_1 or from Tables (Abramowitz and Stegun 1965).

We need the differential or integral properties for Bessel functions to compute the net rate of heat transfer. We discuss these properties in the next section, and then use them to complete the above example.

3.5.4 Properties of Bessel Functions

It is easily verified that all Power Series presented thus far as definitions of Bessel functions are convergent for finite values of x. However, because of the appearance of $\ln(x)$ in the second solutions, only $J_p(x)$ and $I_p(x)$ are finite at $x = 0$ ($p \geq 0$). Thus, near the origin, we have the important results:

$$k = 0, \; J_0(0) = I_0(0) = 1 \tag{3.176}$$

$$k > 0 \text{ (integer)}, \; J_k(0) = I_k(0) = 0 \tag{3.177}$$

$$p > 0, \; J_{-p}(0) = \pm I_{-p}(0) \to \pm \infty \tag{3.178}$$

The sign in the last expression depends on the sign of $\Gamma(m + p + 1)$, as noted in Eqs. 3.149 and 3.161. However, it is sufficient to know that a discontinuity exists at $x = 0$ in order to evaluate the constant of integration. We also observed earlier that $\ln(x)$ appeared in the second solutions, so it is useful to know (e.g., Example 3.5)

$$-Y_k(0) = K_k(0) \to \infty \tag{3.179}$$

hence only $J_k(x)$ and $I_k(x)$ are admissible solutions.

Asymptotic expressions are also useful in taking limits or in finding approximate solutions; for *small values* of x, the approximations are

$$J_p(x) \simeq \frac{1}{2^p} \frac{x^p}{\Gamma(p+1)}; \qquad J_{-p}(x) \simeq \frac{2^p x^{-p}}{\Gamma(1-p)} \qquad (x \ll 1) \tag{3.180}$$

and for integer or zero orders, we have

$$Y_n \simeq -\frac{2^n(n-1)!}{\pi} x^{-n} \qquad (x \ll 1, n \neq 0) \tag{3.181}$$

$$Y_0(x) \simeq \frac{2}{\pi} \ln(x) \qquad (x \ll 1) \tag{3.182}$$

The modified functions for small x are

$$I_p(x) \simeq \frac{x^p}{2^p \Gamma(p+1)}; \qquad I_{-p}(x) \simeq \frac{2^p x^{-p}}{\Gamma(1-p)} \qquad (x \ll 1) \tag{3.183}$$

$$K_n(x) \simeq 2^{n-1}(n-1)! x^{-n} \qquad (x \ll 1, n \neq 0) \tag{3.184}$$

$$K_0(x) \simeq -\ln(x) \tag{3.185}$$

For *large* arguments, the modified functions sustain exponential type behavior and become independent of order (p may be integer or zero):

$$I_p(x) \simeq \exp(x)/\sqrt{2\pi x} \tag{3.186}$$

$$K_p(x) \simeq \exp(-x) \cdot \sqrt{\frac{\pi}{2x}} \tag{3.187}$$

Table 3.1 Selected Values for Bessel Functions

x	$J_0(x)$	$J_1(x)$	$I_0(x)$	$I_1(x)$
0	1.0000	0.0000	1.0000	0.0000
1	0.7652	0.4401	1.266	0.5652
2	0.2239	0.5767	2.280	1.591
3	−0.2601	0.3391	4.881	3.953
4	−0.3971	−0.0660	11.30	9.759
5	−0.1776	−0.3276	27.24	24.34
6	0.1506	−0.2767	67.23	61.34
7	0.3001	−0.0047	168.6	156.0
8	0.1717	0.2346	427.6	399.9
9	−0.0903	0.2453	1094	1031

However, for *large arguments*, $J_p(x)$ and $Y_p(x)$ behave in a transcendental manner:

$$J_p(x) \simeq \sqrt{\frac{2}{\pi x}} \cos\left(x - \frac{\pi}{4} - p\frac{\pi}{2}\right) \tag{3.188}$$

$$Y_p(x) \simeq \sqrt{\frac{2}{\pi x}} \sin\left(x - \frac{\pi}{4} - p\frac{\pi}{2}\right) \tag{3.189}$$

where p can be any real value including integer or zero. It is also clear in the limit $x \to \infty$ that $J_p(x)$ and $Y_p(x)$ tend to zero. The oscillatory behavior causes $J_p(x)$ and $Y_p(x)$ to pass through zero [called zeros of $J_p(x)$] and these are separated by π for large x. Values of several Bessel functions are tabulated in Table 3.1 and some zeros of $J_p(x)$ are shown Table 3.2. Table 3.2 illustrates the zeros for a type III homogeneous boundary condition. The transcendental behavior of $J_p(x)$ plays an important part in finding eigenvalues for partial differential equations expressed in cylindrical coordinates, as we show in Chapter 10.

Table 3.2 Zeros for $J_n(x)$; Values of x to Produce $J_n(x) = 0$

$n = 0$	$n = 1$	$n = 2$
2.4048	3.8371	5.1356
5.5201	7.0156	8.4172
8.6537	10.1735	11.6198
11.7915	13.3237	14.7960
14.9309	16.4706	17.9598
18.0711	19.6159	21.1170

Table 3.3 Values of x to Satisfy: $x \cdot J_1(x) = N \cdot J_0(x)$

↓ N/x →	x_1	x_2	x_3	x_4	x_5
0.1	0.4417	3.8577	7.0298	10.1833	13.3312
1.0	1.2558	4.0795	7.1558	10.2710	13.3984
2.0	1.5994	4.2910	7.2884	10.3658	13.4719
10.0	2.1795	5.0332	7.9569	10.9363	13.9580

It is easy to show by variables transformation that Bessel functions of $1/2$ order are expressible in terms of elementary functions:

$$J_{1/2}(x) = \sqrt{\frac{2}{\pi x}} \, \sin(x) \tag{3.190}$$

$$J_{-1/2}(x) = \sqrt{\frac{2}{\pi x}} \, \cos(x) \tag{3.191}$$

$$I_{1/2}(x) = \sqrt{\frac{2}{\pi x}} \, \sinh(x) \tag{3.192}$$

$$I_{-1/2}(x) = \sqrt{\frac{2}{\pi x}} \, \cosh(x) \tag{3.193}$$

3.5.5   Differential, Integral and Recurrence Relations

The following differential properties may be proved with reference to the defining equations and are of great utility in problem solving (Mickley et al. 1957; Jenson and Jeffreys 1977):

$$\frac{d}{dx}\left[x^p Z_p(\lambda x)\right] = \begin{cases} \lambda x^p Z_{p-1}(\lambda x), & Z = J, Y, I \\ -\lambda x^p Z_{p-1}(\lambda x), & Z = K \end{cases} \tag{3.194}$$

$$\frac{d}{dx}\left[x^{-p} Z_p(\lambda x)\right] = \begin{cases} -\lambda x^{-p} Z_{p+1}(\lambda x), & Z = J, Y, K \\ \lambda x^{-p} Z_{p+1}(\lambda x), & Z = I \end{cases} \tag{3.195}$$

$$\frac{d}{dx}\left[Z_p(\lambda x)\right] = \begin{cases} \lambda Z_{p-1}(\lambda x) - \dfrac{p}{x} Z_p(\lambda x), & Z = J, Y, I \\ -\lambda Z_{p-1}(\lambda x) - \dfrac{p}{x} Z_p(\lambda x), & Z = K \end{cases} \tag{3.196}$$

By applying the recurrence relations, these can also be written in the more useful form

$$\frac{d}{dx}\left[Z_p(\lambda x)\right] = \begin{cases} -\lambda Z_{p+1}(\lambda x) + \dfrac{p}{x} Z_p(\lambda x), & Z = J, Y, K \\ \lambda Z_{p+1}(\lambda x) + \dfrac{p}{x} Z_p(\lambda x), & Z = I \end{cases} \tag{3.197}$$

Most tables of Bessel functions present only positive order values, so the *recurrence relations* are needed to find negative order values

$$Z_p(\lambda x) = \frac{\lambda x}{2p} \big[Z_{p+1}(\lambda x) + Z_{p-1}(\lambda x) \big], \qquad Z = J, Y \qquad \textbf{(3.198)}$$

$$I_p(\lambda x) = \frac{-\lambda x}{2p} \big[I_{p+1}(\lambda x) - I_{p-1}(\lambda x) \big] \qquad \textbf{(3.199)}$$

$$K_p(\lambda x) = \frac{\lambda x}{2p} \big[K_{p+1}(\lambda x) - K_{p-1}(\lambda x) \big] \qquad \textbf{(3.200)}$$

Also, for n integer or zero, the following inversion properties are helpful:

$$J_{-n}(\lambda x) = (-1)^n J_n(\lambda x) \qquad \textbf{(3.201)}$$

$$I_{-n}(\lambda x) = I_n(\lambda x) \qquad \textbf{(3.202)}$$

$$K_{-n}(\lambda x) = K_n(\lambda x) \qquad \textbf{(3.203)}$$

Eqs. 3.194 and 3.195 are exact differentials and yield the key integral properties directly, for example,

$$\int \lambda x^p J_{p-1}(\lambda x)\, dx = x^p J_p(\lambda x) \qquad \textbf{(3.204)}$$

and

$$\int \lambda x^p I_{p-1}(\lambda x)\, dx = x^p I_p(\lambda x) \qquad \textbf{(3.205)}$$

Later, in Chapter 10, we introduce the orthogonality property, which requires the integrals

$$\int_0^x J_k(\lambda \xi) J_k(\beta \xi) \xi\, d\xi$$
$$= \frac{x}{\lambda^2 - \beta^2} \big[\lambda J_k(\lambda x) \cdot J_{k+1}(\beta x) - \beta J_k(\lambda x) J_{k+1}(\beta x) \big] \qquad \textbf{(3.206)}$$

and if $\lambda = \beta$, this gives the useful result

$$\int_0^x [J_k(\lambda \xi)]^2 \xi\, d\xi = \frac{1}{2} x^2 \big[J_k^2(\lambda x) - J_{k-1}(\lambda x) J_{k+1}(\lambda x) \big] \qquad \textbf{(3.207)}$$

For k integer or zero, we saw in Eq. 3.201 that $J_{-k}(\lambda x) = (-1)^k J_k(\lambda x)$, so that if $k = 0$, the right-hand side of Eq. 3.207 may be written as

$$\frac{1}{2} x^2 \big[J_0^2(\lambda x) + J_1^2(\lambda x) \big]$$

EXAMPLE 3.6

We were unable to complete the solution to Example 3.3 in order to find the effectiveness factor for cylindrical catalyst pellets. Thus, we found the expression for composition profile to be

$$\frac{C_A}{C_{As}} = \frac{I_0\left(r\sqrt{\dfrac{k}{D_A}}\right)}{I_0\left(R\sqrt{\dfrac{k}{D_A}}\right)} \tag{3.208}$$

The effectiveness factor η_A is defined as the ratio of actual molar uptake rate to the rate obtainable if all the interior pellet area is exposed to the reactant, without diffusion taking place. Thus, this maximum uptake rate for species A is computed from

$$W_{\max} = \pi R^2 L k C_{As} \tag{3.209}$$

and the actual net positive uptake rate is simply exterior area times flux

$$W_A = 2\pi RL\left(+D_A\frac{dC_A}{dr}\right)_{r=R} \tag{3.210}$$

We thus need to differentiate C_A in Eq. 3.208, and we can use Eq. 3.197 to do this; we find

$$\frac{dC_A}{dr} = \frac{C_{As}}{I_0\left(R\sqrt{\dfrac{k}{D_A}}\right)} \frac{dI_0\left(r\sqrt{\dfrac{k}{D_A}}\right)}{dr} \tag{3.211}$$

and from Eq. 3.197 we see

$$\frac{d}{dr}I_0\left(r\sqrt{\frac{k}{D_A}}\right) = \sqrt{\frac{k}{D_A}}\,I_1\left(r\sqrt{\frac{k}{D_A}}\right) \tag{3.212}$$

so we finally obtain W_A as

$$W_A = 2\pi RLC_{As}\sqrt{kD_A}\,\frac{I_1\left(R\sqrt{\dfrac{k}{D_A}}\right)}{I_0\left(R\sqrt{\dfrac{k}{D_A}}\right)} \tag{3.213}$$

Defining the Thiele modulus as

$$\Lambda = R\sqrt{k/D_A}$$

the effectiveness factor is

$$\eta_A = \frac{2}{\Lambda} \frac{I_1(\Lambda)}{I_0(\Lambda)} = \frac{W_A}{W_{A,\,\max}} \tag{3.214}$$

Curves for η_A versus Λ are presented in Bird et al. (1960).

EXAMPLE 3.7

We were unable to express net heat flux for the Pin-promoters in Example 3.5, in the absence of differential properties of Bessel functions. Thus, the temperature profile obtained in Eq. 3.175 was found to be expressible as a first order, modified Bessel function

$$T(x) = T_a + (T_b - T_a)\sqrt{\frac{H}{x}} \frac{I_1(2\sqrt{\lambda x})}{I_1(2\sqrt{\lambda H})} \tag{3.215}$$

where

$$\lambda = \frac{2h}{\left[k\left(\dfrac{b}{H}\right)\cos\beta\right]}$$

To find the net rate of transfer to a single pin, we need to compute the heat leaving the base (which must be the same as heat loss from the cone surface)

$$Q = (\pi b^2)\left(+k\frac{dT}{dx}\bigg|_{x=H}\right) \tag{3.216}$$

This will require differentiation of $I_1(2\sqrt{(\lambda x)})$. To do this, define $u(x) = 2\sqrt{(\lambda x)}$, so that from Eq. 3.196

$$\frac{dI_1(u)}{dx} = \frac{dI_1(u)}{du}\frac{du}{dx} = \sqrt{\frac{\lambda}{x}}\left[I_0(u) - \frac{1}{u}I_1(u)\right] \tag{3.217}$$

Thus, the net rate is

$$Q = \frac{k\pi b^2(T_b - T_a)}{H}\left[-1 + \sqrt{\lambda}H\frac{I_0(2\sqrt{\lambda H})}{I_1(2\sqrt{\lambda H})}\right] \tag{3.218}$$

It is clear from Table 3.1 that $xI_0(x) > I_1(x)$, so that the net rate is positive as required. To find the effectiveness factor for Pin-Promoters, take the ratio of actual heat rate to the rate obtainable if the entire pin existed at temperature T_b

(base temperature), which is the maximum possible rate (corresponding to ∞ conductivity)

$$Q_{max} = +h \cdot A_{cone} \cdot (T_b - T_a) = h\pi b\sqrt{b^2 + H^2} \cdot (T_b - T_a) \quad (3.219)$$

hence, we find

$$\eta = \frac{Q}{Q_{max}} = \left(\frac{k}{hH}\right) \frac{1}{\sqrt{1 + (H/b)^2}} \left[\frac{\sqrt{\lambda H}\, I_0(2\sqrt{\lambda H})}{I_1(2\sqrt{\lambda H})} - 1 \right] \quad (3.220)$$

For large arguments (λH), we can obtain the asymptotic result, since $I_0 \sim I_1$ [Eq. 3.186]

$$\eta \simeq \sqrt{\left(\frac{k}{hb}\right) \frac{2}{\left(1 + \left(\frac{H}{b}\right)^2\right)\cos\beta}} \quad (3.221)$$

Thus, the Biot number, defined as hb/k, and the geometric ratio (H/b) control effectiveness. Thus, for ambient air conditions, $Bi \sim 2$, and if $H = b$, then $\cos\beta = 1/\sqrt{2}$, hence an effectiveness is estimated to be 0.84 (84% effectiveness). Under these conditions, $\lambda H = 4\sqrt{2}$, which can be considered large enough (see Table 3.1) to use the approximation $I_0 \sim I_1$.

For small values of (λH), we can use the series expansion in Eq. 3.161 to see that

$$I_0(2\sqrt{\lambda H}) \simeq 1 + \lambda H \quad \text{and} \quad I_1(2\sqrt{\lambda H}) \simeq \sqrt{\lambda H} + \left(\sqrt{\lambda H}\right)^3/2!$$

so approximately

$$\eta \sim \frac{k}{hH} \frac{\frac{1}{2}\lambda H}{1 + \frac{1}{2}\lambda H} \frac{1}{\sqrt{1 + (H/b)^2}} \quad (3.222)$$

Moreover, since $\lambda H \ll 1$, then we have finally that $\eta \to 1$ for small λH (to see this, replace $\lambda = 2hH/(kb\cos\beta)$ and $\cos\beta = H/\sqrt{H^2 + b^2}$).

This problem illustrates the usefulness of the asymptotic approximations given in Eqs. 3.186–3.189.

3.6 REFERENCES

1. Abramowitz, M., and I. A. Stegun. *Handbook of Mathematical Functions*. Dover Publications, Inc., New York (1965).
2. Bird, R. B., W. E. Stewart, and E. N. Lightfoot. *Transport Phenomena*. John Wiley and Sons, Inc., New York (1960).
3. Boas, M. L. *Mathematical Methods in Physical Science*. 2nd ed., John Wiley and Sons, Inc., New York (1983), p. 12.

4. Jenson, V. G., and G. V. Jeffreys. *Mathematical Methods in Chemical Engineering*. 2*nd ed.*, Academic Press, New York (1977).

5. Mickley, H. S., T. K. Sherwood, and C. E. Reid. *Applied Mathematics in Chemical Engineering*. McGraw Hill, New York (1957).

6. Villadsen, J., and M. L. Michelsen. *Solution of Differential Equation Models by Polynomial Approximation*. Prentice Hall, Englewood Cliffs, N.J. (1978).

3.7 PROBLEMS

3.1₂. The Taylor series expansion of $f(x)$ around a point $x = x_0$, can be expressed, provided all derivatives of $f(x)$ exist at x_0, by the series

$$y = f(x) = f(x_0) + \frac{\partial f(x_0)}{\partial x}(x - x_0) + \frac{1}{2!}\frac{\partial^2 f(x_0)}{\partial x^2}(x - x_0)^2$$

$$+ \cdots + \frac{1}{n!}\frac{\partial^n f(x_0)}{\partial x^n}(x - x_0)^n$$

Functions that can be expressed this way are said to be "regular."

(a) Expand the function $\sqrt{(1 + x)}$ around the point $x_0 = 0$ by means of Taylor's series.

(b) Use part (a) to deduce the useful approximate result $\sqrt{(1 + x)} \approx 1 + x/2$. What error results when $x = 0.1$?

(c) Complete Example 2.24 for the case $n = 1/2$ by Taylor expansion of $\exp(-E/RT)$ around the point T_0, retaining the first two terms.

3.2₂. Determine the roots of the indicial relationship for the Frobenius method applied to the following equation:

(a)
$$x\frac{d^2y}{dx^2} + \frac{1}{2}\frac{dy}{dx} + y = 0$$

(b)
$$x\frac{d^2y}{dx^2} + 2\frac{dy}{dx} + xy = 0$$

(c)
$$x^2\frac{d^2y}{dx^2} + x\frac{dy}{dx} + (x^2 - 1)y = 0$$

3.3₂. Thin, metallic circular fins of thickness b can be attached to cylindrical pipes as heat transfer promoters. The fins are exposed to an ambient temperature T_a, and the root of each fin contacts the pipe at position $r = R_1$, where the temperature is constant, T_w. The fin loses heat to ambient air through a transfer coefficient h. The metallic fin transmits heat by conduction in the radial direction.

(a) Show that the steady-state heat balance on an elementary annular element of fin yields the equation

$$\frac{1}{r}\frac{d}{dr}\left(r\frac{dT}{dr}\right) - \left(\frac{2h}{bk}\right)(T - T_a) = 0$$

(b) Define a dimensionless radial coordinate as

$$x = r\sqrt{\frac{2h}{bk}}$$

and introduce $y = T - T_a$, and thus show the elementary equation

$$x^2\frac{d^2y}{dx^2} + x\frac{dy}{dx} - x^2y = 0$$

describes the physical situation.

(c) Apply the method of Frobenius and find the roots of the indicial equation to show that $c_1 = c_2 = 0$.

(d) Complete the solution and show that the first few terms of the solution are

$$y = a_0\left[1 + \left(\frac{x}{2}\right)^2 + \left(\frac{x}{2}\right)^4\frac{1}{2!} + \cdots\right]$$

$$+ b_0\left\{\ln(x)\left[1 + \left(\frac{x}{2}\right)^2 + \left(\frac{x}{2}\right)^4\frac{1}{(2!)^2} + \cdots\right]\right.$$

$$\left. - \left[\left(\frac{x}{2}\right)^2 + \frac{3}{2}\left(\frac{x}{2}\right)^4\frac{1}{(2!)^2} + \cdots\right]\right\}$$

3.4₃. The Graetz equation arises in the analysis of heat transfer to fluids in laminar flow

$$x^2\frac{d^2y}{dx^2} + x\frac{dy}{dx} + \lambda^2x^2(1 - x^2)y = 0$$

Apply the method of Frobenius and show that the only solution that is finite at $x = 0$ has the first few terms as

$$y = a_0\left[1 - \left(\frac{\lambda^2}{4}\right)x^2 + \frac{\lambda^2}{16}\left(1 + \frac{\lambda^2}{4}\right)x^4 + \cdots\right]$$

3.5₃. Use the method of Frobenius to find solutions for the equations:

(a) $x(1 - x)\dfrac{d^2y}{dx^2} - 2\dfrac{dy}{dx} + 2y = 0$

(b) $x\dfrac{d^2y}{dx^2} + (1 - 2x)\dfrac{dy}{dx} - y = 0$

3.6₁. Villadsen and Michelsen (1978) define the Jacobi polynomial as solutions of the equation

$$x(1 - x)\frac{d^2y}{dx^2} + [\beta + 1 - (\beta + \alpha + 2)x]\frac{dy}{dx}$$

$$+ n(n + \alpha + \beta + 1)y = 0$$

Show how this can be obtained from the conventional definition of Jacobi's ODE, Eq. 3.145.

3.7₁. In the method of orthogonal collocation to be described later, Villadsen and Michelsen (1978) define Legendre's equation as

$$x(1 - x)\frac{d^2y}{dx^2} + (1 - 2x)\frac{dy}{dx} + n(n + 1)y = 0$$

Use the change of variables $u(x) = 2x - 1$ and show this leads to the usual form of Legendre equation given in Eq. 3.146.

3.8₂. (a) Show that the solution of Problem 3.3 can be conveniently represented by the modified Bessel functions

$$y = AI_0(x) + BK_0(x)$$

(b) Evaluate the arbitrary constants, using the boundary conditions

$$y = T_w - T_a \quad @ \quad x = R_1\sqrt{\frac{2h}{bk}}$$

$$-k\frac{dy}{dr} = hy \quad @ \quad r = R_2 \quad \text{(outer fin radius)}$$

3.9₂. Rodriques' formula is useful to generate Legendre's polynomials for positive integers n

$$P_n(x) = \frac{1}{2^n n!}\frac{d^n(x^2 - 1)^n}{dx^n}$$

Show that the first three Legendre polynomials are

$$P_0(x) = 1; \qquad P_1(x) = x; \qquad P_2(x) = \frac{1}{2}(3x^2 - 1)$$

and then prove these satisfy Legendre's equation.

3.10₃. A wedge-shaped fin is used to cool machine-gun barrels. The fin has a triangular cross section and is L meters high (from tip to base) and W meters wide at the base. This longitudinal fin is ℓ meters long. It loses heat through a constant heat transfer coefficient h to ambient air at temperature T_A. The flat base of the fin sustains a temperature T_H. Show that the temperature variation obeys

$$x\frac{d^2y}{dx^2} + \frac{dy}{dx} - \left(\frac{2hL\sec\theta}{kW}\right)y = 0$$

where

x = distance from tip of fin
$y = T - T_A$
T = local fin temperature
T_A = ambient temperature
h = heat transfer coefficient
k = thermal conductivity of fin material
L = height of fin
W = thickness of fin at base
θ = half wedge angle of fin

(a) Find the general solution of the above equation with the provision that temperature is at least finite at the tip of the fin.

(b) Complete the solution using the condition $T = T_H$ at the base of the fin.

(c) If a fin is 5 mm wide at the base, 5 mm high, and has a total length 71 cm (28″), how much heat is transferred to ambient (desert) air at 20°C? Take the heat transfer coefficient to be 10 Btu/hr-ft²-°F (0.24×10^{-3} cal/cm² · sec · °C) and the barrel temperature (as a design basis) to be 400°C. The conductivity of steel under these conditions is 0.10 cal/(cm · sec · °K). From this, given the rate of heat generation, the number of fins can be specified. Today, circular fins are used in order to reduce bending stresses caused by longitudinal fins.

3.11₂. Porous, cylindrical pellets are used in packed beds for catalytic reactions of the type $A \xrightarrow{k_s}$ Products. Intraparticle diffusion controls the reaction rate owing to the tortuous passage of reactant through the uniform porous structure of the pellet.

(a) Ignoring transport at the end caps, perform a steady material balance for the diffusion-reaction of species A to obtain for linear kinetics

$$D_A \frac{1}{r} \frac{d}{dr}\left(r \frac{dC_A}{dr}\right) - k_s a_s C_A = 0$$

where D_A is pore diffusivity of species A, C_A is molar composition, k_s is surface rate constant, and a_s represents internal area per unit volume of pellet.

(b) If the bulk gas composition is C_{A0}, and the gas velocity is so slow that a finite film resistance exists at the particle boundary (where $r = R$), then introduce the variables transformation

$$y = \frac{C_A}{C_{A0}}, \qquad x = r\sqrt{\frac{k_s a_s}{D_A}}$$

and show the equation in part (a) becomes

$$x^2 \frac{d^2 y}{dx^2} + x\frac{dy}{dx} - x^2 y = 0$$

(c) Apply the boundary condition

$$-D_A \frac{dC_A}{dr}\bigg|_{r=R} = k_c\left(C_A|_{r=R} - C_{A0}\right)$$

and show

$$\frac{C_A}{C_{A0}} = \frac{I_0\left(r\sqrt{\dfrac{k_s a_s}{D_A}}\right)}{\left[\dfrac{\sqrt{D_A k_s a_s}}{k_c} I_1\left(R\sqrt{\dfrac{k_s a_s}{D_A}}\right) + I_0\left(R\sqrt{\dfrac{k_s a_s}{D_A}}\right)\right]}$$

Note: As $k_c \to \infty$, the solution is identical to Example 3.3.

3.12₃. Thin, circular metal fins are used to augment heat transfer from circular pipes, for example, in home heating units. For such thin fins, heat is conducted mainly in the radial direction by the metal, losing heat to the atmosphere by way of a constant heat transfer coefficient h. In Problem 3.3, we introduced the follow variables

$$y = T - T_A$$

$$x = r\sqrt{\frac{2h}{bk}}$$

to obtain for circular fins

$$x^2 \frac{d^2 y}{dx^2} + x\frac{dy}{dx} - x^2 y = 0$$

(a) If the pipe of radius R_p takes a temperature T_p, and the outer rim of the fin at position R exists at ambient temperature T_A, show that the temperature profile is

$$\frac{T - T_A}{T_p - T_A} = \frac{I_0(x)K_0(x_R) - K_0(x)I_0(x_R)}{I_0(x_p)K_0(x_R) - I_0(x_R)K_0(x_p)}$$

where

$$x_R = R\sqrt{\frac{2h}{bk}}$$

$$x_p = R_p\sqrt{\frac{2h}{bk}}$$

(b) For small arguments, such that $x_p < x < x_R < 1$, show that the temperature profile in part (a) reduces to

$$\frac{T - T_A}{T_p - T_A} \approx \frac{\ln\left(\dfrac{r}{R}\right)}{\ln\left(\dfrac{R_p}{R}\right)}$$

3.13*. Darcy's law can be used to represent flow-pressure drop through uniform packed beds

$$V_{0z} = -\frac{\kappa}{\mu}\frac{dp}{dz} \qquad (\kappa \text{ is permeability})$$

where V_{0z} is the superficial velocity along the axial (z) direction. The Brinkman correction has been suggested (Bird et al. 1960), so that pipe wall effects can be accounted for by the modification

$$0 = -\frac{dp}{dz} - \frac{\mu}{\kappa}V_{0z} + \mu\frac{1}{r}\frac{d}{dr}\left(r\frac{dV_{0z}}{dr}\right)$$

(a) For a constant applied pressure gradient such that $-dp/dz = \Delta p/L$, show that a modified Bessel's equation results if we define

$$y = V_{0z} - V_{0z}^*; \quad V_{0z}^* = \frac{\Delta P}{L}\frac{\kappa}{\mu} \quad (\text{Darcy velocity}); \quad x = \frac{r}{\sqrt{\kappa}}$$

hence, obtain

$$x^2\frac{d^2y}{dx^2} + x\frac{dy}{dx} - x^2y = 0$$

(b) Use the symmetry condition to show

$$V_{0z}(r) = \frac{\Delta P}{L}\frac{\kappa}{\mu} + AI_0\left(r/\sqrt{\kappa}\right)$$

What is the remaining boundary condition to find the arbitrary constant A?

Chapter 4

Integral Functions

4.1 INTRODUCTION

The final step in solving differential equations has been shown to be integration. This often produced well-known elementary functions, such as the exponential, logarithmic, and trigonometric types, which show the relationship between dependent (y) and independent variable (x). However, it frequently occurs that the final integral cannot be obtained in closed form. If this occurs often enough, the integral is given a name, and its values are tabulated. Such is the case with integral expressions such as the error function, the Beta function, and so on. It is important to know the properties of these functions and their limiting values, in order to use them to finally close up analytical solutions. We start with the most common integral relationship: the *error function*.

4.2 THE ERROR FUNCTION

This function occurs often in probability theory, and diffusion of heat and mass, and is given the symbolism:

$$\operatorname{erf}(x) = \frac{2}{\sqrt{\pi}} \int_0^x \exp\left(-\beta^2\right) d\beta \tag{4.1}$$

Dummy variables within the integrand are used to forestall possible errors in differentiating $\operatorname{erf}(x)$. Thus, the *Leibnitz formula* for differentiating under the integral sign is written

$$\frac{d}{d\alpha} \int_{u_0(\alpha)}^{u_1(\alpha)} f(x, \alpha)\, dx = f(u_1, \alpha)\frac{du_1}{d\alpha} - f(u_0, \alpha)\frac{du_0}{d\alpha} + \int_{u_0(\alpha)}^{u_1(\alpha)} \frac{\partial f(x, \alpha)}{\partial \alpha}\, dx$$

$$\tag{4.2}$$

Thus, if we wish to find

$$\frac{d}{dx}\operatorname{erf}(x)$$

then

$$\frac{d}{dx} \text{erf}(x) = \frac{2}{\sqrt{\pi}} \frac{dx}{dx} \exp(-x^2) = \frac{2}{\sqrt{\pi}} \exp(-x^2) \qquad (4.3)$$

Fundamentally, $\text{erf}(x)$ is simply the area under the curve of $\exp(-\beta^2)$ between values of $\beta = 0$ and $\beta = x$; thus it depends only on the value of x selected. The normalizing factor $2/\sqrt{\pi}$ is introduced to ensure the function takes a value of unity as $x \rightarrow \infty$ (see Problem 4.10)

$$\text{erf}(\infty) = 1 \qquad (4.4)$$

4.2.1 Properties of Error Function

While the $\text{erf}(x)$ is itself an integral function, nonetheless it can also be integrated again, thus the indefinite integral

$$\int \text{erf}(x)\, dx = ? \qquad (4.5)$$

can be performed by parts, using the differential property given in Eq. 4.3

$$\int \text{erf}(x)\, dx = \int u(x)\, dv = uv - \int v \frac{du}{dx}\, dx \qquad (4.6)$$

where we let $u(x) = \text{erf}(x)$ and $v = x$, so that

$$\int \text{erf}(x)\, dx = x \cdot \text{erf}(x) - \int \frac{2}{\sqrt{\pi}} \exp(-x^2) x\, dx + K \qquad (4.7)$$

Now, since $2x\, dx = d(x^2)$, we can write finally,

$$\int \text{erf}(x)\, dx = x \cdot \text{erf}(x) + \frac{1}{\sqrt{\pi}} \exp(-x^2) + K \qquad (4.8)$$

where K is a constant of integration.

For continuous computation, the $\text{erf}(x)$ can be given an asymptotic expansion (Abramowitz and Stegun 1965) or it can be represented in terms of the Confluent Hypergeometric function after Kummer (e.g., see Example 3.4). A rational approximation given by C. Hastings in Abramowitz and Stegun (1965) is useful for digital computation

$$\text{erf}(x) \simeq 1 - \left(a_1 t + a_2 t^2 + a_3 t^3\right) \exp(-x^2) + \varepsilon \qquad (4.9)$$

where

$$t = 1/(1 + px) \qquad (4.10)$$

$$p = 0.47047, \quad a_1 = 0.34802, \quad a_2 = -0.09587$$

$$a_3 = 0.74785; \quad \varepsilon \le 2.5 \cdot 10^{-5} \qquad (4.11)$$

Table 4.1 Selected Values for Error Function:

$$\text{erf}(x) = \frac{2}{\sqrt{\pi}} \int_0^x \exp(-\beta^2)\, d\beta$$

x	$\text{erf}(x)$
0	0
0.01	0.01128
0.1	0.11246
0.5	0.52049
1.0	0.84270
1.5	0.96610
2.0	0.99532
∞	1.00000

This approximation ensures $\text{erf}(\infty) = 1$. Occasionally, the complementary error function is convenient; it is defined as

$$\text{erfc}(x) = 1 - \text{erf}(x) \tag{4.12}$$

Table 4.1 illustrates some selected values for $\text{erf}(x)$.

4.3   THE GAMMA AND BETA FUNCTIONS

4.3.1   The Gamma Function

The Gamma function was apparently first defined by the Swiss mathematician Euler. In terms of real variable x, it took a product form:

$$\Gamma(x) = \frac{1}{x} \prod_{n=1}^{\infty} \frac{\left(1 + \dfrac{1}{n}\right)^x}{\left(1 + \dfrac{x}{n}\right)} \tag{4.13}$$

The notation Gamma function was first used by Legendre in 1814. From the infinite product, the usual integral form can be derived

$$\Gamma(x) = \int_0^{\infty} t^{x-1} e^{-t}\, dt, \qquad x > 0 \tag{4.14}$$

It is clear from this that $\Gamma(1) = 1$, and moreover, integration by parts shows

$$\Gamma(x+1) = x \cdot \Gamma(x) \tag{4.15}$$

If x is integer, then since

$$\Gamma(n) = (n-1)\Gamma(n-1) \tag{4.16}$$

Table 4.2 Selected Values for Gamma Function:

$$\Gamma(x) = \int_0^\infty t^{x-1}e^{-t}\,dt$$

x	$\Gamma(x)$
1.00	1.00000
1.01	0.99433
1.05	0.97350
1.10	0.95135
1.15	0.93304
1.20	0.91817
1.30	0.89747
1.40	0.88726
1.50	0.88623
1.60	0.89352
1.70	0.90864
1.80	0.93138
1.90	0.96177
2.00	1.00000

we can repeat this to $\Gamma(1)$ to get

$$\Gamma(n) = (n-1)(n-2)\ldots(2)(1)\Gamma(1) = (n-1)! \qquad \textbf{(4.17)}$$

The defining integral, Eq. 4.14, is not valid for negative x, and in fact tabulations are usually only given for the range $1 < x < 2$; see Table 4.2 for selected values. These results can be extended for all positive values of x by using Eq. 4.15. Moreover, we can extend the definition of Gamma function to the realm of negative numbers (exclusive of negative integers) by application of Eq. 4.15. For integer values, we can also see the requirement $(-1)! = \infty$, since $\Gamma(1) = 1$ and from Eq. 4.16

$$\Gamma(1) = 0 \cdot \Gamma(0) = 0 \cdot (-1)! \qquad \textbf{(4.18)}$$

It is clear that all negative integers eventually contain $\Gamma(0)$, since

$$\Gamma(n-1) = \Gamma(n)/(n-1) = \Gamma(n+1)/[n(n-1)],\text{ etc.} \qquad \textbf{(4.19)}$$

hence, $\Gamma(n)$ is always infinite for negative integers.

Occasionally, the range of integration in Eq. 4.14 is not infinite, and this defines the *incomplete Gamma function*

$$\Gamma(x,\tau) = \int_0^\tau t^{x-1}e^{-t}\,dt, \qquad x > 0, \tau > 0 \qquad \textbf{(4.20)}$$

EXAMPLE 4.1

Prove that $\Gamma(1/2) = \sqrt{\pi}$.

This can be done analytically as follows, starting with the definition

$$\Gamma(1/2) = \int_0^\infty t^{-1/2} e^{-t}\, dt \tag{4.21}$$

Let $t = u^2$ so $dt = 2u\, du$ and $\sqrt{t} = u$; hence,

$$\Gamma(1/2) = \int_0^\infty 2 e^{-u^2}\, du \tag{4.22}$$

Thus the right-hand side is a form of the error function, and moreover, since $\mathrm{erf}(\infty) = 1$, then

$$2\int_0^\infty e^{-u^2}\, du = \sqrt{\pi}\, \frac{2}{\sqrt{\pi}} \int_0^\infty e^{-u^2}\, du = \sqrt{\pi} \tag{4.23}$$

This example allows us to write, for integer n

$$\Gamma\left(n + \frac{1}{2}\right) = \frac{(2n-1)(2n-3)(2n-5)\ldots(3)(1)\sqrt{\pi}}{2^n} \tag{4.24}$$

4.3.2   The Beta Function

The Beta function contains two arguments, but these can be connected to Gamma functions, so we define

$$B(x, y) = \int_0^1 t^{x-1}(1-t)^{y-1}\, dt; \qquad x > 0, y > 0 \tag{4.25}$$

It is an easy homework exercise to show

$$B(x, y) = \Gamma(x)\Gamma(y)/\Gamma(x+y) \tag{4.26}$$

From this, it is clear that $B(x, y) = B(y, x)$.

4.4   THE ELLIPTIC INTEGRALS

In our earlier studies of the nonlinear Pendulum problem (Example 2.6), we arrived at an integral expression, which did not appear to be tabulated. The physics of this problem is illustrated in Fig. 4.1, where R denotes the length of (weightless) string attached to a mass m, θ denotes the subtended angle, and g is the acceleration due to gravity. Application of Newton's law along the path s

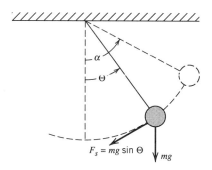

Figure 4.1 Pendulum problem.

yields

$$\frac{d^2\theta}{dt^2} + \omega^2 \sin \theta = 0 \qquad (4.27)$$

where the natural frequency, $\omega = \sqrt{(g/R)}$. Replacing as before $p = d\theta/dt$ yields

$$p\frac{dp}{d\theta} + \omega^2 \sin \theta = 0 \qquad (4.28)$$

hence

$$p^2 = 2\omega^2 \cos \theta + c_1 \qquad (4.29)$$

If the velocity is zero; that is,

$$V = R\frac{d\theta}{dt} = 0$$

at some initial angle α, then $p(\alpha) = 0$, so

$$c_1 = -2\omega^2 \cos(\alpha) \qquad (4.30)$$

We then have, taking square roots

$$p = \frac{d\theta}{dt} = \pm\sqrt{2\omega^2(\cos \theta - \cos \alpha)} \qquad (4.31)$$

hence for positive angles

$$\frac{d\theta}{\sqrt{2\omega^2(\cos \theta - \cos \alpha)}} = dt \qquad (4.32)$$

If we inquire as to the time required to move from angle zero to α, then this is exactly $1/4$ the pendulum period, so that

$$\frac{T}{4} = \frac{1}{\omega\sqrt{2}} \int_0^\alpha \frac{d\theta}{\sqrt{\cos\theta - \cos\alpha}} \tag{4.33}$$

Now, if we define $k = \sin(\alpha/2)$, and using the identities

$$\cos\theta = 1 - 2\sin^2(\theta/2)$$
$$\cos\alpha = 1 - 2\sin^2(\alpha/2) \tag{4.34}$$

gives

$$T = \frac{2}{\omega} \int_0^\alpha \frac{d\theta}{\sqrt{k^2 - \sin^2\left(\dfrac{\theta}{2}\right)}} \tag{4.35}$$

Introducing a new variable φ

$$\sin\frac{\theta}{2} = k \sin\varphi \tag{4.36}$$

then φ ranges from 0 to $\pi/2$ when θ goes from 0 to α. Noting that,

$$d\theta = \frac{2k\cos\varphi\, d\varphi}{\cos\left(\dfrac{\theta}{2}\right)} = \frac{2\sqrt{k^2 - \sin^2\left(\dfrac{\theta}{2}\right)}}{\sqrt{1 - k^2\sin^2\varphi}} \cdot d\varphi \tag{4.37}$$

we finally obtain

$$T = \frac{4}{\omega} \int_0^{\pi/2} \frac{d\varphi}{\sqrt{1 - k^2\sin^2\varphi}} \tag{4.38}$$

The integral term is called the *complete Elliptical integral* of the first kind and is widely tabulated (Byrd and Friedman 1954). In general, the *incomplete* elliptic integrals are defined by

$$F(k,\varphi) = \int_0^\varphi \frac{d\beta}{\sqrt{1 - k^2\sin^2\beta}} \tag{4.39}$$

which is the first kind, and

$$E(k,\varphi) = \int_0^\varphi \sqrt{1 - k^2\sin^2\beta} \cdot d\beta \tag{4.40}$$

which is the second kind. These are said to be "complete" if $\varphi = \pi/2$, and they are denoted under such conditions as simply $F(k)$ and $E(k)$, since they depend only on the parameter k.

The series expansions of the elliptic integrals are useful for computation. For the range $0 < k < 1$, the complete integral expansions are

$$F(k) = \frac{\pi}{2}\left[1 + \left(\frac{1}{2}\right)^2 k^2 + \left(\frac{1 \cdot 3}{2 \cdot 4}\right)^2 k^4 + \left(\frac{1 \cdot 3 \cdot 5}{2 \cdot 4 \cdot 6}\right)^2 k^6 + \cdots\right]$$ (4.41)

and

$$E(k) = \frac{\pi}{2}\left[1 - \left(\frac{1}{2}\right)^2 k^2 - \left(\frac{1 \cdot 3}{2 \cdot 4}\right)^2 \frac{k^4}{3} - \left(\frac{1 \cdot 3 \cdot 5}{2 \cdot 4 \cdot 6}\right)^2 \frac{k^6}{5} + \cdots\right]$$ (4.42)

Table 4.3 gives selected values for the complete integrals in terms of angle φ (degrees), where $k = \sin \varphi$.

It is useful to compare the nonlinear solution of the pendulum problem to an approximate solution; for small angles, $\sin \theta \approx \theta$, hence Eq. 4.27 becomes

$$\frac{d^2\theta}{dt^2} + \omega^2\theta = 0$$ (4.43)

This produces the periodic solution

$$\theta = A \cos(\omega t) + B \sin(\omega t)$$ (4.44)

Now, since it was stated

$$\frac{d\theta}{dt} = 0$$

at time zero, such that $\theta(0) = \alpha$, then we must take $B = 0$. Moreover, we see

Table 4.3 Selected Values for Complete Elliptic Integrals: $k = \sin \varphi$

φ (degrees)	$F(k)$	$E(k)$
0	1.5708	1.5708
1	1.5709	1.5707
5	1.5738	1.5678
10	1.5828	1.5589
20	1.6200	1.5238
30	1.6858	1.4675
40	1.7868	1.3931
50	1.9356	1.3055
60	2.1565	1.2111
70	2.5046	1.1184
80	3.1534	1.0401
90	∞	1.000

that $A = \alpha$, so we find

$$\theta = \alpha \cos(\omega t) \tag{4.45}$$

This can be solved directly for time

$$\omega t = \cos^{-1}(\theta/\alpha) \tag{4.46}$$

The time required for θ to move from 0 to α is again $T/4$ (1/4 the period), so that

$$\frac{T}{4} = \frac{1}{\omega}\left(\frac{\pi}{2}\right) \tag{4.47}$$

We can compare this with the exact result in Eq. 4.38, using the first two terms of the expansion in Eq. 4.41, since it is required $k \ll 1$

$$\frac{T}{4} = \frac{1}{\omega}\frac{\pi}{2}\left[1 + \left(\frac{1}{2}\right)^2 k^2 + \cdots\right] \tag{4.48}$$

Now, suppose $\alpha = 10°$, so that $\sin \alpha/2 = k = 0.087$, hence the exact solution gives approximately

$$\frac{T}{4} \simeq \frac{1}{\omega}\frac{\pi}{2}[1 + 0.001899] \tag{4.49}$$

Thus, the approximate ODE in Eq. 4.43 sustains a very small error (per period) if the original deflected angle α is small (less than 10°). Corrections for bearing friction and form drag on the swinging mass are probably more significant.

4.5   THE EXPONENTIAL AND TRIGONOMETRIC INTEGRALS

Integrals of exponential and trigonometric functions appear so frequently, they have become widely tabulated (Abramowitz and Stegun 1965). These functions also arise in the inversion process for Laplace transforms. The exponential, sine, and cosine integrals are defined according to the relations:

$$Ei(x) = \int_{-\infty}^{x} \frac{\exp(t)}{t} dt \tag{4.50}$$

$$Si(x) = \int_{0}^{x} \frac{\sin(t)}{t} dt \tag{4.51}$$

$$Ci(x) = \int_{\infty}^{x} \frac{\cos(t)}{t} dt \tag{4.52}$$

The exponential integral for negative arguments is represented by

$$-Ei(-x) = \int_{x}^{\infty} \frac{\exp(-t)}{t} dt = E_1(x) \tag{4.53}$$

The behavior of these integral functions is illustrated in Fig. 4.2

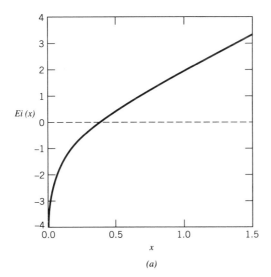

Figure 4.2a Plot of $Ei(x)$ versus x.

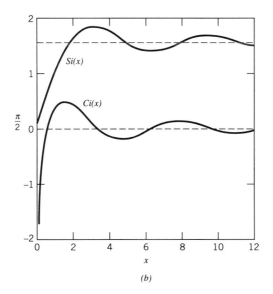

Figure 4.2b Plots of $Si(x)$ and $Ci(x)$ versus x.

Only the sine integral remains finite at the origin, and this information is helpful in evaluating arbitrary constants of integration. It is also important to note that $Ei(x)$ is infinite at $x \to \infty$, hence $Ei(-x)$ is also unbounded as $x \to -\infty$. These boundary values are worth listing

$$Si(0) = 0; \quad Ci(0) = -\infty; \quad Ei(0) = -\infty; \quad E_1(0) = +\infty$$

$$Ei(\infty) = +\infty; \quad E_1(\infty) = 0; \quad E_1(-\infty) = \infty$$

Useful recipes for calculating integral functions are available in Press et al. (1988).

4.6 REFERENCES

1. Abramowitz, A., and I. A. Stegun. *Handbook of Mathematical Functions*. Dover Publications, Inc., New York (1965).
2. Byrd, P. F., and M. D. Friedman. *Handbook of Elliptic Integrals for Engineers and Physicists*. Springer-Verlag, Berlin (1954).
3. Press, W. H., B. P. Flannery, S. A. Teukolsky, and W. T. Vetterling. *Numerical Recipes in C: The Art of Scientific Computing*. Cambridge University Press, Cambridge (1988).

4.7 PROBLEMS

4.1$_3$. One complementary solution of the equation

$$x^2 \frac{d^2y}{dx^2} + (x - 1)\left(x \frac{dy}{dx} - y\right) = x^2 e^{-x}$$

is $y_1 = xC_1$ as seen by inspection.

(a) Show that the second linearly independent, complementary solution can be constructed from the first as

$$y_2 = C_2 x \int \frac{e^{-x}}{x} \, dx$$

(b) Since the indefinite integral is not tabulated, introduce limits and obtain using a new arbitrary constant

$$y_2 = -C_2' x \int_x^\infty \frac{e^{-t}}{t} \, dt$$

Now we see that tabulated functions can be used

$$y_2 = C_2' x Ei(-x)$$

so the complete complementary solution is

$$y_c = C_1 x + C_2' x Ei(-x)$$

(c) Use the method of variation of parameters and prove that a particular solution is

$$y_p = -xe^{-x}$$

so the complete solution is finally,

$$y(x) = x\left[C_1 + C_2' Ei(-x) - e^{-x}\right]$$

4.2$_2$. We wish to find the value of the sine integral

$$Si(\infty) = \int_0^\infty \frac{\sin(t)}{t} \, dt$$

A useful, general approach is to redefine the problem as

$$I(\beta) = \int_0^\infty e^{-\beta t} \frac{\sin(t)}{t} \, dt$$

Then produce the differential

$$\frac{dI(\beta)}{d\beta} = -\int_0^\infty e^{-\beta t} \sin(t) \, dt$$

(a) Use integration by parts twice on the integral expression to show

$$\frac{dI}{d\beta} = -\frac{1}{1 + \beta^2}$$

Hence,

$$I(\beta) = -\tan^{-1}\beta + K$$

where K is an arbitrary constant of integration.

(b) Invoke the condition that $I(\infty) = 0$; hence, show that $K = \pi/2$, and so get the sought-after result

$$I(0) = Si(\infty) = \frac{\pi}{2}$$

4.3$_2$. The complementary error function can be expressed as

$$\text{erfc}(x) = \frac{2}{\sqrt{\pi}} \int_x^\infty e^{-u^2} \, du$$

We wish to develop an asymptotic expression for erfc(x), valid for large arguments.

(a) Use integration by parts to show

$$\int_x^\infty e^{-u^2}\, du = \frac{e^{-x^2}}{2x} - \int_x^\infty \frac{1}{2u^2} e^{-u^2}\, du$$

(b) Repeat this process again on the new integral to get

$$\int_x^\infty \frac{1}{2u^2} e^{-u^2}\, du = \frac{e^{-x^2}}{4x^3} - \frac{3}{4}\int_x^\infty \frac{e^{-u^2}}{u^4}\, du$$

(c) Show that continuation finally yields

$$\int_x^\infty e^{-u^2}\, du \approx \frac{e^{-x^2}}{2x}\left[1 - \frac{1}{2x^2} + \frac{1\cdot 3}{(2x^2)^2} - \frac{1\cdot 3\cdot 5}{(2x^2)^3} + \cdots\right]$$

and thus for large arguments

$$\text{erfc}\,(x) \approx \frac{e^{-x^2}}{x\sqrt{\pi}}$$

4.4$_2$. The Beta function can be expressed in alternative form by substituting

$$t = \sin^2\theta; \qquad 1 - t = \cos^2\theta$$

(a) Show that

$$B(x, y) = 2\int_0^{\pi/2} \sin^{2x-1}\theta \cdot \cos^{2y-1}\theta\, d\theta$$

(b) Now, use the substitution $t = \cos^2\theta$, repeat the process in (a) and thereby prove

$$B(x, y) = B(y, x)$$

4.5$_3$. Use the method illustrated in Problem 8.2 to evaluate

$$I(\alpha) = \int_0^\infty e^{-x^2}\frac{\sin(2\alpha x)}{x}\, dx \tag{I}$$

and thus obtain the useful result

$$I(\alpha) = \frac{1}{2}\pi\, \text{erf}\,(\alpha) \tag{II}$$

(a) Apply two successive differentiations with respect to α and show that

$$\frac{d^2 I}{d\alpha^2} + 2\alpha \frac{dI}{d\alpha} = 0$$

(b) Replace $dI/d\alpha$ with p and obtain

$$p = A \exp(-\alpha^2)$$

and then deduce that $A = \sqrt{\pi}$.

(c) Integrate again and show that

$$I(\alpha) = \frac{1}{2}\pi \operatorname{erf}(\alpha) + B$$

Since the defining integral shows $I(0) = 0$, then $B = 0$, hence obtain the elementary result in (II) above.

4.6₂. Evaluate the derivative of the Gamma function and see

$$\Gamma'(x) = \frac{d\Gamma(x)}{dx} = \int_0^\infty t^{x-1}(\ln t) e^{-t}\, dt$$

and then show that at $x = 1$

$$\Gamma'(1) = -\gamma$$

where $\gamma \approx 0.5772$ is Euler's constant

$$\gamma = \lim_{n \to \infty} \left(1 + \frac{1}{2} + \frac{1}{3} + \frac{1}{4} + \cdots + \frac{1}{n} - \ln(n)\right)$$

4.7₂. An approximation to $\Gamma(x + 1)$ for large x can be obtained by an elementary change of variables.

(a) Starting with

$$\Gamma(x + 1) = \int_0^\infty t^x e^{-t}\, dt$$

change variables to

$$\tau = \frac{t - x}{\sqrt{x}}$$

and show that

$$\frac{\Gamma(x + 1)}{e^{-x} x^{x+1/2}} = \int_{-\sqrt{x}}^\infty e^{-\tau\sqrt{x}} \left(1 + \frac{\tau}{\sqrt{x}}\right)^x d\tau$$

(b) Rearrange the integral by noting

$$\left(1 + \frac{\tau}{\sqrt{x}}\right)^x = \exp\left[x \ln\left(1 + \frac{\tau}{\sqrt{x}}\right)\right]$$

then use the Taylor expansion for $\ln(1 + \tau/\sqrt{x})$ around the point $\tau = 0$; that is,

$$\ln(1 + x) = \ln(1) + \frac{\partial \ln(1+x)}{\partial x}\bigg|_{x=0} \cdot x + \frac{1}{2!}\frac{\partial^2 \ln(1+x)}{\partial x^2}\bigg|_{x=0} \cdot x^2$$

(this is called the Maclauren expansion) and thus shows

$$\frac{\Gamma(x+1)}{e^{-x}x^{x+1/2}} = \int_{-\sqrt{x}}^{\sqrt{x}} \exp\left[-\tau\sqrt{x} + x\left(\frac{\tau}{\sqrt{x}} - \frac{\tau^2}{2x} + \cdots\right)\right] d\tau$$

$$+ \int_{\sqrt{x}}^{\infty} \exp\left[-\tau\sqrt{x} + x \ln\left(1 + \frac{\tau}{\sqrt{x}}\right)\right] d\tau$$

(c) For large x, the second integral has a narrow range of integration, so ignore it and hence show

$$\frac{\Gamma(x+1)}{e^{-x}x^{x+1/2}} \approx \sqrt{2\pi}$$

When $x = n$ (an integer), we have $\Gamma(n + 1) = n!$, so the above yields the useful result

$$n! \approx \sqrt{2\pi}\, e^{-n} n^{n+1/2}$$

This is known as *Stirling's formula*. It is quite accurate even for relatively small n; for example, for $n = 3$, $n! = 6$, while Stirling's formula gives $3! = 5.836$.

4.8₁. Evaluate the integral

$$I = \int_0^{\pi/6} \frac{d\phi}{\sqrt{1 - 4\sin^2\phi}}$$

and show that

$$I = \frac{1}{2}F\left(\frac{1}{2}\right)$$

4.9₂. It is useful to obtain a formula to calculate elliptic integrals outside the range normally provided in tables (0 to $\pi/2$ radians). Use the relations

$$F(k, \pi) = 2F(k)$$
$$E(k, \pi) = 2E(k)$$

to show

$$F(k, \phi + m\pi) = 2mF(k) + F(k, \phi)$$
$$E(k, \phi + m\pi) = 2mE(k) + E(k, \phi)$$

$m = 0, 1, 2, 3, \ldots$

4.10$_2$. Show that

$$\int_0^\infty x^a e^{-bx^c} \, dx = \frac{\Gamma\left(\dfrac{a+1}{c}\right)}{cb^{(a+1)/c}}$$

and then use this result to show the important probability integral has the value

$$\int_0^\infty e^{-x^2} \, dx = \frac{\sqrt{\pi}}{2}$$

Chapter **5**

Staged-Process Models: The Calculus of Finite Differences

5.1 INTRODUCTION

Chemical processing often requires connecting finite stages in series fashion. Thus, it is useful to develop a direct mathematical language to describe interaction between finite stages. Chemical engineers have demonstrated considerable ingenuity in designing systems to cause intimate contact between countercurrent flowing phases, within a battery of stages. Classic examples of their clever contrivances include plate-to-plate distillation, mixer-settler systems for solvent extraction, leaching batteries, and stage-wise reactor trains. Thus, very small local driving forces are considerably amplified by imposition of multiple stages. In the early days, stage-to-stage calculations were performed with highly visual graphical methods, using elementary principles of geometry and algebra. Modern methods use the speed and capacity of digital computation. Analytic techniques to exploit finite-difference calculus were first published by Tiller and Tour (1944) for steady-state calculations, and this was followed by treatment of unsteady problems by Marshall and Pigford (1947).

We have seen in Chapter 3 that finite difference equations also arise in Power Series solutions of ODEs by the Method of Frobenius; the recurrence relations obtained there are in fact finite-difference equations. In Chapters 7 and 8, we show how finite-difference equations also arise naturally in the numerical solutions of differential equations.

In this chapter, we develop analytical solution methods, which have very close analogs with methods used for linear ODEs. A few nonlinear difference equations can be reduced to linear form (the Riccati analog) and the analogous Euler-Equidimensional finite-difference equation also exists. For linear equations, we again exploit the property of superposition. Thus, our general solutions will be composed of a linear combination of complementary and particular solutions.

5.1.1 Modeling Multiple Stages

Certain assumptions arise in stage calculations, especially when contacting immiscible phases. The key assumptions relate to the intensity of mixing and the attainment of thermodynamic equilibrium. Thus, we often model stages using the following idealizations:

- *CSTR Assumption (Continuous Stirred Tank Reactor)* implies that the composition everywhere within the highly mixed fluid inside the vessel is exactly the same as the composition leaving the vessel.
- *Ideal Equilibrium Stage Assumption* is the common reference to an "ideal" stage, simply implying that all departing streams are in a state of thermodynamic equilibrium.

As we show by example, there are widely accepted methods to account for practical inefficiencies that arise, by introducing, for example, Murphree Stage efficiency, and so on.

To illustrate how finite-difference equations arise, consider the countercurrent liquid–liquid extraction battery shown in Fig. 5.1. We first assume the phases are completely immiscible (e.g., water and kerosene). The heavy underflow phase has a continuous mass flow L (water, kg/sec), and the light solvent phase flows at a rate V (kerosene, kg/sec). Under steady-state operation, we wish to extract a solute X_0 (e.g., acetic acid) from the heavy phase, and transfer it to the light phase (kerosene), using a nearly pure solvent with composition Y_{N+1}. Since the solvent flows are constant, it is convenient in writing the solute balance to use mass ratios

$$Y = \text{mass solute}/\text{mass extracting solvent} \tag{5.1}$$

$$X = \text{mass solute}/\text{mass carrier solvent} \tag{5.2}$$

The equilibrium relation is taken to be linear, so that

$$Y_n = KX_n \tag{5.3}$$

With a slight change in the meaning of the symbols, we could also use this model to describe countercurrent distillation; here V would be the vapor phase, and L would denote the downflowing liquid. We illustrate distillation through an example later as a case for nonlinear equations.

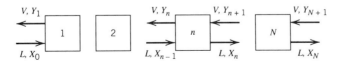

Figure 5.1 Continuous countercurrent extraction cascade.

A material balance on the nth stage can be written, since accumulation is nil (steady state)

$$LX_{n-1} + VY_{n+1} - LX_n - VY_n = 0 \tag{5.4}$$

We can eliminate either X or Y, using Eq. 5.3. Since we are most concerned with the concentration of the light product phase (Y_1), we choose to eliminate X; hence,

$$\left(\frac{L}{K}\right)Y_{n-1} + VY_{n+1} - \left(\frac{L}{K}\right)Y_n - VY_n = 0 \tag{5.5}$$

Dividing through by V yields a single parameter, so we have finally,

$$Y_{n+1} - (\beta + 1)Y_n + \beta Y_{n-1} = 0 \tag{5.6}$$

where

$$\beta = \frac{L}{VK}$$

This equation could be incremented upward (replace n with $n + 1$) to readily see that a second order difference equation is evident

$$Y_{n+2} - (\beta + 1)Y_{n+1} + \beta Y_n = 0 \tag{5.7}$$

Thus, the *order* of a difference equation is simply the difference between the highest and lowest subscripts appearing on the *dependent* variable (Y in the present case). Thus, we treat n as an independent variable, which takes on only integer values.

5.2   SOLUTION METHODS FOR LINEAR FINITE DIFFERENCE EQUATIONS

It was stated at the outset that analytical methods for linear difference equations are quite similar to those applied to linear ODE. Thus, we first find the complementary solution to the homogeneous (unforced) equation, and then add the particular solution to this. We shall use the methods of Undetermined Coefficients and Inverse Operators to find particular solutions.

The general linear finite difference equation of kth order can be written just as we did in Section 2.5 for ODE

$$Y_{n+k} + a_{k-1}(n) \cdot Y_{n+k-1} + \cdots + a_0(n)Y_n = f(n) \tag{5.8}$$

It frequently occurs that the coefficients a_k are independent of n and take constant values. Moreover, as we have seen, the second order equation arises most frequently, so we illustrate the complementary solution for this case

$$Y_{n+2} + a_1 Y_{n+1} + a_0 Y_n = 0 \tag{5.9}$$

Staged processes usually yield a negative value for a_1, and by comparing with the extraction battery, Eq. 5.7, we require $a_1 = -(\beta + 1)$ and $a_0 = \beta$.

5.2.1   Complementary Solutions

In solving ODE, we assumed the existence of solutions of the form $y = A \exp(rx)$ where r is the characteristic root, obtainable from the characteristic equation. In a similar manner, we assume that linear, homogeneous finite difference equations have solutions of the form, for example, in the previous extraction problem

$$Y_n = A(r)^n \tag{5.10}$$

where A is an arbitrary constant obtainable from end conditions (i.e., where $n = 0$, $n = N + 1$, etc.). Thus, inserting this into this second order Eq. 5.9, with $a_1 = -(\beta + 1)$ and $a_0 = \beta$ yields the characteristic equation for the extraction battery problem

$$r^2 - (\beta + 1)r + \beta = 0 \tag{5.11}$$

This implies two characteristic values obtainable as

$$r_{1,2} = \frac{(\beta + 1) \pm \sqrt{(\beta + 1)^2 - 4\beta}}{2} = \frac{(\beta + 1) \pm (\beta - 1)}{2} \tag{5.12}$$

which gives two distinct roots

$$r_1 = \beta; \qquad r_2 = 1$$

Linear superposition then requires the sum of the two solutions of the form in Eq. 5.10, so that

$$Y_n = A(r_1)^n + B(r_2)^n = A(\beta)^n + B(1)^n \tag{5.13}$$

We can now complete the solution for the extraction battery. Thus, taking the number of stages N to be known, it is usual to prescribe the feed composition, X_0, and the solvent composition Y_{N+1}. From these, it is possible to find the constants A and B by defining the fictitious quantity $Y_0 = KX_0$; hence

$$KX_0 = A(\beta)^0 + B(1)^0 = A + B \tag{5.14}$$

$$Y_{N+1} = A(\beta)^{N+1} + B \tag{5.15}$$

so we can solve for A and B

$$A = \frac{KX_0 - Y_{N+1}}{1 - \beta^{N+1}} \tag{5.16}$$

and

$$B = \frac{Y_{N+1} - KX_0\beta^{N+1}}{1 - \beta^{N+1}} \tag{5.17}$$

The general solution then becomes

$$Y_n = \left(\frac{KX_0 - Y_{N+1}}{1 - \beta^{N+1}} \right) \beta^n + \left(\frac{Y_{N+1} - KX_0 \beta^{N+1}}{1 - \beta^{N+1}} \right) \tag{5.18}$$

and the exit composition is obtained by taking $n = 1$; hence,

$$Y_1 = \frac{\left[KX_0 \beta (1 - \beta^N) + Y_{N+1}(1 - \beta) \right]}{\left[1 - \beta^{N+1} \right]} \tag{5.19}$$

It is possible, if Y_1, Y_{N+1} and β are specified, to rearrange Eq. 5.19 to solve for the required number of stages to effect a specified level of enrichment. Thus, rearranging to solve for β^{N+1} gives

$$\beta^{N+1} = \frac{\left[1 - \dfrac{Y_{N+1}}{Y_1}(1 - \beta) - \dfrac{KX_0}{Y_1}\beta \right]}{\left[1 - \dfrac{KX_0}{Y_1} \right]} = \psi \tag{5.20}$$

If we denote the ratio on the right-hand side as ψ, then taking logarithms gives

$$(N + 1) = \frac{\log(\psi)}{\log(\beta)} \tag{5.21}$$

In the case just considered, the characteristic roots $r_{1,2}$ were distinct. However, it is also possible for the roots to be *equal*. For this case, with $r_1 = r_2 = r$, we proceed exactly as for continuous ODE, and write the second linearly independent solution as nr^n; hence,

$$Y_n = (A + Bn)r^n \tag{5.22}$$

It is also possible for the roots to take on complex values, and these will always occur in conjugate pairs

$$r_1 = \sigma + i\omega; \qquad r_2 = \sigma - i\omega \tag{5.23}$$

As we illustrate more fully in Chapter 9, such complex numbers can always be written in polar form

$$r_1 = \sigma + i\omega = |r| \exp(i\varphi) \tag{5.24}$$

$$r_2 = \sigma - i\omega = |r| \exp(-i\varphi) \tag{5.25}$$

where

$$|r| = \sqrt{\sigma^2 + \omega^2}; \qquad \varphi = \tan^{-1}(\omega/\sigma)$$

Here, $|r|$ is called the modulus, and φ is the phase angle. It is now clear that the Euler formula can be used to good effect

$$|r| \exp(i\varphi) = |r|[\cos(\varphi) + i \sin(\varphi)] \tag{5.26}$$

so we insert the two complex roots and rearrange to get

$$Y_n = |r|^n [A \cos(n\varphi) + B \sin(n\varphi)] \tag{5.27}$$

since

$$\exp(in\varphi) = \cos(n\varphi) + i \sin(n\varphi)$$

EXAMPLE 5.1

It is desired to find the required number of ideal stages in the extraction cascade illustrated in Fig. 5.1. Take the feed-to-solvent ratio (L/V) and equilibrium distribution constant (K) to be unity, then $\beta = L/(KV) = 1$. Pure solvent is specified, so $Y_{N+1} = 0$ kg solute/kg solvent, and the feed is $X_0 = 1$ kg solute/kg carrier. It is desired to produce a rich extract product such that $Y_1 = 0.9$ kg solute/kg solvent.

   This problem can be solved by classical methods, using graphical construction as illustrated in Fig. 5.2. The stage-to-stage calculations require linear connections between equilibrium line (i.e., departing streams) and operating line. The passing streams are related by way of a material balance between any stage n and the first stage, and this relationship is called the *operating line*; thus, we

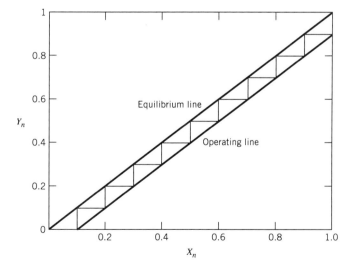

Figure 5.2 Graphical stage-to-stage calculations.

write the material balance:

$$VY_{n+1} + LX_0 = VY_1 + LX_n \tag{5.28}$$

and rearrange to find $Y_{n+1} = f(X_n)$ which is the operating line

$$Y_{n+1} = \left(\frac{L}{V}\right)X_n + \left(Y_1 - X_0\frac{L}{V}\right) \tag{5.29}$$

with slope L/V and intercept $(Y_1 - X_0 L/V)$. This line is plotted along with the equilibrium curve $(Y_n = KX_n)$ in Fig. 5.2, and stages are stepped-off as illustrated to yield $N = 9$ stages.

A direct computation could have been performed using Eq. 5.21. However, since $\beta = 1$, a difficulty arises, since

$$\log(\psi)/\log(\beta) = \log(1)/\log(1) = 0/0$$

which is indeterminate. To resolve this difficulty, we can perform the limiting operation

$$(N + 1) = \lim_{\beta \to 1} \frac{\log(\psi)}{\log(\beta)} \tag{5.30}$$

To perform this limit, we can use the series expansion for logarithms

$$\log(x) = (x - 1) - \frac{1}{2}(x - 1)^2 + \cdots \tag{5.31}$$

so for $x \sim 1$, use the first term to get

$$(N + 1) = \lim_{\beta \to 1} \left[\frac{Y_{N+1}(\beta - 1) - KX_0(\beta - 1)}{(Y_1 - KX_0)(\beta - 1)}\right] \tag{5.32}$$

This gives a finite limit, since $(\beta - 1)$ cancels and we finally get

$$(N + 1) = \frac{KX_0 - Y_{N+1}}{KX_0 - Y_1} \tag{5.33}$$

valid for $\beta = 1$. Inserting the given parameters yields

$$(N + 1) = \frac{1}{(1 - 0.9)} = 10 \qquad \therefore N = 9 \tag{5.34}$$

We could also have resolved this indeterminacy using L'Hôpital's rule.

In Section 2.4.2, we illustrated how to reduce a certain class of variable coefficient ODE to an elementary constant coefficient form; such equation forms were called Equidimensional or Euler equations. The analog of this also exists for finite difference equations; for example,

$$(n + 2)(n + 1)Y_{n+2} + A(n + 1)Y_{n+1} + BY_n = 0 \qquad (5.35)$$

This can be reduced to constant coefficient status by substituting

$$Y_n = \frac{z_n}{n!} \qquad (5.36)$$

hence, we have in terms of z

$$z_{n+2} + Az_{n+1} + Bz_n = 0 \qquad (5.37)$$

EXAMPLE 5.2

Find the analytical solution for a_n in the recurrence relation given in Eq. 3.9, subject only to the condition a_0 is different from 0

$$a_{n+2}(n + 2)(n + 1) - 2a_{n+1}(n + 1) + a_n = 0 \qquad (5.38)$$

As suggested in Eq. 5.36, let

$$a_n = \frac{\alpha_n}{n!} \qquad (5.39)$$

so we get a relation for α_n

$$\alpha_{n+2} - 2\alpha_{n+1} + \alpha_n = 0 \qquad (5.40)$$

Now, assume a solution for this constant coefficient case to be

$$\alpha_n = Ar^n \qquad (5.41)$$

so the characteristic equation is

$$r^2 - 2r + 1 = 0 \qquad (5.42)$$

Hence, we have equal roots: $r_1 = r_2 = 1$. According to Eq. 5.22, the solution is

$$\alpha_n = (A + Bn)r^n = (A + Bn) \qquad (5.43)$$

and so from Eq. 5.38,

$$a_n = \frac{(A + Bn)}{n!} \qquad (5.44)$$

Now, the only boundary condition we have stipulated is that a_0 is different from 0, so one of the constants A or B must be set to zero. If we set $A = 0$, then we would have

$$a_n = \frac{B}{(n-1)!} \tag{5.45}$$

But when $n = 0$, then $(-1)! = \infty$. This says $a_0 = 0$. Thus, we must set $B = 0$ and write $a_n = A/n!$. Now, since a_0 is different from zero, set $n = 0$ to see that $A = a_0$, since $0! = 1$. This reproduces the result given by Eq. 3.10, with $r = 1$

$$a_n = \frac{a_0}{n!} \tag{5.46}$$

5.3   PARTICULAR SOLUTION METHODS

We shall discuss two techniques to find particular solutions for finite difference equations, both having analogs with continuous ODE solution methods:

1. Method of Undetermined Coefficients
2. Method of Inverse Operators

A Variation of Parameters analog was published by Fort (1948).

5.3.1   Method of Undetermined Coefficients

Under forced conditions, the constant coefficient, second order finite difference equation can be written, following Eq. 5.8

$$Y_{n+2} + a_1 Y_{n+1} + a_0 Y_n = f(n) \tag{5.47}$$

where $f(n)$ is usually a polynomial function of n. Thus, if the forcing function is of the form

$$f(n) = b_0 + b_1 n + b_2 n^2 + \cdots \tag{5.48}$$

then the particular solution is also assumed to be of the same form

$$Y_n^p = \alpha_0 + \alpha_1 n + \alpha_2 n^2 + \cdots \tag{5.49}$$

where α_i are the undetermined coefficients. The particular solution is inserted into the defining equation, yielding a series of algebraic relations obtained by equating coefficients of like powers. Difficulties arise if the forcing function has the same form as one of the complementary solutions. Thus, linear independence of the resulting solutions must be guaranteed, as we illustrated earlier, for example, in Eq. 5.22 (for the occurrence of equal roots). We illustrate the method by way of examples in the following.

EXAMPLE 5.3

Find the particular solution for the forced finite difference equation

$$y_{n+1} - 2y_n - 3y_{n-1} = (n + 1) \tag{5.50}$$

Inserting an assumed solution of form

$$y_n = \alpha_0 + \alpha_1 n \qquad (5.51)$$

into Eq. 5.50 yields

$$\alpha_0 + \alpha_1(n + 1) - 2(\alpha_0 + \alpha_1 n) - 3[\alpha_0 + \alpha_1(n - 1)] = n + 1 \quad (5.52)$$

Equating coefficients of unity and n gives, first for coefficient of unity

$$\alpha_0 + \alpha_1 - 2\alpha_0 - 3\alpha_0 + 3\alpha_1 = 1 \qquad (5.53)$$

hence,

$$4\alpha_1 - 4\alpha_0 = 1 \qquad (5.54)$$

Next, coefficients of n give

$$\alpha_1 - 2\alpha_1 - 3\alpha_1 = 1 \qquad \therefore \alpha_1 = -\frac{1}{4} \qquad (5.55)$$

We can now solve Eq. 5.54 for α_0

$$\alpha_0 = -\frac{1}{2} \qquad (5.56)$$

The particular solution can now be written

$$y_n^p = -\frac{1}{2} - \frac{1}{4}n \qquad (5.57)$$

To check linear independence with the complementary parts, first find the characteristic roots from the homogeneous equation to see

$$r^2 - 2r - 3 = 0 \qquad \therefore r_{1,2} = 3, -1 \qquad (5.58)$$

$$y_n^c = A(3)^n + B(-1)^n \qquad (5.59)$$

and these are obviously independent of the particular solutions. The general solution is written as before: $y_n = y_n^c + y_n^p$. The coefficients A and B are found for the general solution using end conditions.

EXAMPLE 5.4

Repeat the previous problem, except the forcing function, $f(n) = (3)^n$.
  It is clear that an assumed solution of the form

$$y_n^p = \alpha(3)^n \qquad (5.60)$$

will not be linearly independent of one of the complementary solutions.

Therefore, we assume a solution form, which is clearly independent

$$y_n^p = \beta n(3)^n \tag{5.61}$$

Inserting this gives

$$\beta(n + 1)3^{n+1} - 2\beta n 3^n - 3\beta(n - 1)3^{n-1} = 3^n \tag{5.62}$$

Dividing through by 3^n gives

$$3\beta(n + 1) - 2\beta n - \beta(n - 1) = 1 \tag{5.63}$$

Equating coefficients of unity allows the undetermined coefficient β to be found

$$3\beta + \beta = 1 \qquad \therefore \beta = \frac{1}{4} \tag{5.64}$$

Coefficients of n are exactly balanced, since

$$3\beta n - 2\beta n - 1\beta n = 0 \tag{5.65}$$

as required. Thus, the linearly independent particular solution is

$$y_n^p = n\frac{(3)^n}{4} \tag{5.66}$$

and the general solution is

$$y_n = A(3)^n + B(-1)^n + n\frac{(3)^n}{4} \tag{5.67}$$

5.3.2 Inverse Operator Method

This method parallels the Heaviside operator method used in Chapter 2. We first define the incrementing operator as

$$Ey_0 = y_1 \tag{5.68}$$

$$E(y_{n-1}) = y_n \tag{5.69}$$

$$E(Ey_{n-2}) = E^2 y_{n-2} = y_n \tag{5.70}$$

In fact, operating n times shows

$$y_n = E^n y_0 \tag{5.71}$$

The exponent n can take positive or negative integer values; if n is negative, we increment downward. The equivalent of Rule 1 in Chapter 2 for arbitrary c is

$$Ec^n = c^{n+1} = cc^n \tag{5.72}$$

$$E^m c^n = c^{m+n} = c^m c^n \tag{5.73}$$

and as before, for any polynomial of E, Rule 1 is

$$P(E)c^n = P(c)c^n \tag{5.74}$$

This has obvious applications to find particular solutions when $f(n) = c^n$. The analogous form for Rule 2 is

$$P(E)(c^n f_n) = c^n P(cE) f_n \tag{5.75}$$

Here, we see that cE replaces E. This product rule has little practical value in problem solving.

Thus, we can rewrite Eq. 5.47 as

$$\left(E^2 + a_1 E + a_0\right) y_n = f(n) \tag{5.76}$$

Now, if $f(n)$ takes the form Kc^n, then we can solve directly for the particular solution

$$y_n^p = \frac{1}{\left(E^2 + a_1 E + a_0\right)} Kc^n \tag{5.77}$$

With the usual provision that the inverse operator can be expanded in series, we have, using Rule 1,

$$y_n^p = \frac{Kc^n}{\left[c^2 + a_1 c + a_0\right]} \tag{5.78}$$

provided $c^2 + a_1 c + a_0$ is different from zero.

EXAMPLE 5.5

Find the particular solution for

$$y_{n+2} - 2y_{n+1} - 8y_n = e^n \tag{5.79}$$

Writing in operator notation, we have in one step

$$y_n^p = \frac{1}{[E^2 - 2E - 8]} e^n = e^n \frac{1}{[e^2 - 2e - 8]} \tag{5.80}$$

This would have required considerably more algebra using the method of undetermined coefficients. To check for linear independence, the roots of the homogeneous equation are

$$r^2 - 2r - 8 = 0 \qquad \therefore r_{1,2} = 4, -2 \tag{5.81}$$

so the general solution is

$$y_n = A(4)^n + B(-2)^n + \frac{e^n}{[e^2 - 2e - 8]} \tag{5.82}$$

5.4 NONLINEAR EQUATIONS (RICCATI EQUATION)

Very few nonlinear equations yield analytical solutions, so graphical or trial–error solution methods are often used. There are a few nonlinear finite difference equations, which can be reduced to linear form by elementary variable transformation. Foremost among these is the famous Riccati equation

$$y_{n+1}y_n + Ay_{n+1} + By_n + C = 0 \tag{5.83}$$

We translate the coordinates by letting

$$y_n = z_n + \delta \tag{5.84}$$

Inserting this into Eq. 5.83 yields

$$z_{n+1}z_n + (A + \delta)z_{n+1} + (B + \delta)z_n + \delta^2 + (A + B)\delta + C = 0 \tag{5.85}$$

We use the last group of terms to define δ

$$\delta^2 + (A + B)\delta + C = 0 \tag{5.86}$$

The remainder can be made linear by dividing by $z_{n+1}z_n$ and introducing a new variable

$$v_n = \frac{1}{z_n} = \frac{1}{y_n - \delta} \tag{5.87}$$

$$(B + \delta)v_{n+1} + (A + \delta)v_n + 1 = 0 \tag{5.88}$$

This is an elementary first order linear equation with forcing by a constant. The characteristic root is simply

$$r = -\frac{A + \delta}{B + \delta} \tag{5.89}$$

and the particular solution, taking $f(n) = -1 \cdot 1^n$, is simply (using inverse operators)

$$v_n^p = -\frac{1}{[A + B + 2\delta]} \tag{5.90}$$

so the general solution is, replacing $v_n = 1/[y_n - \delta]$

$$\frac{1}{y_n - \delta} = K \left[-\frac{A + \delta}{B + \delta} \right]^n - \frac{1}{[A + B + 2\delta]} \tag{5.91}$$

where K is an arbitrary constant.

EXAMPLE 5.6

The cascade shown in Fig. 5.1 could also represent a plate-to-plate distillation operation if we denote y_n as solute mole fraction in vapor and x_n is the fraction in liquid. For constant molar flow rates, L and V are then constant. Now, y_{N+1} represents hot vapor feed, and x_0 represents desired product recycle, which is rich in the volatile solute. For the high concentration expected, the relative volatility (α) is taken to be constant, so the equilibrium relation can be taken as

$$y_n = \frac{\alpha x_n}{[1 + (\alpha - 1)x_n]} \tag{5.92}$$

This is obtained from the usual definition of relative volatility for binary systems

$$\alpha = \frac{\left(\dfrac{y_n}{x_n} \right)}{\dfrac{(1 - y_n)}{(1 - x_n)}} \tag{5.93}$$

The material balance between the nth and first plate is the same as in Eq. 5.28, except replace Y_n with y_n, and so on

$$y_{n+1} = (L/V)x_n + (y_1 - x_0 L/V) \tag{5.94}$$

This equation represents a general plate in the enriching section of a binary distillation column. In the usual case, a total condenser is used, so $V = L + D$, and $L/D = R$ is the recycle ratio, D being the distillate product removed. In the present case, it is the liquid composition that is monitored, so we proceed to eliminate y_n using the equilibrium relation, Eq. 5.92

$$\frac{\alpha x_{n+1}}{[1 + (\alpha - 1)x_{n+1}]} = \left(\frac{L}{V} \right) x_n + \left(y_1 - x_0 \frac{L}{V} \right) \tag{5.95}$$

Since a total condenser is used, $y_1 = x_0$ and $L/V = R/(R + 1)$, so the intercept term becomes $x_0/(R + 1)$. Multiplying through by $[1 + (\alpha - 1)x_{n+1}]$ yields a form of the Riccati equation

$$\frac{R}{R + 1}(\alpha - 1)x_n x_{n+1} + \left[\frac{(\alpha - 1)x_0}{R + 1} - \alpha \right] x_{n+1}$$

$$+ \frac{R}{R + 1} x_n + \frac{x_0}{R + 1} = 0 \tag{5.96}$$

Rearranging this to the familiar form gives

$$x_n x_{n+1} + A x_{n+1} + B x_n + C = 0 \tag{5.97}$$

$$A = \frac{x_0(\alpha - 1) - \alpha(R + 1)}{(\alpha - 1)R} \tag{5.98}$$

$$B = \frac{1}{(\alpha - 1)}$$

$$C = \frac{x_0}{R(\alpha - 1)}$$

The solution has already been worked out in Eq. 5.91, so write

$$\frac{1}{x_n - \delta} = K\left[-\frac{A + \delta}{B + \delta} \right]^n - \frac{1}{[A + B + 2\delta]} \tag{5.99}$$

The parameter δ can be found by solving the quadratic in Eq. 5.86. However, a geometric basis was presented in Mickley et al. (1957) by the following expedient. Let the intersection of the operating and equilibrium curve be denoted by y, x. Thus, ignoring subscripts, we search for the intersection

$$y = \frac{R}{R + 1}x + \frac{x_0}{R + 1} \tag{5.100}$$

and

$$y = \frac{\alpha x}{[1 + (\alpha - 1)x]} \tag{5.101}$$

Eliminating y, one can obtain

$$x^2 + (A + B)x + C = 0 \tag{5.102}$$

which is identical to the equation for δ. Thus, the parameter δ represents the x-value of the point of intersection for the operating line and the equilibrium curve. The arbitrary constant K can be found if the composition x_0 is known (corresponding to $n = 0$). Thus, it is seen that

$$K = \frac{1}{x_0 - \delta} + \frac{1}{A + B + 2\delta} \tag{5.103}$$

To find the number of stages N in the upper (enriching) section, the composition y_{N+1} or x_N must be known. Rearrangement, followed by taking logarithm (as in Eqs. 5.20 and 5.21) allows N to be computed.

5.5 REFERENCES

1. Fort, T. *Finite Differences and Difference Equations in the Real Domain*. Oxford University Press, New York (1948).
2. Foust, A. S., L. A. Wenzel, C. W. Clump, L. Mans, and L. B. Andersen. *Principles of Unit Operations*. p. 117, John Wiley & Sons, Inc., New York (1980).
3. Marshall, W. R., and R. L. Pigford. *The Application of Differential Equations to Chemical Engineering Problems*. University of Delaware, Newark, Delaware (1947).
4. Mickley, H. S., T. K. Sherwood, and C. E. Reid. *Applied Mathematics in Chemical Engineering*. McGraw Hill, New York (1957).
5. Pigford, R. L., B. Burke and D. E. Blum. "An Equilibrium Theory of the Parametric Pump," *Ind. Eng. Chem. Fundam.*, 8, 144 (1969).
6. Tiller, F. M., and R. S. Tour, "Stagewise Operations—Applications of the Calculus of Finite Differences to Chemical Engineering," *Trans. AIChE*, 40, 317–332 (1944).

5.6 PROBLEMS

5.1$_2$. Find the analytical solutions for the following finite difference equations

(a)
$$y_{n+3} - 3y_{n+2} + 2y_{n+1} = 0$$

(b)
$$y_{n+3} - 6y_{n+2} + 11y_{n+1} - 6y_n = 0$$

(c)
$$y_{n+2}y_n = y_{n+1}^2$$

Hint: Use logarithms

(d)
$$R\bar{x}_{n-1} - R\bar{x}_n = \nu s\bar{x}_n$$

when

$$\bar{x}_0 = \frac{x_0}{s}$$

is initial state.

5.2*. A continuous cascade of stirred tank reactors consists of N tanks in series as shown below.

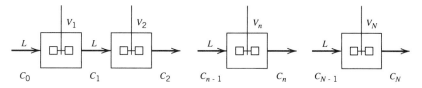

The feed to the first tank contains C_0 moles/cc of component A. Each tank undergoes the irreversible reaction A $\rightarrow$ B so that the reaction rate is $R_n = kC_n$ (moles/cc/sec). The volume rate of flow in the cascade is constant at a value L (cc/sec).

(a) Show that the material balance for the $(n + 1)$th tank yields, at steady state

$$C_{n+1} = \frac{C_n}{k\theta_{n+1} + 1}$$

where $\theta_n = V_n/L$ is residence time.

(b) Show that the relationship between first and last stages is

$$C_N = \frac{C_0}{(k\theta_1 + 1)(k\theta_2 + 1) \cdots (k\theta_N + 1)}$$

(c) If the rate constant k is invariant, show that the highest concentration of B, which may be produced by the cascade, occurs when all volumes are equal: $V_1 = V_2 = V_N$, with only the provision that total volume $(V_T = V_1 + V_2 + \cdots + V_N)$ remains constant.

Hint: The maximum in B occurs when A is minimum, so search for the minimum in C_N using

$$dC_N = \sum_{n=1}^{N} \frac{\partial C_N}{\partial \theta_n} d\theta_n = 0$$

with the constraint that

$$\sum_{n=1}^{N} \theta_n = \frac{V_T}{L}$$

5.3$_3$. Pigford et al. (1969) developed a simple equilibrium model to predict separations in a closed packed column. The process, called "parametric pumping," uses the synchronization of fluid motion and adsorption to effect a separation. Thus, on upflow the column is made hot (low adsorption on solid), while on downflow the column is made cold (high adsorption on solid). The bottom and top compositions depend on cycle number n according to

$$\langle y_B \rangle_n = \langle y_B \rangle_{n-1} \left(\frac{1-b}{1+b} \right); \qquad b < 1$$

$$\langle y_T \rangle_n = \langle y_T \rangle_{n-1} + \langle y_B \rangle_{n-3} \left(\frac{2b}{1+b} \right)$$

where the dimensionless parameter b reflects the difference in adsorption between hot and cold conditions. The starting conditions were given as

$$\langle y_B \rangle_{n=0} = y_0$$

Show that the separation factor for this process is

$$\frac{\langle y_T \rangle_n}{\langle y_B \rangle_n} = \alpha_n = \left(2 + \frac{2b}{1+b}\right)\left(\frac{1+b}{1-b}\right)^n - \left(\frac{1+b}{1-b}\right)^2; \qquad n > 0$$

$$\langle y_T \rangle_2 = y_0\left(1 + \frac{2b}{1+b}\right)$$

Is the separation factor bounded as $n \to \infty$?

5.4∗. A gas–liquid chemical reaction is carried out in a cascade of shallow bubble columns, which are assumed to be well mixed owing to the mixing caused by gas bubbles. Each column contains H moles of liquid composed mainly of inert, nonvolatile solvent with a small amount of dissolved, nonvolatile catalyst. Liquid does not flow from stage to stage. Pure gas A is fed to the first column at a rate G (moles/sec). The exit gas from stage one goes to stage two, and so on. On each stage, some A dissolves and reacts to form a volatile product B by way of the reversible reaction

$$A \underset{k_B}{\overset{k_A}{\rightleftharpoons}} B$$

A linear rate law is assumed to exist for each pathway. The B formed by reaction is stripped and carried off by the gas bubbling through the liquid mixture. Overall absorption, desorption efficiencies are predictable using Murphree's vapor rule

$$E_A = \frac{y_{n-1}^A - y_n^A}{y_{n-1}^A - \left(y_n^A\right)^*}$$

$$E_B = \frac{y_{n-1}^B - y_n^B}{y_{n-1}^B - \left(y_n^B\right)^*}$$

where the equilibrium solubilities obey Henry's law

$$\left(y_n^A\right)^* = m_A x_n^A$$

$$\left(y_n^B\right)^* = m_B x_n^B$$

Here, y_n and x_n denote mole fractions, for the nth stage, and in the vapor phase, only A and B exist, so that $y_n^A + y_n^B = 1$.

(a) The moles of A or B lost to the liquid phase can be represented in terms of stage efficiency; thus, for component A

$$G\left(y_{n-1}^A - y_n^A\right) = GE_A\left[y_{n-1}^A - \left(y_n^A\right)^*\right] = GE_A\left(y_{n-1}^A - m_A x_n^A\right)$$

Write the steady-state material balances for the liquid phase and show that

$$x_n^A = \beta y_{n-1}^A + \alpha$$

where

$$\alpha = \frac{1}{F}\left(-\frac{k_B H}{GE_A}\right)$$

$$\beta = \frac{-m_B + \dfrac{k_B H \Delta}{G}}{F}$$

$$\Delta = \frac{E_B - E_A}{E_A E_B}$$

$$F = \left(m_A + \frac{k_A H}{GE_A}\right)\left(-m_B + \frac{k_B H \Delta}{G}\right) - \frac{k_B H}{GE_A}\left(m_A + \frac{k_A H \Delta}{G}\right)$$

(b) Find the relationship to predict y_n^A as a function of n, where $y_0^A = 1$.
(c) The statement of the problem implies $m_B \gg m_A$ (B is volatile) $E_B \geq E_A$ (volatiles have higher efficiencies), and $k_A > k_B$. Are these the appropriate conditions on physical properties to maximize production of B? Compute the composition y_n^B, leaving the 5th stage for the following conditions:

$$m_B = 10; \quad m_A = 1; \quad \frac{G}{H} = 0.1 \text{ sec}^{-1}; \quad E_A = E_B = 0.5;$$

$$k_A = 2 \text{ sec}^{-1}; \quad k_B = 1 \text{ sec}$$

Ans: 0.92

5.5₃. A hot vapor stream containing 0.4 mole fraction ammonia and 0.6 mole fraction water is to be enriched in a distillation column consisting of enriching section and total condenser. The saturated vapor at 6.8 atm pressure (100 psia) is injected at a rate 100 moles/hour at the bottom of the column. The liquid distillate product withdrawn from the total condenser has a composition 0.9 mole fraction NH_3. Part of the distillate is returned as reflux, so that 85% of the NH_3 charged must be recovered as distillate product.

(a) Complete the material balance and show that $x_N = 0.096$ (mole fraction NH_3 in liquid leaving Nth tray) and that the required reflux ratio is $R = 1.65$.
(b) The NH_3-H_2O system is highly nonideal, and hence, the relative volatility is not constant, so that the assumption of constant molar overflow is invalid. Nonetheless, we wish to estimate the number of ideal stages necessary, using the single available piece of vapor–liquid equilibria data at 6.8 atm pressure: $y(NH_3) = 0.8$ when $x(NH_3) = 0.2$. This suggests $\alpha \approx 16$. Use this information and the results from Example 5.6 to find the required number of ideal stages (an enthalpy-composition plot shows that exactly two stages are needed).

5.6₃. Acetone can be removed from acetone–air mixtures using simple counter-current cascades, by adsorption onto charcoal (Foust et al. 1985). We wish to find the required number of equilibrium stages to reduce a gas stream carrying 0.222 kg acetone per kg air to a value 0.0202 kg acetone per kg air. Clean charcoal ($X_0 = 0$) enters the system at 2.5 kg/sec, and the air rate is constant at 3.5 kg/sec. Equilibrium between the solid and gas can be taken to obey the Langmuir-type relationship

$$Y_n = \frac{KX_n}{1 + KX_n}; \qquad K = 0.5$$

where
Y_n = kg acetone/kg air
X_n = kg acetone/kg charcoal

(a) Write the material balance between the first stage (where X_0 enters) and the nth stage, and use the equilibrium relationship to derive the finite difference equation for X_n.
(b) Rearrange the expression in part (a) to show that the Riccati equation arises

$$X_n X_{n+1} + AX_{n+1} + BX_n + C = 0$$

What are the appropriate values for A, B, and C?
(c) Use the condition $X_0 = 0$ (clean entering charcoal) to evaluate the arbitrary constant in the general solution from part (b) and thus obtain a relationship to predict $X_n = f(n)$.
(d) Use an overall material balance to find X_N and use this to calculate the number of equilibrium stages (N) required.

Ans: (d) 2.9 stages

Chapter **6**

Approximate Solution Methods for ODE: Perturbation Methods

In Chapters 2 and 3, various analytical techniques were given for solving ordinary differential equations. In this chapter, we develop an approximate solution technique called the *perturbation* method. This method is particularly useful for model equations that contain a small parameter, and the equation is analytically solvable when that small parameter is set to zero. We begin with a brief introduction into the technique. Following this, we teach the technique using a number of examples, from algebraic relations to differential equations. It is in the class of nonlinear differential equations that the perturbation method finds the most fruitful application, since numerical solutions to such equations are often mathematically intractable.

The perturbation method complements solution of ordinary differential equations by numerical methods, which are taught in Chapters 7 and 8 for initial and boundary type of problems, respectively.

6.1 PERTURBATION METHODS

6.1.1 Introduction

Modeling problems nearly always contain parameters, which are connected to the physicochemical dynamics of the system. These parameters may take a range of values. The solution obtained when the parameter is zero is called the *base case*. If one of the parameters is small, the behavior of the system either deviates slightly from the base case, or it can take a trajectory that is remote from the base case. The analysis of systems having the former behavior is called *regular perturbation*, whereas that of the latter is referred to as *singular perturbation*.

Perturbation methods involve series expansions, which are called *asymptotic* expansions in terms of *a small parameter*. This is sometimes called *parameter* perturbation. The term asymptotic implies that the solution can be adequately described by only a few terms of the expansion, which is different from the power series taught in Chapter 3. Moreover, perturbation schemes can be applied to nonlinear systems, where they find their greatest utility.

The perturbation method can be applied to algebraic equations as well as differential equations. For example, we may wish to find an approximate solution for the algebraic equation containing a small parameter

$$\varepsilon y^2 - y + a = 0 \tag{6.1}$$

An example of a differential equation containing a small parameter is

$$\frac{dy}{dx} + \varepsilon y = 0 \tag{6.2}$$

In the above two examples, ε is the small parameter and the perturbation method will be carried out in terms of this small parameter. For a general problem, there are many parameters, but we assume that at least one of them is small and the rest are of order of unity and we denote that small parameter as ε.

If the equation in question is a differential equation, there must exist boundary or initial conditions. For example, conditions may take the following form

$$y(0) = 1; \quad \frac{dy}{dx}\bigg|_{x=0} = 0 \tag{6.3}$$

These conditions may also contain the parameter ε in the general case such as

$$y(1) + \varepsilon \frac{dy(1)}{dx} = 1 + 2\varepsilon \tag{6.4}$$

Let us further assume that the solution to the problem posed is not possible by analytical means for a nonzero value of the parameter ε, but an analytical solution is readily available when that parameter ε is equal to zero. If this is the case, we will endeavor to find a solution (asymptotic form) in terms of ε using the following series expansion

$$y(x; \varepsilon) = y_0(x) + \varepsilon y_1(x) + \varepsilon^2 y_2(x) + \cdots + \varepsilon^n y_n(x) + \cdots \tag{6.5}$$

Thus, if all the *coefficients* $y_j(x)$ are of the same order of magnitude, then it is quite clear that the higher order terms are getting smaller since ε^n becomes increasingly smaller. This is, however, not always the case in some practical problems, since it is possible that the higher order coefficients could become larger than the leading coefficients. Usually one stops the expansion with a few terms, generally with two terms, and as we will see the asymptotic expansions with two terms can often describe the solution very closely, relative to the exact solution.

To start, we substitute the series expansion (Eq. 6.5) into the governing equation and the boundary condition (if the equation is a differential equation), and then apply a Taylor expansion to the equation and the boundary condition. Now, since the coefficients of each power of ε are independent of ε, a set of identities will be produced. This leads to a simpler set of equations, which may have analytical solutions. The solution to this set of simple subproblems is done *sequentially*, that is, the zero order solution $y_0(x)$ is obtained first, then the next solution $y_1(x)$, and so on. If analytical solutions cannot be obtained easily for the first few leading coefficients (usually two), there is no value in using the perturbation methods. In such circumstances, a numerical solution may be sought.

EXAMPLE 6.1

To demonstrate the elementary concept, let us first study the simple nonlinear algebraic equation

$$y = a + \varepsilon y^2 \tag{6.6}$$

where ε is a small number. The solution for this problem when ε is zero (hereafter called the base case) is simply

$$y_0 = a \tag{6.7}$$

Now, we assume that there exists an asymptotic expansion of the form

$$y = y_0 + \varepsilon y_1 + \varepsilon^2 y_2 + \cdots \tag{6.8}$$

Substituting this into the original equation (Eq. 6.6) yields

$$y_0 + \varepsilon y_1 + \varepsilon^2 y_2 + \cdots = a + \varepsilon [y_0 + \varepsilon y_1 + \cdots]^2 \tag{6.9}$$

Expanding the squared function gives

$$y_0 + \varepsilon y_1 + \varepsilon^2 y_2 + \cdots = a + \varepsilon y_0^2 \left[1 + 2\varepsilon \frac{y_1}{y_0} + \cdots \right] \tag{6.10}$$

Matching the coefficients of unity and each of the powers of ε yields

$$0(1): y_0 = a \tag{6.11}$$

$$0(\varepsilon): y_1 = y_0^2 \tag{6.12}$$

$$0(\varepsilon^2): y_2 = 2y_1 y_0 \tag{6.13}$$

Hence, solutions for the coefficients of the asymptotic expansion (Eq. 6.8) are obtained sequentially, starting with the zero order solution

$$y_0 = a \tag{6.14a}$$

$$y_1 = a^2 \tag{6.14b}$$

$$y_2 = 2a^3 \tag{6.14c}$$

Thus, the required asymptotic expansion is

$$y(\varepsilon) = a + \varepsilon a^2 + 2\varepsilon^2 a^3 + \cdots \tag{6.15}$$

Inspection of the original solution, which is a quadratic, suggests there must exist two solutions to the original problem. The asymptotic solution presented above assumed that the base solution is finite. There is, however, another asymptotic expansion of the form

$$y(\varepsilon) = \frac{1}{\varepsilon}[u_0 + \varepsilon u_1 + \cdots] \tag{6.16}$$

which indicates the second solution is of order of $1/\varepsilon$. Substituting this expansion into Eq. 6.6 yields

$$\frac{1}{\varepsilon}[u_0 + \varepsilon u_1 + \cdots] = a + \varepsilon \left\{ \frac{1}{\varepsilon^2}[u_0 + \varepsilon u_1 + \cdots]^2 \right\}$$

Next, matching multiples of 1, ε, ε^2, and so on gives

$$u_0 = u_0^2 \therefore u_0 = 1 \tag{6.17}$$

$$u_1 = a + 2u_0 u_1 \tag{6.18}$$

Knowing $u_0 = 1$, the solution for u_1 is

$$u_1 = -a \tag{6.19}$$

Thus, the second solution has the following asymptotic form

$$y(\varepsilon) = \frac{1}{\varepsilon}[1 - \varepsilon a + \cdots] \tag{6.20}$$

We have two asymptotic solutions, given in Equations 6.15 and 6.20. Now we know that the exact solution to the original quadratic equation is simply

$$y = \frac{1 \pm \sqrt{1 - 4\varepsilon a}}{2\varepsilon} \tag{6.21}$$

It is not difficult to show that the two asymptotic expansions above (Eqs. 6.15 and 6.20) are in fact the Taylor series expansions in terms of ε around the point

$\varepsilon = 0$; that is,

$$y = \frac{1}{2\varepsilon} f(\varepsilon) = \frac{1}{2\varepsilon} \left[1 \pm \sqrt{1 - 4\varepsilon a} \right]$$

$$= \frac{1}{2\varepsilon} \left[f(0) + f'(0)\varepsilon + f''(0)\frac{\varepsilon^2}{2!} + f'''(0)\frac{\varepsilon^3}{3!} + \cdots \right]$$

for the two exact solutions.

The asymptotic solution (6.15) differs only slightly from the base solution, which is the solution when $\varepsilon = 0$. This is the regular perturbation solution. The asymptotic solution (6.20) deviates substantially from the base solution; it is the singular perturbation solution.

This algebraic example illustrates the salient features of the perturbation method. In the next example, we apply the method to an elementary differential equation.

EXAMPLE 6.2

The basic principles underlying perturbation methods can be taught using the elementary first order equation discussed in Chapter 2

$$\frac{dy}{dx} + \varepsilon y = 0; \qquad y(0) = 1; \qquad \varepsilon \ll 1 \tag{6.22}$$

The complete analytical solution is known to be

$$y = \exp(-\varepsilon x) = 1 - \varepsilon x + \frac{\varepsilon^2 x^2}{2} + \cdots \tag{6.23}$$

Suppose we search for an approximate solution of the form

$$y = y_0(x) + \varepsilon y_1(x) + \varepsilon^2 y_2(x) + \cdots \tag{6.24}$$

Inserting this into the defining equation (Eq. 6.22) yields

$$\left[\frac{dy_0}{dx} + \varepsilon \frac{dy_1}{dx} + \varepsilon^2 \frac{dy_2}{dx} + \cdots \right] + \varepsilon \left[y_0 + \varepsilon y_1 + \varepsilon^2 y_2 + \cdots \right] = 0 \tag{6.25}$$

Next, we stipulate the following identities, by matching like multiples of unity, ε, ε^2, and so on.

$$\frac{dy_0}{dx} = 0 \therefore y_0 = K_0 \tag{6.26a}$$

$$\frac{dy_1}{dx} = -y_0 \therefore y_1 = -K_0 x + K_1 \tag{6.26b}$$

$$\frac{dy_2}{dx} = -y_1 \therefore y_2 = \frac{K_0 x^2}{2} - K_1 x + K_2 \tag{6.26c}$$

where K_0, K_1, and K_2 are constants of integration.

Obviously, the solution $y_0(x)$ corresponds to the base case (when $\varepsilon = 0$), so we shall stipulate that

$$y(0) = y_0(0) = 1 \therefore K_0 = 1 \qquad (6.27)$$

This is a critical component of regular perturbation methods, since *only the base case carries the primary boundary conditions*; hence, by implication, we must have for the other solutions

$$y_1(0) = y_2(0) = \cdots = 0 \qquad (6.28)$$

This allows the determination of K_1, K_2, K_3, and so on in sequence

$$K_1 = 0; \qquad K_2 = 0; \qquad \text{etc.} \qquad (6.29)$$

We finally see to the order ε^2

$$y(x) = 1 - \varepsilon x + \varepsilon^2 \frac{x^2}{2} + \cdots \qquad (6.30)$$

This is identical to the first three terms of the analytical solution (Eq. 6.23), as one may have expected if the technique is to be useful. This example illustrates *regular perturbation*. The power of the method is most useful for nonlinear systems, where an analytical solution is not easily obtainable. The method is also quite useful in providing a simplified form of an unwieldy analytical solution. We take up the issue of singular perturbations for differential equations in the next section following some preliminary concepts.

6.2 THE BASIC CONCEPTS

Over the years, users of perturbation methods have evolved a shorthand language to express ideas. This reduces repetition and allows compact illustration. We first present the *gauge functions*, which are used to compare the size of functions, and then we present the *order concept*, which is convenient in expressing the order of a function (i.e., the speed it moves when ε tends small). Finally, we discuss asymptotic expansions and sequences, and the sources of nonuniformity, which cause the solution for $\varepsilon \neq 0$ to behave differently from the base case.

6.2.1 Gauge Functions

Let us consider a function $f(\varepsilon)$ containing a small parameter ε. There are three possible behavior patterns that $f(\varepsilon)$ can take as ε tends to zero. These are

$$f(\varepsilon) \to 0$$
$$f(\varepsilon) \to \alpha$$
$$f(\varepsilon) \to \infty$$

where α is finite. The second possibility needs no further explanation. The other two require careful inspection. Mainly, we need to analyze how fast the magnitude of $f(\varepsilon)$ moves as ε tends to zero. To help us with this, we use so-called gauge functions, that is, we compare $f(\varepsilon)$ with *known functions* called gauge functions. Examples of a class of gauge functions are

$$\ldots, \varepsilon^{-n}, \varepsilon^{-n+1}, \ldots \varepsilon^{-1}, 1, \varepsilon, \varepsilon^2, \ldots, \varepsilon^n, \ldots.$$

In some cases, the following gauge functions are useful (mostly in fluid flow problems)

$$\log(\varepsilon^{-1}), \log(\log(\varepsilon^{-1})), \ldots$$

Other classes of gauge functions are discussed in Van Dyke (1975) and Nayfeh (1973).

To compare the behavior of a function $f(\varepsilon)$ relative to a gauge function, we need to define the order symbols, O and o.

6.2.2 Order Symbols

The notation

$$f(\varepsilon) = O(g(\varepsilon)) \tag{6.31}$$

expresses boundedness, and implies the sense of a limit

$$\lim_{\varepsilon \to 0} \frac{f(\varepsilon)}{g(\varepsilon)} < \infty \tag{6.32}$$

For example,

$$\sin(\varepsilon) = O(\varepsilon)$$
$$\cos(\varepsilon) = O(1)$$
$$\coth(\varepsilon) = O(\varepsilon^{-1})$$

The first simply expresses the fact that

$$\lim_{\varepsilon \to 0} \frac{\sin \varepsilon}{\varepsilon} = 1 \quad \text{(bounded)} \tag{6.33}$$

that is, the function $f(\varepsilon) = \sin(\varepsilon)$ behaves like ε for small values of ε.

The o notation

$$f(\varepsilon) = o(g(\varepsilon)) \tag{6.34}$$

implies a zero limit

$$\lim_{\varepsilon \to 0} \left| \frac{f(\varepsilon)}{g(\varepsilon)} \right| = 0 \tag{6.35}$$

This means that $f(\varepsilon)$ decreases to zero faster than the function $g(\varepsilon)$. For example,

$$\sin(\varepsilon) = o(1) \qquad \therefore \lim_{\varepsilon \to 0} \frac{\sin(\varepsilon)}{1} = 0$$

In summary, the O implies finite boundedness, while the o implies zero in the limit $\varepsilon \to 0$.

6.2.3 Asymptotic Expansions and Sequences

In the presentation of an asymptotic expansion, we need not restrict ourselves to power series $(1, \varepsilon, \varepsilon^2, \varepsilon^3,$ etc.$)$, such as the previous examples, but we could also use a general sequence of functions $\{\delta_n\}$ such that

$$\delta_{n+1}(\varepsilon) = o\big(\delta_n(\varepsilon)\big) \therefore \lim_{\varepsilon \to 0} \frac{\delta_{n+1}(\varepsilon)}{\delta_n(\varepsilon)} \to 0 \qquad (6.36)$$

as ε approaches zero. Such a sequence is called an *asymptotic* sequence. Using this asymptotic sequence, we can write the *asymptotic expansion*

$$y(\varepsilon) = y_0 \delta_0(\varepsilon) + y_1 \delta_1(\varepsilon) + \cdots = \sum_{j=0}^{\infty} y_j \delta_j(\varepsilon) \qquad (6.37)$$

as ε approaches zero, where y_j are independent of ε. We could truncate the series and form the asymptotic expansion as

$$y(\varepsilon) = \sum_{j=0}^{N} y_j \delta_j(\varepsilon) + O\big[\delta_{N+1}(\varepsilon)\big] \qquad (6.38)$$

The second term in the RHS of Eq. 6.38 means that the error of the result of the series truncation to N terms has the order of magnitude as the sequence δ_{N+1}; that is,

$$\lim_{\varepsilon \to 0} \frac{y(\varepsilon) - \sum_{j=0}^{N} y_j \delta(\varepsilon)}{\delta_{N+1}(\varepsilon)} < \infty$$

A function y can have many asymptotic expansions simply because there are many sets of asymptotic sequences $\{\delta_n\}$ that could be selected. However, for a *given* asymptotic sequence, the asymptotic expansion is unique, and the coefficients y_j are determined as follows. First divide Eq. 6.37 by δ_0 to see

$$\frac{y(\varepsilon)}{\delta_0(\varepsilon)} = y_0 + \sum_{j=1}^{\infty} y_j \frac{\delta_j(\varepsilon)}{\delta_0(\varepsilon)} \qquad (6.39)$$

Now take the limit of the above equation when ε approaches zero, and make use of the asymptotic sequence property (Eq. 6.36), so we have

$$y_0 = \lim_{\varepsilon \to 0} \frac{y(\varepsilon)}{\delta_0(\varepsilon)} \qquad (6.40)$$

since

$$\lim_{\varepsilon \to 0} \frac{\delta_j(\varepsilon)}{\delta_0(\varepsilon)} \to 0 \qquad \text{for } j \geq 1$$

Knowing y_0, we can rearrange the asymptotic expansion as follows

$$\frac{y(\varepsilon) - y_0 \delta_0(\varepsilon)}{\delta_1(\varepsilon)} = y_1 + \sum_{j=2}^{\infty} y_j \frac{\delta_j(\varepsilon)}{\delta_1(\varepsilon)} \qquad (6.41)$$

Taking its limit when ε approaches zero, we have the following expression for y_1

$$y_1 = \lim_{\varepsilon \to 0} \frac{y(\varepsilon) - y_0 \delta_0(\varepsilon)}{\delta_1(\varepsilon)} \qquad (6.42)$$

Similarly, we can prove that

$$y_n = \lim_{\varepsilon \to 0} \frac{y(\varepsilon) - \displaystyle\sum_{j=0}^{n-1} y_j \delta_j(\varepsilon)}{\delta_n(\varepsilon)} \qquad (6.43)$$

In solving practical problems, the function y usually involves another variable in addition to the small parameter ε. If we denote that variable as t, then the asymptotic expansion for a given asymptotic sequence $\{\delta_n(\varepsilon)\}$ is

$$y(t; \varepsilon) = \sum_{j=0}^{\infty} y_j(t) \delta_j(\varepsilon) \qquad (6.44)$$

where $y_j(t)$ is only a function of t. This asymptotic expansion is said to be *uniformly valid* over the entire domain of definition of t if

$$y(t; \varepsilon) = \sum_{j=0}^{N} y_j(t) \delta_j(\varepsilon) + R_{N+1}(t; \varepsilon) \qquad (6.45)$$

and

$$R_{N+1}(t; \varepsilon) = O[\delta_{N+1}(\varepsilon)] \qquad \text{or} \qquad \lim_{\varepsilon \to 0} \frac{R_{N+1}(t; \varepsilon)}{\delta_{N+1}(\varepsilon)} < \infty \qquad (6.46)$$

for *all* t in the domain of interest. Otherwise, the asymptotic expansion is said to be nonuniformly valid (sometimes called a singular perturbation expansion).

For the uniformity condition to be valid, the next term in the asymptotic expansion must be smaller than the preceding term; hence,

$$\delta_{n+1}(\varepsilon) = o[\delta_n(\varepsilon)] \qquad \text{or} \qquad \lim_{\varepsilon \to 0} \frac{\delta_{n+1}}{\delta_n} = 0 \qquad (6.47)$$

and we require that the coefficient $y_{n+1}(t)$ be no more singular than the preceding coefficient $y_n(t)$. In other words, each term is a small correction to the preceding term irrespective of the value of t.

6.2.4 Sources of Nonuniformity

There are several sources of nonuniformity that might give rise to the singular perturbation expansions. Some of these are the following:

1. Infinite domain (either time or space).
2. Small parameter multiplying the highest derivative.
3. Type change of a partial differential equation (e.g., from parabolic to hyperbolic equation).
4. Nonlinearity (e.g., the algebraic equation of Example 6.1).

Examples of the nonuniformity source for infinite domain come from problems involving $\cos(t)$ and $\sin(t)$ functions, and the higher order terms involve $t \sin(t)$ or $t \cos(t)$, which would make the higher order terms more secular (i.e., more unbounded) than the preceding term.

The second source of nonuniformity may seem obvious. For example, when the highest derivative is removed (i.e., when the small parameter is set to zero) the differential equation is one order less, and hence, one boundary condition becomes redundant. To invoke this boundary condition, there must exist a *boundary layer* wherein a boundary condition becomes redundant when the parameter ε is set to zero.

EXAMPLE 6.3

To demonstrate sources of nonuniformity, let us study the following simple second order differential equation with a small parameter multiplying the second order derivative

$$\varepsilon \frac{d^2 y}{dx^2} + \frac{dy}{dx} + y = 0; \qquad \varepsilon \ll 1 \tag{6.48}$$

subject to

$$x = 0; \qquad y = \alpha \tag{6.49a}$$

$$x = 1; \qquad y = \beta \tag{6.49b}$$

This problem has been treated by Latta (1964).

Now, if we formally attempt the following asymptotic expansion

$$y(x; \varepsilon) = y_0(x) + \varepsilon y_1(x) + \cdots \tag{6.50}$$

by placing it into the differential equation (Eq. 6.48), equate the coefficients of like powers of ε to zero, then we obtain the following first two subproblems:

$$O(1): \frac{dy_0}{dx} + y_0 = 0 \tag{6.51}$$

$$O(\varepsilon): \frac{dy_1}{dx} + y_1 = -\frac{d^2y_0}{dx^2} \tag{6.52}$$

The first subproblem is a *first order* differential equation, which is one order less than the original differential equations (Eq. 6.48), and therefore, it cannot satisfy two boundary conditions. So one condition must be dropped. It will be argued later that the boundary condition at $x = 0$ must be dropped. The solution for y_0 which satisfies $y(1) = \beta$, is

$$y_0(x) = \beta e^{1-x} \tag{6.53}$$

Substitute this into the equation for y_1 and solve for y_1 where as before we require all cases beyond zero order to have null boundaries; that is, $y_1(1) = 0$, $y_2(1) = 0$, and so on.

$$y_1(x) = \beta(1 - x)e^{1-x} \tag{6.54}$$

Therefore, the asymptotic expansion is

$$y(x;\varepsilon) = \beta e^{1-x} + \varepsilon\beta(1 - x)e^{1-x} + O(\varepsilon^2) \tag{6.55}$$

At $x = 0$, the above asymptotic expansion gives

$$y(0;\varepsilon) = \beta(1 + \varepsilon)e^1 \neq \alpha \tag{6.56}$$

which is, in general, different from α. Thus, the asymptotic expansion is not uniformly valid over the whole domain of interest $[0, 1]$. Figure 6.1 shows computations for the asymptotic and the exact solutions for $\varepsilon = 0.05$. The exact solution takes the following form:

$$y_{\text{exact}} = \frac{(\alpha e^{s_1} - \beta)e^{s_1 x} + (\beta - \alpha e^{s_2})e^{s_2 x}}{e^{s_2} - e^{s_1}} \tag{6.57}$$

where

$$s_{1,2} = \frac{-1 \pm \sqrt{1 - 4\varepsilon}}{2\varepsilon} \tag{6.58}$$

The zero order asymptotic solution (Eq. 6.53) agrees reasonably well with the exact solution over most of the domain $[0, 1]$ except close to the origin. The first order asymptotic expansion (Eq. 6.55), even though it agrees with the exact solution better, also breaks down near the origin. This suggests that a boundary layer solution must be constructed near the origin to take care of this nonuniformity. We will come back to this point later.

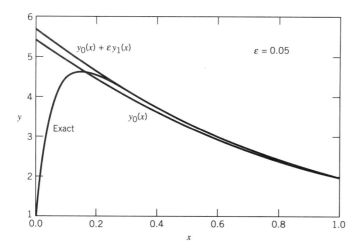

Figure 6.1 Plots of the exact solution and the zero order outer solution (Eq. 6.53) and the first order outer solution (Eq. 6.55); $\alpha = 0, \beta = 2$.

6.3   THE METHOD OF MATCHED ASYMPTOTIC EXPANSION

There are a number of variations for the perturbation technique. Among them, the method of matched asymptotic expansion is the easiest to apply. The method is useful for obtaining expansions from separate domains of validity.

Using the straightforward expansion, such as the last example, we have obtained what is called the *outer* solution; that is, the solution that is valid over *most* of the domain of interest. This solution, however, fails in a small region, which we call the boundary layer. In this thin boundary layer, there is a sharp change from a boundary point to the outer solution. To study this sharp change in the boundary layer, we need to *magnify* the scale; the choice of this magnified variable changes from problem to problem. We will demonstrate this procedure in the example of Eqs. 6.48 and 6.49.

Outer Solutions

Let us consider the problem in the last section (Eq. 6.48). The straightforward expansion we obtained is the outer solution (Eq. 6.55); that is, it is valid over most of the domain, except close to the origin (see Fig. 6.1). The outer solution is

$$y^{(0)}(x;\varepsilon) = y_0^{(0)}(x) + \varepsilon y_1^{(0)}(x) + \cdots \tag{6.59}$$

where the superscript (0) denotes outer solution and we have already found

$$y_0^{(0)}(x) = \beta e^{1-x} \tag{6.60}$$

$$y_1^{(0)} = \beta(1 - x)e^{1-x} \tag{6.61}$$

Inner Solutions

The previous section clearly indicated that this outer solution is not valid near the origin. So there is a sharp change in the solution behavior as the trajectory moves away from the outer region. To describe this sharp change, we need to magnify the variable as follows

$$x^* = \frac{x}{\varepsilon^n} \tag{6.62}$$

where ε^n represents the thickness of the boundary layer. In terms of this new magnified variable, the differential equation (Eq. 6.48) becomes

$$\frac{d^2y}{dx^{*2}} + \varepsilon^{n-1}\frac{dy}{dx^*} + \varepsilon^{2n-1}y = 0 \tag{6.63}$$

Note that n is unknown at this stage, since we have no preconceived notion of the thickness of the boundary. Next, we assume that $y(x^*; \varepsilon)$ has the following asymptotic expansion (called *inner solution*)

$$y^{(i)}(x^*; \varepsilon) = y_0^{(i)}(x^*) + \varepsilon y_1^{(i)}(x^*) + \cdots \tag{6.64}$$

where the superscript (i) denotes inner solution.

Since the second order derivative did not appear in the zero order equation of the outer expansion (Eq. 6.51), we must *ensure* that it appears in the leading order equation of the inner solution (i.e., the zero order inner solution must be second order). Observing the differential equation, we have two possibilities. One is that the second order derivative term balances with the first order term, and the second possibility is that the second order derivative term balances with the remaining third term.

First, let us balance the second order derivative term with the third term. To do this, we insert Eq. 6.64 into 6.63 and by inspection select

$$2n - 1 = 0; \quad \text{i.e., } n = \frac{1}{2} \tag{6.65}$$

With this value of n, the original differential equation becomes

$$\frac{d^2y}{d(x^*)^2} + \varepsilon^{-1/2}\frac{dy}{dx^*} + y = 0 \tag{6.66}$$

So, when we substitute the inner expansion into the above equation, the zero order equation is

$$\frac{dy_0^{(i)}}{dx^*} = 0 \tag{6.67}$$

which means that $y_0^{(i)}(x^*)$ is a constant. This is not acceptable because the boundary layer solution must sustain sharp behavior from the boundary point to the outer solution. Also, it is clear that our aim of getting a second order ODE for $y_0^{(i)}$ was not accomplished. This then leads us to the second possibility, that is, we balance the second order derivative term with the first order derivative term. This implies setting

$$n - 1 = 0; \quad \text{i.e., } n = 1 \tag{6.68}$$

Thus, the region of nonuniformity near the origin has thickness of order of ε. With $n = 1$, the differential equation becomes

$$\frac{d^2 y}{d(x^*)^2} + \frac{dy}{dx^*} + \varepsilon y = 0 \tag{6.69}$$

Substituting the inner expansion (Eq. 6.64) into the above equation, we have the following two leading subproblems

$$\frac{d^2 y_0^{(i)}}{d(x^*)^2} + \frac{dy_0^{(i)}}{dx^*} = 0 \tag{6.70}$$

and

$$\frac{d^2 y_1^{(i)}}{d(x^*)^2} + \frac{dy_1^{(i)}}{dx^*} + y_0^{(i)} = 0 \tag{6.71}$$

Let us have a look at the zero order subproblem. Because this solution is valid in the boundary layer around $x = 0$, it must satisfy the condition at $x = 0$; that is,

$$x = 0; \quad y_0^{(i)}(0) = \alpha \tag{6.72}$$

The solution of this leading order subproblem is

$$y_0^{(i)} = C + (\alpha - C)e^{-x^*} \tag{6.73}$$

where C is the only constant of integration since we have used the condition (6.72).

Matching

The inner solution obtained in Eq. 6.73 still contains a constant of integration. To find a value for this constant, we need to match the inner solution (Eq. 6.73) with the outer solution (Eq. 6.60). The matching principle is (Van Dyke 1975)

The inner limit of the outer solution must match with the outer limit of the inner solution.

The matching principle is mathematically equivalent to

$$\lim_{x \to 0} y^{(0)}(x;\varepsilon) = \lim_{x^* \to \infty} y^{(i)}(x^*;\varepsilon) \tag{6.74}$$

Therefore, for the zero order solution, we have

$$y_0^{(0)}(0) = y_0^{(i)}(\infty) \tag{6.75}$$

From Eq. 6.73, this gives

$$y_0^{(i)}(\infty) = C \tag{6.76}$$

and from Eq. 6.60

$$y_0^{(0)}(0) = \beta e^1 \tag{6.77}$$

Therefore, substitution of Eqs. 6.76 and 6.77 into Eq. 6.75 gives

$$C = \beta e^1 \tag{6.78}$$

Hence, the leading order inner solution is

$$y_0^{(i)}(x^*) = \beta e^1 + (\alpha - \beta e^1)e^{-x^*} \tag{6.79}$$

Composite Solutions

What we have done so far is to obtain an inner expansion and an outer expansion. Each is valid in their respective regions. We now wish to find a solution that is valid over the whole domain. This solution is called the *composite solution*. It is found simply by adding the inner solution and the outer solution and subtracting the common parts of the two solutions (because we don't want to count the common parts of the two solutions twice). The common parts are simply the terms that arise in the matching process. For the zero order solution, this common part is Eq. 6.77

$$y_{\text{com}} = \beta e^1 \tag{6.80}$$

Thus, the composite solution is

$$y = \beta e^{1-x} + (\alpha - \beta e^1)e^{-x/\varepsilon} + O(\varepsilon) \tag{6.81}$$

Figure 6.2 shows computations of the zero order composite solution for a value of $\varepsilon = 0.05$. The exact solution is also shown, and it is seen that the agreement is fairly good. Note that this composite solution is the zero order composite solution, that is, it has an error of order of ε. We next inspect the first order composite solution, which has an error of order ε^2. But before we can achieve this goal, we have to develop a matching principle in general, beyond the principle we have in Eq. 6.75, which is applicable for only zero order solutions.

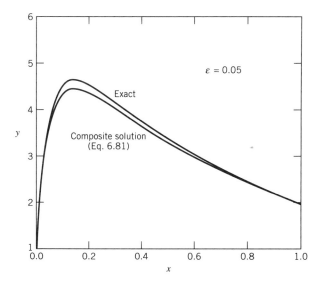

Figure 6.2 Plots of the exact solution and the zero order composite solution (Eq. 6.81).

General Matching Principle

The success of the matching is due to the existence of the *overlapping* region. In this overlapping region, one could define an intermediate variable as

$$x_\delta = \frac{x}{\delta(\varepsilon)} \tag{6.82}$$

where

$$\lim_{\varepsilon \to 0} \delta(\varepsilon) = 0 \tag{6.83}$$

and $\delta(\varepsilon)$ is such that

$$\lim_{\varepsilon \to 0} \frac{\delta(\varepsilon)}{\varepsilon} = \infty \tag{6.84}$$

This means that the overlapping region lies between the boundary layer (which has a thickness of order of ε) and the outer region (which has a thickness of order of unity). In this overlapping region, the *intermediate variable is of order of unity*. Now, we write the outer variable x and the inner variable x^* in terms of this intermediate variable

$$x = \delta(\varepsilon) x_\delta \tag{6.85}$$

and

$$x^* = \frac{\delta(\varepsilon)}{\varepsilon} x_\delta \tag{6.86}$$

Thus, in *the overlapping region* (intermediate region), x_δ is of order of unity, and when ε approaches zero, we will have

$$x \to 0; \qquad x^* \to \infty \qquad\qquad (6.87)$$

Equation 6.87 simply states that for a very large distance from the origin (outer solution), the intermediate region appears to have zero thickness, whereas the intermediate region would appear to be extremely far away when viewed from the origin. This proves Eq. 6.74.

   Note, we previously obtained both the inner and the outer solutions for the zero order solutions. Hence, the composite solution obtained has an error of order of ε. To find a composite solution, which has an error of the order of ε^2, we need to find the first order inner solution, as we did for the outer solution (Eq. 6.61), and then we make use of the matching principle using the intermediate variable of Eq. 6.82 to achieve our goal of composite solution having an error of ε^2.

Composite Solution of Higher Order

Solving Eq. 6.71 for the first order inner solution we obtain

$$y_1^{(i)}(x^*) = C_1(1 - e^{-x^*}) - \left[\alpha - (\alpha - \beta e^1)(1 + e^{-x^*})\right]x^* \qquad (6.88)$$

in which we have used the boundary condition at $x = 0$, $y_1^{(i)}(0) = 0$. Thus, the inner solution order having error of $0(\varepsilon^2)$ is

$$y^{(i)}(x^*; \varepsilon) = y_0^{(i)}(x^*) + \varepsilon y_1^{(i)}(x^*) + O(\varepsilon^2) \qquad (6.89)$$

To solve for the constant C_1, we resort to the matching principle and utilize the intermediate variable x_δ. The inner and outer variables are related to this intermediate variable as shown in Eqs. 6.85 and 6.86. Thus, in the intermediate region, the inner solution (Eq. 6.89) and the outer solution (Eq. 6.59) behave like

$$y^{(i)}(x_\delta) = \beta e^1 - \beta e^1 \delta x_\delta + \varepsilon C_1 + o(\varepsilon) + o(\delta) \qquad (6.90)$$

$$y^{(0)}(x_\delta) = \beta e^1 - \beta e^1 \delta x_\delta + \varepsilon \beta e^1 + o(\varepsilon) + o(\delta) \qquad (6.91)$$

In the intermediate region, these two solutions match. Hence, the constant C_1 must take the value

$$C_1 = \beta e^1 \qquad\qquad (6.92)$$

Hence, the inner solution that has an error of order of ε^2 is

$$y^{(i)}(x^*) = \beta e^1 + (\alpha - \beta e^1)e^{-x^*}$$

$$+ \varepsilon\{\beta e^1(1 - e^{-x^*}) - [\beta e^1 - (\alpha - \beta e^1)e^{-x^*}]x^*\} + O(\varepsilon^2)$$

(6.93)

The composite solution having an error of the order of ε^2 is the summation of the inner and outer solutions minus the common parts, which are the matching terms given in Eq. 6.91; that is, the composite solution is

$$y_{\text{comp}} = \beta[1 + \varepsilon(1 - x)]e^{1-x} + [(\alpha - \beta e^1)(1 + x) - \varepsilon\beta e^1]e^{-x/\varepsilon} + O(\varepsilon^2)$$

(6.94)

Equation 6.94 is the composite solution including the first order correction terms, in contrast to the composite solution (Eq. 6.81), which only includes the zero order terms. The zero order composite solution (Eq. 6.81) has an error of order of ε, whereas the first order composite solution (Eq. 6.94) has an error of order ε^2. Figure 6.3 shows plots of the composite solution (Eq. 6.94) and the exact solution for $\varepsilon = 0.05$. It is remarkable that these two solutions practically overlap each other. Even when the parameter ε is increased to 0.2, these two solutions agree very well.

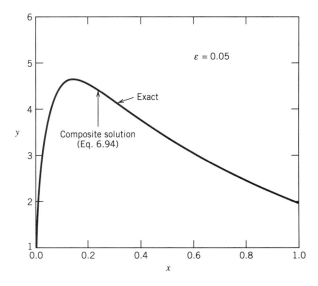

Figure 6.3 Plots of the exact solution and the first order composite solution (Eq. 6.94).

6.3.1 Matched Asymptotic Expansions for Coupled Equations

We have shown the basic steps for the method of matched asymptotic expansion. The matching principle is based on the use of the overlapping region and an intermediate variable in that region. In what follows is another simple method of matching for a class of problems, which takes the following form

$$\varepsilon \frac{dx}{dt} = f(x, y; \varepsilon) \tag{6.95}$$

and

$$\frac{dy}{dt} = g(x, y; \varepsilon) \tag{6.96}$$

subject to the initial conditions

$$t = 0; \quad x(0; \varepsilon) = \alpha(\varepsilon); \quad y(0; \varepsilon) = \beta(\varepsilon) \tag{6.97}$$

These types of coupled first order equations arise frequently in chemical engineering, where there are two dependent variables and their dynamics are coupled through the nonlinear functions f and g, but the dynamic evolution of one variable is faster than the other (in this case, x is faster).

The outer variable for this problem is t, which is sometimes called *slow time* in the literature. The appropriate measure of the change of the dependent variable in the initial short period is an inner variable (sometimes called *fast time*), which is defined as

$$t^* = \frac{t}{\varepsilon} \tag{6.98}$$

In effect, the time variable is broken into two regions, the outer region and the inner region. In the inner region, the evolution of the x-variable is observed, whereas the evolution of the variable y is only seen in the outer region. The expansions associated with the inner and outer regions are called the inner and outer expansions, respectively.

Outer Expansion

First, let us construct the outer expansion with the independent variable t. The outer asymptotic expansions are assumed to take the following form

$$x^{(0)}(t; \varepsilon) = x_0^{(0)}(t) + \varepsilon x_1^{(0)}(t) + \cdots \tag{6.99a}$$

$$y^{(0)}(t; \varepsilon) = y_0^{(0)}(t) + \varepsilon y_1^{(0)}(t) + \cdots \tag{6.99b}$$

where the superscript (o) denotes the outer solution.

Next, we substitute the above expansions into Eqs. 6.95 and 6.96 and equate coefficients of the like powers of ε^j. We then obtain a set of subproblems; the zero-order for these are

$$f\left(x_0^{(0)}, y_0^{(0)}; 0\right) = 0 \tag{6.100a}$$

and

$$\frac{dy_0^{(0)}}{dt} = g\left(x_0^{(0)}, y_0^{(0)}; 0\right) \tag{6.100b}$$

The curve (expressed as y versus x) defined in Eq. 6.100a relates the *algebraic* relationship between the two state variables, that is, if y takes some value, x then responds instantaneously to the value of y according to the algebraic relationship. Let us now assume that we can solve the subproblems for the outer region, that is, $y_j^{(0)}$ are obtained. There are constants of integration carried by outer solutions, simply because the outer solutions are not valid in the inner region, and hence, they *do not* satisfy the imposed initial conditions (Eq. 6.97).

Inner Expansion

In terms of the inner variable t^* (Eq. 6.98), the governing equations (Eqs. 6.95, 6.96) can be seen to be

$$\frac{dx}{dt^*} = f(x, y; \varepsilon) \tag{6.101a}$$

and

$$\frac{dy}{dt^*} = \varepsilon g(x, y; \varepsilon) \tag{6.101b}$$

Now we assume that the inner expansions take the following forms

$$x^{(i)}(t^*; \varepsilon) = x_0^{(i)}(t^*) + \varepsilon x_1^{(i)}(t^*) + \cdots \tag{6.102a}$$

$$y^{(i)}(t^*; \varepsilon) = y_0^{(i)}(t^*) + \varepsilon y_1^{(i)}(t^*) + \cdots \tag{6.102b}$$

where the superscript (i) denotes inner solution.

We assume that the initial conditions also have expansions

$$t^* = 0; \quad x(0; \varepsilon) = \alpha_0 + \varepsilon \alpha_1 + \cdots \tag{6.103a}$$

$$t^* = 0; \quad y(0; \varepsilon) = \beta_0 + \varepsilon \beta_1 + \cdots \tag{6.103b}$$

Now, the corresponding initial conditions for the coefficients of the inner expansions are

$$t^* = 0; \quad x_j^{(i)}(0) = \alpha_j; \quad y_j^{(i)}(0) = \beta_j \tag{6.104}$$

for $j = 1, 2, \ldots$.

If we substitute the inner expansions (Eq. 6.102) into Eqs. 6.101 and then equate the coefficients of like powers of ε^j, we obtain a set of subproblems, and we shall assume that their solutions can be obtained by some analytical means. The inner expansions are completely known because their initial conditions are given (Eqs. 6.104). Unlike the inner expansions, the outer expansions still contain unknown constants of integration. These must be found from the matching between the inner expansions and the outer expansions.

Matching

To carry out the matching procedure between the inner and outer expansions, the outer solutions are first written in terms of the inner variable, t^*, and then are expanded using a Taylor series with respect to t^* around the point $t^* = 0$. The final stage is to equate to the Taylor expanded outer solution to the inner expansions in the limit when t^* tends to infinity. The terms that are matched between the inner and outer expansions are called the *common parts*.

The matching of the derivatives in the Taylor expansions is straightforward and yields the following common parts (Vasileva 1963)

$$\sum_{j=0}^{n} \frac{(t^*)^j}{(j)!} \frac{d^j x_{n-j}^{(0)}(0)}{d(t^*)^j} = x_n^{(i)}(t^*) \tag{6.105a}$$

$$\sum_{j=0}^{n} \frac{(t^*)^j}{(j)!} \frac{d^j y_{n-j}^{(0)}(0)}{d(t^*)^j} = y_n^{(i)}(t^*) \tag{6.105b}$$

as $t^* \to \infty$.

The terms given in the above equations are the common parts and will be used in the construction of the composite solution.

The initial conditions for the outer solution in terms of the completely determined inner solutions are

$$y_n^{(0)}(0) = \lim_{t^* \to \infty} \left[y_n^{(i)}(t^*) + \sum_{k=1}^{n} \frac{(-t^*)^k}{k!} \frac{d^k y_n^{(i)}(t^*)}{d(t^*)^k} \right] \tag{6.106a}$$

$$x_n^{(0)}(0) = \lim_{t^* \to \infty} \left[x_n^{(i)}(t^*) + \sum_{k=1}^{n} \frac{(-t^*)^k}{k!} \frac{d^k x_n^{(i)}(t^*)}{d(t^*)^k} \right] \tag{6.106b}$$

EXAMPLE 6.4

To illustrate the procedure for coupled equations, we apply the above technique to a CSTR problem experiencing a slow catalyst decay, so that the deactivation rate is proportional to the reactant concentration (parallel deactivation). Let x be the reactant concentration and y be the catalyst activity. The reaction rate is represented by the product xy and the rate of catalyst decay is given by εxy ($\varepsilon \ll 1$), which is taken to be slower than the main reaction rate.

The mass balance equations are

$$\frac{dx}{dt^*} = 1 - x - xy \tag{6.107a}$$

and

$$\frac{dy}{dt^*} = -\varepsilon xy \tag{6.107b}$$

subject to

$$t^* = 0: \quad x = 0; \quad y = 1 \tag{6.108}$$

We can take the inner solutions to have the following expansions

$$x^{(i)}(t^*; \varepsilon) = x_0^{(i)}(t^*) + \varepsilon x_1^{(i)}(t^*) + \cdots \tag{6.109a}$$

$$y^{(i)}(t^*; \varepsilon) = y_0^{(i)}(t^*) + \varepsilon y_1^{(i)}(t^*) + \cdots \tag{6.109b}$$

When we substitute these expansions into Eqs. 6.107 and follow the usual procedure, we obtain a set of subproblems. Solving the first two, we have

$$x_0^{(i)}(t^*) = \frac{(1 - e^{-2t^*})}{2} \tag{6.110a}$$

$$y_0^{(i)}(t^*) = 1 \tag{6.110b}$$

$$x_1^{(i)}(t^*) = \frac{t^*}{8} + \left(\frac{t^*}{4}\right)e^{-2t^*} - \left(\frac{(t^*)^2}{8}\right)e^{-2t^*}$$

$$- \left(\frac{1}{8}\right)(1 - e^{-2t^*}) - e^{-2t^*}\frac{(1 - e^{-2t^*})}{16} \tag{6.111a}$$

$$y_1^{(i)}(t^*) = -\frac{t^*}{2} + \frac{(1 - e^{-2t^*})}{4} \tag{6.111b}$$

We see that the inner solutions are completely defined.

Now we turn to the outer solutions. The time variable for the outer region is taken as

$$t = \varepsilon t^* \tag{6.112}$$

the mass balance equations (Eq. 6.107) become, in terms of t,

$$\varepsilon \frac{dx}{dt} = 1 - x - xy \tag{6.113a}$$

$$\frac{dy}{dt} = -xy \tag{6.113b}$$

We can write the outer expansions in the following form

$$x^{(0)}(t;\varepsilon) = x_0^{(0)}(t) + \varepsilon x_1^{(0)}(t) + \cdots \qquad (6.114a)$$

$$y^{(0)}(t;\varepsilon) = y_0^{(0)}(t) + \varepsilon y_1^{(0)}(t) + \cdots \qquad (6.114b)$$

The substitution of these expansions into Eqs. 6.113 will yield a set of subproblems. To find the initial conditions for $y_j^{(0)}(0)$ we apply Eq. 6.106, and get

$$y_0^{(0)}(0) = \lim_{t^* \to \infty} y_0^{(i)}(t^*) = 1 \qquad (6.115a)$$

$$y_1^{(0)}(0) = \lim_{t^* \to \infty} \left[y_1^{(i)}(t^*) - t^* \frac{dy_1^{(i)}(t^*)}{dt^*} \right] = \frac{1}{4} \qquad (6.115b)$$

Knowing the initial conditions, the outer solutions can be completely obtained

$$x_0^{(0)} = \frac{1}{\left(1 + y_0^{(0)}\right)} \qquad (6.116a)$$

$$\left(1 - y_0^{(0)}\right) + \ln\left(\frac{1}{y_0^{(0)}}\right) = t \qquad (6.116b)$$

$$x_1^{(0)} = -\frac{y_0^{(0)}}{\left(1 + y_0^{(0)}\right)^4} - \frac{y_0^{(0)}}{\left(1 + y_0^{(0)}\right)^4} \qquad (6.117a)$$

$$y_1^{(0)} = \frac{y_0^{(0)}}{\left(1 + y_0^{(0)}\right)^2} \qquad (6.117b)$$

Now that the inner and outer solutions are known, the composite solutions can be obtained by adding the inner and outer solutions and subtracting the common parts, which are given in Eq. 6.105. The first order composite solutions are

$$y^c = y_0^{(0)} + \varepsilon \left[\frac{y_0^{(0)}}{\left(1 + y_0^{(0)}\right)^2} - \left(\frac{1}{4}\right)e^{-2t} \right] + O(\varepsilon^2) \qquad (6.118a)$$

$$x^c = \left[\frac{1}{1 + y_0^{(0)}} - \left(\frac{1}{2}\right)e^{-2t} \right]$$

$$\qquad\qquad (6.118b)$$

$$+ \varepsilon \left[\left(\frac{t}{4}\right)e^{-2t} - \left(\frac{t^2}{8}\right)e^{-2t} + \left(\frac{1}{8}\right)e^{-2t} - \frac{e^{2t}}{16}(1 - e^{-2t}) \right.$$

$$\left. - \frac{2y_0^{(0)}}{\left(1 + y_0^{(0)}\right)^4} \right] + O(\varepsilon^2)$$

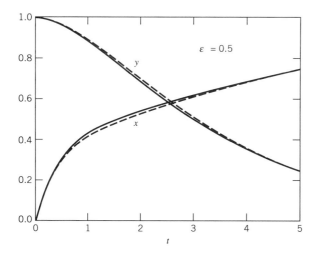

Figure 6.4 Plots of the numerically exact solution (continuous line) and the first order composite solutions (Eqs. 6.118) for $\varepsilon = 0.5$.

Figure 6.4 shows the comparison between the composite asymptotic solution and the numerically exact solution for $\varepsilon = 0.5$. It is useful to note that the asymptotic solution agrees quite well with the exact solution even when $\varepsilon = 0.5$.

We have shown, using the method of matched asymptotic expansions, that in the outer domain there is an adjustable variable, and in the inner region there is another such variable. The composite solution is, therefore, a function of these two variables. Exploitation of this function is the essential idea behind the Multiple Time Scale method. Interested readers should refer to Nayfeh (1973) for exposition of this technique.

6.4   REFERENCES

1. Abramowitz, A., and I. A. Stegun, *Handbook of Mathematical Functions*, U.S. Government Printing Office, Washington, D.C. (1964).
2. Aziz, A., and T. Y. Na. *Perturbation Methods in Heat Transfer*. Springer-Verlag, Berlin (1984).
3. Latta, G. E., "Advanced Ordinary Differential Equations," Lecture Notes, Stanford University (1964).
4. Lighthill, M. J., "A Technique for Rendering Approximate Solutions to Physical Problems Uniformly Valid," *Philosophical Mag.*, 40, 1179–1201 (1949).
5. Nayfeh, A. H. *Perturbation Methods*. John Wiley & Sons, Inc., New York (1973).
6. Van Dyke, M., *Perturbation Methods in Fluid Mechanics*. The Parabolic Press, Stanford, California (1975).
7. Vasil'eva, A. B., "Asymptotic Behavior of Solutions to Certain Problems Involving Nonlinear Differential Equations. Containing a Small Parameter Multiplying the Highest Derivatives," *Russian Mathematical Surveys*, 18, 13–84 (1963).

6.5 PROBLEMS

6.1$_2$. To solve the heat conduction problem for a slab geometry in Example 11.4 by the method of finite integral transform (or alternately by the Laplace transform or the separation of variables method), it was necessary to find eigenvalues for the transcendental equation given by Eq. 11.88b, rewritten here for completeness

$$\xi \tan(\xi) = Bi$$

where Bi is the Biot number for heat transfer, and ξ is the eigenvalue.
(a) For the small Biot number ($Bi \ll 1$), assume that ξ has the asymptotic expansion,

$$\xi = \xi_0 + Bi\xi_1 + Bi^2\xi_2 + Bi^3\xi_3 + \cdots$$

Then substitute this expansion into the transcendental equation and make use of the following Taylor series for the trigonometric tan function

$$\tan(x) = x + \frac{x^3}{3} + \frac{2}{15}x^5 + \frac{17}{315}x^7 + \cdots \qquad \text{for } x \ll 1$$

to show that the zero order solution is

$$\tan(\xi_0) = 0$$

hence, $\xi = n\pi$.
(b) Solve for the higher order subproblems to show that ξ has the asymptotic solution,

$$\xi = \xi_0 + Bi\left(\frac{1}{\xi_0}\right) + Bi^2\left(-\frac{1}{\xi_0^3}\right) + Bi^3\left(\frac{2}{\xi_0^5} - \frac{1}{3\xi_0^3}\right) + O(Bi^4)$$

This asymptotic expansion works for all eigenvalues ($n\pi$), except the first one, that is, $\xi_0 = 0$. For this eigenvalue, the second and subsequent terms are more singular than the first one. This problem of growing in singularity is a common problem in perturbation methods and will be dealt with in the next homework problem.

6.2$_3$. The solution to the last homework problem works well for all eigenvalues except the first one. This homework will consider the solution to this first eigenvalue. For small values of Bi, it is expected that the first eigenvalue is a small number as indicated by Problem 6.1.
(a) To solve for the first eigenvalue when the Biot number is small, set

$$\xi = Bi^\nu \xi_0$$

where ξ_0 has an order of unity and Bi^ν represents the magnitude of the smallness of the first eigenvalue. Subsitute this into the transcendental equation, $\xi \tan(\xi) = Bi$, to show that $\nu = 1/2$ and $\xi_0 = 1$. This means that the leading order solution for the first eigenvalue is

$$\xi = Bi^{1/2}$$

(b) To obtain the asymptotic solution for the first eigenvalue beyond the leading order solution given in part (a), assume that it has the asymptotic expansion in terms of the small parameter Bi

$$\xi = Bi^{1/2}\left[1 + Bi\xi_1 + Bi^2\xi_2 + Bi^3\xi_3 + \cdots\right]$$

Again substitute this asymptotic expansion into the transcendental equation, and equate the like powers of Bi, and then show that

$$\xi_1 = -\frac{1}{6}; \qquad \xi_2 = \frac{11}{360}; \qquad \xi_3 = -\frac{1}{432}$$

Compute the approximate first eigenvalue for $Bi = 0.01$, 0.1, and 1.

(c) To compare the approximate solution with the "exact" solution, we wish to solve the transcendental equation numerically using Newton–Raphson to obtain the exact solution. Rewrite the transcendental equation in terms of sin and cos as

$$f = \xi \sin(\xi) - Bi \cos(\xi) = 0$$

We do this because the $\sin(\varepsilon)$ and $\cos(\varepsilon)$ functions do not tend to infinity, as does the $\tan(\varepsilon)$ function. Use the Newton–Raphson formula given in Appendix A to show that the iteration equation to obtain the "numerically exact" solution for the eigenvalue is

$$\xi^{(k+1)} = \xi^{(k)} - \frac{f^{(k)}}{f'^{(k)}}$$

where $\xi^{(k)}$ and $\xi^{(k+1)}$ are the kth and $(k + 1)$th iterated eigenvalues, respectively, and

$$f'^{(k)} = \xi^{(k)} \cos\left(\xi^{(k)}\right) + (1 + Bi)\sin\left(\xi^{(k)}\right)$$

Write a program and perform the computations for the "numerically exact" solutions to give values of the first eigenvalue for $Bi = 0.01$, 0.1, and 1 equal to 0.099833639, 0.3110528, and 0.86033, respectively, using relative percentage error of less than 1.d-06, that is, the Nth iterated solution will be accepted as the "numerically exact" solution when

$$\left|\frac{\xi^{(N)} - \xi^{(N-1)}}{\xi^{(N)}}\right| \times 100 \ll 0.000001$$

(d) Compare the "numerically exact" solution obtained for $Bi = 1$ and the approximate solution obtained in part (b) to show that the relative percentage error is 0.14%. This demonstrates that the asymptotic solution, derived for small Bi, can be useful even when Bi is of order of unity.

(e) Calculate the approximate solution when $Bi = 5$. Use the approximate solution obtained in part (b) as the initial guess in the Newton–Raphson scheme to show that the "numerically exact" solution can be achieved within *two* iterations!

6.3₃. The last two problems solve the asymptotic solution for the transcendental equation $\xi \tan(\xi) = Bi$ when Bi is a small number.

(a) To deal with the same transcendental equation when Bi is a large number, use the usual perturbation approach with the asymptotic expansion

$$\xi = \xi_0 + \frac{1}{Bi}\xi_1 + \frac{1}{Bi^2}\xi_2 + \frac{1}{Bi^3}\xi_3 + \cdots$$

and substitute this expansion into the governing equation to show that the coefficients are

$$\xi_0 = \left(n + \frac{1}{2}\right)\pi; \quad \xi_1 = -\xi_0; \quad \xi_2 = \xi_0;$$

$$\xi_3 = \frac{\xi_0\left(\xi_0^2 - 3\right)}{3}; \quad \xi_4 = -\frac{\xi_0\left(4\xi_0^2 - 3\right)}{3}$$

(b) Use the Newton–Raphson program of Problem 6.2 to compute the first eigenvalue when $Bi = 100$, 10, and 1, and compare the asymptotic solutions obtained in part (a) with the corresponding exact solutions.

6.4₃. Repeat Problems 6.1, 6.2, and 6.3 with the transcendental equation

$$\xi \cot(\xi) - 1 = -Bi$$

obtained when solving the heat conduction problem in a spherical object by the method of separation of variables.

Show that for small Bi number, the first eigenvalue is

$$\xi = Bi\left(3 - Bi\frac{3}{5} + \cdots\right)$$

Hint: The following Taylor series expansion for cot is useful:

$$\cot(\xi) = \frac{1}{\xi} - \frac{\xi}{3} - \frac{\xi^2}{45} + \cdots$$

6.5₂. Use the perturbation method to find the root of the cubic equation

$$\varepsilon z^3 = a - z$$

where a is a constant of order of unity, and ε is a small number ($\varepsilon \ll 1$).
(a) Use the asympotic expansion for z

$$z = z_0 + \varepsilon z_1 + \varepsilon^2 z_2 + \cdots$$

and substitute it into the cubic equation to show that the coefficients are

$$z_0 = a; \quad z_1 = -a^3; \quad z_2 = 3a^5$$

The regular perturbation yields only one real root to the cubic equation. The other two roots are not found because of the simple fact that the cubic term was not retained in the solution for the zero-order coefficient, z_0, explaining why the regular perturbation method fails to locate the other two roots.
(b) To find the other two roots, it is essential to retain the cubic term in the zero-order subproblem. To do this, start with the expansion

$$z = \frac{1}{\varepsilon^n} [y_0 + \varepsilon^m y_1 + \cdots]$$

where $1/\varepsilon^n$ represents the magnitude of the two missing roots. At this stage, the exponents n and m are unknown. Substitute this expansion into the cubic equation and show that in order to keep the cubic term in the zero-order solution, which is essential to find the two missing roots, the exponents n and m must be

$$n = m = \frac{1}{2}$$

(c) Solve the zero-order and first-order subproblems to show that the coefficients y_0 and y_1 are to take the form

$$y_0 = \pm i; \qquad y_1 = -\frac{a}{2}$$

where i is the imaginary number, $i = \sqrt{-1}$.
Thus, the missing two roots are complex conjugates.
(d) Compare the asymptotic solutions with the exact solutions[1] for $\varepsilon = 0.01$ and 0.1. Discuss your results.

---

[1]For the cubic equation of the form (Abramowitz and Stegun 1964)

$$z^3 + a_2 z^2 + a_1 z + a_0 = 0$$

6.6₂. Repeat Problem 6.5 with the cubic equation

$$\varepsilon z^3 = z - a \qquad (\varepsilon \ll 1)$$

and show that the three roots to this cubic equation have the following asymptotic expansions:

$$z_1 = a + \varepsilon(a^3) + \varepsilon^2(3a^5) + \cdots$$

$$z_2 = \frac{1}{\sqrt{\varepsilon}}\left[1 + \sqrt{\varepsilon}\left(-\frac{a}{2}\right) + \cdots\right]$$

$$z_3 = \frac{1}{\sqrt{\varepsilon}}\left[-1 + \sqrt{\varepsilon}\left(-\frac{a}{2}\right) + \cdots\right]$$

6.7₂. Modeling of diffusion coupled with a reaction of order p inside a catalyst particle poses difficulties owing to the nonlinear reaction term (e.g., see Example 2.7).

(a) If the particle has a slab geometry, set up the coordinate frame with the origin being at the center of the particle. Show that the differential mass balance equation for the reactant is

$$D_e \frac{d^2C}{dr^2} - kC^p = 0$$

where C represents reactant concentration

$$C \xrightarrow{k} \text{Products}$$

If the external film resistance surrounding the particle is negligible

---

let

$$q = \frac{a_1}{3} - \frac{a_2^2}{9}; \quad r = \frac{(a_1 a_2 - 3a_0)}{6} - \frac{a_2^3}{27}; \quad \Delta = q^3 + r^2$$

If $\Delta > 0$, we have one real root and a pair of complex conjugate roots.
If $\Delta = 0$, all roots are real and at least two are equal.
If $\Delta < 0$, all roots are real.

The three roots are

$$z_1 = (s_1 + s_2) - \frac{a_2}{3}$$

$$z_2 = -\frac{(s_1 + s_2)}{2} - \frac{a_2}{3} + \frac{i\sqrt{3}}{2}(s_1 - s_2)$$

$$z_3 = -\frac{(s_1 + s_2)}{2} - \frac{a_2}{3} - \frac{i\sqrt{3}}{2}(s_1 - s_2)$$

where

$$s_1 = \left[r + \sqrt{\Delta}\right]^{1/3}; \quad s_2 = \left[r - \sqrt{\Delta}\right]^{1/3}$$

compared to the internal diffusion, show that the boundary conditions for the above mass balance equation are

$$r = 0; \qquad \frac{dC}{dr} = 0$$

$$r = R; \qquad C = C_0$$

where R is the half thickness of the slab catalyst and C_0 is the external bulk concentration, which can be assumed constant.

(b) Convert the mass balance equation and its boundary conditions to the following nondimensional form

$$\frac{d^2 y}{dx^2} - \phi^2 y^p = 0$$

$$x = 0; \qquad \frac{dy}{dx} = 0$$

$$x = 1; \qquad y = 1$$

What are the required definitions of the nondimensional variables x, y, and ϕ?

(c) The quantity of interest in analyzing the relative importance between diffusional resistance and chemical reaction is called the effectiveness factor, η, which is given by

$$\eta = \frac{1}{\phi^2} \frac{dy}{dx}\bigg|_{x=1}$$

The effectiveness factor is the ratio between the actual reaction rate per unit catalyst particle and the ideal reaction rate when there is no diffusional resistance. Derive the above expression for the effectiveness factor.

(d) For a first order (linear) reaction, show the following solutions for y and η are obtained

$$y(x) = \frac{\cosh(\phi x)}{\cosh(\phi)}; \qquad \eta = \frac{\tanh(\phi)}{\phi}$$

(e) For a general reaction order, the explicit analytical solution for the effectiveness factor is not possible, but perturbation methods can be applied here to derive asymptotic solutions when ϕ is either small or large. When ϕ is small, the problem can be handled by the regular perturbation method. Assume that the solution for y possesses the asymptotic expansion

$$y(x; \phi) = y_0(x) + \phi y_1(x) + \phi^2 y_2(x) + \cdots$$

Show that the effectiveness factor for a reaction of order p is

$$\eta = 1 - \frac{p}{3}\phi^2 + \frac{2p^2 + p(p-1)}{15}\phi^4 + O(\phi^6)$$

(f) When the parameter ϕ is large, a singular perturbation problem arises. This is caused by the small parameter multiplying the second derivative term in the mass balance equation, which is a source of nonuniformity, as discussed in Section 6.2.4. Let

$$\varepsilon = \frac{1}{\phi}$$

so the mass balance equation will become

$$\varepsilon^2 \frac{d^2 y}{dx^2} - y^p = 0$$

Show that the straightforward application of the expansion

$$y(x; \varepsilon) = y_0 + \varepsilon y_1 + \cdots$$

will yield the solution

$$y_0 = y_1 = 0$$

This solution clearly does not satisfy the boundary condition at the particle surface. However, the above solution is valid for most of the interior region of the catalyst, except for a thin zone very close to the particle surface. This solution, therefore, is called the *outer solution*.

(g) To find the solution that is valid close to the particle surface, the coordinate must be stretched (magnified) near the surface as

$$\xi = \frac{1 - x}{\varepsilon^n}$$

where ε^n is the boundary layer thickness, and n is unknown at this stage and must be deduced based on further analysis. In terms of the new stretched variable, show that the new mass balance equation is

$$\frac{\varepsilon^2}{\varepsilon^{2n}} \frac{d^2 y}{d\xi^2} = y^p$$

Since the inner solution is valid near the particle surface, it is essential to retain the second order derivative term in the zero order solution; this means we must select $n = 1$. The new equation now becomes

$$\frac{d^2 y}{d\xi^2} = y^p$$

Show that the proper boundary conditions for this new equation valid in the boundary layer are

$$\xi = 0; \qquad y = 1$$

$$\xi \to \infty; \qquad y = \frac{dy}{dx} = 0$$

Hint: The second boundary condition is a result of matching between the inner and the outer solution.

(h) If the solution valid in the boundary layer is to have the expansion

$$y^{(i)}(\xi; \varepsilon) = y_0^{(i)}(\xi) + \varepsilon y_1^{(i)}(\xi) + \cdots$$

then show that the equation for the leading order solution is

$$\frac{d^2 y_0^{(i)}}{d\xi^2} = \left[y_0^{(i)} \right]^p$$

and its associated boundary conditions are

$$\xi = 0; \qquad y_0^{(i)} = 1$$

$$\xi \to \infty; \qquad y_0^{(i)} = \frac{dy_0^{(i)}}{d\xi} = 0$$

(i) Integrate the relation in part (h) (use the method in Example 2.7) and show that the solution for the effectiveness factor for large ϕ is

$$\eta = \frac{1}{\phi} \sqrt{\frac{2}{p + 1}}$$

6.8₃. If the catalyst particle in Problem 6.7 is cylindrical, and if the transfer to end caps is ignored,

(a) Show that when the reaction is first order, the solutions for the concentration distribution and the effectiveness factor are (c.f. Examples 3.3 and 3.6)

$$y(x) = \frac{I_0(\phi x)}{I_0(\phi)}$$

$$\eta = \frac{2 I_1(\phi)}{\phi I_0(\phi)}$$

(b) When the reaction is of order p, show that the solutions for effectiveness factor for small and large ϕ are

$$\eta = 1 - \frac{p}{8}\phi^2 + \frac{2p^2 + p(p-1)}{96}\phi^4 + \cdots$$

and

$$\eta = \frac{2}{\phi}\sqrt{\frac{2}{p+1}}$$

respectively.

6.9₃. If we redo Problem 6.7 for spherical particles, show that the solutions for the effectiveness factor for small ϕ and large ϕ are

$$\eta = 1 - \frac{p}{15}\phi^2 + \frac{2p^2 + p(p-1)}{315}\phi^4 + \cdots$$

and

$$\eta = \frac{3}{\phi}\sqrt{\frac{2}{p+1}}$$

respectively.

6.10₃. Apply the regular perturbation method to solve the following first order ordinary differential equation

$$(1 + \varepsilon y)\frac{dy}{dx} + y = 0$$

subject to

$$x = 1; \qquad y = 1$$

(a) Show that the asymptotic solution has the form

$$y = e^{1-x} + \varepsilon[e^{1-x} - e^{2(1-x)}] + \cdots$$

(b) Compare the approximate solution with the exact solution for $\varepsilon = 0.01$ and 0.1

$$\ln(y) + \varepsilon y = -x - (1 + \varepsilon)$$

6.11₃. Modeling of mass transport through a membrane with variable diffusivity usually gives rise to the nondimensional form

$$\frac{d}{dx}\left[f(y)\frac{dy}{dx}\right] = 0$$

subject to the following boundary conditions:

$$x = 0; \quad y = 0$$
$$x = 1; \quad y = 1$$

(a) For $f(y) = 1 + \varepsilon(y - 1)$, show by the method of regular perturbation that the asymptotic solution is

$$y = 1 - x + \varepsilon\left(\frac{x}{2} - \frac{x^2}{2}\right) + \cdots$$

(b) Compare this asymptotic solution with the following exact solution for $\varepsilon = 0.01$ and 0.1,

$$\frac{y(1 - \varepsilon) + \frac{\varepsilon}{2}y^2}{1 - \frac{\varepsilon}{2}} = x$$

6.12₃. Use the regular perturbation method to solve the nonlinear ordinary differential equation

$$(x + \varepsilon y)\frac{dy}{dx} + y = 0$$

subject to

$$x = 1; \quad y = 1$$

(a) Show that the asymptotic solution is

$$y = \frac{1}{x} + \varepsilon\left(\frac{x^2 - 1}{2x^3}\right) + \varepsilon^2\left(-\frac{x^2 - 1}{2x^5}\right) + \cdots$$

(b) The exact solution is given by

$$xy = 1 + \frac{\varepsilon}{2}(1 - y^2)$$

Compare the asymptotic solution obtained in part (a) with the exact solution for $\varepsilon = 0.1$. Discuss the diverging behavior of the asymptotic solution near $x = 0$. Does the asymptotic solution behave better near $x = 0$ as more terms are retained? Observing the asymptotic solution, it shows that the second term is more singular (secular) than the first term, and the third term is more singular than the second term. Thus, it is seen that just like the cubic equations dealt with in Problems 6.5 and 6.6, where the cubic term is multiplied by a small parameter, this differential equation also suffers the same growth in singular behavior.

(c) If the equation is now rearranged such that x is treated as the dependent variable and y as the independent variable, that is, their roles are reversed, show that the governing equation is

$$y\frac{dy}{dx} = -(x + \varepsilon y)$$

Applying the regular perturbation to this new equation to show that the asymptotic solution will have the form

$$x = \frac{1}{y} + \varepsilon\frac{(1 - y^2)}{2y}$$

which is in fact the exact solution, a pure coincidence.

6.13₃. The differential equation having the type dealt with in Problem 6.12 can be solved by a variation of the regular perturbation method, called the *strained coordinates* method. The idea was initially due to Lighthill (1949). Basically, this idea is to expand the dependent as well as the independent variables in terms of a new variable. The coefficients of the expansion of the independent variable are called the *straining functions*. Regular perturbation is then applied by substituting the two expansions of the independent and dependent variables into the equation, and the crucial requirement in finding the straining function is that the higher order terms are no more singular than the preceding term. This is called *Lighthill's rule*, after Lighthill (1949). Let us now reinvestigate the equation dealt with in Problem 6.12 and expand the independent variable as

$$x = x_0 + \varepsilon x_1 + \cdots$$

Here, we allow the first coefficient, x_0, to behave as the new independent variable, and the subsequent coefficients will be a function of x_0; that is, the independent variable x is strained very slightly

$$x = x_0 + \varepsilon x_1(x_0) + \cdots$$

If x_1 is set to zero, the strained coordinates method will become the traditional regular perturbation method.

Assume that the dependent variable y has the following asymptotic expansion with coefficents being the function of the new independent variable x_0; that is,

$$y = y_0(x_0) + \varepsilon y_1(x_0) + \cdots$$

(a) Show by the chain rule of differentiation that the derivative will have the asymptotic expansion

$$\frac{dy}{dx} = \frac{dy}{dx_0} \cdot \frac{dx_0}{dx} = \frac{dy_0}{dx_0} + \varepsilon\left(\frac{dy_1}{dx_0} - \frac{dy_0}{dx_0}\frac{dx_1}{dx_0}\right) + \cdots$$

(b) Substitute the expansions for x, y, and dy/dx into the governing equation and equate like powers of ε to show that the zero-order and first-order subproblems are

$$\varepsilon^0: \qquad x_0\frac{dy_0}{dx_0} + y_0 = 0$$

and

$$\varepsilon^1: \qquad x_0\frac{dy_1}{d_0} + y_1 = -(x_1 + y_0)\frac{dy_0}{dx_0} + x_0\frac{dy_0}{dx_0}\frac{dx_1}{dx_0}$$

(c) In principle, we can solve for y_0 and y_1 from the above two subproblems with x_1 chosen such that the coefficient y_1 is not more singular than the preceding coefficient y_0. To completely solve for y_0 and y_1, we need to find their conditions. Use the given initial condition $x = 1$, $y = 1$ to the expansion for x to show that

$$1 = x_0^* + \varepsilon x_1(x_0^*)$$

The value x_0^* is the value of x_0 at $x = 1$. Since x_0^* is very close to unity, assume x_0^* to have the asymptotic expansion

$$x_0^* = 1 + \varepsilon\alpha$$

and substitute this into the initial condition to show that

$$\alpha = -x_1(1)$$

Hence,

$$x_0^* = 1 - \varepsilon x_1(1)$$

(d) To find the condition for y_0 and y_1, use the initial condition and then substitute it into the expansion for y to show that

$$y_0(1) = 1 \qquad \text{and} \qquad y_1(1) = x_1(1)\frac{dy_0(1)}{dx_0}$$

Note that the initial condition for y_1 depends on the straining function x_1, which is still unknown at this stage. This will be determined during the course of analysis, and the Lighthill's requirement is that the subsequent term not be more singular than the preceeding term.

(e) Show that the solution for y_0 is

$$y_0 = \frac{1}{x_0}$$

(f) Use this solution for y_0 into the first-order subproblem to show that the equation for y_1 takes the form

$$\frac{d(x_0 y_1)}{dx_0} = -\frac{d}{dx_0}\left(\frac{x_1}{x_0} + \frac{1}{2x_0^2}\right)$$

then show that the solution for y_1 is

$$y_1 = \frac{K}{x_0} - \frac{1}{x_0^2}\left(x_1 + \frac{1}{2x_0}\right)$$

where K is the constant of integration.

(g) Show that the simplest form of x_1, taken to ensure that the function y_1 is not more singular than the function y_0, is

$$x_1 = -\frac{1}{2x_0}$$

then show that the initial condition for y_1 is

$$y_1(1) = \frac{1}{2}$$

and then prove that the complete solution for y_1 is

$$y_1 = \frac{1}{2x_0}$$

Thus, the complete solutions are

$$y = \frac{1}{x_0} + \varepsilon\frac{1}{2x_0}; \qquad x = x_0 - \varepsilon\frac{1}{2x_0}$$

which are implicit because of the presence of x_0.

(h) Eliminate x_0 between the two solutions in part (g) to show that the final implicit asymptotic solution is

$$\frac{2\varepsilon y^2}{2 + \varepsilon} + 2xy - (2 + \varepsilon) = 0$$

Compute this solution for $\varepsilon = 0.1$ and 1 and compare the results with the exact solution. Discuss your observations especially in the neighborhood of $x = 0$.

Further exposition of this method of strained coordinates can be found in Nayfeh (1973) and Aziz and Na (1984).

6.14*. A well–stirred reactor containing solid catalysts can be modelled as a perfect stirred tank reactor, that is, concentration is the same everywhere inside the reactor. When such catalysts are very small, the diffusional resistance inside the catalyst can be ignored. The catalyst is slowly deactivated by the presence of the reactant (this is called *parallel deactivation*). The rate of the reaction per unit volume of the reactor is taken to follow a second order kinetics with respect to the reactant concentration as

$$R = kC^2a$$

where C is the reactant concentration, k is the reaction rate constant, and a is the catalyst activity, which is unity (i.e., fresh catalyst) at $t = 0$.
(a) Show that the mass balance equation for the reactant is

$$V\frac{dC}{dt} = F(C_0 - C) - VkC^2a$$

where F is the volumetric flow rate, V is the reactor volume, and C_0 is the inlet reactant concentration.
  The catalyst activity declines with time, and the equation describing such change is assumed to take the form

$$\frac{da}{dt} = -k_d Ca$$

where k_d is the rate constant for deactivation.
(b) Show that the mass balance equations in nondimensional form can be cast to the following form

$$\frac{dy}{d\tau} = \alpha(1 - y) - y^2 a$$

and

$$\frac{da}{d\tau} = -\varepsilon y a$$

where y is the nondimensional concentration and τ is the nondimensional time given as

$$y = \frac{C}{C_0}; \qquad \tau = (kC_0)t$$

and ε (small parameter) and α are

$$\varepsilon = \frac{k_d}{k} \ll 1; \qquad \alpha = \frac{F}{kVC_0}$$

(c) With respect to the time scale τ, use the perturbation method to obtain the solutions for y and a, given below

$$a_0^{(0)}(\tau) = 1$$

$$y_0^{(0)}(\tau) = \frac{m_1\left[1 - \exp\left(-\tau\sqrt{\alpha^2 + 4\alpha}\right)\right]}{2\left[1 + \left(\dfrac{m_1}{m_2}\right)\exp\left(-\tau\sqrt{\alpha^2 + 4\alpha}\right)\right]}$$

where m_1 and m_2 are given by

$$m_1 = \sqrt{\alpha^2 + 4\alpha} - \alpha; \qquad m_2 = \sqrt{\alpha^2 + 4\alpha} + \alpha$$

This represents the start-up problem (inner solution), that is, the catalyst is still fresh ($a = 1$).

(d) To obtain the solution behavior beyond the start-up period, define a slower time scale, which is slower than the time scale τ, as

$$\bar{t} = \varepsilon\tau$$

and show that the mass balance equations in terms of this new time scale are

$$\varepsilon\frac{dy}{d\bar{t}} = \alpha(1 - y) - y^2 a$$

and

$$\frac{da}{d\bar{t}} = -ya$$

(e) Apply the perturbation method on this new set of equations to show that the solutions for the slow time scale are

$$a_0^{(0)} = \frac{\alpha\left[1 - y_0^{(0)}\right]}{\left[y_0^{(0)}\right]^2}$$

and

$$\ln\left\{\frac{y_0^{(0)}(\bar{t})\left[1 - y_0^{(0)}(0)\right]}{y_0^{(0)}(0)\left[1 - y_0^{(0)}(\bar{t})\right]}\right\} + 2\left[\frac{1}{y_0^{(0)}(0)} - \frac{1}{y_0^{(0)}(\bar{t})}\right] = \bar{t}$$

where $y_0^{(o)}(0)$ is the initial condition of the zero order outer solution. This initial condition is found by matching the inner solutions obtained in part (b), and shows that

$$y_0^{(0)}(0) = \frac{1}{2}\left[\sqrt{\alpha^2 + 4\alpha} - \alpha\right]$$

(f) Knowing the inner solutions in part (b) and the outer solutions in part (d), obtain the composite solutions that are valid over the whole time domain.

PART TWO

With them the seed of wisdom did I sow,
And with my own hand wrought to make it grow;
And this was all the Harvest that I reap'd—
"I came like water, and like Wind I go."

        Rubáiyát of Omar Khayyám, XXXI.

Chapter 7

Numerical Solution Methods (Initial Value Problems)

In the previous chapter, we discussed analytical and approximate methods to obtain solutions to ordinary differential equations. When these approaches fail, the only remaining course of action is a numerical solution. This chapter and the next consider numerical methods for solving ODEs, which may not be obtainable by techniques presented in Chapters 2, 3, and 6. ODEs of initial value type (i.e., conditions are specified at one boundary in time or space) will be considered in this chapter, whereas ODEs of boundary value type (conditions are specified at two boundary points) will be considered in Chapter 8.

7.1 INTRODUCTION

Modeling of process equipment often requires a careful inspection of start-up problems that lead to differential equations of initial value type, for example, modeling of chemical kinetics in a batch reactor and modeling of a plug flow reactor. For the first example with two reactions in series

$$A \xrightarrow{k_1} B \xrightarrow{k_2} C$$

the model equations for constant volume are

$$\frac{dC_A}{dt} = -k_1 C_A^n$$

$$\frac{dC_B}{dt} = +k_1 C_A^n - k_2 C_B^m$$

and

$$\frac{dC_C}{dt} = +k_2 C_B^m$$

The batch reactor is initially filled with A, B, and C of concentrations C_{A0}, C_{B0}, and C_{C0}. Mathematically, one can write:

$$t = 0; \quad C_A = C_{A0}, \quad C_B = C_{B0}, \quad C_C = C_{C0}$$

The last equation is the condition that we impose on the differential condition at $t = 0$, that is, before the reactions are allowed to start. These are called the *initial* conditions, and the mathematical problem attendant to such conditions is called the *initial value problem* (IVP).

The plug flow reactor also gives rise to an IVP. If we write down a mass balance equation for the thin element at the position z with a thickness of Δz (see the control volume of Fig. 1.1b) and then allow the element thickness to approach zero, we obtain the following equations for the two reactions in series:

$$u \frac{dC_A}{dz} = -k_1 C_A^n$$

$$u \frac{dC_B}{dz} = +k_1 C_A^n - k_2 C_B^m$$

$$u \frac{dC_C}{dz} = +k_2 C_B^m$$

where u is the superficial velocity. The conditions imposed on these equations are the inlet concentrations of these three species at the entrance of the reactor; that is,

$$z = 0; \quad C_A = C_{A0}, \quad C_B = C_{B0}, \quad C_C = C_{C0}$$

Again, in the second example the conditions are specified at one point (the entrance), and hence the problem is the initial value problem. The similarity of the two systems can be seen more readily if we introduce the local residence time to the plug flow problem, $\tau = z/u$, we thus see the above equations are identical to the batch start-up problem, except t is replaced by τ.

Noting the form of the model equations of the last two examples, we can write the standard form for first order ODE as follows:

$$\frac{dy_i}{dt} = f_i(y_1, y_2, \ldots, y_N) \tag{7.1}$$

for $i = 1, 2, \ldots, N$, and where N is the number of equations. The independent variable t does not appear explicitly in Eq. 7.1; such a system is called *autonomous*. Otherwise, the system is called *nonautonomous* [see Eq. (7.3a)].

The initial conditions for the above equations are values of $y_i (i = 1, 2, \ldots, N)$ for an initial instant of time (usually, $t = 0$)

$$t = 0; \quad y_i(0) = \alpha_i \quad (\text{known}) \tag{7.2}$$

Equation 7.1 includes all first order nonlinear ODEs, provided the independent variable t does not appear in the RHS. Equations of the last two examples fall into the general format of Eq. 7.1.

If the argument in the RHS of the coupled ordinary differential equations contains t, such as

$$\frac{dy_i}{dt} = f_i(y_1, y_2, \ldots, y_N, t) \qquad \text{for} \qquad i = 1, 2, \ldots, N \qquad (7.3a)$$

$$t = 0; \qquad y_i(0) = \alpha_i \qquad \text{(known)} \qquad (7.3b)$$

we can nonetheless redefine the problem such that the relation of the form of Eq. 7.1 is recovered.

For such a case, we simply introduce one more dependent variable by replacing t with y_{N+1}, a new dependent variable, for which the differential equation is defined as

$$\frac{dy_{N+1}}{dt} = 1 \qquad (7.4a)$$

and

$$t = 0; \qquad y_{N+1}(0) = 0 \qquad (7.4b)$$

or in the general case,

$$t = t_0; \qquad y_{N+1}(t_0) = t_0 \qquad (7.4c)$$

Thus, we see that the new "dependent variable" is exactly equal to t.

With the introduction of the new dependent variable y_{N+1}, Eq. 7.3a can now be cast into the form

$$\frac{dy_i}{dt} = f_i(y_1, y_2, \ldots, y_N, y_{N+1}) \qquad (7.5)$$

for $i = 1, 2, \ldots, N$.

The new set now has $N + 1$ coupled ordinary differential equations (Eqs. 7.5 and 7.4). Thus, the standard form of Eq. 7.1 is recovered, and we are not constrained by the time appearing explicitly or implicitly. In this way, numerical algorithms are developed only to deal with autonomous systems.

EXAMPLE 7.1

We illustrate this with an example of heating of an oil bath with a time-dependent heat source. Let T and $Q(t)$ represent the bath temperature and heat rate, respectively. We shall take the bath to be well mixed so the temperature is uniform throughout. A heat balance on the bath gives

$$V\rho C_p \frac{dT}{dt} = Ah(T - T_a) - VQ(t)$$

where V is the bath volume and A is the heat transfer area for heat loss to ambient air at temperature T_a.

The initial temperature of the bath is T_0, before the heater is turned on; that is:

$$t = 0; \qquad T = T_0$$

The balance equation has a time variable in the RHS, so to convert it to the format of Eq. 7.1 we simply define

$$y_1 = T$$
$$y_2 = t$$

As suggested by Eq. 7.4a, the differential equation for y_2 is

$$\frac{dy_2}{dt} = 1 \qquad \therefore y_2 = t$$

and the equation for y_1 is simply the heat balance equation with t being replaced by y_2

$$V\rho C_p \frac{dy_1}{dt} = Ah(y_1 - T_a) - VQ(y_2)$$

The initial conditions for y_1 and y_2 are

$$t = 0; \qquad y_1 = T_0, \qquad y_2 = 0$$

Thus, the standard form of Eq. 7.1 is recovered.

Occasionally, we encounter problems involving higher order derivatives. To render the governing equations for these problems to the standard form of Eq. 7.1, we can proceed as follows. For illustrative purposes, we shall consider a second order differential equation, but the procedure is general and can be applied to higher order systems. If the governing equation contains second derivatives of the general form

$$y'' + F(y', y) = 0 \tag{7.6}$$

we can recover the form of Eq. 7.1 by simply defining, as discussed in Chapter 2

$$y_1 = y \tag{7.7a}$$

$$y_2 = \frac{dy}{dt} \tag{7.7b}$$

With the definition (7.7), Eq. 7.6 becomes

$$\frac{dy_2}{dt} = -F(y_2, y_1) \tag{7.8}$$

Thus, the new set of equations written in the format of Eq. 7.1 is

$$\frac{dy_1}{dt} = y_2 = f_1(y_1, y_2)$$

$$\frac{dy_2}{dt} = -F(y_2, y_1) = f_2(y_1, y_2) \tag{7.9}$$

Thus, any order equation can be expressed in the format of Eq. 7.1. We next address the issue of initial conditions. If the initial conditions of Eq. 7.6 are

$$H_i(y'(0), y(0)) = 0; \quad i = 1, 2 \tag{7.10}$$

then we can use the definition of Eq. 7.7 to convert this initial condition (7.10) in terms of the new dependent variables

$$H_i(y_2(0), y_1(0)) = 0; \quad i = 1, 2 \tag{7.11}$$

Equation 7.11 represents a set of two possibly nonlinear algebraic equations in terms of $y_1(0)$ and $y_2(0)$. This algebraic problem can be solved by trial and error, using for example the Newton–Raphson technique (Appendix A). This will yield $y_1(0)$ and $y_2(0)$, which will form the initial conditions for Eq. 7.9. At this point, we need to assume that the numerical (approximate) solution of Eq. 7.11 will give rise to an initial condition, which will produce a trajectory that is arbitrarily close to the one with the exact initial condition.

EXAMPLE 7.2

Convert the following second order differential equation

$$a(y, t)\frac{d^2 y}{dt^2} + b(y, t)\frac{dy}{dt} + c(y, t)y = f(t)$$

to the standard format of Eq. 7.1.

This is a challenging example, since the equation is second order and nonlinear. We first take care of the independent variable time by introducing y_1 as the time variable; that is,

$$\frac{dy_1}{dt} = 1$$

with the initial condition

$$t = 0; \quad y_1 = 0$$

To account for the second order derivative, we define

$$y_2 = y \qquad \text{and} \qquad y_3 = \frac{dy_2}{dt}$$

With these definitions the original differential equation becomes

$$a(y_2, y_1) \frac{dy_3}{dt} + b(y_2, y_1) y_3 + c(y_2, y_1) y_2 = f(y_1)$$

Next, we rearrange the differential equations for y_1, y_2, and y_3 as follows

$$\frac{dy_1}{dt} = 1$$

$$\frac{dy_2}{dt} = y_3$$

$$\frac{dy_3}{dt} = -\frac{b(y_2, y_1)}{a(y_2, y_1)} y_3 - \frac{c(y_2, y_1)}{a(y_2, y_1)} y_2 + \frac{f(y_1)}{a(y_2, y_1)}$$

provided that $a(y_2, y_1)$ is not equal to zero. Now, we see that the standard form of Eq. 7.1 is recovered.

The same procedure presented above for second order differential equations can be extended to an nth order differential equation and to coupled nth order differential equations, as the reader will see in the homework problems.

Thus, no matter whether the independent variable, t, is in the RHS of the differential equation or the governing equation involves higher order derivatives, we can perform elementary transformations, as illustrated in the last two examples, to convert these equations to the standard form of Eq. 7.1. The compact language of vector representation could also be used to express Eq. 7.1

$$\frac{d\mathbf{y}}{dt} = \mathbf{f}(\mathbf{y}) \tag{7.12a}$$

and

$$t = 0; \qquad \mathbf{y} = \boldsymbol{\alpha} \qquad (\text{known vector}) \tag{7.12b}$$

where the vectors are

$$\mathbf{y} = [y_1 y_2, \ldots, y_N]^T$$
$$\mathbf{f} = [f_1, f_2, \ldots, f_N]^T \tag{7.13}$$
$$\boldsymbol{\alpha} = [\alpha_1, \alpha_2, \ldots, \alpha_N]^T$$

7.2 TYPE OF METHOD

Arranging equations in a systematic format is time well spent, before the act of computation takes place. The computation methods are generally of two types: *explicit* and *implicit*. By explicit methods, we mean that if we know the value of **y** at the instant of time t_n, the calculation of the vector **y** at the next time t_{n+1} requires only the *known* values of the vector **y** and its derivatives $d\mathbf{y}/dt = \mathbf{f}(\mathbf{y})$ at the time t_n and previous times. The implicit methods, however, involve solving equations for the unknown value of the vector **y** at the time t_{n+1}.

The simplest example of the explicit type is the *Euler method* (unfortunately, it is unstable and the most inaccurate method), in which the recursive formula to calculate the vector **y** at time t_{n+1} is

$$\mathbf{y}(t_{n+1}) = \mathbf{y}(t_n) + h\mathbf{f}(\mathbf{y}(t_n)) \tag{7.14}$$

Here, we have represented the differential with

$$\frac{d\mathbf{y}}{dt} \approx \frac{\mathbf{y}(t_{n+1}) - \mathbf{y}(t_n)}{\Delta t} \tag{7.15a}$$

where

$$\Delta t = h \tag{7.15b}$$

and the average value for **f** over the interval h is taken as $\mathbf{f}(\mathbf{y}(t_n))$.

Thus, we see that the evaluation of the vector **y** at t_{n+1} only requires the known vector **y** at t_n and its derivatives at t_n, that is, $\mathbf{f}(\mathbf{y}(t_n))$. It is the simplest method and unfortunately it is unstable if the step size h is not properly chosen. This problem justifies more sophisticated methods to *ensure* numerical stability, a topic that will be discussed later.

Note that Eq. 7.14 is basically the Taylor series of $y(t_{n+1})$, where $\mathbf{y}_{n+1} = y(t_{n+1})$

$$\mathbf{y}(t_{n+1}) = \mathbf{y}(t_n) + h\frac{d\mathbf{y}(t_n)}{dt} + O(h^2)$$

$$= \mathbf{y}(t_n) + \mathbf{f}(\mathbf{y}(t_n)) + O(h^2)$$

where the order symbol O is defined in 6.2.2, and simply means that "left-out" terms have a size magnitude of order h^2 (the smaller h, the better). Comparing the above Taylor series and Eq. 7.14, the explicit Euler equation has a local truncation error of $O(h^2)$, with the local truncation error being the error incurred by the approximation over a *single* step. As time increases, the *overall* error increases. Since the number of calculations is inversely proportional to the step size, the actual accuracy of the explicit Euler method is $O(h)$.

To simplify the notation of the recursive formula, we denote the numerical value of **y** at the times t_n and t_{n+1} as $\mathbf{y}(t_n) = \mathbf{y}_n$ and $\mathbf{y}(t_{n+1}) = \mathbf{y}_{n+1}$, and the explicit Euler formula (7.14) is rewritten

$$\mathbf{y}_{n+1} = \mathbf{y}_n + h\mathbf{f}(\mathbf{y}_n) \tag{7.16}$$

The *implicit method*, to be introduced next, alleviates the stability problem inherent in the explicit method. An example of the implicit formalism is the *trapezoidal method*. Here, the derivative $d\mathbf{y}/dt$ at the time t_n is calculated using the *Trapezoidal rule*; that is,

$$\left[\frac{d\mathbf{y}}{dt}\right]_{t=t_n} = \frac{1}{2}\left[\mathbf{f}(\mathbf{y}_n) + \mathbf{f}(\mathbf{y}_{n+1})\right] \tag{7.17}$$

where the RHS is simply the average $\mathbf{f}$ at two successive times.

The recursive formula for $\mathbf{y}(t_{n+1})$ for the Trapezoidal rule is obtained by replacing as before

$$\frac{d\mathbf{y}}{dt} \approx \frac{\mathbf{y}_{n+1} - \mathbf{y}_n}{h}$$

so that:

$$\mathbf{y}_{n+1} = \mathbf{y}_n + \frac{h}{2}\left[\mathbf{f}(\mathbf{y}_n) + \mathbf{f}(\mathbf{y}_{n+1})\right] \tag{7.18}$$

Equation (7.18) represents a set of N nonlinear algebraic equations, which must be solved by a trial and error method such as Newton–Raphson or successive substitution for $\mathbf{y}_{n+1}$ (Appendix A). This is the characteristic difference between the implicit and explicit types of solution; the easier explicit method allows sequential solution one at a time, while the implicit method requires simultaneous solutions of sets of equations; hence, an iterative solution at a given time t_{n+1} is required.

The *backward Euler method* yields a simpler implicit formula, and its recursive relation to calculate $\mathbf{y}_{n+1}$ is given as

$$\mathbf{y}_{n+1} = \mathbf{y}_n + h\mathbf{f}(\mathbf{y}_{n+1}) \tag{7.19}$$

where the average $\mathbf{f}$ over the interval h is taken to be $\mathbf{f}(\mathbf{y}_{n+1})$. Again, the recursive formula is a set of nonlinear algebraic equations.

In terms of stability, the explicit Euler method is unstable if the step size is not properly chosen. The implicit methods, such as the backward Euler and the trapezoidal methods, are stable, but the solution may oscillate if the step size is not chosen small enough. The illustrative example in the next section reveals the source of the stability problem.

7.3   STABILITY

Numerical integration of a problem usually gives rise to results that are unusual in the sense that often the computed values "blow up." The best example of this so-called stability problem is illustrated in the numerical integration of $dy/dt = -y$ using the Euler method with a very large step size.

Let us investigate the phenomena using the following decay equation

$$\frac{dy}{dt} = -\lambda y \qquad (7.20a)$$

with

$$t = 0; \qquad y = 1 \qquad (7.20b)$$

If we denote y_e as the exact solution, then the numerically calculated solution y can be expressed as a deviation from this exact value

$$y = y_e + \varepsilon \qquad (7.21)$$

where ε represents the error, which is a function of time. Substitution of Eq. 7.21 into Eq. 7.20a gives

$$\frac{d\varepsilon}{dt} = -\lambda\varepsilon \qquad (7.22)$$

The method is considered to be stable if the error decays with time.

If the explicit Euler method is applied to Eq. 7.22 from t_n to t_{n+1}, we obtain

$$\varepsilon_{n+1} = \varepsilon_n + h(-\lambda\varepsilon_n) = \varepsilon_n(1 - h\lambda) \qquad (7.23)$$

If stability is enforced, we need to have the error at t_{n+1} to be smaller than that at t_n; that is,

$$\left|\frac{\varepsilon_{n+1}}{\varepsilon_n}\right| \leq 1 \qquad (7.24)$$

Substituting Eq. 7.23 into Eq. 7.24, we have

$$|1 - h\lambda| \leq 1 \qquad (7.25)$$

This implies

$$0 \leq h\lambda \leq 2 \qquad (7.26)$$

The step size for the explicit Euler method must be smaller than $2/\lambda$ to ensure stability.

If we apply the trapezoidal rule (Eq. 7.18) to Eq. 7.22, we find

$$\varepsilon_{n+1} = \varepsilon_n + \frac{h}{2}(-\lambda\varepsilon_n - \lambda\varepsilon_{n+1}) \qquad (7.27)$$

from which

$$\frac{\varepsilon_{n+1}}{\varepsilon_n} = \frac{1 - h\lambda/2}{1 + h\lambda/2} \qquad (7.28)$$

Table 7.1 Comparison of Stability Behavior

Method	Stable and No Oscillation	Stable and Oscillation	Unstable
Euler	$0 < h\lambda < 1$	$1 < h\lambda < 2$	$h\lambda > 2$
Trapezoid	$0 < h\lambda < 2$	$2 < h\lambda < \infty$	none
Backward Euler	$0 < h\lambda < \infty$	none	none

Now, to ensure stability for this case, we conclude the criterion is $|\varepsilon_{n+1}/\varepsilon_n| < 1$ is satisfied for any step size $h(h > 0)$. Thus, the trapezoidal rule is always stable. Nonetheless, the error oscillates around zero if $h\lambda > 2$.

If we next apply the backward Euler method (Eq. 7.19) to Eq. 7.22, we have

$$\varepsilon_{n+1} = \varepsilon_n + h(-\lambda\varepsilon_{n+1}) \tag{7.29}$$

which yields

$$\frac{\varepsilon_{n+1}}{\varepsilon_n} = \frac{1}{1 + h\lambda} \tag{7.30}$$

It is clear from Eq. 7.30 that the method is always stable, and moreover, the problem of oscillation around the exact solution disappears.

Table 7.1 highlights behavior of the three techniques.

The explicit Euler method is stable only when the absolute value of $h\lambda$ is less than 2, and is not stable when it is greater than or equal to 2. The backward Euler is always stable and does not oscillate, but it is not particularly accurate (because it is a first order method). The trapezoid method is of second order, but it oscillates for large $|h\lambda|$. Analysis of the Order of Errors in the various numerical schemes will be discussed in Section 7.6.

EXAMPLE 7.3

To illustrate the Euler and trapezoidal methods, we apply them to the following problem of chemical decomposition of nitrogen dioxide in a plug flow reactor. The chemical reaction rate is second order with respect to the nitrogen dioxide concentration; that is,

$$R_{\text{rxn}} = kC^2 \tag{7.31}$$

The rate constant at 383°C is 5030 cc/mole/sec.

Here, we assume that there is no axial diffusion along the reactor, and the velocity profile is taken to be plug shaped. We wish to study the steady-state behavior at constant temperature. With these assumptions, we can set up a mass balance around a thin element having a thickness of Δz (Fig. 7.1).

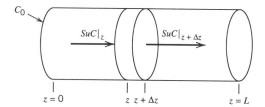

Figure 7.1 Schematic diagram of the plug flow reactor.

The mass balance equation is

$$SuC|_z - SuC|_{2+\Delta z} - (S\,\Delta z)R_{rxn} = 0 \qquad (7.32)$$

where u is the velocity, and S is the cross-sectional area of the reactor.

Dividing Eq. 7.32 by $S\Delta z$ and allowing the element to shrink to zero, we obtain the following ordinary differential equation describing the nitrogen dioxide concentration variation along the reactor

$$u\frac{dC}{dz} = -R_{rxn} = -kC^2 \qquad (7.33)$$

This equation is a first order ordinary differential equation; hence, one boundary condition is needed for the complete formulation of the problem. This is known since the concentration of nitrogen dioxide at the entrance is given as C_0. Mathematically, we write

$$z = 0; \quad C = C_0 \qquad (7.34)$$

This set of problems can be solved analytically using the methods of Chapter 2. The analytical solution is

$$\frac{C}{C_0} = \frac{1}{\left[1 + \left(\dfrac{kC_0 L}{u}\right)\right]} \qquad (7.35)$$

which we shall use as a basis for comparison among the several numerical integration schemes. If we denote the bracketed term in the denominator of Eq. 7.35 as

$$B = \frac{kC_0 L}{u} = \frac{kC_0 LS}{uS} = \frac{kC_0 V}{F} \qquad (7.36)$$

where V is the reactor volume and F is the volumetric flow rate, Eq. 7.35 then becomes

$$\frac{C}{C_0} = \frac{1}{1 + B} \tag{7.37}$$

Hence, the conversion is

$$X = 1 - \frac{C}{C_0} = \frac{B}{1 + B} \tag{7.38}$$

Thus, the reactor performance is governed by a simple equation (Eq. 7.37), or in terms of design, it is governed by a simple *parameter group B*. If we wish to design a reactor with a given conversion, the parameter group B is then calculated from Eq. 7.38. For example, to achieve a 50% conversion, we need to have the design parameter group B to be 1. So, if the rate constant is known ($k = 5030$ cc/mole/sec at $383°$ C), the group C_0V/F can be calculated

$$\frac{C_0V}{F} = \frac{B}{k} = \frac{1}{k} = \frac{1}{5030 \text{ cc/mole/sec}} = 1.99 \times 10^{-4} \frac{\text{mole sec}}{\text{cc}}$$

Therefore, if the inlet concentration, C_0, and the volumetric flow rate are known (which is usually the case in design), the reactor volume then can be calculated. So, for example, if

$$C_0 = 2 \times 10^{-8} \frac{\text{mole}}{\text{cc}}; \qquad F = 72,000 \frac{\text{cc}}{\text{hr}}$$

the reactor volume is

$$V = 199 \text{ liters}$$

Now let us return to the original equation and attempt to solve it numerically. First, we multiply the numerator and denominator of the LHS of Eq. 7.33 by the cross-sectional area, S, to obtain

$$F\frac{dC}{dV} = -kC^2 \tag{7.39}$$

We can define the following nondimensional variables to simplify the form for numerical integration

$$y = \frac{C}{C_0}; \qquad \tau = \left(\frac{kC_0V}{F}\right) \tag{7.40}$$

Equation 7.39 then becomes

$$\frac{dy}{d\tau} = -y^2 \tag{7.41}$$

with the initial condition

$$\tau = 0; \quad y = 1 \tag{7.42}$$

If we use the explicit Euler formula (Eq. 7.14), the recurrence formula for this problem (Eq. 7.41) is

$$y_{n+1} = y_n + h\left(-y_n^2\right) \tag{7.43}$$

where h is the step size (to be chosen) in the numerical integration; also we note $y_0 = 1$.

Using the implicit formulas (the backward Euler and the trapezoidal), we find the following recursive equations; first, the backward Euler

$$y_{n+1} = y_n + h\left(-y_{n+1}^2\right) \tag{7.44}$$

and the trapezoidal is

$$y_{n+1} = y_n + h\left[\frac{\left(-y_n^2\right) + \left(-y_{n+1}^2\right)}{2}\right] \tag{7.45}$$

We note again that the recurrence formula for the implicit methods are nonlinear algebraic equations. This means that at every time step, t_{n+1}, these nonlinear algebraic equations (Eqs. 7.44 or 7.45) must be solved iteratively for y at that time. Appendix A reviews a number of methods for solving nonlinear algebraic equations, and here we use the Newton–Raphson to solve the nonlinear algebraic equations. Let us begin the demonstration using the backward Euler formula (Eq. 7.44).

We rearrange the recurrence formula for the backward Euler formula (Eq. 7.44) to the following standard format, suitable for the application of the Newton–Raphson trial-error technique

$$F = hy_{n+1}^2 + y_{n+1} - y_n = 0 \tag{7.46}$$

Although this equation is a solvable quadratic (owing to the second order reaction), we will solve it nonetheless by numerical means to demonstrate the general procedure of the implicit integration method. The iteration scheme for the Newton–Raphson procedure is (Appendix A)

$$y_{n+1}^{(k+1)} = y_{n+1}^{(k)} - \frac{F^{(k)}}{\dfrac{\partial F^{(k)}}{\partial y_{n+1}}} \tag{7.47}$$

hence, performing the indicated differentiation

$$y_{n+1}^{(k+1)} = y_{n+1}^{(k)} - \frac{hy_{n+1}^{(k)2} + y_{n+1}^{(k)} - y_n}{1 + 2hy_{n+1}^{(k)}} \tag{7.48}$$

where the superscript k denotes the iteration number.

With this iteration equation, at any time t_{n+1} the unknown y_{n+1} must be solved iteratively using Eq. 7.48 until some, as yet unspecified, convergence criterion is satisfied. The following relative error is commonly used to stop the iteration process

$$\left| \frac{y_{n+1}^{(k+1)} - y_{n+1}^{(k)}}{y_{n+1}^{(k+1)}} \right| < \varepsilon \tag{7.49}$$

The initial guess for y_{n+1} is taken as

$$y_{n+1}^{(0)} = y_n \tag{7.50}$$

For small step sizes, this is a prudent initial guess. Alternatively, one can use the explicit Euler formula to provide *a better initial guess* for the iteration process for y_{n+1} in the implicit scheme; that is,

$$y_{n+1}^{(0)} = y_n + h\left(-y_n^2\right) \tag{7.51}$$

Now, we consider the trapezoidal rule. We rearrange the recurrence formula for the trapezoidal rule (Eq. 7.45) as

$$F = \frac{h}{2} y_{n+1}^2 + y_{n+1} + \frac{h}{2} y_n^2 - y_n \tag{7.52}$$

Again, we use the Newton–Raphson formula for this implicit procedure

$$y_{n+1}^{(k+1)} = y_{n+1}^{(k)} - \frac{\frac{h}{2} y_{n+1}^{(k)2} + y_{n+1}^{(k)} + \frac{h}{2} y_n^2 - y_n}{1 + h y_{n+1}^{(k)}} \tag{7.53}$$

As in the case of the backward Euler method, the initial guess for y_{n+1} is either chosen as in Eq. 7.50 or Eq. 7.51.

We have now laid the foundation for a numerical computation using three schemes. First, let us study the worst case, which is the explicit Euler formula (Eq. 7.43). The calculation for y_{n+1} at time $t_{n+1} = nh$ is explicit in Eq. 7.43. For example, if we use $h = 0.2$, the value of y_1 at time $t_1 = 0.2$ is

$$y_1 = y_0 + h\left(-y_0^2\right) = 1 - (0.2)(1)^2 = 0.8 \tag{7.54}$$

and similarly, the value of y_2 at $t_2 = 0.4$ can be obtained as

$$y_2 = y_1 + (0.2)\left(-y_1^2\right) = 0.8 - (0.2)(0.8)^2 = 0.672 \tag{7.55}$$

The same procedure can be carried out sequentially for $t = 0.6, 0.8$, and so on. This can be done quite easily with a hand calculator, which makes it attractive. Figure 7.2 illustrates computation of y versus τ for several step sizes, 0.1, 0.2, 0.5, 1, and 2. The exact solution (Eq. 7.37) is also shown in the figure. We see

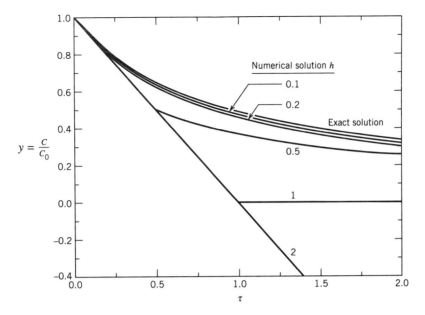

Figure 7.2 Plots of numerical solutions using the explicit Euler method.

that the explicit Euler integration scheme starts to fail (i.e., the integration becomes divergent) when $h = 1$. To apply the stability theorem from the earlier decay equation (Eq. 7.20*a*), it was necessary to linearize the nonlinear equation (Eq. 7.41) around the point $y = 1$ (i.e., at zero time). The linearized version of Eq. 7.41 is obtained by applying the Taylor series around y_0; so if $f(y) = y^2$

$$f(y) = f(y_0) + \frac{\partial f}{\partial y}\bigg|_{y=1}(y - y_0) + \cdots \approx y_0^2 + (2y_0)(y - y_0)$$

so the linearized equivalent to the decay equation is given by

$$\frac{dy}{d\tau} = -y_0^2 - (2y_0)(y - y_0) = -1 - (2)(1)(y - 1) = -2y + 1 \quad \textbf{(7.56)}$$

This means that the decay constant λ is equal to 2 *at zero time*. For nonlinear systems, the decay constant changes with time. So with the decay constant of 2 at zero time, the step size that will make the explicit Euler integration unstable is

$$h = \frac{2}{\lambda} = \frac{2}{2} = 1$$

and so we observe such instability in Fig. 7.2, where the numerical solutions became unstable when we used $h = 1$ and 2.

Thus, we see problems arise when we apply a step size criterion based on linear analysis to a nonlinear problem, leading to a time-variable decay constant. Let us choose the step size as 0.2 (less than 1, so that the solution integration from time zero is stable), so the numerical solution for y at $t = 0.2$ is 0.8 (Eq. 7.54). Now we linearize the nonlinear function (Eq. 7.41) around this point ($t = 0.2$, $y = 0.8$), and the governing equation becomes

$$\frac{dy}{d\tau} = -y_1^2 - (2y_1)(y - y_1) = -1.6y + 0.64 \qquad (7.57)$$

The decay constant at time $t = 0.2$ is now 1.6 instead of 2 calculated earlier (at time zero). This simply implies that the maximum step size for the explicit Euler formula to be stable at time $t = 0.2$ is

$$h = \frac{2}{\lambda} = \frac{2}{1.6} = 1.25$$

instead of the value $h = 1$ computed at zero time. This is a typical pattern for systems exhibiting decay behavior; that is, the maximum allowable step size increases as time progresses. This is, indeed, a salient feature utilized by good integration packages on the market, that is, to vary the step size in order to reduce the total number of steps to reach a certain time. This increases computation speed.

Up to now, we have placed great emphasis on the maximum allowable step size to ensure integration stability. Good stability does not necessarily mean good accuracy, as we observed in Fig. 7.2, where we saw that a step size of 0.5 gave a stable integration, but the accuracy was not very good. To this end, we need to use smaller step sizes as we illustrated using $h = 0.2$ or $h = 0.1$. But the inevitable question arises, how small is small? In practice, we usually do not possess analytical solutions with which to compare.

We next consider an implicit formula, the backward Euler formula (Eq. 7.44). The value y_{n+1} at time t_{n+1} is obtained *iteratively* using the Newton–Raphson equation (Eq. 7.48). This can also be done with a personal computer or a programmable calculator. To demonstrate this iteration process, we choose a step size of 0.2, so that the next calculated value for y_1 is at $t_1 = 0.2$. The Newton–Raphson iteration equation for y_1 is

$$y_1^{(k+1)} = y_1^{(k)} - \frac{h\left(y_1^{(k)}\right)^2 + y_1^{(k)} - y_0}{1 + 2hy_1^{(k)}} = y_1^{(k)} - \frac{(0.2)\left(y_1^{(k)}\right)^2 + y_1^{(k)} - 1}{1 + 2(0.2)y_1^{(k)}} \qquad (7.58)$$

If we now choose the initial guess for y_1 as $y_0 = 1$, the first iterated solution for y_1 is

$$y_1^{(1)} = y_1^{(0)} - \frac{(0.2)\left(y_1^{(0)}\right)^2 + y_1^{(0)} - y_0}{1 + 2(0.2)y_1^{(0)}} = 1 - \frac{(0.2)1^2 + 1 - 1}{1 + 2(0.2)1} = 0.857142857$$

Using this first iterated solution for y_1, we then calculate its second iterated solution as

$$y_1^{(2)} = y_1^{(1)} - \frac{(0.2)\left(y_1^{(1)}\right)^2 + y_1^{(1)} - y_0}{1 + 2(0.2)y_1^{(1)}}$$

$$= 0.857 \cdots - \frac{(0.2)(0.857\cdots)^2 + (0.857\cdots) - 1}{1 + 2(0.2)(0.857\cdots)} = 0.854103343$$

We can see that the solution converges very quickly. The third iterated solution is calculated in a similar way, and we have

$$y_1^{(3)} = 0.854101966$$

Thus the relative error calculated using Eq. 7.49 is

$$\text{Rel. Error} = \left| \frac{(0.854101966) - (0.854103343)}{0.854101966} \right| = 0.000001612$$

This error is indeed small enough for us to accept the solution for y_1, which is 0.8541. The few iterations needed show the fast convergence of the Newton–Raphson method.

Knowing the solution for y_1 at $t_1 = 0.2$, we can proceed in a similar fashion to obtain y_2, y_3, and so on. Figure 7.3 illustrates computations for a number of step sizes. No instability is observed for the case of backward Euler method.

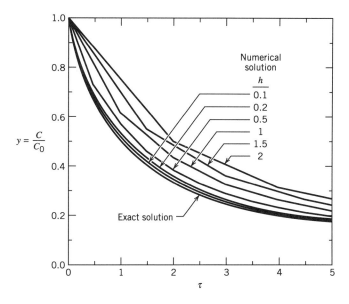

Figure 7.3 Plots of the numerical solutions for the backward Euler method.

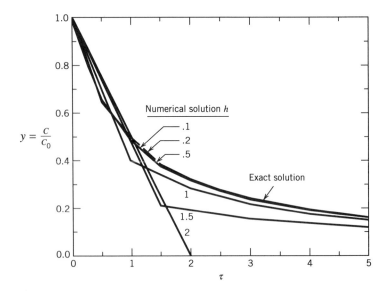

Figure 7.4 Computations of the numerical solutions for the trapezoidal method.

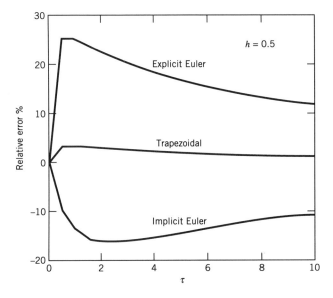

Figure 7.5 Relative errors between the numerical solutions and the exact solution versus τ.

Notice also that the solution does not oscillate, as one might expect in light of the stability analysis in the previous section.

Again, we use Newton–Raphson in Eq. 7.53 to obtain the numerical solution for the trapezoidal rule, and Fig. 7.4 illustrates these computations for various step sizes. We note that oscillation of the numerical solution occurs around the exact solution for $h = 1.5$ and 2. This arises because the decay constant (λ) at time zero is 2 and for the two step sizes ($h = 1.5$ and 2) producing oscillation, we see that $h\lambda > 2$. Under such circumstances, oscillations will always occur for the trapezoidal rule.

To summarize the comparison of methods, we illustrate in Fig. 7.5 the percentage relative error between the numerical solutions and the exact solution for a fixed step size of 0.5. With this step size all three methods, explicit Euler, backward Euler, and trapezoid, produced integration stability, but the relative error is certainly unacceptable.

7.4 STIFFNESS

A characteristic of initial value type ordinary differential equations is the so-called stiffness. It is easiest to illustrate stiffness by considering again the simple decay equation (or first order kinetics equation)

$$\frac{dy}{dt} = -\lambda y \qquad (7.59a)$$

$$t = 0; \qquad y = y_0 \qquad (7.59b)$$

The exact solution for this equation is

$$y = y_0 e^{-\lambda t} \qquad (7.60)$$

Let us assume that we would like to solve Eqs. 7.59 by an explicit method from $t = 0$ to a time, say t_1. The step size that we could use to maintain numerical stability is

$$\lambda \, \Delta t \leq p \qquad (7.61)$$

If the explicit Euler method is used, the value of p is 2 (see section 7.3 on stability for more details), and hence the maximum step size to maintain numerical stability is

$$(\Delta t)_{\max} = \frac{2}{\lambda} \qquad (7.62)$$

Thus, to integrate the problem up to time t_1, the number of steps needed is

$$\text{Number of steps} = \frac{\lambda t_1}{2} \qquad (7.63)$$

It is clear from Eq. 7.63 that the number of steps required for numerical stability increases with the constant λ (which is sometimes called the characteristic value).

Next, we consider the following coupled ordinary differential equations, represented in vector form (see Appendix B for a review of this subject)

$$\frac{d\mathbf{y}}{dt} = \begin{bmatrix} -500.5 & 499.5 \\ 499.5 & -500.5 \end{bmatrix} \mathbf{y} = \mathbf{Ay} \qquad (7.64a)$$

with

$$\mathbf{y}(0) = [2, 1]^T \qquad (7.64b)$$

The solutions of Eqs. 7.64 are

$$y_1 = 1.5e^{-t} + 0.5e^{-1000t} \qquad (7.65a)$$

$$y_2 = 1.5e^{-t} - 0.5e^{-1000t} \qquad (7.65b)$$

The characteristic values (roots) or eigenvalues for this system of equations are $\lambda_1 = -1$ and $\lambda_2 = -1000$. The corresponding time constants are simply the inverse of these eigenvalues. To ensure the stability of an explicit method (such as the explicit Euler), one needs to ensure that the step size should be less than

$$\Delta t \le \frac{2}{|\lambda_{max}|} \qquad (7.66)$$

where λ_{max} is the maximum eigenvalue for the matrix $\mathbf{A}$. Thus, the step size has to be less than 0.002 to achieve integration stability.

Inspection of the exact solution (7.65) shows that the problem is controlled by the smallest eigenvalue, in this case $\lambda = -1$. The curves of y_1 and y_2 versus time are shown in Fig. 7.6.

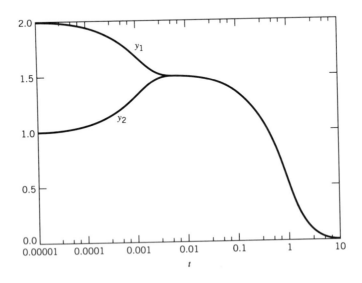

Figure 7.6 Plots of y_1 and y_2 (Eqs. 7.65) versus time.

A logarithmic time scale is used to illustrate the importance of the two distinctly separate eigenvalues, $\lambda_1 = -1$ and $\lambda_2 = -1000$. The fast change of y_1 and y_2 occurs over a very short time scale. We therefore have a situation where the step size is controlled by the maximum eigenvalue (i.e., very small step size), whereas the full evolution is controlled by the smallest eigenvalue. This type of problem is called a *stiff* problem. This occurs when small and large time constants occur in the same system. The small time constant controls earlier response, whereas the large one controls tailing.

To measure the degree of stiffness, one can introduce the following stiffness ratio:

$$SR = \frac{\max |\boldsymbol{\lambda}|}{\min |\boldsymbol{\lambda}|} \tag{7.67}$$

When $SR < 20$ the problem is not stiff, when $SR = 1000$ the problem is classified stiff, and when $SR => 1{,}000{,}000$ the problem is very stiff (Finlayson 1980).

The example we used to demonstrate stiffness is a linear problem. For a nonlinear problem of the following type

$$\frac{d\mathbf{y}}{dt} = \mathbf{f}(\mathbf{y}) \tag{7.68}$$

we linearize the above equation around time $t_n(y_n)$ using a Taylor expansion, retaining only the first two terms

$$\frac{d\mathbf{y}}{dt} \approx \mathbf{f}(\mathbf{y}_n) + \mathbf{J}(t_n) \cdot (\mathbf{y} - \mathbf{y}_n) \tag{7.69}$$

where

$$\mathbf{J}(t_n) = \left\{ a_{ij} = \left[\frac{\partial f_i(y)}{\partial y_j} \right]_{t_n} \right\} \tag{7.70}$$

which is the Jacobian matrix for the problem at $t = t_n$. The definition of stiffness in Eq. 7.67 utilizes the eigenvalues obtained from the Jacobian, and since this Jacobian matrix changes with time, the *stiffness of the problem also changes with time.*

EXAMPLE 7.4

In this example, we illustrate computations using the Jacobian with a system of two coupled differential equations

$$\frac{dy_1}{dt} = (1 - y_1) - 10y_1^2 y_2$$

$$\frac{dy_2}{dt} = -0.05 y_1^2 y_2$$

The initial conditions are

$$t = 0; \qquad y_1 = 0.2, \qquad y_2 = 1$$

For a pair of coupled differential equations, the Jacobian will be a 2×2 matrix. For the above problem, the four elements of the Jacobian are computed to be

$$\frac{\partial f_1}{\partial y_1} = -1 - 20 y_1 y_2; \qquad \frac{\partial f_1}{\partial y_2} = -10 y_1^2$$

$$\frac{\partial f_2}{\partial y_1} = -0.1 y_1 y_2; \qquad \frac{\partial f_2}{\partial y_2} = -0.05 y_1^2$$

To find the Jacobian at time $t = 0$, we simply replace y_1 and y_2 by their values at $t = 0$, that is, the four elements will be

$$\frac{\partial f_1}{\partial y_1} = -0.2 - 20(0.2)(1) = -4.2$$

$$\frac{\partial f_1}{\partial y_2} = -10(0.2)^2 = -0.4$$

$$\frac{\partial f_2}{\partial y_1} = -0.1(0.2)(1) = -0.02$$

$$\frac{\partial f_2}{\partial y_2} = -0.05(0.2)^2 = -0.002$$

The eigenvalues of this Jacobian are seen to be -4.2019 and 9.52×10^{-5}. The absolute ratio of these eigenvalues is 44,140, which indicates that the present problem is quite stiff at time $t = 0$.

In the next two sections, we discuss the basic theories underlying the essential differences between explicit and implicit methods. To help with this, we need to recall a few elementary steps such as the Newton interpolation formula. Details on the various interpolation theories can be found elsewhere (Burden and Faires 1981).

7.5   INTERPOLATION AND QUADRATURE

To facilitate the development of explicit and implicit methods, it is necessary to briefly consider the origins of interpolation and quadrature formulas (i.e., numerical approximation to integration). There are essentially two methods for performing the differencing operation (as a means to approximate differentiation); one is the forward difference, and the other is the backward difference. Only the backward difference is of use in the development of explicit and implicit methods.

   Let us assume that we have a set of data points at *equally* spaced times, ..., $t_{n-1}, t_n, t_{n+1}, \ldots$, and let $y(t_n) = y_n$ (i.e., y_n are values of y at those equally spaced times). The forward difference in finite difference terms is

$$\Delta y_n = y_{n+1} - y_n \tag{7.71}$$

which can be related to the first order differentiation

$$\frac{dy}{dt} \approx \frac{\Delta y_n}{\Delta t} = \frac{y_{n+1} - y_n}{h} \tag{7.72}$$

For second order differentiation, we shall need

$$\Delta^2 y_n = \Delta y_{n+1} - \Delta y_n = (y_{n+2} - y_{n+1}) - (y_{n+1} - y_n)$$

that is:

$$\Delta^2 y_n = y_{n+2} - 2y_{n+1} + y_n \tag{7.73}$$

To see this, write

$$\frac{d^2 y}{dt^2} \approx \frac{\Delta^2 y}{\Delta t^2} = \frac{\Delta(y_{n+1} - y_n)}{h^2} = \frac{1}{h^2}(y_{n+2} - 2y_{n+1} + y_n) \tag{7.74}$$

The same procedure can be applied for higher order differences.
  The backward difference is defined as

$$\nabla y_n = y_n - y_{n-1} \tag{7.75}$$

From this definition, the second and third order differences are

$$\nabla^2 y_n = \nabla y_n - \nabla y_{n-1}$$
$$= (y_n - y_{n-1}) - (y_{n-1} - y_{n-2}) \tag{7.76}$$
$$\nabla^2 y_n = y_n - 2y_{n+1} + y_{n-2}$$

and

$$\nabla^3 y_n = y_n - 3y_{n-1} + 3y_{n-2} - y_{n-3} \tag{7.77}$$

Having defined the necessary backward difference relations, we present the *Newton backward interpolation formula*, written generally as

$$y(\alpha) = y_n + \alpha \nabla y_n + \frac{\alpha(\alpha+1)}{2!} \nabla^2 y_n + \cdots + \frac{\alpha(\alpha+1)\cdots(\alpha+j-1)}{j!} \nabla^j y_n \tag{7.78}$$

where

$$\alpha = \frac{(t - t_n)}{h}; \qquad h = \Delta t \tag{7.79}$$

Since n in Eq. 7.78 is arbitrary, one can write another equation similar to Eq. 7.78 by replacing n by $n + 1$.

$$y(\alpha) = y_{n+1} + \alpha \nabla y_{n+1} + \frac{\alpha(\alpha + 1)}{2!} \nabla^2 y_{n+1}$$

$$+ \cdots + \frac{\alpha(\alpha + 1)(\alpha + 2) \cdots (\alpha + j - 1)}{j!} \nabla^j y_{n+1} \qquad (7.80)$$

where

$$\alpha = \frac{(t - t_{n+1})}{h}; \qquad h = \Delta t \qquad (7.81)$$

The polynomials (7.78) or (7.80) are continuous and can be differentiated and integrated.

If we keep two terms in the RHS of Eq. 7.78, the resulting formula is equivalent to a linear equation passing through two points (t_n, y_n) and (t_{n-1}, y_{n-1}). Similarly, if three terms are retained, the resulting formula corresponds to a parabolic curve passing through three data points (t_n, y_n), (t_{n-1}, y_{n-1}) and (t_{n-2}, y_{n-2}). Note that these latter points are prior to (t_n, y_n).

The polynomial (7.78) is a continuous function in terms of t (through the variable α defined in Eq. 7.79; hence, it can be differentiated as well as integrated. Taking the derivative of Eq. 7.78 and evaluating at $t = t_n$ (i.e., $\alpha = 0$), we obtain

$$h \frac{dy}{dt} \bigg|_{t=t_n} = (y_n - y_{n-1}) - \frac{1}{2}(y_n - 2y_{n-1} + y_{n-2}) + \cdots \qquad (7.82)$$

Equation 7.82 simply states that knowing the values of $y_n, y_{n-1}, y_{n-2}, \ldots$, we can calculate the derivative at $t = t_n$ using just these values.

Similarly, we can use Eq. 7.78 to evaluate higher order derivatives at $t = t_n$, and also perform integration.

Also, if the data points are composed of derivatives of y at equally spaced times (i.e., $y'_n, y'_{n-1}, y'_{n-2}, \ldots$), we can write the backward interpolation formula in terms of derivatives

$$y'(\alpha) = y'_n + \alpha \nabla y'_n + \frac{\alpha(\alpha + 1)}{2!} \nabla^2 y'_n + \frac{\alpha(\alpha + 1)(\alpha + 2)}{3!} \nabla^3 y'_n$$

$$+ \cdots + \frac{\alpha(\alpha + 1) \cdots (\alpha + j - 1)}{j!} \nabla^j y'_n \qquad (7.83)$$

where

$$\alpha = \frac{(t - t_n)}{h}; \qquad h = \Delta t \qquad (7.84)$$

Since n in Eq. 7.83 is arbitrary, one can write a relation similar to Eq. 7.83 by replacing n by $n + 1$

$$y'(\alpha) = y'_{n+1} + \alpha \nabla y'_{n+1} + \frac{\alpha(\alpha + 1)}{2!} \nabla^2 y'_{n+1}$$

$$+ \cdots + \frac{\alpha(\alpha + 1)(\alpha + 2) \cdots (\alpha + j - 1)}{j!} \nabla^j y'_{n+1} \quad (7.85)$$

where

$$\alpha = \frac{(t - t_{n+1})}{h}; \qquad h = \Delta t \quad (7.86)$$

The above expansions now allow representation for functions and derivatives to a higher order than previously used.

7.6 EXPLICIT INTEGRATION METHODS

The interpolation formula (Eqs. 7.83 and 7.85) presented in the previous section can be utilized to derive a method for integration. Let us consider a single scalar ODE:

$$\frac{dy}{dt} = f(y) \quad (7.87)$$

subject to some known initial condition, say $y(0) = y_0$.

A straightforward integration of Eq. 7.87 with respect to t from t_n to t_{n+1} would yield

$$y_{n+1} - y_n = \int_{t_n}^{t_{n+1}} f(y) \, dt \quad (7.88)$$

or:

$$y_{n+1} = y_n + \int_{t_n}^{t_{n+1}} y' \, dt \quad (7.89)$$

By defining $\alpha = (t - t_n)/h$, where $h = (t_{n+1} - t_n)$, Eq. 7.89 becomes

$$y_{n+1} = y_n + h \int_0^1 y'(\alpha) \, d\alpha \quad (7.90)$$

Now, we will show how integration schemes can be generated. Numerical integration (quadrature) formulas can be developed by approximating the integrand $y'(\alpha)$ of Eq. 7.90 with any interpolation polynomials. This is done by substituting any of several interpolation formulas into Eq. 7.90. For example, if we use the backward interpolation formula using values from t_n going backward

(Eq. 7.83) into the RHS of Eq. 7.90, we have

$$y_{n+1} = y_n + h \int_0^1 \left[y_n' + \alpha \nabla y_n' + \frac{\alpha(\alpha + 1)}{2!} \nabla^2 y_n' + \cdots \right] d\alpha \quad (7.91)$$

Note that $y_{n+1}', \nabla y_{n+1}', \ldots$ are numbers; hence, they can be moved outside the integration sign, and we have

$$y_{n+1} = y_n + h \left[y_n' \int_0^1 d\alpha + \nabla y_n' \int_0^1 \alpha \, d\alpha + \nabla^2 y_n' \int_0^1 \frac{\alpha(\alpha + 1)}{2!} d\alpha + \cdots \right] \quad (7.92)$$

In simplifying, the integration becomes a sum

$$y_{n+1} = y_n + h \sum_{i=0}^{M} \delta_i \nabla^i y_n' \quad (7.93)$$

where M is an arbitrary integer, and δ_i is defined by the following relation

$$\delta_i = \int_0^1 \frac{\alpha(\alpha + 1) \cdots (\alpha + i - 1)}{i!} d\alpha \quad (7.94)$$

Explicit evaluation of δ_i from Eq. 7.94 and substitution of the result into Eq. 7.93 gives

$$y_{n+1} = y_n + h \left(1 + \frac{1}{2} \nabla + \frac{5}{12} \nabla^2 + \cdots \right) y_n' \quad (7.95)$$

Keeping only the first term in the bracket, we have

$$y_{n+1} = y_n + hy_n' + O(h^2) \quad (7.96)$$

but $y_n' = f(y_n)$, so we get

$$y_{n+1} = y_n + hf(y_n) + O(h^2) \quad (7.97)$$

This formula is simply the *explicit Euler method*. The *local truncation error* is of order of $O(h^2)$, that is, the error is proportional to h^2 if all the previous values, $y_n, y_{n-1}, \ldots$ are exact. However, in the integration from time $t = 0$, the error of the method at time t_{n+1} is accurate up to $O(h)$ because the number of integration step is inversely proportional to h.

The explicit Euler integration method is simply the linear extrapolation from the point at $t = t_n$ to t_{n+1} using the slope of the curve at t_n. Figure 7.7 shows this Euler method with the local and global errors.

If we now keep two terms in the bracket of Eq. 7.95, we have

$$y_{n+1} = y_n + h \left(y_n' + \frac{1}{2} \nabla y_n' \right) \quad (7.98)$$

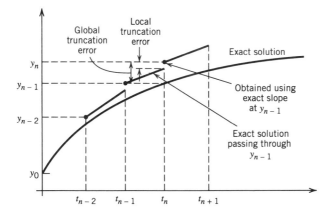

Figure 7.7 Graphical representation of Euler method, and local and global truncation error.

and since the backward difference is defined as $\nabla y'_n = y'_n - y'_{n-1}$, we have

$$y_{n+1} = y_n + h\left[y'_n + \frac{1}{2}(y'_n - y'_{n-1}) \right] \tag{7.99}$$

or

$$y_{n+1} = y_n + h\left(\frac{3}{2}y'_n - \frac{1}{2}y'_{n-1} \right) \tag{7.100}$$

Using $y' = dy/dt = f(y)$, Eq. 7.100 becomes

$$y_{n+1} = y_n + h\left[\frac{3}{2}f(y_n) - \frac{1}{2}f(y_{n-1}) \right] \tag{7.101}$$

This formula is sometimes called the *second-order Adam–Bashford method*, and it requires two values of the function f at t_n and t_{n-1}. The accuracy of the method is $O(h^2)$.

If four terms are retained in Eq. 7.95, we obtain the following *fourth order Adams–Bashford method*.

$$y_{n+1} = y_n + \frac{h}{24}\left[55f(y_n) - 59f(y_{n-1}) + 37f(y_{n-2}) - 9f(y_{n-3}) \right] + O(h^5) \tag{7.102}$$

The accuracy of the method is $O(h^4)$. The Adams–Bashford method is among the most attractive methods used for solving nonstiff, coupled ordinary differential equations.

It is noted that using the higher order Adams–Bashford methods, we must have previous points other than the current point, y_n, to execute the RHS of Eq. 7.101 or Eq. 7.102. At the starting point (i.e., $t = 0$), only one point is available (y_0); hence, lower order methods such as the Euler method or Runge–Kutta methods (which are presented later), which require only one point, must be used to generate enough points before higher order methods can be applied.

7.7 IMPLICIT INTEGRATION METHODS

Section 7.6 gave a number of explicit integration formulas, and in this section we present implicit methods, which are more stable than the explicit formula because they utilize the information about the unknown point, t_{n+1}, in the integration formula.

To derive implicit integration methods, we start with the Newton backward difference interpolation formula starting from the point t_{n+1} backward (Eq. 7.85) rather than from the point t_n backward as used in the generation of the explicit methods. With a single equation of the type shown by Eq. 7.87, we have

$$y_{n+1} = y_n + \int_{t_n}^{t_{n+1}} y' \, dt \tag{7.103}$$

or

$$y_{n+1} = y_n + h \int_{-1}^{0} y'(\alpha) \, d\alpha \tag{7.104}$$

where $\alpha = (t - t_{n+1})/h$ with $h = t_{n+1} - t_n$. Using the incremented interpolation formula for the function y' (Eq. 7.85), we substitute it into Eq. 7.104 and carry out the integration with respect to α to obtain

$$y_{n+1} = y_n + h\left(1 - \frac{1}{2}\nabla - \frac{1}{12}\nabla^2 - \frac{1}{24}\nabla^3 - \cdots\right)y'_{n+1} \tag{7.105}$$

If we keep the first term in the bracket of Eq. 7.105, we have

$$y_{n+1} = y_n + hy'_{n+1} + O(h^2) \tag{7.106}$$

Since $y'_{n+1} = f(y_{n+1})$, Eq. 7.106 becomes

$$y_{n+1} = y_n + hf(y_{n+1}) \tag{7.107}$$

Equation 7.107 is the *implicit Euler method*, which is in a similar form to the explicit Euler method, except that the evaluation of the function f is done at the *unknown point y_{n+1}*. Hence, Eq. 7.107 is a nonlinear algebraic equation and must be solved by trial to find y_{n+1}. Example 7.1 demonstrated this iteration process in solving nonlinear algebraic equations.

We proceed further with the generation of more implicit schemes. If the second term in the RHS of Eq. 7.105 is retained, we have

$$y_{n+1} = y_n + h\left(1 - \frac{1}{2}\nabla\right)y'_{n+1} \tag{7.108}$$

Using the definition of the backward difference, $\nabla y'_{n+1} = y'_{n+1} - y'_n$, we have

$$y_{n+1} = y_n + h\left[y'_{n+1} - \frac{1}{2}(y'_{n+1} - y'_n)\right] \qquad (7.109)$$

that is,

$$y_{n+1} = y_n + \frac{h}{2}[f(y_n) + f(y_{n+1})] \qquad (7.110)$$

in which we have used Eq. 7.87. Like the implicit Euler method (Eq. 7.107), Eq. 7.110 is a nonlinear algebraic equation and it must be solved by trial methods to find y_{n+1}. This representation in Eq. 7.110 is called the trapezoidal method or Crank–Nicolson method, and is also sometimes called the second order implicit method. For simple problems, this method is very attractive, and as we have shown in Example 7.1, it is more accurate than the implicit Euler scheme. This is because the trapezoidal technique has an accuracy of $O(h^2)$, whereas for the implicit Euler, it is $O(h)$.

If we truncate up to the fourth term in the RHS of Eq. 7.105, we obtain the following *fourth order Adams–Moulton method*.

$$y_{n+1} = y_n + \frac{h}{24}[9f(y_{n+1}) + 19f(y_n) - 5f(y_{n-1}) + f(y_{n-2})] \qquad (7.111)$$

The common factor in the implicit Euler, the trapezoidal (Crank–Nicolson), and the Adams–Moulton methods is simply their recursive nature, which are nonlinear algebraic equations with respect to y_{n+1} and hence must be solved numerically; this is done in practice by using some variant of the Newton–Raphson method or the successive substitution technique (Appendix A).

7.8 PREDICTOR-CORRECTOR METHODS AND RUNGE–KUTTA METHODS

7.8.1 Predictor-Corrector Methods

A compromise between the explicit and implicit methods is the predictor-corrector technique, where the explicit method is used to obtain a first estimate of y_{n+1} and this estimated y_{n+1} is then used in the RHS of the implicit formula. The result is the corrected y_{n+1}, which should be a better estimate to the true y_{n+1} than the first estimate. The corrector formula may be applied several times (i.e., successive substitution) until the convergence criterion (7.49) is achieved. Generally, predictor-corrector pairs are chosen such that they have truncation errors of approximately the same degree in h but with a difference in sign, so that the truncation errors compensate one another.

One of the most popular predictor-corrector methods is the fourth order Adams–Bashford and Adams–Moulton formula.

Adams–Bashford (for prediction)

$$y_{n+1} = y_n + \frac{h}{24}[55f(y_n) - 59f(y_{n-1}) + 37f(y_{n-2}) - 9f(y_{n-3})] \quad (7.112a)$$

Adams–Moulton (for correction)

$$y_{n+1} = y_n + \frac{h}{24}[9f(y_{n+1}) + 19f(y_n) - 5f(y_{n-1}) + f(y_{n-2})] \quad (7.112b)$$

7.8.2 Runge–Kutta Methods

Runge–Kutta methods are among the most popular methods for integrating differential equations. Let us start with the derivation of a second order Runge–Kutta method and then generalize to a pth order method. The second order Runge–Kutta integration formula for integrating equation

$$\frac{dy}{dt} = f(t, y) \quad (7.113)$$

takes the form

$$y_{n+1} = y_n + w_1 k_1 + w_2 k_2 \quad (7.114a)$$

where k_1 and k_2 are given by

$$k_1 = hf(t_n, y_n) = hf_n \quad (7.114b)$$

$$k_2 = hf(t_n + ch, y_n + ak_1) \quad (7.114c)$$

where w_1, w_2, c, and a are constants.

First, we apply the Taylor series to $y_{n+1} = y(t_n + h)$

$$y_{n+1} = y(t_{n+1}) = y(t_n + h) = y(t_n) + \frac{\partial y(t_n)}{\partial t}h + \frac{1}{2}\frac{\partial^2 y(t_n)}{\partial t^2}h^2 + \cdots$$

$$(7.115)$$

Noting that

$$\frac{\partial y}{\partial t} = f(t, y) \quad (7.116a)$$

and

$$\frac{\partial^2 y}{\partial t^2} = \frac{\partial f}{\partial t} + \frac{\partial f}{\partial y} \cdot \frac{\partial y}{\partial t} \quad (7.116b)$$

then Eq. 7.115 becomes, using the notation $\partial f/\partial t = f_t$, $\partial f/\partial y = f_y$, and so on

$$y_{n+1} = y_n + hf_n + \frac{h^2}{2}(f_t + ff_y)_n + \cdots \tag{7.117}$$

Next, we take the Taylor series for k_1 and k_2 and obtain

$$k_1 = hf_n$$
$$k_2 = hf_n + h^2[cf_t + aff_y]_n + \cdots \tag{7.118}$$

Next, substitute Eq. 7.118 into Eq. 7.114a, and obtain

$$y_{n+1} = y_n + (w_1 + w_2)hf_n + w_2 h^2(cf_t + aff_y)_n + \cdots$$

Comparing this equation with Eq. 7.117 yields the following equalities

$$w_1 + w_2 = 1$$
$$cw_2 = 0.5$$
$$aw_2 = 0.5$$

Here, we have four unknowns but only three equations. So, we can specify one constant and solve for the other three. If we take $c = 0.5$, we have

$$w_1 = 0; \qquad w_2 = 1; \qquad a = 0.5$$

the integration formula then becomes

$$y_{n+1} = y_n + hf\left(t_n + \frac{h}{2}, y_n + \frac{h}{2}f_n\right) \tag{7.119}$$

This integration is basically the midpoint scheme, where the midpoint is used to calculate the unknown point at t_{n+1}. Note that the argument $y_n + (h/2)f_n$ is the slope at $t_n + (h/2)$, the midpoint between t_n and t_{n+1}.

One can also choose $c = 1$, from which we calculate the other three unknowns

$$w_1 = w_2 = \frac{1}{2}; \qquad a = 1$$

With these values, the integration formula is

$$y_{n+1} = y_n + \frac{h}{2}[f_n + f(t_n + h, y_n + hf_n)] \tag{7.120}$$

which is the Euler predictor-trapezoid corrector scheme.

We have presented the essence of the Runge–Kutta scheme, and now it is possible to generalize the scheme to a pth order. The Runge–Kutta methods are explicit methods, and they involve the evaluation of derivatives at various points between t_n and t_{n+1}. The Runge–Kutta formula is of the following general form

$$y_{n+1} = y_n + \sum_{i=1}^{p} w_i k_i \qquad (7.121a)$$

where p is the order of the Runge–Kutta method and

$$k_i = hf\left(t_n + c_i h, y_n + \sum_{j=1}^{i-1} a_{ij} k_j\right) \qquad (7.121b)$$

and

$$c_1 = 0 \qquad (7.121c)$$

The Runge–Kutta technique is different from the predictor-corrector method in the sense that, instead of using a number of points from t_{n+1} backward it uses a number of functional evaluations between the current point t_n and the desired point t_{n+1}.

Next, we apply the Taylor series to k_i (Eq. 7.121b) and obtain

$$k_1 = hf_n$$

$$k_2 = hf_n + h^2\left[c_2(f_t)_n + a_{21} f_n(f_y)_n\right] + \cdots$$

and so on.

Finally substituting these equations and Eq. 7.117 into the formula (7.121a) and matching term by term we obtain

$$w_1 + w_2 + \cdots + w_p = 1 \qquad (7.122a)$$

$$w_2 a_{21} + \cdots = 0.5 \qquad (7.122b)$$

For the second order Runge–Kutta method (i.e., $p = 2$), we recover the formula obtained earlier.

The following Runge–Kutta–Gill method (proposed by Gill in 1951, to reduce computer storage requirements) is fourth order and is among the most widely used integration schemes (written in vector form for coupled ODEs with N being the number of equations)

$$\mathbf{y}_{n+1} = \mathbf{y}_n + \frac{1}{6}(\mathbf{k}_1 + \mathbf{k}_4) + \frac{1}{3}(b\mathbf{k}_2 + d\mathbf{k}_3) \qquad (7.123)$$

where

$$\mathbf{k}_1 = h\mathbf{f}(t_n, \mathbf{y}_n) \qquad (7.124a)$$

$$\mathbf{k}_2 = h\mathbf{f}\left(t_n + \frac{h}{2}, \mathbf{y}_n + \frac{1}{2}\mathbf{k}_1\right) \qquad (7.124b)$$

$$\mathbf{k}_3 = h\mathbf{f}\left(t_n + \frac{h}{2}, \mathbf{y}_n + a\mathbf{k}_1 + b\mathbf{k}_2\right) \qquad (7.124c)$$

$$\mathbf{k}_4 = h\mathbf{f}(t_n + h, \mathbf{y}_n + c\mathbf{k}_2 + d\mathbf{k}_3) \qquad (7.124d)$$

with

$$a = \frac{\sqrt{2} - 1}{2}, b = \frac{2 - \sqrt{2}}{2}, c = -\frac{\sqrt{2}}{2}, d = 1 + \frac{\sqrt{2}}{2} \qquad (7.124e)$$

where

$$\mathbf{y} = [y_1, y_2, y_3, \ldots, y_N]^T; \qquad \mathbf{f} = [f_1, f_2, f_3, \ldots, f_N]^T \qquad (7.124f)$$

The following algorithm for semi-implicit third order Runge–Kutta is suggested by Michelsen (1976) for the following equation

$$\frac{d\mathbf{y}}{dt} = \mathbf{f}(\mathbf{y}) \qquad (7.125)$$

The integration formula is

$$\mathbf{y}_{n+1} = \mathbf{y}_n + w_1\mathbf{k}_1 + w_2\mathbf{k}_2 + w_3\mathbf{k}_3 \qquad (7.126)$$

$$\mathbf{k}_1 = h[\mathbf{I} - ha_1\mathbf{J}(\mathbf{y}_n)]^{-1}\mathbf{f}(\mathbf{y}_n) \qquad (7.127a)$$

$$\mathbf{k}_2 = h[\mathbf{I} - ha_1\mathbf{J}(\mathbf{y}_n)]^{-1}(\mathbf{f}(\mathbf{y}_n + b_2\mathbf{k}_1)) \qquad (7.127b)$$

$$\mathbf{k}_3 = h[\mathbf{I} - ha_1\mathbf{J}(\mathbf{y}_n)]^{-1}(b_{31}\mathbf{k}_1 + b_{32}\mathbf{k}_2) \qquad (7.127c)$$

where the various constants are defined as

$$a_1 = 0.43586659 \qquad b_2 = 0.75$$

$$b_{32} = \frac{2}{9a_1}(6a_1^2 - 6a_1 + 1) = -0.24233788$$

$$b_{31} = -\frac{1}{6a_1}(8a_1^2 - 2a_1 + 1) = -0.63020212$$

$$w_1 = \frac{11}{27} - b_{31} = 1.037609527$$

$$w_2 = \frac{16}{27} - b_{32} = 0.834930473; \qquad w_3 = 1$$

Michelsen's third order semi-implicit Runge–Kutta method is a modified version of the method originally proposed by Caillaud and Padmanabhan (1971). This third-order semi–implicit method is an improvement over the original version of semi–implicit methods proposed in 1963 by Rosenbrock.

7.9   EXTRAPOLATION

Information from truncation error may be utilized to obtain a better estimate for y_{n+1}. If the step size h is used, the calculated y_{n+1} is equal to the exact solution y_{n+1} plus an error, which is proportional to the truncation error. So if the Euler method is used, denoting exact solutions as hat variable, we have

$$y_{n+1}(h) = \hat{y}_{n+1} + \alpha h \tag{7.128}$$

If the step size is halved, we have

$$y_{n+1}\left(\frac{h}{2}\right) = \hat{y}_{n+1} + \alpha \frac{h}{2} \tag{7.129}$$

Eliminating the proportional constant α from Eqs. 7.128 and 7.129 we have

$$\hat{y}_{n+1} = 2 y_{n+1}\left(\frac{h}{2}\right) - y_{n+1}(h) \tag{7.130}$$

Equation 7.130 gives the exact solution for y at $t = t_{n+1}$ provided the error formula (7.128) and (7.129) are exact. Because of the approximate nature of Eqs. 7.128 and 7.129, Eq. 7.130 will not give the exact solution, but rather it gives a better estimate to the solution at t_{n+1} because the error is proportional to h^2 rather than h.

If a higher order method is used, say the trapezoidal rule method (Eq. 7.110), we then have

$$y_{n+1}(h) = \hat{y}_{n+1} + \alpha h^2 \tag{7.131}$$

and

$$y_{n+1}\left(\frac{h}{2}\right) = \hat{y}_{n+1} + \alpha \left(\frac{h}{2}\right)^2 \tag{7.132}$$

Eliminating α between the above two equations gives

$$\hat{y}_{n+1} = \frac{1}{3}\left[4 y_{n+1}\left(\frac{h}{2}\right) - y_{n+1}(h)\right] \tag{7.133}$$

Equation 7.133 will give a better estimate to the solution at $t = t_{n+1}$.

7.10   STEP SIZE CONTROL

There are various ways of controlling the step size. One method suggested by Bailey (1969) is as follows. For $\mathbf{y} = (y_1, y_2, y_3, \ldots, y_N)^T$ and $\mathbf{y}(t_n) = \mathbf{y}_n$, we calculate

$$\Delta y = |y(t_{n+1}) - y(t_n)| \tag{7.134}$$

If any component Δy is less than 0.001, we ignore that component in the following crude but practical tests.

- If $\Delta y_i / y_i < 0.01$, the step size is doubled.
- If $\Delta y_i / y_i > 0.1$, the step size is halved.
- Otherwise, the old step size is retained.

This method is applicable for *any* integration method.

Michelsen used Eq. 7.125, which is a third order method, and performed two calculations at every time step. One calculation uses a step size of h and the other uses the step size of $h/2$. The error at the time t_{n+1} is then defined

$$e_{n+1} = y_{n+1}\left(\frac{h}{2}\right) - y_{n+1}(h) \tag{7.135}$$

Then for a tolerance of ε, the step size is accepted when the following *ratio is less than unity*.

$$q = \max_i \left| \frac{e_{n+1}}{\varepsilon} \right|_i \tag{7.136}$$

Knowing the error e_{n+1}, the accepted solution at t_{n+1} is

$$y_{n+1} = y_{n+1}\left(\frac{h}{2}\right) + \frac{1}{7}e_{n+1} \tag{7.137}$$

where e_{n+1} is calculated from Eq. 7.135. Equation 7.137 is derived as follows. Because Michelsen used a third order method, the truncation error is $O(h^3)$; hence,

$$y_{n+1}(h) = \hat{y}_{n+1} + \alpha(h)^3 \tag{7.138a}$$

and

$$y_{n+1}\left(\frac{h}{2}\right) = \hat{y}_{n+1} + \alpha\left(\frac{h}{2}\right)^3 \tag{7.138b}$$

Elimination of α between Eq. 7.138a and 7.138b yields the improved estimate for y_{n+1} as given in Eq. 7.137.

Once the better estimate for y_{n+1} is obtained (Eq. 7.137) the next step size is chosen as

$$h_{n+1} = h_n \min\left[(4q)^{-1/4}, 3\right] \tag{7.139}$$

The exponent $-1/4$ comes from the fourth order method, resulting from the extrapolation.

Note that the step size selection given in Eq. 7.139 is valid only when $q < 1$ (defined in Eq. 7.136). When q is greater than unity, the step size at t_n is halved and the method is recalculated at the time step of t_n until q is less than unity.

Unlike the strategy of Bailey, the technique after Michelsen utilizes a user-provided tolerance (ε).

7.11 HIGHER ORDER INTEGRATION METHODS

We have shown a number of integration techniques that are suitable to handle nonstiff coupled ODEs. The semi-implicit third order Runge–Kutta method (Michelsen, 1976) presented so far is the only method presented that can handle the stiff ODEs. More details for methods in solving stiff problems can be found in Gear (1971). Weimer and Clough (1979) performed a comparison between Michelsen's semi-implicit third order Runge–Kutta method and the Gear method on 18 different problems, and concluded that the Gear method is more computationally efficient than Michelsen's method. Moreover, in Michelsen's method the Jacobian must be evaluated exactly, whereas in the Gear method it can be approximated numerically.

Large-order chemical engineering problems, when solved numerically, usually give rise to a set of coupled differential-algebraic equations (DAE). This type of coupling is more difficult to deal with compared to coupled algebraic equations or coupled ODEs. The interested reader should refer to Brenan et al. (1989) for the exposition of methods for solving DAE.

7.12 REFERENCES

1. Bailey, H. E., "Numerical Integration of the Equation Governing the One-Dimensional Flow of a Chemically Reactive Gas," *Phys. Fluid* 12, 2292–2300 (1969).
2. Brenan, K. E., S. L. Campbell, and L. R. Petzold, *Numerical Solution of Initial Value Problems in Differential-Algebraic Equations*, Elsevier, New York (1989).
3. Burden, R. L., and J. D. Faires, *Numerical Analysis*, PWS Publishers, Boston (1981).
4. Caillaud, J. B., and L. Padmanabhan, "An Improved Semi-Implicit Runge–Kutta Method for Stiff Systems," *Chem. Eng. J.* 2, 227–232 (1971).
5. Finlayson, B. A., *Nonlinear Analysis in Chemical Engineering*, McGraw Hill, New York (1980).
6. Gear, C. W., *Numerical Initial Value Problems in Ordinary Differential Equations*, Prentice Hall, New Jersey (1971).
7. Gill, S., "A Process for the Step-by-Step Integration of Differential Equations in an Automatic Digital Computing Machine," *Proc. Camb. Philo. Soc.* 47, 96–108 (1951).
8. Michelsen, M. L., "An Efficient General Purposes Method for the Integration of Stiff Ordinary Differential Equations," *AIChEJ.* 22, 594–597 (1976).
9. Rosenbrock, H. H., "Some General Implicit Processes for the Numerical Solution of Differential Equation," *Comput. J.* 5, 329–340 (1963).
10. Weimer, A. W., and D. E. Clough, "A Critical Evaluation of the Semi-Implicit Runge–Kutta Methods for Stiff Systems," *AIChEJ.* 25, 730–732 (1979).

7.13 PROBLEMS

7.1$_2$. Convert the following nth order differential equation to the standard format of Eq. 7.1:

$$y^{(n)} + F\left(y^{(n-1)}, y^{(n-2)}, \ldots, y'', y', y\right) = 0$$

7.2$_2$. Convert the following differential equations to the standard format of Eq. 7.1:

(a) $$y'' + y' = \cos(t)$$

(b) $$y'' + y'z' = 3t^2$$

$$z' + z^2 + y = t$$

7.3$_2$. Runge–Kutta is one of the most popular methods for solving nonstiff ordinary differential equations because it is a self-starting method, that is, it needs only a condition at one point to start the integration, which is in contrast to the Adams family methods where values of dependent variables at several values of time are needed before they can be used. For the following equation

$$\frac{dy}{dt} = f(t, y)$$

subject to

$$t = t_0; \qquad y = y_0$$

(a) show that the Taylor series of $y_{n+1} = y(t_{n+1}) = y(t_n + h)$ is

$$y_{n+1} \approx y_n + f_n h + \left(\frac{\partial f_n}{\partial t} + \frac{\partial f_n}{\partial y} f_n\right)\frac{h^2}{2} + \cdots$$

where

$$y_{n+1} = y(t_{n+1}); \qquad y_n = y(t_n);$$

$$\frac{\partial f_n}{\partial t} = \frac{\partial f(t_n, y_n)}{\partial t}; \qquad \frac{\partial f_n}{\partial y} = \frac{\partial f(t_n, y_n)}{\partial y}$$

(b) The Runge–Kutta of order 2 is assumed to take the form (Eq. 7.121)

$$y_{n+1} = y_n + w_1 h f_n + w_2 h f(t_n + c_1 h, y_n + a_1 h f_n)$$

Expand the Runge–Kutta formula for h and compare with the Taylor series in part (a) to show that one possible solution for w_1, w_2, c_1, and a_1 is

$$w_1 = w_2 = \frac{1}{2}; \qquad c_1 = a_1 = 1$$

Show that this second order Runge–Kutta is the Euler predictor-corrector method.

(c) Another possible solution for w_1, w_2, c_1, and a_1 is $w_1 = w_2 = 1/2$ and $c_1 = a_1 = 1/2$.

7.4$_3$. (a) Starting from the general formula for the third order Runge–Kutta

$$y_{n+1} = y_n + \sum_{i=1}^{3} w_i k_i$$

where

$$k_i = hf\left(t_n + c_i h, y_n + \sum_{j=1}^{i-1} a_{ij} k_j\right); \qquad c_1 = 0$$

obtain a number of third order Runge–Kutta formula to integrate the equation

$$\frac{dy}{dt} = f(t, y)$$

(b) Show that one of the possible Runge–Kutta formulas is

$$k_1 = hf(t_n, y_n)$$

$$k_2 = hf\left(t_n + \frac{1}{2}h, y_n + \frac{1}{2}k_1\right)$$

$$k_3 = hf(t_n + h, y_n + 2k_2 - k_1)$$

and $w_1 = w_2 = w_3 = 1/6$.

7.5$_2$. Use the third order Runge–Kutta formula obtained in Problem 7.4 to derive the integration formulas for the coupled ordinary differential equations

$$\frac{dy_1}{dt} = f_1(t, y_1, y_2)$$

$$\frac{dy_2}{dt} = f_2(t, y_1, y_2)$$

subject to

$$t = 0; \qquad y_1 = y_{10}, \qquad y_2 = y_{20}$$

7.6$_2$. For a fourth order method, use the extrapolation procedure to show that the better estimate for $y(t_{n+1})$ is

$$\hat{y}_{n+1} = \frac{1}{15}\left[16 y_{n+1}\left(\frac{h}{2}\right) - y_{n+1}(h)\right]$$

7.7*. The following describes the catalytic cracking of gas oils to gasoline. The reaction network is as follows: Gas oil is cracked catalytically to gasoline according a second order chemical kinetics, and in a parallel reaction it is also cracked to give light gases and coke with second order chemical kinetics. Gasoline, once formed from the gas oil, is also cracked to give light gases and coke with a first order chemical kinetics.

(a) If we denote A, G for gas oil and gasoline, and a single symbol C for either coke or light gas, show that the kinetic equations describing the change of these species (expressed in weight fraction) are

(i)
$$\frac{dy_A}{dt} = -k_1 y_A^2 - k_3 y_A^2$$

(ii)
$$\frac{dy_G}{dt} = +k_1 y_A^2 - k_2 y_G$$

(iii)
$$\frac{dy_C}{dt} = +k_3 y_A^2 + k_2 y_G$$

The initial conditions for these species are

$$t = 0; \qquad y_A = 1; \qquad y_G = y_C = 0$$

(b) Before attempting numerical solutions, it is tedious to show that an analytical solution exists, with which we can compare the numerical solutions. To start this, note that the equations are not independent, but rather they are related by the overall mass balance equation. After adding the equations, show that

$$\frac{dy_A}{dt} + \frac{dy_G}{dt} + \frac{dy_C}{dt} = 0$$

Show that integration of this exact differential yields

$$y_A + y_G + y_C = 1$$

(c) Divide the kinetic equation for A by that for G to show

$$\frac{dy_G}{dy_A} = -\frac{k_1}{k_0} + \frac{k_2}{k_0} \frac{y_G}{y_A^2}$$

where

$$k_0 = k_1 + k_3$$

Next, show that the integration of this equation leads to the analytical result

$$y_G = \left(\frac{k_1 k_2}{k_0^2}\right) \exp(-\eta) \left[\frac{k_0}{k_2} \exp\left(\frac{k_2}{k_0}\right) - \frac{1}{\eta} \exp(\eta) + Ei(\eta) - Ei\left(\frac{k_2}{k_0}\right)\right]$$

where

$$\eta = \frac{k_2}{k_0 y_A}$$

and *Ei* is the tabulated exponential integral defined in Chapter 4 as

$$Ei(x) = \int_{-\infty}^{x} e^s \frac{ds}{s}$$

The weight fraction for A is obtained from direct integration of Eq. i to yield

$$y_A = \frac{1}{1 + k_0 t}$$

(d) It is clear that the analytical result is quite tedious to use, so that it may be easier to use a numerical solution. Use the explicit and implicit Euler methods to integrate the full kinetic equations. Use step size as your variable, and determine the minimum step size for the explicit Euler to remain stable. Compare your numerical results obtained by these techniques with the exact solutions obtained in part (c).

(e) Use the trapezoidal rule and show numerically that for a given step size the trapezoidal scheme is more accurate than the Euler methods. Explain why.

(f) Implement the Runge–Kutta–Gill scheme to numerically solve this problem. Use the same step size as in previous parts in order to compare the results.

7.8*. Perform the integration in Problem 7.5 using the semi-implicit formula of Michelsen (Eq. 7.126) with a constant step size.

7.9*. Repeat Problem 7.6, but instead of a constant step size, use the step size control described in Section 7.10 for the Michelsen method.

7.10*. (a) Reduce the following Van der Pol equation to the standard form

$$\frac{d^2 y}{dt^2} - \mu(1 - y^2)\frac{dy}{dt} + y = 0$$

subject to

$$t = 0; \qquad y = 1, \qquad y' = 0$$

(b) For $\mu = 0.1$, 1, and 10, solve the Van der Pol equation using the Runge–Kutta–Gill method and observe the behavior of the solution as μ varies. For large μ, study the solution behavior as step size is changed in the integration. Explain why the choice of step size is critical in this case.

(c) Repeat part (b) using the method of Michelsen.

7.11₃. Solve the differential equation

$$\frac{dy}{dt} + y = t; \qquad t = 0, y = 1$$

by using the following iteration scheme.

(a) First, put the nonderivative terms to the RHS and show that the solution can be put in the form of an implicit integral equation

$$y(t) = 1 + \int_0^t [s - y(s)] \, ds$$

where s is the dummy integration variable. The equation is implicit because of the presence of y in the integrand of the integral.

(b) The solution for y can be facilitated by using the iteration scheme

$$y^{(k+1)}(t) = 1 + \int_0^t [s - y^{(k)}(s)] \, ds$$

where $y^{(0)}$ is taken as the initial condition, that is, $y^{(0)} = 1$. Show that the next five iterated solutions are

$$y^{(1)}(t) = 1 - x + \frac{x^2}{2}$$

$$y^{(2)}(t) = 1 - x + x^2 - \frac{x^3}{6}$$

$$y^{(3)}(t) = 1 - x + x^2 - \frac{x^3}{3} + \frac{x^4}{24}$$

$$y^{(4)}(t) = 1 - x + x^2 - \frac{x^3}{3} + \frac{x^4}{12} - \frac{x^5}{120}$$

$$y^{(5)}(t) = 1 - x + x^2 - \frac{x^3}{3} + \frac{x^4}{12} - \frac{x^5}{60} + \frac{x^6}{720}$$

(c) Obtain the exact solution to the original differential equation, and compare the iterated solutions with the Taylor series expansion of the exact solution.

This iteration method is fundamentally different from all other methods discussed in Chapter 7. It basically produces iterated solutions that are valid over the whole domain of interest. As the iteration increases, the iterated solution is (hopefully) getting closer to the exact solution. This method is known as the *Picard* method. The fundamental disadvantage of this technique, which mitigates its usefulness, is that the iterated solutions must be found analytically, so that they can be applied to subsequent iteration steps (which involves the evaluation of an integral).

7.12₃. (a) Show that the modeling of a second order chemical reaction in a fixed bed reactor would give rise to the following differential equation

$$D\frac{d^2C}{dx^2} - u\frac{dC}{dx} - kC^2 = 0$$

where D is the axial diffusion coefficient, u is the superficial velocity, and k is the chemical reaction rate constant based on unit volume of reactor.

The typical Danckwerts boundary conditions are

$$x = 0; \qquad D\frac{dC}{dx} = u(C - C_0)$$

$$x = L; \qquad \frac{dC}{dx} = 0$$

where L is the reactor length and C_0 is the inlet concentration.

(b) Convert the above differential equation to the standard form, and show that it takes the form

$$\frac{dC}{dx} = p$$

$$\frac{dp}{dx} = \frac{u}{D}p + \frac{k}{D}C^2$$

(c) To solve the equations in part (b) using any of the integration techniques described in this chapter, conditions at *one* point must be specified, that is, either C and p $(= dC/dx)$ are specified at $x = 0$ or they are specified at $x = L$. Unfortunately, this problem has conditions specified at two end points. This means that we shall need to guess one of them at one point. For example, we can guess C at $x = 0$ and then p at the same point can be calculated from the condition at $x = 0$, or we can guess C at $x = L$ since p at $x = L$ is already specified. Use the latter option by specifying

$$x = L; \qquad \frac{dC}{dx} = 0$$

and

$$C = \alpha$$

where α is the guessing value of C at $x = L$. Use the third order Runge–Kutta scheme of Problem 7.4 to develop the integration formula for the above set of standard equations.

(d) It is obvious that after the integration from $x = L$ back to $x = 0$, the calculated C and p at $x = 0$ will not satisfy the specified condition at $x = 0$ (unless by coincidence the choice of α was the correct value of

C at $x = L$), so we expect the inequality to arise

$$Dp(0) \neq u[C(0) - C_0]$$

If the condition at $x = 0$ is not satisfied, a different value of α is chosen and the integration process is repeated until the above relation is finally satisfied within some prespecified error. This method is commonly called the *shooting* method. Use the secant method of Appendix A to show that the next guess of α should satisfy the following equation

$$\alpha^{(k+1)} = \alpha^{(k)} - \frac{f(\alpha^{(k)})}{\dfrac{f(\alpha^{(k)}) - f(\alpha^{(k-1)})}{\alpha^{(k)} - \alpha^{(k-1)}}}$$

where

$$f(\alpha^{(k)}) = Dp(0; \alpha^{(k)}) - uC(0; \alpha^{(k)}) + uC_0$$

Chapter **8**

Approximate Methods for Boundary Value Problems: Weighted Residuals

In the previous chapter we presented numerical techniques to solve differential equations of the initial value type, that is, the type where conditions are specified at only one position or time, such as $t = 0$ in the time domain. In this chapter we discuss methods to handle ordinary differential equations of the boundary value type, that is, conditions are specified at two different points in the domain, such as conditions at the two ends of a fixed bed reactor or conditions at two sides of a membrane. We start the chapter by discussing general aspects of the underlying basis for the method: weighted residuals. As a subset of this, we pay particular attention to the orthogonal collocation method, which enjoys wide popularity for current applications in chemical engineering research problems. It is particularly attractive for solving nonlinear problems which heretofore have defied analytical treatment.

8.1 THE METHOD OF WEIGHTED RESIDUALS

The method of weighted residuals has been used in solving a variety of boundary value problems, ranging from fluid flow to heat and mass transfer problems. It is popular because of the interactive nature of the first step, that is, the user provides a first guess at the solution and this is then forced to satisfy the governing equations along with the conditions imposed at the boundaries. The left-over terms, called *residuals*, arise because the chosen form of solution does not exactly satisfy either the equation or the boundary conditions. How these residual terms are minimized provides the basis for parameter or function selection. Of course, the optimum solution depends on the intelligent selection of a proposed solution.

To illustrate the salient features of the method, we first consider the following boundary value problem in an abstract form, and then later attempt an elementary example of diffusion and reaction in a slab of catalyst material. We shall assume there exists an operator of the type discussed in Chapter 2 (Section 2.5) so that in compact form, we can write

$$L(y) = 0 \tag{8.1}$$

where L is some differential operator. Examples of Eq. 8.1 are

$$L(y) = \frac{d^2y}{dx^2} - 100y^2 = 0$$

$$L(y) = \frac{d}{dx}\left[(1 + 10y)\frac{dy}{dx}\right] = 0$$

The differential equation (Eq. 8.1) is subject to the boundary condition

$$M(y) = 0 \tag{8.2}$$

These boundary values could be initial values, but any boundary placement is allowed, for example $dy(0)/dx = 0$, or $y(1) = 0$, where $M(y)$ is simply a general representation of the operation on y as $dy(0)/dx = 0$ or $y(1) = 0$, respectively.

The essential idea of the method of weighted residuals is to *construct* an approximate solution and denote it as y_a. Because of the approximate nature of the estimated solution, it *may not*, in general, satisfy the equation and the boundary conditions; that is:

$$L(y_a) = R \neq 0 \tag{8.3}$$

and

$$M(y_a) = R_b \neq 0 \tag{8.4}$$

where the *residuals* R and R_b are not identically zero. If the approximate solution is constructed such that the differential equation is satisfied exactly (i.e., $R = 0$) the method is called the *boundary method*. However, if it is constructed such that the boundary conditions are satisfied exactly (i.e., $R_b = 0$) the method is called the *interior method*. If neither the differential equation nor the boundary conditions are satisfied exactly, it is referred to as the *mixed method*.

The method of weighted residuals will require two types of *known functions*. One is called the *trial* function, and the other is called the *test* function. The former is used to construct the trial solution, and the latter is used as a basis (criterion) to make the residual R small (a *small* residual leads to a small error in the approximate solution). To minimize the residual, which is usually a function of x, we need a means to convert this into a scalar quantity so that a minimization can be performed. This is done by way of some form of *averaging*, which we call an *inner product*. This can be regarded as a measure of distance between the two functions; that is, between the residual function and the test function, as we show later.

The approximate solution to the governing equation Eq. 8.1 can be written as a polynomial, for example,

$$y_a(x) = y_0(x) + \sum_{i=1}^{N} a_i \phi_i(x) \tag{8.5}$$

where y_0 is suitably chosen to satisfy the boundary conditions exactly, and generally it is a function of x. In the following discussion, we shall use the interior method (i.e., the approximate solution satisfies the boundary conditions exactly). The *trial functions* ϕ_i chosen by the analyst must satisfy the boundary conditions, which are usually of the homogeneous type (Chapter 1). The coefficients a_i are unknown and will be determined by the method of residuals to force a close matching of y_a with the proposed equations. Thus, the solution of the governing equation (Eq. 8.1) is reduced to the determination of N coefficients, a_i, in the assumed approximate solution (Eq. 8.5).

Substituting this trial solution y_a into the differential equation (Eq. 8.1), we see

$$R(x) = L\left[y_0(x) + \sum_{i=1}^{N} a_i \phi_i(x) \right] \tag{8.6}$$

The residual R is in general nonzero over the whole domain of interest, so that it will be dependent on x, in the usual case.

Since the residual R is a function of x, we shall need to minimize it over the whole domain of interest. To do this, we need to define some form of averaging. For example, the following integral over the whole domain may be used as a means of averaging

$$\int_V R(x) w_k(x) \, dx \tag{8.7}$$

where V is the domain of interest, and w_k is some selected set of independent functions ($k = 1, 2, \ldots, N$), which are called the test functions. Such an integral is called an inner product, and we denote this averaging process as

$$(R, w_k) \tag{8.8a}$$

This notation is analogous to the dot product used in the analysis of vectors in Euclidean space. The dot product is an operation that maps two vectors into a scalar. Here, in the context of functions, the inner product defined in Eqs. 8.7 or 8.8a will map two functions into a scalar, which will be used in the process of minimization of the residual R. This minimization of the residual intuitively implies a small error in the approximate solution, $y_a(x)$.

Since we have N unknown coefficients a_i in the trial solution (Eq. 8.5), we will take the inner product (defined in Eq. 8.8a) of the residual with the first N test functions and set them to zero, and as a result we will have the following set of N nonlinear *algebraic* equations

$$(R, w_k) = 0 \qquad \text{for} \qquad k = 1, 2, 3, \ldots, N \qquad (8.8b)$$

which can be solved by using any of the algebraic solvers discussed in Appendix A to obtain the coefficients a_i ($i = 1, 2, \ldots, N$) for the approximate solution (Eq. 8.5).

This completes a brief overview of the approximation routine. To apply the technique, specific decisions must be made regarding the selection of test function and a definition of an inner product (Eq. 8.7 being only one possible choice).

8.1.1 Variations on a Theme of Weighted Residuals

There are five widely used variations of the method of weighted residuals for engineering and science applications. They are distinguished by the choice of the test functions, used in the minimization of the residuals (Eq. 8.8). These five methods are

1. The collocation method
2. The subdomain method
3. The least square method
4. The moment method
5. The Galerkin method

Each of these methods have attractive features, which we discuss as follows. Later, we shall concentrate on the collocation method, because it is easy to apply and it can also give good accuracy. The Galerkin method gives better accuracy, but it is somewhat intractable for higher order problems, as we illustrate in the examples to follow.

1. *The collocation method* In this method the test function is the Dirac delta function at N *interior* points (called collocation points) within the domain of interest, say $0 < x < L$:

$$w_k = \delta(x - x_k) \qquad (8.9)$$

where x_k is the kth collocation point.

The useful property of the Dirac's delta function is

$$\int_{x_k^-}^{x_k^+} f(x)\delta(x - x_k)\,dx = f(x_k) \qquad (8.10)$$

If these N interior collocation points are chosen as roots of an orthogonal Jacobi polynomial of Nth degree, the method is called the orthogonal collocation method (Villadsen and Michelsen 1978). It is possible to use other orthogonal functions, but Jacobi is popular because it is compact and contains only a few terms. Another attractive feature is that the solution can be derived in terms of *the dependent variable y at the collocation points*. This will be illustrated in Section 8.4.

2. *The subdomain method* In this method the domain V of the boundary value problem is split into N subdomains V_i; hence, the origin of the name "subdomain method." The test function is chosen such that

$$w_k = 1 \qquad (8.11)$$

in the subdomain V_k and is zero elsewhere.

3. *The least square method* In this method, the test function is chosen as

$$w_k = \frac{\partial R}{\partial a_k} \qquad (8.12)$$

With this definition, Eq. 8.8 becomes

$$\left(R, \frac{\partial R}{\partial a_k} \right) = \frac{1}{2} \frac{\partial}{\partial a_k} (R, R) = 0 \qquad (8.13)$$

Thus, if the inner product is defined as an integral such as in Eq. 8.7, Eq. 8.13 can be written explicitly as

$$\int_V R \frac{\partial R}{\partial a_k} \, dx = \frac{1}{2} \frac{\partial}{\partial a_k} \int_V R^2 \, dx = 0 \qquad (8.14)$$

This means that the coefficients a_k are found as the minimum of (R, R). The least squares result is the most well-known criterion function for weighted residuals. The test function for this technique is more complicated, owing to the requirement of differentiation in Eq. 8.12.

4. *The moment method* In this method, the test function is chosen as

$$w_k = x^{k-1} \qquad \text{for} \qquad k = 1, 2, \ldots, N \qquad (8.15)$$

5. *The Galerkin method* In this method, the weighting function is chosen from the same family as the trial functions, ϕ_k, that is:

$$w_k = \phi_k(x) \qquad (8.16)$$

For all five methods just discussed, the only restriction on the trial functions is that they must belong to a complete set of linearly independent functions. These functions need not be orthogonal to each other.[1] The trial and test

---

[1] The definition of orthogonality in the sense used here is discussed in Chapter 10, Section 10.5.

functions must also be chosen as the first N members of that set of independent functions. This will improve the efficiency of the methods of weighted residuals. For the Galerkin method, if the trial and test functions are chosen based on the knowledge of the form of the exact solution of a closely related problem, the efficiency of the method is enhanced (Fletcher 1984). It is noted here that an orthogonal set can always be created from the given set of independent functions.

EXAMPLE 8.1

We illustrate the above five variations of weighted residuals with the following example of diffusion and first order chemical reaction in a slab catalyst (Fig. 8.1). We choose the first order reaction here to illustrate the five methods of weighted residual. In principle, these techniques can apply equally well to nonlinear problems, however, with the exception of the collocation method, the integration of the form (8.7) may need to be done numerically.

Carrying out the mass balance over a thin shell at the position r with a thickness of Δr, we have

$$SJ|_r - SJ|_{r+\Delta r} - S\,\Delta r(kC) = 0 \qquad (8.17)$$

where S is the cross-sectional area of the catalyst, J is the diffusion flux, defined as moles per unit total area per time, and kC is the chemical reaction rate per unit volume of the catalyst.

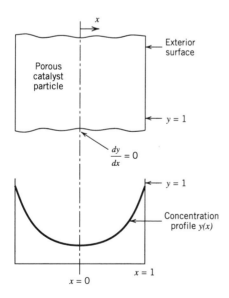

Figure 8.1 Diffusion and reaction in a slab catalyst.

Dividing the equation by $S\Delta r$, we have

$$-\frac{J|_{r+\Delta r} - J|_r}{\Delta r} - kC = 0 \tag{8.18}$$

If we now allow the shell as thin as possible; that is, in the limit of Δr approaching zero, we have

$$-\lim_{\Delta r \to 0} \frac{J|_{r+\Delta r} - J|_r}{\Delta r} - kC = 0 \tag{8.19}$$

Using the definition of the derivative, the first two terms in the LHS of Eq. 8.19 become the first derivative of J; that is,

$$-\frac{dJ}{dr} - kC = 0 \tag{8.20}$$

The flux into the catalyst particle is taken to be proportional to the concentration gradient

$$J = -D_e \frac{dC}{dr} \tag{8.21}$$

where D_e is the effective diffusivity of the reactant within the catalyst particle, and it is a function of the structure of the catalyst.

Substituting the expression for the diffusion flux (Eq. 8.21) into the mass balance equation (Eq. 8.20) yields

$$D_e \frac{d^2C}{dr^2} - kC = 0 \tag{8.22}$$

This is a second order differential equation. The boundary conditions for this problem are

$$r = 0; \qquad \frac{dC}{dr} = 0 \tag{8.23a}$$

$$r = R; \qquad C = C_0 \tag{8.23b}$$

The first condition indicates that there is no flux at the center of the catalyst particle. This condition also says that the reactant concentration profile is symmetrical at the center. This situation is commonly referred to as the symmetry condition. The second boundary condition corresponds to high velocity at the boundary since the reactant concentration at the surface is taken equal to that of the bulk surrounding the particle, which is taken to be invariant in the present problem.

It is convenient to cast the mass balance equation and the boundary conditions into dimensionless form, with the independent variable having the domain from 0 to 1. By defining the following nondimensional variables and parameters

$$y = \frac{C}{C_0}; \qquad x = \frac{r}{R}; \qquad \phi^2 = \frac{kR^2}{D_e} \qquad (8.24)$$

the mass balance equation and the boundary conditions take the following clean form

$$\frac{d^2y}{dx^2} - \phi^2 y = 0 \qquad (8.25a)$$

$$x = 0; \qquad \frac{dy}{dx} = 0 \qquad (8.25b)$$

$$x = 1; \qquad y = 1 \qquad (8.25c)$$

The quantity of interest is the overall reaction rate, which is the observed rate. Starting with the thin shell, the chemical reaction rate in the shell is

$$\Delta R_{r \times n} = (S \, \Delta r)(kC) \qquad (8.26)$$

Thus, the overall reaction rate is obtained by summing all thin elements, which leads to the integral

$$R_{r \times n} = \int_{-R}^{R} SkC \, dr = 2 \int_{0}^{R} SkC \, dr \qquad (8.27)$$

Hence, the overall reaction rate per unit volume of the catalyst slab is

$$\frac{R_{r \times n}}{V_p} = \frac{1}{R} \int_{0}^{R} kC \, dr \qquad (8.28)$$

Written in terms of the nondimensional variables, the overall reaction rate per unit volume is

$$\frac{R_{r \times n}}{V_p} = kC_0 \int_{0}^{1} y \, dx \qquad (8.29)$$

If we denote

$$\eta = \int_{0}^{1} y \, dx \qquad (8.30)$$

which is called the effectiveness factor, the overall reaction rate per unit catalyst volume is given by

$$\frac{R_{r \times n}}{V_p} = kC_0 \cdot \eta \qquad (8.31)$$

Thus, it is seen now that the overall reaction rate per unit volume is equal to the intrinsic reaction rate per unit volume (kC_0) modified by a factor, called the effectiveness factor, η. The presence of this factor accounts for the fact that the chemical reaction in the catalyst particle is affected by a diffusional resistance. If this diffusional resistance is negligible compared to the reaction rate, the overall reaction rate must be equal to the intrinsic reaction rate. If this resistance is strong, we would expect the overall reaction rate to be less than the intrinsic reaction rate; that is, $\eta < 1$.

Let us start the illustration with the situation where $\phi = 1$, that is, the rate of reaction is comparable to the rate of diffusion. Using the techniques taught in Chapters 2 and 3, the exact solution to Eqs. 8.25 (which is needed later as a basis for comparison with approximate solutions) is

$$y = \frac{\cosh(\phi x)}{\cosh(\phi)} \tag{8.32}$$

and for $\phi = 1$, this yields

$$y = \frac{\cosh(x)}{\cosh(1)} \tag{8.32a}$$

Knowing the dimensionless concentration inside the porous catalyst, the dimensionless reaction rate is given by the integral (Eq. 8.30), which in general is

$$\eta = \frac{\tanh(\phi)}{\phi} \tag{8.33a}$$

and when $\phi = 1$ it is

$$\eta = \int_0^1 y\, dx = \tanh(1) = 0.7616 \tag{8.33b}$$

Now we undertake to find approximate solutions using the several variations of weighted residuals. Central to all methods is the choice of the trial solution, y_a. By noting the boundary conditions at $x = 0$ and $x = 1$, a parabolic function seems to be a good choice so we write our first guess as

$$y_a = 1 + a_1(1 - x^2) \tag{8.34}$$

Here, the function $y_0(x) = 1$, which satisfies the boundary conditions (Eqs. 8.25), and the trial function $\phi_1(x) = 1 - x^2$, which satisfies the homogeneous boundary conditions

$$x = 0; \quad \frac{d\phi_1}{dx} = 0 \quad \text{and} \quad x = 1; \quad \phi_1 = 0$$

Hence, the trial solution $y_a(x)$ satisfies the boundary conditions (Eqs. 8.25) exactly; that is,

$$x = 0; \quad \frac{dy_a}{dx} = 0$$

and

$$x = 1; \quad y_a = 1 \tag{8.35}$$

Next, substituting the trial function into the differential equation (Eq. 8.25a), we obtain the residual, defined here as

$$Ly_a = \frac{d^2 y_a}{dx^2} - y_a = R \tag{8.36}$$

or specifically

$$R(x) = -2a_1 - \left[1 + a_1(1 - x^2)\right] \neq 0 \tag{8.37}$$

Note that the residual is *a function of x*. To find the coefficient a_1, we need to "transform" this x-dependent function into a quantity, which is not x dependent. This is the central idea of the inner product, which we have discussed earlier. Since we are dealing with functions, the inner product should be defined in the form of an integral, to eliminate dependence on x. For the present problem, the inner product between the residual and a test function w_1 we only need one, since there is only one coefficient, a_1, in the trial solution, (Eq. 8.34) is

$$(R, w_1) = \int_0^1 R w_1 \, dx = 0 \tag{8.38}$$

Observing the inner product of Eq. 8.38, we note, following integration, an algebraic relationship is produced, the form and complexity of which depends on the selection of the test function w_1. This is the only difference among the various methods of weighted residuals. Let us demonstrate this with various methods one by one. We start with the collocation method.

(a) *The collocation method*   In this method, the test function was stipulated earlier to be

$$w_1(x) = \delta(x - x_1) \tag{8.39}$$

where x_1 is a (as yet unknown) collocation point chosen in the domain $[0, 1]$. With one collocation point, the method is called *single point* collocation, a convenient tool for assessing system behavior.

With this test function, Eq. 8.38 can be integrated directly and we obtain the following algebraic equation

$$\begin{aligned}(R, w_1) &= \int_0^1 \left\{-2a_1 - \left[1 + a_1(1 - x^2)\right]\right\} \delta(x - x_1) \, dx \\ &= -2a_1 - \left[1 + a_1(1 - x_1^2)\right] = 0\end{aligned} \tag{8.40}$$

where we have used the property of the Dirac delta function (Eq. 8.10). Equation 8.40 is simply the residual at $x = x_1$, and the collocation method gets its name from the fact that it forces the residual to be zero at particular points, called collocation points.

From Eq. 8.40, the unknown coefficient a_1 can be readily determined in terms of the point x_1 (which is the single collocation point)

$$a_1 = - \frac{1}{\left[2 + \left(1 - x_1^2\right)\right]} \tag{8.41}$$

We note that x_1 is unspecified, and it must be chosen in the range: $0 < x_1 < 1$.
The trial solution, which approximates the exact solution, is simply

$$y_a = 1 - \frac{\left(1 - x^2\right)}{\left[2 + \left(1 - x_1^2\right)\right]} \tag{8.42}$$

The quantity of interest is the overall reaction rate, and it takes the dimensionless form

$$\eta = \int_0^1 y \, dx \approx \int_0^1 y_a \, dx \tag{8.43}$$

If we substitute the trial solution of Eq. 8.42 into Eq. 8.43, we have

$$\eta_a = 1 - \frac{2}{3\left[2 + \left(1 - x_1^2\right)\right]} \tag{8.44}$$

Thus, if we choose the collocation point as the midpoint of the domain $[0, 1]$ so that $x_1 = 1/2$, then the integral of Eq. 8.44 will take the value

$$\eta_a = 1 - \frac{2}{3\left\{2 + \left[1 - (0.5)^2\right]\right\}} = 1 - \frac{8}{33} = 0.7576 \tag{8.45}$$

We see that the approximate solution obtained by the collocation method agrees fairly well with the exact solution ($\eta = 0.7616$). Note here that the selection of x_1 was arbitrary, and intuitive. Other choices are possible, as we shall see.

(b) *The subdomain method* Here, since we have only one unknown coefficient in the trial solution, only one subdomain is dealt with, and it is the full domain of the problem, that is, $[0, 1]$; hence, the test function is

$$w_1(x) = 1 \qquad \text{for} \qquad 0 < x < 1 \tag{8.46}$$

The inner product is defined as before (Eq. 8.38). Substituting the residual of Eq. 8.37 and the test function (Eq. 8.46) into the inner product (Eq. 8.38) we have

$$(R, w_1) = \int_0^1 R w_1 \, dx = \int_0^1 \left\{-2a_1 - \left[1 + a_1(1 - x^2)\right]\right\}(1) \, dx = 0 \tag{8.47}$$

Integrating the above equation, we finally obtain the following solution for a_1

$$a_1 = -\frac{3}{8} \tag{8.48}$$

Hence, the trial solution by the subdomain method is

$$y_a = 1 - \frac{3}{8}(1 - x^2) \tag{8.49}$$

and the approximate nondimensional chemical reaction rate is

$$\eta_a = \int_0^1 y_a \, dx = \int_0^1 \left[1 - \frac{3}{8}(1 - x^2)\right] dx = \frac{3}{4} = 0.75 \tag{8.50}$$

which also compares well with the exact solution $\eta_{exact} = 0.7616$.

(c) *The least square method*   The test function for the least square approach is

$$w_1 = \frac{\partial R}{\partial a_1} \tag{8.51}$$

The inner product (Eq. 8.38) for this method is

$$(R, w_1) = \int_0^1 R \frac{\partial R}{\partial a_1} \, dx = \frac{1}{2} \frac{\partial}{\partial a_1} \int_0^1 R^2 \, dx = 0 \tag{8.52}$$

that is,

$$(R, w_1) = \frac{1}{2} \frac{\partial}{\partial a_1} \int_0^1 \left\{-2a_1 - \left[1 + a_1(1 - x^2)\right]\right\}^2 dx = 0 \tag{8.53}$$

Integrating with respect to x and then differentiating with respect to a_1 yields

$$(R, w_1) = \frac{1}{2} \left[\frac{216}{15} a_1 + \frac{16}{3}\right] = 0 \tag{8.54}$$

Hence, solving for a_1, we have

$$a_1 = -\frac{10}{27} \tag{8.55}$$

The trial solution for the least square method is

$$y_a = 1 - \frac{10}{27}(1 - x^2) \tag{8.56}$$

and the approximate nondimensional chemical reaction rate is

$$\eta_a = \int_0^1 y_a \, dx = \int_0^1 \left[1 - \frac{10}{27}(1 - x^2)\right] dx = \frac{61}{81} = 0.75309 \qquad (8.57)$$

compared to the exact solution of $\tanh(1) = 0.7616$.

(d) *The moment method* The test function for the moment method is

$$w_1 = x^0 = 1 \qquad (8.58)$$

which is identical to the test function of the subdomain method. Thus, the solution of the moment method is the same as that of the subdomain method. This is true because only one term is retained in the trial solution.

(e) *The Galerkin method* The test function is the same as the trial solution; that is,

$$w_1 = (1 - x^2) \qquad (8.59)$$

Thus, the inner product (Eq. 8.38) becomes

$$(R, w_1) = \int_0^1 \left\{-2a_1 - \left[1 + a_1(1 - x^2)\right]\right\}(1 - x^2) \, dx = 0 \qquad (8.60)$$

that is,

$$(R, w_1) = -\frac{2}{3}(1 + 2a_1) - \frac{8}{15}a_1 = 0 \qquad (8.61)$$

Solving for a_1 gives

$$a_1 = -\frac{10}{28} \qquad (8.62)$$

Thus, the trial solution obtained by the Galerkin method is

$$y_a = 1 - \frac{10}{28}(1 - x^2) \qquad (8.63)$$

and the approximate nondimensional chemical reaction rate, η_a, is

$$\eta_a = \int_0^1 y_a \, dx = \int_0^1 \left[1 - \frac{10}{28}(1 - x^2)\right] dx = \frac{64}{84} = 0.7619 \qquad (8.64)$$

Table 8.1 provides a summary of the five methods using only one term.

It is seen from Table 8.1 that Galerkin appears to be the most accurate method for this specific problem. However, when more terms are used in the trial solution, the Galerkin method presents more analytical difficulties than the collocation method. As a matter of fact, all the weighted residual procedures

Table 8.1 Comparison of Accuracy for Approximate Solutions

Method	$\eta = \int_0^1 y_a \, dx$	Relative Percentage Error
Collocation	25/33	0.53
Subdomain	3/4	1.5
Least square	61/81	1.1
Moment	3/4	1.5
Galerkin	64/84	0.041

require an integration of the form of Eq. 8.7 or 8.8 and hence, with the exception of the collocation method, may require numerical evaluation of this integral if analytical integration is impossible. The ease of performing integrals with the Dirac delta function is an enormous advantage for the collocation technique.

All the methods tried so far show very low relative errors, in spite of the fact that only one term was retained in the trial solution (Eq. 8.34). This arises because the diffusion-reaction problem used for illustration has a slow reaction rate. This implies the concentration profile inside the particle is *very shallow* and can be easily described with only one term in the series of parabolic trial functions.

EXAMPLE 8.2

If we allow higher reaction rates in Eq. 8.25a, one can see that the concentration profile becomes rather steep and the trial solution with only one term obviously will become inadequate, as we shall show.

Let us take $\phi = 10$ (i.e., fast reaction relative to diffusion), hence the exact solution of Eq. 8.25 now becomes

$$y = \frac{\cosh(10x)}{\cosh(10)} \tag{8.65}$$

The nondimensional chemical reaction rate is obtained from the integral

$$\eta = \int_0^1 y \, dx = \int_0^1 \frac{\cosh(10x)}{\cosh(10)} \, dx = \frac{\tanh(10)}{10} = 0.1 \tag{8.66}$$

If we use only one term in Eq. 8.34 and substitute it into Eq. 8.25a with $\phi = 10$, we obtain the following residual

$$R = -2a_1 - 100\left[1 + a_1(1 - x^2)\right] \tag{8.67}$$

We now apply the five variations of the weighted residuals and follow the procedure as presented in the last example.

1. Collocation method Using this method, we obtain

$$a_1 = -\frac{100}{\left[2 + 100\left(1 - x_1^2\right)\right]} \tag{8.68a}$$

$$\eta_a = \int_0^1 y_a \, dx = 1 - \frac{200}{3\left[2 + 100\left(1 - x_1^2\right)\right]} \tag{8.68b}$$

For $x_1 = 0.5$, the nondimensional reaction rate is

$$\eta_a = 1 - \frac{200}{3\left[2 + 100\left(1 - (0.5)^2\right)\right]} = 0.1342 \tag{8.68c}$$

Comparing with the exact solution of $\eta = 0.1$, the relative error is 34%.

2. Subdomain method For this method, we have

$$a_1 = -\frac{300}{206} \quad \text{and} \quad \eta_a = 0.02913 \tag{8.69}$$

The relative error between the approximate solution η_a and the exact solution is 71%.

3. Least square method We have

$$a_1 = -\frac{41,200}{33,624} \quad \text{and} \quad \eta_a = 0.183123 \tag{8.70}$$

The relative error is 83%.

4. Moment method same as the subdomain method.

5. Galerkin method We have

$$a_1 = -\frac{1000}{820} \quad \text{and} \quad \eta_a = 0.18699 \tag{8.71}$$

The relative error is 87%.

We noted that in the case of high reaction rate, the collocation technique seems to be superior to the others. However, the relative errors between all approximate solutions and the exact solution are unacceptably high. The reason for the high error arises from the sharpness of the concentration profiles, as illustrated in Fig. 8.2. Moreover, the approximate solutions yield negative concentration over some parts of the domain $[0, 1]$.

To improve the accuracy in this case, we need to retain more terms in the trial solutions. In general, the more terms retained, the better the accuracy. This

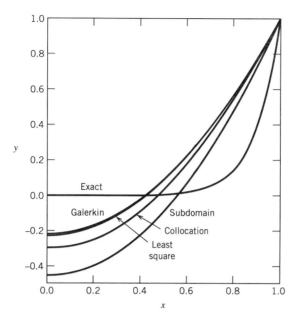

Figure 8.2 Comparison of concentration profiles for weighted residual methods.

can be carried out in a straightforward way with a computer. To demonstrate how this is done, we shall now retain two terms in the general expansion

$$y_a = 1 + \sum_{j=1}^{N} a_j x^{2(j-1)}(1 - x^2)$$

which is simply

$$y_a = 1 + a_1(1 - x^2) + a_2 x^2(1 - x^2) \tag{8.72}$$

Because of the symmetry property discussed earlier, our trial functions should be chosen to be even functions. This saves time, since a selection of odd functions would ultimately lead to a zero value for the multiplicative coefficient.

Inserting y_a into the defining equation yields the residual

$$R = -2a_1 - 10a_2 - 100$$
$$+ (12a_2 - 100a_1)(1 - x^2) - 100a_2 x^2(1 - x^2) \tag{8.73}$$

We now apply the collocation method to this residual, using the following two test functions

$$w_1 = \delta(x - x_1) \quad \text{and} \quad w_2 = \delta(x - x_2) \tag{8.74}$$

where x_1 and x_2 are two separate collocation points, chosen in the domain $[0, 1]$.

Next, we evaluate the following two inner products

$$(R, w_1) = \int_0^1 R w_1 \, dx = 0 \quad \text{and} \quad (R, w_2) = \int_0^1 R w_2 \, dx = 0 \quad \textbf{(8.75)}$$

to obtain two algebraic equations, solvable for unknowns a_1 and a_2

$$a_1 = -0.932787 \quad \text{and} \quad a_2 = -1.652576 \quad \textbf{(8.76)}$$

for two equally spaced collocation points chosen at $x_1 = 1/3$ and $x_2 = 2/3$.

Knowing the coefficients and the trial solution as in Eq. 8.72, we can now evaluate the integral

$$\eta_a = \int_0^1 y_a \, dx = 0.1578 \quad \textbf{(8.77)}$$

This means that the error relative to the exact solution is 58%, which is even higher than the error when we kept only one term in the trial solution! This is a somewhat surprising result; however, we have made no effort to select the collocation points in any optimum way. Nonetheless, if we retained more terms in the trial solutions (say 10 terms), the accuracy becomes better, as we might expect.

Let us return to the trial solution with two terms and note that we used two collocation points at $1/3$ and $2/3$, that is, we used *equal spacing collocation points*. Now, let us try two different collocation points $x_1 = 0.285231517$ and $x_2 = 0.765055325$, which lead to the coefficients $a_1 = -0.925134$ and $a_2 = -2.040737$. Knowing these two coefficients, we can evaluate η_a to obtain only 11% relative error, in contrast to 58% using equal spacing. This means that the choice of the collocation points is critical, and cannot be undertaken in an arbitrary way. The full potential of the collocation method can only be realized by judicious selection of the collocation points. Moreover, the choice of functions is critical. Orthogonal functions, such as Jacobi polynomials, are particularly attractive, since they are compact and contain only a few terms. One cannot expect good results with *any* orthogonal polynomial. The last choices of the two collocation points were in fact roots of the Jacobi polynomial. They gave good result because the weighting function of the Jacobi polynomial is

$$x^\beta (1 - x)^\alpha$$

and the effectiveness factor also has a weighting factor of a similar form.

This previous discussion highlights the essential features of the *orthogonal collocation method*. The word orthogonal implies the Jacobi polynomials are orthogonal in the sense of the integral shown in Eq. 8.83. We give more detailed discussion in Section 8.2. Also, we again emphasize the origin of orthogonality relates to the Sturm–Liouville equation, discussed in Section 10.5.

Among the many variations of weighted residuals, the orthogonal collocation is the easiest to use because its formulation is straightforward (owing to the Dirac delta function), the accuracy is good, and if many collocation points are used, the accuracy becomes excellent. With high-speed personal computers, the computation time required for many such terms is minimal. The subdomain and least square methods are more tedious to use. The Galerkin is used by applied mathematicians in some specific cases, but it is not as popular as the orthogonal collocation technique.

Because it has become so widely used, we devote the remainder of the chapter to orthogonal collocation. Before discussing additional details, we introduce a number of preliminary steps, which are needed for further development. These steps include a discussion of Jacobi polynomials (its choice has been explained above) and the Lagrangian interpolation polynomials. The Lagrangian interpolation polynomial is chosen as a convenient vehicle for interpolation between collocation points.

The Jacobi polynomials are important in providing the optimum positions of the collocation points. Since these collocation points are not equally spaced, the Lagrangian interpolation polynomials are useful to effect an approximate solution.

8.2 JACOBI POLYNOMIALS

Since all finite domains can be expressed in the range $[0, 1]$ through a linear transformation, we will consider the Jacobi polynomials defined in this domain. This is critical because the orthogonality condition for the polynomials depends on the choice of the domain. The Jacobi polynomial is a solution to a class of second order differential equation defined by Eq. 3.145.

The Jacobi polynomial of degree N has the power series representation

$$J_N^{(\alpha, \beta)}(x) = \sum_{i=0}^{N} (-1)^{N-i} \gamma_{N,i} x^i \tag{8.78}$$

with $\gamma_{N,0} = 1$. Note, the series contains a finite number of terms (N), and is therefore compact (not carried to infinity).

Here, $\gamma_{N,i}$ are constant coefficients, and α and β are parameters characterizing the polynomials, as shown in Eq. 8.83. That is, $J_N^{(\alpha, \beta)}(x)$ is the polynomial orthogonal with respect to the weighting function $x^\beta (1 - x)^\alpha$. The term $(-1)^{N-i}$ is introduced in the series to insure the coefficients γ are always positive. Note that the Jacobi function $J_N^{(\alpha, \beta)}(x)$ is a polynomial of degree N, since the summation in Eq. 8.78 is bounded and not infinite as with other orthogonal functions, such as Bessel's.

8.2.1 Rodrigues Formula

The Jacobi polynomials are given explicitly by the Rodrigues formula

$$J_N^{(\alpha, \beta)}(x)\left[x^\beta(1 - x)^\alpha\right] = \frac{(-1)^N \Gamma(\beta + 1)}{\Gamma(N + \beta + 1)} \frac{d^N}{dx^N}\left[x^{N+\beta}(1 - x)^{N+\alpha}\right] \tag{8.79}$$

where Γ is the Gamma function and its definition and properties are detailed in Chapter 4. For many applications, $\alpha = \beta = 0$, and we may conveniently drop the superscripts, that is, $J_N^{(0,0)} = J_N$.

EXAMPLE 8.3

For $\alpha = \beta = 0$, we have the following three Jacobi polynomials ($N = 1, 2, 3$) using Eq. 8.79

$$J_1(x) = -1 + 2x \qquad (8.80)$$

$$J_2(x) = 6x^2 - 6x + 1 \qquad (8.81)$$

$$J_3(x) = 20x^3 - 30x^2 + 12x - 1 \qquad (8.82)$$

The curves for these three Jacobi polynomials are shown in Fig. 8.3 over the domain [0, 1]. It is important to note that J_1 has one zero, J_2 has two zeros, and J_3 has three zeros *within* the domain [0, 1]; zeros are the values of x, which cause $J_N(x) = 0$. These zeros will be used later as the interior collocation points for the orthogonal collocation method.

8.2.2 Orthogonality Conditions

Since the Jacobi polynomials belong to a class of orthogonal polynomials, they satisfy the orthogonality condition

$$\int_0^1 \left[x^\beta (1 - x)^\alpha \right] J_j^{(\alpha, \beta)}(x) J_N^{(\alpha, \beta)}(x) \, dx = 0 \qquad (8.83)$$

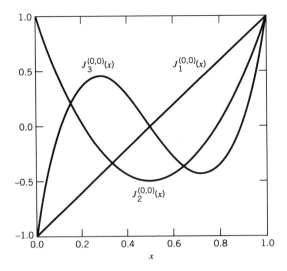

Figure 8.3 Plots of three Jacobi polynomials.

for $j = 0, 1, 2, \ldots, (N - 1)$, that is, all Jacobi polynomials are orthogonal to each other except to itself (i.e., when $j = N$). This condition arises by consideration of the Sturm–Liouville equation, the detailed discussion of which is given in Section 10.5.1.

The integration is defined in the domain $[0, 1]$. Outside this domain, orthogonality cannot be guaranteed. Any physical systems having a finite domain can be easily scaled to reduce to the domain $[0, 1]$.

The weighting function for this particular orthogonality condition defined with reference to the Sturm–Liouville equation is

$$W(x) = x^{\beta}(1 - x)^{\alpha} \tag{8.84}$$

The exponents α and β are seen to dictate the nature of the orthogonal Jacobi polynomials.

There are N equations of orthogonality (Eq. 8.83) because $j = 0, 1, 2, \ldots,$ $N - 1$, and there are exactly N unknown coefficients of the Jacobi polynomial of degree N to be determined

$$\gamma_{N,1}, \gamma_{N,2}, \gamma_{N,3}, \ldots, \gamma_{N,N}$$

Note that $\gamma_{N,0} = 1$.

Solving these N linear equations for N unknown coefficients, the following explicit solution is obtained for γ (Villadsen 1970)

$$\gamma_{N,i} = \binom{N}{i} \frac{\Gamma(N + i + \alpha + \beta + 1)\Gamma(\beta + 1)}{\Gamma(N + \alpha + \beta + 1)\Gamma(i + \beta + 1)} \tag{8.85a}$$

where the representation for the lead term is

$$\binom{N}{i} = \frac{N!}{i!(N - i)!} \tag{8.85b}$$

The above equation provides the explicit formula for the coefficients. During computation, it is easier to evaluate coefficients using the following recurrence formula

$$\frac{\gamma_{N,i}}{\gamma_{N,i-1}} = \frac{N - i + 1}{i} \cdot \frac{N + i + \alpha + \beta}{i + \beta}$$

starting with $\gamma_{N,0} = 1$.

Using the formula obtained above for the coefficients, we can evaluate the first four Jacobi polynomials, with $\alpha = \beta = 0$

$$J_0 = 1, \quad J_1 = 2x - 1, \quad J_2 = 6x^2 - 6x + 1, \quad J_3 = 20x^3 - 30x^2 + 12x - 1$$

On inspection of these, we note that (except the first one, which is equal to unity) all have *zeros within the domain from 0 to 1* (Fig. 8.3). Here, zeros are used to denote roots of $J_N(x) = 0$, so that $J_N(x)$ has N values of x causing it to become zero.

Since the orthogonal collocation method will require roots of the Jacobi polynomial, we shall need to discuss methods for the computation of zeros of $J_N^{(\alpha,\beta)}(x)$.

It has been proved by Villadsen and Michelsen (1978), using the orthogonality condition in Eq. 8.83, that $J_N^{(\alpha,\beta)}$ has N distinct, real-valued zeros in the domain $[0, 1]$.

If the Newton–Raphson method is applied using an initial guess $x = 0$, the first root found will be x_1. Once this root is obtained, we can obtain the next root by suppressing the previously determined zero at x_1. In general, if $x_1, x_2, \ldots, x_k$ are previously determined zeros, we can suppress these roots by constructing the following function

$$G_{N-k} = \frac{p_N(x)}{\displaystyle\prod_{i=1}^{k}(x - x_i)} \tag{8.86a}$$

where $p_N(x)$ is the rescaled polynomial

$$p_N(x) = \frac{J_N^{(\alpha,\beta)}(x)}{\gamma_{N,N}} \tag{8.86b}$$

The Newton–Raphson formula to determine the root x_{k+1} at the ith iteration is

$$x_{k+1}^{(i)} = x_{k+1}^{(i-1)} - \left[\frac{G_{N-k}(x)}{G'_{N-k}(x)}\right]_{x_{k+1}^{(i-1)}} \tag{8.87a}$$

for $i = 1, 2, \ldots$, and the initial guess for x_{k+1} is

$$x_{k+1}^{(0)} = x_k + \varepsilon \tag{8.87b}$$

where ε is a small number and a good starting value of 1×10^{-4} is recommended.

The function in the bracket of Eq. 8.87a is obtained from Eq. 8.86a and can be written explicitly as

$$\frac{G_{N-k}(x)}{G'_{N-k}(x)} = \frac{\left[\dfrac{p_N(x)}{p'_N(x)}\right]}{1 - \left[\dfrac{p_N(x)}{p'_N(x)}\right]\displaystyle\sum_{i=1}^{k}\frac{1}{(1 - x_i)}} \tag{8.88a}$$

with $p_N(x)$ and $p'_N(x)$ determined from the following recursive formula for computation

$$p_N(x) = (x - g_N)p_{N-1} - h_N p_{N-2} \tag{8.88b}$$

$$p'_N(x) = p_{N-1} + (x - g_N)p'_{N-1} - h_N p'_{N-2} \tag{8.88c}$$

where

$$g_1 = \frac{\beta + 1}{\alpha + \beta + 2};$$

$$g_N = \frac{1}{2}\left[1 - \frac{\alpha^2 - \beta^2}{(2N + \alpha + \beta - 1)^2 - 1}\right] \quad \text{for} \quad N > 1 \tag{8.88d}$$

$$h_1 = 0; \quad h_2 = \frac{(\alpha + 1)(\beta + 1)}{(\alpha + \beta + 2)^2(\alpha + \beta + 3)} \tag{8.88e}$$

$$h_N = \frac{(N - 1)(N + \alpha - 1)(N + \beta - 1)(N + \alpha + \beta - 1)}{(2N + \alpha + \beta - 1)(2N + \alpha + \beta - 2)^2(2N + \alpha + \beta - 3)} \quad \text{for} \quad N > 2$$

$$\tag{8.88f}$$

The recursive formula for $p_N(x)$ and $p'_N(x)$ are started with $N = 1$ and

$$p_0 = 1; \quad p'_0 = 0; \quad p_{-1} \quad \text{and} \quad p'_{-1} \quad \text{are arbitrary}$$

The choice of p_{-1} and p'_{-1} is immaterial because h_1 is zero in Eq. 8.88b and c.

8.3   LAGRANGE INTERPOLATION POLYNOMIALS

We have discussed a class of orthogonal functions called Jacobi polynomials, which have been found to be quite useful in the development of the choice of the interior collocation points for the orthogonal collocation method.

For a given set of data points $(x_1, y_1), (x_2, y_2), \ldots, (x_N, y_N)$ and (x_{N+1}, y_{N+1}), an interpolation formula passing through all $(N + 1)$ points is an Nth degree polynomial. We shall call this an *interpolation polynomial*, and it is expressed as

$$y_N(x) = \sum_{i=1}^{N+1} y_i l_i(x) \tag{8.89}$$

where y_N is the Nth degree polynomial, y_i is the value of y at the point x_i, and $l_i(x)$ is called the Lagrange interpolation polynomial. It is defined as

$$l_i(x_j) = \begin{cases} 0 & i \neq j \\ 1 & i = j \end{cases} \tag{8.90a}$$

The Lagrange interpolation polynomial is a useful building block. There are $(N + 1)$ building blocks, which are Nth degree polynomials. The building blocks are given as

$$l_i(x) = \prod_{\substack{j=1 \\ j \neq i}}^{N+1} \frac{(x - x_j)}{(x_i - x_j)} = \frac{p_{N+1}(x)}{(x - x_i)\left[\dfrac{dp_{N+1}(x_i)}{dx}\right]} \tag{8.90b}$$

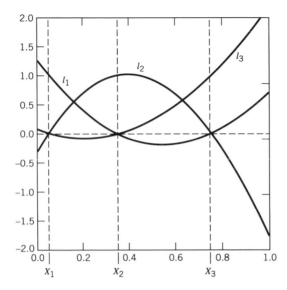

Figure 8.4 Typical plots of the Lagrangian interpolation polynomials.

where $p_{N+1}(x)$ is called the node polynomial. It is a $N + 1$ degree polynomial and is defined as

$$p_{N+1}(x) = (x - x_1)(x - x_2) \ldots (x - x_N)(x - x_{N+1}) \qquad (8.91)$$

where x_i $(i = 1, 2, \ldots, N, N + 1)$ are locations of the data set. The $p_{N+1}(x)$ is called node polynomial because it passes through all the nodes x_i $(i = 1, 2, \ldots, N + 1)$.

Figure 8.4 shows typical plots of the Lagrangian interpolation polynomials for $x_1 = 0.044$, $x_2 = 0.35$, $x_3 = 0.76$. Note that they satisfy Eq. 8.90, for example, $l_1(x)$ is unity at $x = x_1$ and zero at $x = x_2$ and $x = x_3$.

The construction of the Lagrange interpolation polynomial proceeds as follows. First, the $N + 1$ interpolation points are chosen, then the $N + 1$ building blocks $l_i(x)$ can be constructed (Eq. 8.90). If the functional values of y at those $N + 1$ points are known, the interpolation polynomial is given in Eq. 8.89. Hence, the value of y at any point including the interpolation points, say x^*, is given by

$$y_N(x^*) = \sum_{i=1}^{N+1} y_i l_i(x^*) \qquad (8.92)$$

8.4 ORTHOGONAL COLLOCATION METHOD

The previous development on Jacobi and Lagrangian polynomials allows us to proceed directly to computations for the orthogonal collocation method.

8.4.1 Differentiation of a Lagrange Interpolation Polynomial

The interpolation polynomial defined in Eq. 8.89 is a continuous function and therefore can be differentiated as well as integrated.

Taking the first and second derivatives of the interpolation polynomial (Eq. 8.89), we obtain

$$\frac{dy_N(x)}{dx} = \sum_{i=1}^{N+1} y_i \frac{dl_i(x)}{dx} \tag{8.93}$$

$$\frac{d^2 y_N(x)}{dx^2} = \sum_{i=1}^{N+1} y_i \frac{d^2 l_i(x)}{dx^2} \tag{8.94}$$

In most practical problems, only the first two derivatives are required, so these are presented here. However, if higher derivatives are needed, the Lagrange interpolation polynomial can be differentiated further.

In particular, if we are interested in obtaining the derivative *at the interpolation points*, we have

$$\frac{dy_N(x_i)}{dx} = \sum_{j=1}^{N+1} \frac{dl_j(x_i)}{dx} y_j \tag{8.95}$$

for $i = 1, 2, \ldots, N, N + 1$.

Similarly, the second derivative is obtained as

$$\frac{d^2 y_N(x_i)}{dx^2} = \sum_{j=1}^{N+1} \frac{d^2 l_j(x_i)}{dx^2} y_j \tag{8.96}$$

for $i = 1, 2, 3, \ldots, N, N + 1$.

The summation format in the RHS of Eqs. 8.95 and 8.96 suggests the use of a vector representation for compactness, as we will show next.

We define the first derivative vector, composed of $(N + 1)$ first derivatives at the $N + 1$ interpolation points, as

$$\mathbf{y}_N' = \left[\frac{dy_N(x_1)}{dx}, \frac{dy_N(x_2)}{dx}, \ldots, \frac{dy_N(x_N)}{dx}, \frac{dy_N(x_{N+1})}{dx} \right]^T \tag{8.97}$$

Similarly, the second derivative vector is defined as

$$\mathbf{y}_N'' = \left[\frac{d^2 y_N(x_1)}{dx^2}, \frac{d^2 y_N(x_2)}{dx^2}, \ldots, \frac{d^2 y_N(x_N)}{dx^2}, \frac{d^2 y_N(x_{N+1})}{dx^2} \right]^T \tag{8.98}$$

The function vector is defined as values of y at $N + 1$ collocation points as

$$\mathbf{y} = [y_1, y_2, y_3, \ldots, y_N, y_{N+1}]^T \tag{8.99}$$

With these definitions of vectors $\mathbf{y}$ and derivative vectors, the first and second derivative vectors can be written in terms of the function vector $\mathbf{y}$ using matrix notation

$$\mathbf{y}' = \mathbf{A} \cdot \mathbf{y} \tag{8.100}$$

$$\mathbf{y}'' = \mathbf{B} \cdot \mathbf{y} \tag{8.101}$$

where the matrices $\mathbf{A}$ and $\mathbf{B}$ are defined as

$$\mathbf{A} = \left\{ a_{ij} = \frac{dl_j(x_i)}{dx}; \quad i, j = 1, 2, \ldots, N, N + 1 \right\} \tag{8.102a}$$

and

$$\mathbf{B} = \left\{ b_{ij} = \frac{d^2 l_j(x_i)}{dx^2}; \quad i, j = 1, 2, \ldots, N, N + 1 \right\} \tag{8.102b}$$

The matrices $\mathbf{A}$ and $\mathbf{B}$ are $(N + 1, N + 1)$ square matrices. Once the $N + 1$ interpolation points are chosen, then all the Lagrangian building blocks $l_j(x)$ are completely known (Eq. 8.90), and thus the matrices $\mathbf{A}$ and $\mathbf{B}$ are also known. For computation purposes, a_{ij} and b_{ij} are calculated from

$$a_{ij} = \frac{dl_j(x_i)}{dx} = \begin{cases} \dfrac{1}{2} \dfrac{p_{N+1}^{(2)}(x_i)}{p_{N+1}^{(1)}(x_i)} & j = i \\[4mm] \dfrac{1}{(x_i - x_j)} \dfrac{p_{N+1}^{(1)}(x_i)}{p_{N+1}^{(1)}(x_j)} & i \ne j \end{cases} \tag{8.103a}$$

and

$$b_{ij} = \frac{d^2 l_j(x_i)}{dx^2} = \begin{cases} \dfrac{1}{3} \dfrac{p_{N+1}^{(3)}(x_i)}{p_{N+1}^{(1)}(x_i)} & j = i \\[4mm] 2a_{ij}\left[a_{ii} - \dfrac{1}{(x_i - x_j)} \right] & j \ne i \end{cases} \tag{8.103b}$$

where $p_{N+1}^{(1)}$, $p_{N+1}^{(2)}$ and $p_{N+1}^{(3)}$ are calculated from the following recurrence formula

$$p_0(x) = 1$$

$$p_j(x) = (x - x_j)p_{j-1}(x); \quad j = 1, 2, \ldots, N + 1$$

$$p_j^{(1)}(x) = (x - x_j)p_{j-1}^{(1)}(x) + p_{j-1}(x)$$

$$p_j^{(2)}(x) = (x - x_j)p_{j-1}^{(2)}(x) + 2p_{j-1}^{(1)}(x)$$

$$p_j^{(3)}(x) = (x - x_j)p_{j-1}^{(3)}(x) + 3p_{j-1}^{(2)}(x)$$

with

$$p_0^{(1)}(x) = p_0^{(2)}(x) = p_0^{(3)}(x) = 0$$

8.4.2 Gauss–Jacobi Quadrature

We have discussed the differentiation of the interpolation polynomial. Now, we turn to the process of quadrature.[2] This is often needed in solutions, such as chemical reaction rates obtained as an integration of a concentration profile.

Let y_{N-1} be the interpolation polynomial of degree $(N-1)$, passing through N points $(x_1, y_1), (x_2, y_2), \ldots (x_N, y_N)$ presented as

$$y_{N-1}(x) = \sum_{j=1}^{N} y_j l_j(x) \tag{8.104}$$

where $l_j(x)$ is the $(N-1)$th degree Lagrange building block polynomial, and it is given by

$$l_j(x) = \prod_{\substack{i=1 \\ i \neq j}}^{N} \frac{(x - x_i)}{(x_j - x_i)} = \frac{p_N(x)}{(x - x_j)\left[\dfrac{dp_N(x_j)}{dx}\right]}; \qquad P_N(x) = \prod_{i=1}^{N}(x - x_i)$$

The interpolation polynomial, $y_{N-1}(x)$ of Eq. 8.104, is continuous, and therefore we can integrate it with respect to x, using a weighting function $W(x)$, as

$$\int_0^1 W(x)y_{N-1}(x)\,dx = \int_0^1 W(x)\left[\sum_{j=1}^{N} y_j l_j(x)\right]dx \tag{8.105}$$

When we interchange the summation sign and the integral sign, the so-called quadrature relation arises

$$\int_0^1 W(x)y_{N-1}(x)\,dx = \sum_{j=1}^{N} y_j\left[\int_0^1 W(x)l_j(x)\,dx\right] \tag{8.106}$$

---

[2]Quadrature defines the process of expressing the continuous integral

$$\int_a^b W(x)f(x)\,dx$$

as an approximate sum of terms

$$\int_a^b W(x)f(x)\,dx \approx \sum_{k=1}^{N} w_k f(x_k)$$

where x_k are suitably chosen.

Next, if we define

$$w_j = \int_0^1 W(x)l_j(x)\, dx \tag{8.107}$$

which is called the *quadrature weights*, the above quadrature (Eq. 8.106) becomes

$$\int_0^1 W(x)y_{N-1}(x)\, dx = \sum_{j=1}^N w_j y_j \tag{8.108}$$

For a specific choice of weighting function, say the Jacobi weight

$$W(x) = x^\beta (1-x)^\alpha \tag{8.109}$$

the quadrature weight then becomes

$$w_j = \int_0^1 x^\beta (1-x)^\alpha l_j(x)\, dx$$

If the N interpolation points are chosen as N zeroes of the Jacobian polynomial of degree N, the quadrature is called the Gauss–Jacobi quadrature, and w_j are called the Gauss–Jacobi quadrature weights.

For computational purposes, the following formula for the Gauss–Jacobi quadrature weights can be obtained using the properties of the Lagrangian interpolation polynomials $l_j(x)$

$$w_i = \frac{(2N + \alpha + \beta + 1)c_N^{(\alpha,\beta)}}{x_i(1-x_i)\left[\dfrac{dp_N(x_i)}{dx}\right]^2} \tag{8.110a}$$

where $dp_N(x_i)/dx$ is calculated from the recurrence formula

$$p_j^{(1)} = (x - x_j)p_{j-1}^{(1)}(x) + p_{j-1}(x)$$
$$p_j(x) = (x - x_j)p_{j-1}(x)$$

with $p_0(x) = 1$ and $p_0^{(1)}(x) = 0$. Here, $c_N^{(\alpha,\beta)}$ is given by Villadsen (1970)

$$c_N^{(\alpha,\beta)} = \frac{1}{\gamma_{N,N}^2} \frac{\Gamma^2(\beta+1)N!\Gamma(N+\alpha+1)}{\Gamma(N+\beta+1)\Gamma(N+\alpha+\beta+1)(2N+\alpha+\beta+1)} \tag{8.110b}$$

where $\gamma_{N,N}$ is given in Eq. 8.85.

Of practical interest in problem solving, the following situation often arises. Suppose $x_1, x_2, \ldots, x_N$ are N roots of an Nth degree Jacobi polynomial $J_N^{(\alpha,\beta)}$. Now we choose the $(N+1)$th point to be the end point of the domain (i.e., $x_{N+1} = 1$). The interpolation polynomial passing through these $N+1$ points is the Nth degree polynomial, defined as

$$y_N = \sum_{j=1}^{N+1} y_j l_j(x) \tag{8.111}$$

Now if we need to evaluate the quadrature integral

$$\int_0^1 x^\beta (1-x)^\alpha y_N(x)\, dx = \sum_{j=1}^{N+1} w_j y_j \tag{8.112}$$

it is found that the first N quadrature weights are identical to the quadrature weights obtained earlier (Eq. 8.110a) and moreover $w_{N+1} = 0$. This means that adding one more point to the interpolation process will not increase the accuracy of the quadrature, if the N interior interpolation points are chosen as zeros of the Nth degree Jacobian polynomial $J_N^{(\alpha,\beta)}(x)$. For improving the accuracy in the evaluation of the integral (8.112) when one or two boundary points are used as extra interpolation points (in addition to the interior collocation points) we need to use different quadrature formula and this will be addressed in the next section.

8.4.3 Radau and Lobatto Quadrature

To improve the accuracy of the quadrature

$$\int_0^1 x^\beta (1-x)^\alpha y_N(x)\, dx \tag{8.113}$$

when one extra interpolation point is added (say, $x_{N+1} = 1$), the first N interior interpolation points *must be chosen as roots* of the Jacobi polynomial

$$J_N^{(\alpha+1,\beta)}(x) \tag{8.114}$$

rather than as roots of $J_N^{(\alpha,\beta)}$.

For practical computations, the following weighting formula was derived by Villadsen and Michelsen (1978)

$$w_i = \frac{(2N + \alpha + \beta + 2)c_N^{(\alpha+1,\beta)}}{x_i \left[\dfrac{dp_{N+1}(x_i)}{dx} \right]^2} \cdot K; \qquad P_{N+1} = \prod_{j=1}^{N+1}(x - x_j)$$

where $K = 1$ for $i = 1, 2, \ldots, N$, and $K = 1/(\alpha + 1)$ for $i = N + 1$. The coefficient $c_N^{(\alpha+1,\beta)}$ is evaluated using Eq. 8.110b with α being replaced by $\alpha + 1$.

Similarly, when the boundary point at $x = 0$ is added to the N interior interpolation points, the interior points must be chosen as roots of the following Nth degree polynomial

$$J_N^{(\alpha,\beta+1)}(x) \tag{8.115}$$

For computations, the formula for w_i is

$$w_i = \frac{(2N + \alpha + \beta + 2)c_N^{(\alpha,\beta+1)}}{(1 - x_i)\left[\dfrac{dp_{N+1}(x_i)}{dx}\right]^2} \cdot K; \qquad P_{N+1} = x\prod_{j=1}^{N}(x - x_j)$$

where $K = 1/(\beta + 1)$ for $i = 0$, and $K = 1$ for $i = 1, 2, \ldots, N$. The coefficient $c_N^{(\alpha,\beta+1)}$ is evaluated using Eq. 8.110b with β being replaced by $\beta + 1$.

Finally, if both the end points (i.e., $x = 0$ and $x = 1$) are included in the evaluation of the quadrature (Eq. 8.113), the N interior interpolation points must be chosen as roots of the following Nth degree polynomial

$$J_N^{(\alpha+1,\beta+1)}(x) \tag{8.116}$$

The computational formula for w_i is

$$w_i = \frac{(2N + \alpha + \beta + 3)c_N^{(\alpha+1,\beta+1)}}{\left[\dfrac{dp_{N+2}(x_i)}{dx}\right]^2} \cdot K; \qquad P_{N+2} = x(1 - x)\prod_{j=1}^{N}(x - x_j)$$

where $K = 1/(\beta + 1)$ for $i = 0$, $K = 1$ for $i = 1, 2, \ldots, N$, and $K = 1/(\alpha + 1)$ for $i = N + 1$. The coefficient $c_N^{(\alpha+1,\beta+1)}$ is evaluated using Eq. 8.110b with α and β being replaced by $\alpha + 1$ and $\beta + 1$, respectively.

8.5 LINEAR BOUNDARY VALUE PROBLEM—DIRICHLET BOUNDARY CONDITION

The diffusion-reaction problem for slab catalyst particles is a classic problem used to illustrate the orthogonal collocation method. We consider this problem next.

EXAMPLE 8.4

The problem of a slab catalyst particle, sustaining linear reaction kinetics, was posed earlier and the dimensionless material balance equations were given in Eqs. 8.25.

Note, we must ensure the independent variable x has a domain from 0 to 1. Here we note that the problem is symmetrical at $x = 0$, so the following transformation is convenient

$$u = x^2 \tag{8.117}$$

With this transformation, we have

$$\frac{dy}{dx} = \frac{dy}{du} \cdot \frac{du}{dx} = 2\sqrt{u}\,\frac{dy}{du} \tag{8.118}$$

$$\frac{d^2y}{dx^2} = \frac{d}{dx}\left(2\sqrt{u}\,\frac{dy}{du}\right) = \frac{d}{du}\left(2\sqrt{u}\,\frac{dy}{du}\right) \cdot \frac{du}{dx} = 2\frac{dy}{du} + 4u\frac{d^2y}{du^2} \tag{8.119}$$

Using these relations, the mass balance equation (Eq. 8.25a) and the boundary condition at the catalyst surface (Eq. 8.25c) become

$$4u\frac{d^2y}{du^2} + 2\frac{dy}{du} - \phi^2 y = 0 \tag{8.120}$$

$$u = 1; \quad y = 1 \tag{8.121}$$

At first glance, it may appear that we have made the problem more difficult. However, the boundary condition at $x = 0$ is no longer needed owing to the transformation $(u = x^2)$, which makes y always an even function.

Since it is our ultimate objective to evaluate the effectiveness factor (Eq. 8.30), we transform the integral in terms of the u variable, that is,

$$\eta = \int_0^1 y \, dx = \frac{1}{2}\int_0^1 u^{-1/2} y \, du \tag{8.122}$$

The weighting function for the above integral, by comparison with Eq. 8.109, is simply

$$W(u) = u^{-1/2}(1 - u)^0 \tag{8.123a}$$

so we conclude

$$\alpha = 0; \quad \beta = -\frac{1}{2} \tag{8.123b}$$

Now, if we choose $(N + 1)$ interpolation points as N *interior collocation points in the domain* [0, 1] *and the boundary point at* $u = 1$ to evaluate the integral of the form of Eq. 8.122, the N interior collocation points $(u_1, u_2, u_3, \ldots, u_N)$ must be chosen as roots of the Jacobi polynomial $J_N^{(\alpha+1,\beta)} = J_N^{(1,-1/2)}$ (see Eq. 8.114). The $(N + 1)$th interpolation point u_{N+1} is 1.

The interpolation polynomial for this problem is

$$y_N(u) = \sum_{j=1}^{N+1} y_j l_j(u) = l_{N+1}(u) + \sum_{j=1}^{N} y_j l_j(u) \tag{8.124}$$

where the building blocks l_j are defined in Eq. 8.90. Comparing Eq. 8.124 with the trial solution formula (8.5), we see the correspondence

$$l_{N+1} = y_0 \quad \text{and} \quad l_j = \phi_j$$

Since the mass balance equation (Eq. 8.120) is valid at any point inside the domain [0, 1], we evaluate it at the ith *interior collocation point* as follows (note, as a reminder, the residual is zero at the collocation points)

$$R(u_i) = \left[4u\frac{d^2y}{du^2}\right]_i + \left[2\frac{dy}{du}\right]_i - [\phi^2 y]_i = 0 \tag{8.125}$$

for $i = 1, 2, \ldots, N$.

But the derivatives at the point i are given by (Eqs. 8.95 and 8.96)

$$\left[\frac{dy}{du}\right]_i = \sum_{j=1}^{N+1} A_{ij} y_j \tag{8.126}$$

and

$$\left[\frac{d^2 y}{du^2}\right]_i = \sum_{j=1}^{N+1} B_{ij} y_j \tag{8.127}$$

where y_j is the unknown value of y at the interpolation point u_j. Now that the $(N + 1)$ interpolation points are chosen, the matrices $\mathbf{A}$ and $\mathbf{B}$ are completely known.

Substituting these derivatives (Eqs. 8.126 and 8.127) into Eq. 8.125 yields

$$4u_i \sum_{j=1}^{N+1} B_{ij} y_j + 2 \sum_{j=1}^{N+1} A_{ij} y_j - \phi^2 y_i = 0 \tag{8.128}$$

for $i = 1, 2, \ldots, N$.

Because we know the value for y at the interpolation point $u_{N+1} = 1$, we can remove the last term from each of the two series as

$$4u_i \left[\sum_{j=1}^{N} B_{ij} y_j + B_{i, N+1} y_{N+1}\right] + 2\left[\sum_{j=1}^{N} A_{ij} y_j + A_{i, N+1} y_{N+1}\right] - \phi^2 y_i = 0 \tag{8.129}$$

for $i = 1, 2, \ldots, N$.

But the value of y at the boundary $u = 1$ is $y_{N+1} = 1$, so the above equation becomes

$$4u_i \sum_{j=1}^{N} B_{ij} y_j + 2 \sum_{j=1}^{N} A_{ij} y_j - \phi^2 y_i = -4u_i B_{i, N+1} - 2A_{i, N+1} \tag{8.130}$$

for $i = 1, 2, \ldots, N$.

Equation 8.130 represents N coupled algebraic equations, with N unknowns $(y_1, y_2, \ldots, y_N)$, which are functional values of y at N interior collocation points. Techniques for solving large systems of algebraic equations are given in Appendix A. However, in the present problem the algebraic equations are linear and hence they are amenable to solution by matrix methods. By defining the following known matrix $\mathbf{C}$ and vector $\mathbf{b}$ as

$$\mathbf{C} = \left\{C_{ij} = 4u_i B_{ij} + 2A_{ij} - \phi^2 \delta_{ij}; \quad i, j = 1, 2, \ldots, N\right\} \tag{8.131a}$$

$$\mathbf{b} = \left\{b_i = -4u_i B_{i, N+1} - 2A_{i, N+1}; \quad i = 1, 2, \ldots, N\right\} \tag{8.131b}$$

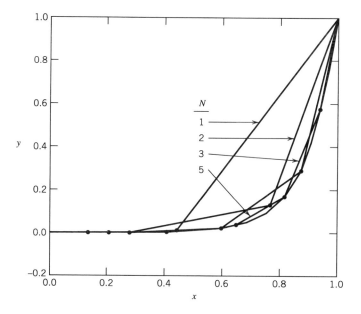

Figure 8.5 Concentration profiles for $\phi = 10$, illustrating advantages of additional collocation points.

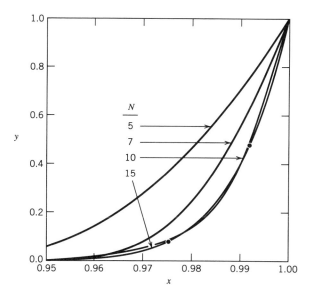

Figure 8.6 Concentration profiles for $\phi = 100$, with number of collocation points as parameter.

where

$$\delta_{ij} = \begin{cases} 1 & i = j \\ 0 & i \neq j \end{cases} \tag{8.132}$$

and the unknown **y** as

$$\mathbf{y} = [y_1, y_2, y_3, \ldots, y_N]^T \tag{8.133}$$

then Eq. 8.130 can be put into the following compact vector form

$$\mathbf{C} \cdot \mathbf{y} = \mathbf{b} \tag{8.134}$$

from which the solution is simply

$$\mathbf{y} = \mathbf{C}^{-1} \cdot \mathbf{b} \tag{8.135}$$

where $\mathbf{C}^{-1}$ is the inverse of the matrix $\mathbf{C}$.

Since the solution for **y** is known, the effectiveness factor, η, can be obtained from Eq. 8.122. Substituting Eq. 8.124 into Eq. 8.122, the quadrature for the integral representing effectiveness factor is

$$\eta = \frac{1}{2} \sum_{j=1}^{N+1} w_j y_j \tag{8.136}$$

Figure 8.5 illustrates the evolution of concentration profiles for $\phi = 10$ with the number of collocation points as parameter. It is seen that when the reaction rate is high, the concentration profile inside the particle is very sharp, so that about 5 interior collocation points are needed to get a satisfactory result. Figure 8.6 treats the case for even faster rates, such that $\phi = 100$. The extreme

Table 8.2 Computations Using Orthogonal Collocation: Diffusion in Catalyst Particle

Number of Interior Collocation Point, N	ϕ	I	Percentage Relative Error
1	10	0.186992	87
2	10	0.111146	11
3	10	0.100917	1
5	10	0.100001	.001
1	100	0.166875	1569
2	100	0.067179	572
5	100	0.017304	73
7	100	0.012006	20
10	100	0.010203	2
15	100	0.010001	.01

sharpness of the profile requires about 10 or more interior collocation points to yield reasonable accuracy.

Table 8.2 summarizes computation of the effectiveness factor using the orthogonal collocation method. Also shown in the table is the relative error between the calculated effectiveness factor with the exact solution, given by

$$\eta = \frac{\tanh(\phi)}{\phi}$$

8.6 LINEAR BOUNDARY VALUE PROBLEM— ROBIN BOUNDARY CONDITION

EXAMPLE 8.5

In this example, we reconsider the catalyst problem in the previous example, but the bulk fluid moves slowly, so that finite film resistance exists. The boundary condition at the catalyst surface (Eq. 8.23b) is replaced by

$$r = R; \qquad -\left[D_e \frac{dC}{dr}\right]_R = k_c(C|_R - C_0) \tag{8.137}$$

which is simply a balance of flux to the solid phase and through the film surrounding the exterior surface.

In nondimensional form, this boundary condition becomes

$$x = 1; \qquad \left[\frac{dy}{dx}\right]_1 = Bi(1 - y|_1) \tag{8.138}$$

where $Bi = k_c R/D_e$.

Thus, the mass balance equation (Eqs. 8.25a and 8.138) written in terms of the variable $u(u = x^2)$ as before

$$4u\frac{d^2y}{du^2} + 2\frac{dy}{du} - \phi^2 y = 0 \tag{8.139}$$

$$u = 1; \qquad \left[\frac{dy}{du}\right]_1 = \frac{Bi}{2}(1 - y|_1) \tag{8.140}$$

Again, our objective here is to calculate the overall reaction rate per unit volume, and from this to obtain the effectiveness factor. Therefore, the $N + 1$ interpolation points are chosen with the first N points being interior collocation points in the catalyst particle and the $(N + 1)$th interpolation point being the boundary point $(u_{N+1} = 1)$. The N interior points are chosen as roots of the Jacobian polynomial $J_N^{(0, -1/2)}$. The optimal choice of N interior points in this example as well as the last one was studied by Michelsen and Villadsen (1980). This is done by using the quadrature approach to the calculation of the integral (Eq. 8.7) in the Galerkin method. A summary of this approach is presented in Section 8.9.

The mass balance equation is discretized at the ith interior collocation point as before, and we have

$$4u_i \left[\sum_{j=1}^{N} B_{ij} y_j + B_{i, N+1} y_{N+1} \right] + 2 \left[\sum_{j=1}^{N} A_{ij} y_j + A_{i, N+1} y_{N+1} \right] - \phi^2 y_i = 0$$

(8.141)

for $i = 1, 2, \ldots, N$.

In this case, unlike the last example where $y_{N+1} = 1$, the value of y at the boundary point is not equal to unity, but is governed by the boundary condition (Eq. 8.140). At the boundary (i.e., at the point u_{N+1}), we have

$$\left[\frac{dy}{du} \right]_{u = u_{N+1}} = \frac{Bi}{2} (1 - y_{N+1})$$

(8.142)

The first derivative at the point u_{N+1} is given by (Eq. 8.95)

$$\left[\frac{dy}{du} \right]_{u = u_{N+1}} = \sum_{j=1}^{N+1} A_{N+1, j} y_j$$

(8.143)

When this is substituted into the boundary equation (Eq. 8.142), we obtain

$$\sum_{j=1}^{N} A_{N+1, j} y_j + A_{N+1, N+1} y_{N+1} = \frac{Bi}{2} (1 - y_{N+1})$$

(8.144)

where we removed the last term from the series in the LHS of this equation.

Solving for y_{N+1}, we have

$$y_{N+1} = \frac{1}{1 + \dfrac{2 A_{N+1, N+1}}{Bi}} \left[1 - \frac{2}{Bi} \sum_{j=1}^{N} A_{N+1, j} y_j \right]$$

(8.145)

Thus, we see that when Bi is extremely large (minuscule film resistance), the above equation reduces to $y_{N+1} = 1$ as required.

Next, substitute the equation for y_{N+1} (Eq. 8.145 into Eq. 8.141), and so obtain the following linear equation in terms of $\mathbf{y}$

$$\mathbf{D} \cdot \mathbf{y} = \mathbf{b}$$

(8.146)

where

$$\mathbf{D} = \left\{ D_{ij} = C_{ij} - \frac{\dfrac{2}{Bi} C_{i, N+1} A_{N+1, j}}{1 + \dfrac{2}{Bi} A_{N+1, N+1}} - \phi^2 \delta_{ij}; \ i, j = 1, 2, \ldots, N \right\}$$

(8.147)

$$\mathbf{C} = \{ C_{ij} = 4u_i B_{ij} + 2 A_{ij}; \ i, j = 1, 2, \ldots, N \}$$

(8.148)

$$\mathbf{b} = \left\{ b_i = - \frac{C_{i, N+1}}{1 + \dfrac{2}{Bi} A_{N+1, N+1}}; \ i = 1, 2, \ldots, N \right\}$$

(8.149)

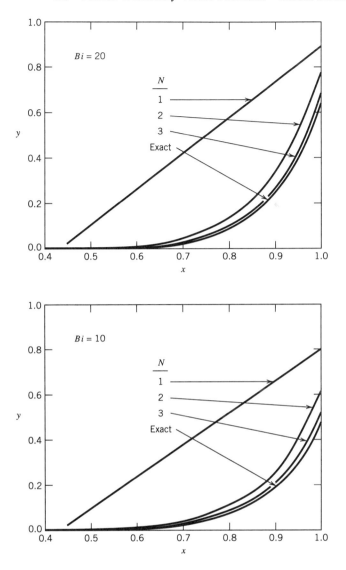

Figure 8.7 Plots of concentration profiles for $\phi = 10$.

Here, we have used a vector-matrix format to achieve compactness. The inverse of Eq. 8.146 will yield the vector **y**, that is, the concentrations y_j at all interior collocation points. Knowing the concentrations at all the interior collocation points, the surface concentration y_{N+1} is calculated from Eq. 8.145. Figure 8.7 presents the concentration profiles for $\phi = 10$, with the number of interior collocation point and the Biot number being parameters.

Table 8.3 Computations Using Orthogonal Collocation: Effect of Boundary Resistance

N	ϕ	Bi	η	Relative Error %
1	10	10	0.150327	200
2	10	10	0.069764	40
3	10	10	0.054334	9
5	10	10	0.050143	0.3
10	10	10	0.050000	0
1	10	20	0.166667	150
2	10	20	0.085722	29
3	10	20	0.070637	6
5	10	20	0.066794	0.2
10	10	20	0.066667	0

The exact solutions to this problem are

$$y = \frac{\cosh(\phi x)}{\left[\cosh(\phi) + \frac{\phi}{Bi} \sinh(\phi)\right]}$$

$$\eta = \frac{\tanh(\phi)}{\phi\left[1 + \frac{\phi}{Bi} \tanh(\phi)\right]}$$

Table 8.3 compares the numerical solution with the exact solution. Even for sharp profiles, the collocation solutions are comparable to the exact solution using only five collocation points.

8.7 NONLINEAR BOUNDARY VALUE PROBLEM— DIRICHLET BOUNDARY CONDITION

EXAMPLE 8.6

We wish to consider the catalyst particle for nonlinear conditions so the local reaction rate is given by

$$R_{\text{local}} = g(C) \tag{8.150}$$

where $g(C)$ is some nonlinear function of concentration.

Setting the shell balance at the position r, we obtain the following mass balance equation

$$D_e \frac{d^2 C}{dr^2} - g(C) = 0 \tag{8.151a}$$

subject to the boundary conditions

$$r = 0; \qquad \frac{dC}{dr} = 0 \tag{8.151b}$$

$$r = R; \qquad C = C_0 \tag{8.151c}$$

The effectiveness factor for a general nonlinear reaction rate is defined as

$$\eta = \frac{\int g(C)\, dV}{\int g(C_0)\, dV} \tag{8.152a}$$

For the slab geometry, $dV = A\,dr$, where A is the cross-sectional area of the catalyst. Hence, Eq. 8.152a becomes for a slab catalyst

$$\eta = \frac{1}{Rg(C_0)} \int_0^R g(C)\, dr \tag{8.152b}$$

Thus, when we know the distribution of C, we replace $g(C)$ in the integrand of Eq. 8.152b and evaluate the integral numerically. Alternately, we can calculate the effectiveness factor in the following way.

Multiplying Eq. 8.151a by dr and integrating the result from 0 to R gives

$$\int_0^R g(C)\, dr = D_e \frac{dC}{dr}\Big|_R$$

When we substitute this result into Eq. 8.152b, we obtain

$$\eta = \frac{D_e}{Rg(C_0)} \frac{dC}{dr}\Big|_R \tag{8.152c}$$

This means that the effectiveness factor can be calculated using the derivative of concentration distribution at the exterior surface of the particle.

By defining the following dimensionless variables and parameters

$$y = \frac{C}{C_0}; \quad x = \frac{r}{R}; \quad \phi^2 = \frac{g(C_0)R^2}{D_e C_0}; \quad G(y) = \frac{g(C_0 y)}{g(C_0)} \tag{8.153}$$

the mass balance equation (Eqs. 8.151) becomes

$$\frac{d^2 y}{dx^2} - \phi^2 G(y) = 0 \tag{8.154a}$$

$$x = 0; \qquad \frac{dy}{dx} = 0 \tag{8.154b}$$

$$x = 1; \qquad y = 1 \tag{8.154c}$$

In terms of the dimensionless variables and parameters (Eq. 8.153), the effectiveness factor given in Eq. 8.152c becomes

$$\eta = \frac{C_0 D_e}{R^2 g(C_0)} \frac{dy}{dx}\bigg|_1 = \frac{1}{\phi^2} \frac{dy}{dx}\bigg|_1 \qquad (8.154d)$$

Thus, when $y(x)$ is numerically determined from Eqs. 8.154a to c, the effectiveness factor is readily evaluated from Eq. 8.154d.

We introduce the u transformation as before ($u = x^2$), and the mass balance equation now becomes

$$4u \frac{d^2 y}{du^2} + 2 \frac{dy}{du} - \phi^2 G(y) = 0 \qquad (8.155a)$$

subject to

$$u = 1; \qquad y = 1 \qquad (8.155b)$$

The reader is reminded again that when the symmetry transformation ($u = x^2$) is introduced, the boundary condition at $x = 0$ (Eq. 8.154b) is automatically satisfied.

If we now discretize the mass balance equation (Eq. 8.155a) at the interior point i, we have

$$4u_i \left[\sum_{j=1}^{N} B_{ij} y_j + B_{i,N+1} y_{N+1} \right] + 2 \left[\sum_{j=1}^{N} A_{ij} y_j + A_{i,N+1} y_{N+1} \right] \qquad (8.156)$$
$$- \phi^2 G(y_i) = 0$$

for $i = 1, 2, \ldots, N$.

Since $y_{N+1} = 1$ (Eq. 8.155b), the above equation becomes

$$4u_i \left[\sum_{j=1}^{N} B_{ij} y_j + B_{i,N+1} \right] + 2 \left[\sum_{j=1}^{N} A_{ij} y_j + A_{i,N+1} \right] \qquad (8.157)$$
$$- \phi^2 G(y_i) = 0$$

for $i = 1, 2, \ldots, N$. This equation represents a set of N nonlinear coupled algebraic equations in terms of $y_1, y_2, \ldots, y_N$. They can be solved by one of several nonlinear algebraic solvers, such as the Newton–Raphson method (see Appendix A).

To solve Eq. 8.157 by the Newton–Raphson procedure, we define

$$F_i(\mathbf{y}) = 4u_i \sum_{j=1}^{N} B_{ij}y_j + B_{i,\,N+1} + 2\left[\sum_{j=1}^{N} A_{ij}y_j + A_{i,\,N+1} \right] - \phi^2 G(y_i) = 0$$

$$(8.158)$$

for $i = 1, 2, \ldots, N$, and where

$$\mathbf{y} = [\, y_1, y_2, \ldots, y_N \,]^T \tag{8.159}$$

The iteration scheme for the Newton–Raphson is

$$\mathbf{y}^{(k+1)} = \mathbf{y}^{(k)} - \mathbf{d}^{(k)} \tag{8.160a}$$

where

$$\mathbf{J}(\mathbf{y}^{(k)})\mathbf{d}^{(k)} = \mathbf{F}(\mathbf{y}^{(k)}) \tag{8.160b}$$

$$\mathbf{F} = \left[F_1(\mathbf{y}), F_2(\mathbf{y}), \ldots, F_N(\mathbf{y}) \right]^T \tag{8.160c}$$

$$\mathbf{J} = \left\{ \frac{\partial F_i}{\partial y_j} = 4u_i B_{ij} + 2A_{ij} - \delta_{ij}\phi^2 \frac{\partial G(y_i)}{\partial y} ;\quad i, j = 1, 2, \ldots, N \right\} \tag{8.160d}$$

where δ_{ij} represents the Krocknecker delta function, defined as

$$\delta_{ij} = \begin{cases} 1 & i = j \\ 0 & i \neq j \end{cases} \tag{8.161}$$

To solve for the concentration vector $\mathbf{y}$ by the Newton–Raphson technique, we need to select an initial set for $\mathbf{y}^{(0)}$. With this initial guess, the function vector $\mathbf{F}$ and the Jacobian $\mathbf{J}$ can be evaluated (Eqs. 8.160c, d). Using any standard linear equation solver, the vector $\mathbf{d}$ can be calculated from Eq. 8.160b and hence the first iterated solution $\mathbf{y}^{(1)}$ is given in Eq. 8.160a. The process is repeated until a convergence criterion is satisfied. One can choose either of the following criteria for stopping the iteration

$$\sum_{j=1}^{N} \frac{\left| y_j^{(k+1)} - y_j^{(k)} \right|}{\left| y_j^{(k)} \right|} < \varepsilon$$

or

$$\max_{j} \frac{\left| y_j^{(k+1)} - y_j^{(k)} \right|}{\left| y_j^{(k)} \right|} < \varepsilon$$

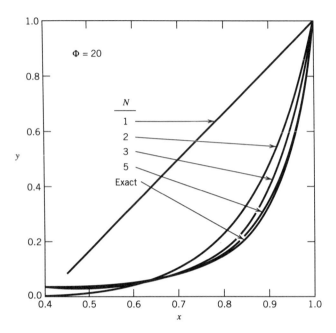

Figure 8.8 Approximate concentration profiles for second order chemical kinetics.

If y is small, the following stopping criterion is recommended

$$\sum_j \left| y_j^{(k+1)} - y_j^{(k)} \right| < \varepsilon$$

Figure 8.8 presents computations of concentration profiles for the case of second order chemical kinetics. The first stopping criterion was used to generate these plots, and ε was 0.001. Again, just as in previous examples, the collocation solutions generated with five or more interior collocation points agree fairly well with the exact solution, which in the present case is taken as the solution generated by using 19 interior collocation points. For this problem of slab geometry, the analytical solution is given as

$$\int_y^1 \frac{ds}{\left[2 \int_{y_0}^s G(m)\, dm \right]^{1/2}} = \phi(1 - x)$$

where y_0 is the value of y at $x = 0$ and is given by

$$\int_{y_0}^1 \frac{ds}{\left[2 \int_{y_0}^s G(m)\, dm \right]^{1/2}} = \phi$$

Table 8.4 Computations Using Orthogonal Collocation: Nonlinear Reaction Kinetics

N	ϕ	η	Relative Error %
1	20	0.230000	109
2	20	0.130439	18
3	20	0.117599	7
5	20	0.111027	0.61
10	20	0.110356	0
19	20	0.110355	—

Table 8.4 shows the numerically calculated effectiveness factor and the relative error as function of the number of interior collocation point.

8.8   ONE-POINT COLLOCATION

The orthogonal collocation method, as we have attempted to illustrate in previous examples, sustains an accuracy, which will increase with the number of points used. Occasionally, one is interested in the approximate behavior of the system instead of the computer intensive exact behavior. To this end, we simply use only one collocation point, and the result is a simplified equation, which allows us to quickly investigate the behavior of solutions, for example, to see how the solution would change when a particular parameter is changed, or to determine whether the solution exhibits multiplicity. Once this is done, detailed analysis can be carried out with more collocation points.

EXAMPLE 8.7

We illuminate these attractive features by considering the difficult problems of diffusion and reaction in a slab catalyst sustaining highly nonlinear Hinshelwood kinetics. The mass balance equations written in nondimensional form are taken to be

$$\frac{d^2y}{dx^2} - \frac{\phi^2 y}{1 + \delta y + \gamma y^2} = 0 \qquad (8.162a)$$

$$x = 0; \qquad \frac{dy}{dx} = 0 \qquad (8.162b)$$

$$x = 1; \qquad y = 1 \qquad (8.162c)$$

Noting the symmetry of this problem, we make the usual substitution $u = x^2$, and the mass balance equations become

$$4u\frac{d^2y}{du^2} + 2\frac{dy}{du} - \frac{\phi^2 y}{1 + \delta y + \gamma y^2} = 0 \qquad (8.163)$$

$$u = 1; \qquad y = 1 \qquad (8.164)$$

Now we choose one collocation point, u_1, in the domain $[0, 1]$, and since we know the value of y at the surface of the catalyst, we will use it as the second interpolation point, that is, $u_2 = 1$. For these two interpolation points, we have two Lagrangian interpolation polynomials, $l_1(u)$ and $l_2(u)$, given as

$$l_1(u) = \frac{u - u_2}{u_1 - u_2}; \qquad l_2(u) = \frac{u - u_1}{u_2 - u_1} \qquad (8.165)$$

where u_1 is the collocation point chosen in the domain $[0, 1]$, and $u_2 = 1$.

Using the Lagrangian interpolation polynomials $l_1(u)$ and $l_2(u)$, the approximate solution for $y(u)$ can be written as

$$y = l_1(u) y_1 + l_2(u) y_2 = l_1(u) y_1 + l_2(u) \qquad (8.166)$$

because $y_2 = y(u_2) = y(1) = 1$.

Next, we substitute the approximate solution 8.166 into the differential equation 8.163, and obtain the following residual

$$R = 4u \frac{d^2 y}{du^2} + 2 \frac{dy}{du} - \frac{\phi^2 y}{1 + \delta y + \gamma y^2} \qquad (8.167)$$

We now have only one unknown, which is the value of y at the collocation point u_1, and the test function for the collocation method is

$$w_1 = \delta(u - u_1) \qquad (8.168)$$

Averaging the residual with the test function w_1 is carried out using the integral

$$\int_0^1 R(u) w_1(u)\, du = \int_0^1 \left(4u \frac{d^2 y}{du^2} + 2 \frac{dy}{du} - \frac{\phi^2 y}{1 + \delta y + \gamma y^2} \right) \delta(u - u_1)\, du = 0$$

that is,

$$4u_1 \frac{d^2 y}{du^2} \bigg|_{u_1} + 2 \frac{dy}{du} \bigg|_{u_1} - \frac{\phi^2 y_1}{1 + \delta y_1 + \gamma y_1^2} = 0 \qquad (8.169)$$

But

$$\frac{dy}{du} \bigg|_{u_1} = \frac{dl_1(u_1)}{du} y_1 + \frac{dl_2(u_1)}{du} = A_{11} y_1 + A_{12} \qquad (8.170a)$$

and

$$\frac{d^2 y}{du^2} \bigg|_{u_1} = \frac{d^2 l_1(u_1)}{du^2} y_1 + \frac{d^2 l_2(u_1)}{du^2} = B_{11} y_1 + B_{12} \qquad (8.170b)$$

When we substitute Eqs. 8.170 into Eq. 8.169, we get

$$C_{11}y_1 + C_{12} - \frac{\phi^2 y_1}{1 + \delta y_1 + \gamma y_1^2} = 0 \tag{8.171}$$

where

$$C_{11} = 4u_1 B_{11} + 2A_{11} \quad \text{and} \quad C_{12} = 4u_1 B_{12} + 2A_{12}$$

Equation 8.171 is a cubic equation in terms of y_1; hence, depending on the values of ϕ, δ, and γ, there may exist three solutions (multiple steady states) for this problem. A knowledge of this will then help the comprehensive computation using more collocation points.

This example serves as a means to allow workers to quickly study the topology of a system before more time is spent on the detailed computation of the governing equations.

8.9   SUMMARY OF COLLOCATION METHODS

We have presented a family of approximate methods, called weighted residuals, which are quite effective in dealing with boundary value problems. The name suggests that we need to generate residuals obtained when the approximate solution is substituted into the governing equation. Then, we try to minimize the residuals or force it to be asymptotically close to zero at certain points. A number of methods have appeared, depending on how we minimize this residual.

Among the five methods studied in this chapter, the orthogonal collocation and the Galerkin methods seem to provide the best approximate routes. The Galerkin method provides solution with good accuracy, while the collocation method is easy to apply and to program, owing to its mechanical structure. The accuracy of collocation is comparable to Galerkin if the collocation points are properly chosen. Because of this attribute, the collocation method has found wide applications in chemical engineering and other branches of engineering.

We now summarize the steps taken in the application of the orthogonal collocation procedure.

Step 1   For the given problem, normalize the range of the independent variable to $(0, 1)$. Any domain (a, b) can be transformed to a $(0, 1)$ by the transformation

$$x = \frac{z - a}{b - a}$$

where x is the new independent variable lying in the domain $(0, 1)$.

Step 2   Next, observe the boundary conditions and if there is symmetry at $x = 0$, make use of the transformation $u = x^2$. Similarly, if the problem is symmetrical at $x = 1$, use the transformation $u = (1 - x)^2$.

Step 3   Assume an approximate solution of the form in Eq. 8.5, where $y_0(x)$ satisfies the boundary conditions exactly and the trial functions, $\phi_i(x)$, satisfy the homogeneous boundary conditions. The problem at this point is then reduced to the problem of solving for N unknown coefficients. If the trial functions are chosen as the Lagrange interpolation polynomials, $l_i(x)$ (Section 8.3), the coefficients a_i then become the values of y at the interpolation points x_i. Interpolation points are those used to generate the Lagrange interpolation polynomials. It is noted here that the Lagrange interpolation polynomial is used here as a convenient vehicle to obtain a solution. Any linearly independent set of trial functions can be used in the collocation method.

Step 4   Substitute the approximate solution prescribed in Step 3 into the governing equation to form a residual R (Eq. 8.6), which is a function of x as well as N coefficients a_i.

Step 5   This step is the most crucial step of the orthogonal collocation method. If the Galerkin method is used for minimizing the residual (remember that the Galerkin is the best method among the many weighted residual methods to provide solution of good accuracy), the following function of N coefficients a_i is created

$$\int_0^1 R(a_1, a_2, \ldots, a_N; x)\phi_j(x)\, dx = 0 \qquad \text{for} \qquad j = 1, 2, \ldots, N$$

These N integrals, in general, cannot be integrated analytically; hence, it must be done numerically by a quadrature method, such as the Gaussian quadrature described in Appendix E. Before doing this, extract a common factor of the form $W(x) = x^\beta(1 - x)^\alpha$ from the integrand of the above integral. This factor **must** be the same for all values of j. The integral can then be written as

$$\int_0^1 R(a_1, a_2, \ldots, a_N; x)\phi_j(x)\, dx$$

$$= \int_0^1 \left[x^\beta(1 - x)^\alpha\right] Q_j(x)\, dx \quad \text{for} \quad j = 1, 2, \ldots, N$$

where

$$Q_j(x) = \frac{R(a_1, a_2, \ldots, a_N; x)\phi_j(x)}{x^\beta(1 - x)^\alpha}$$

If the Gaussian quadrature of N quadrature points is applied to evaluate the above integral approximately, the optimal quadrature formula would be

$$\int_0^1 \left[x^\beta(1 - x)^\alpha\right] Q_j(x)\, dx = \sum_{k=1}^{N} w_k Q_j(x_k) \quad \text{for} \quad j = 1, 2, \ldots, N$$

where w_k are the quadrature weights and x_k are the quadrature points, which are zeros of the Jacobi polynomial $J_N^{(\alpha,\beta)}(x)$. Note that the number of quadrature points used is the same as the number of unknown coefficients a_i. It should also be pointed out that the above quadrature approximation will be exact if the polynomial Q_j is of degree less than or equal to $2N - 1$ (see Appendix E).

Step 6   It is then clear from Step 5 that the *optimal approximation* to the Galerkin method is simply to choose

$$R(a_1, a_2, \ldots, a_N; x_k) = 0 \qquad \text{for} \qquad k = 1, 2, \ldots, N$$

Thus, if the collocation method is used, with the collocation points being zeros of the Jacobi polynomial $J_N^{(\alpha,\beta)}(x)$, then the collocation method will closely approximate the Galerkin method.

These outline the pedagogical steps for undertaking the orthogonal collocation method. They are chosen so that the collocation method closely approximates the Galerkin method. Problem 8.9 illustrates in a practical way how to apply these steps to an engineering problem.

8.10   CONCLUDING REMARKS

Boundary value problems are encountered so frequently in modelling of engineering problems that they deserve special treatment because of their importance. To handle such problems, we have devoted this chapter exclusively to the methods of weighted residual, with special emphasis on orthogonal collocation. The one-point collocation method is often used as the first step to quickly assess the behavior of the system. Other methods can also be used to treat boundary value problems, such as the *finite difference* method. This technique is considered in Chapter 12, where we use this method to solve boundary value problems and partial differential equations.

8.11   REFERENCES

1. Fletcher, C. A. J., *Computational Galerkin Methods*, Springer-Verlag, New York (1984).
2. Michelsen, M. L., and J. Villadsen, "Polynomial Solution of Differential Equations," Proceedings of the First Conference on Computer Aided Process Design held at Henniker, New Hampshire, July 6–11, 1980.
3. Villadsen, J., and M. L. Michelsen, *Solution of Differential Equation Models by Polynomial Approximation*, Prentice Hall, Engelwood Cliffs, New Jersey (1978).
4. Villadsen, J., *Selected Approximation Methods for Chemical Engineering Problems*, Reproset, Copenhagen (1970).

8.12 PROBLEMS

8.1₁. The Lagrangian polynomials $l_j(x)$ defined for $N + 1$ interpolation points are given as

$$l_j(x) = \prod_{\substack{k=1 \\ k \neq j}}^{N+1} \frac{x - x_k}{x_j - x_k} \qquad \text{for} \qquad j = 1, 2 \cdots, N + 1$$

(a) Show that $l_j(x_j) = 1$ and $l_j(x_k) = 0$ for all $k \neq j$.

(b) Prove that the definition of the Lagrangian polynomial is equivalent to the expression

$$l_j(x) = \frac{p_{N+1}(x)}{(x - x_j)p'_{N+1}(x_j)}$$

where $p_{N+1}(x)$ is called the node polynomial (it is called *node* because it becomes zero at the interpolation points) and is defined as

$$p_{N+1}(x) = (x - x_1)(x - x_2)(x - x_3) \cdots (x - x_N)(x - x_{N+1})$$

and $p'_{N+1} = dp_{N+1}/dx$.

8.2₂. For any set of Lagrangian polynomial $l_j(x)$ of degree N, defined as in Problem 8.1, show that the sum of these $N + 1$ Lagrangian polynomials is unity; that is,

$$\sum_{j=1}^{N+1} l_j(x) = 1$$

Hint: Start with the following function $y_N(x)$ representing the polynomial passing through $N + 1$ points $(x_1, y_1), (x_2, y_2), \ldots, (x_N, y_N), (x_{N+1}, y_{N+1})$

$$y_N(x) = \sum_{j=1}^{N+1} l_j(x)y_j$$

8.3₃. With the Lagrangian polynomials defined in part (b) of Problem 8.1, show that they are orthogonal to each other with respect to the weighting function $W(x) = x^\beta(1 - x)^\alpha$ if the $N + 1$ interpolation points $x_1, x_2, x_3, \ldots, x_N, x_{N+1}$ are chosen as roots of the Jacobi polynomial $J_{N+1}^{(\alpha, \beta)}(x) = 0$; that is,

$$\int_0^1 \left[x^\beta(1 - x)^\alpha\right]l_k(x)l_j(x)\, dx = 0 \qquad \text{for} \qquad k \neq j$$

Hint: Use the following orthogonality property of the Jacobi polynomial

$$\int_0^1 \left[x^\beta(1-x)^\alpha\right] x^j J_{N+1}^{(\alpha,\beta)}(x)\, dx = 0 \qquad \text{for} \qquad j = 0,1,2,\ldots, N$$

8.4$_3$. Let $x_1, x_2, x_3, \ldots, x_N$ be N interpolation points chosen as roots of the Jacobi polynomial $J_N^{(\alpha,\beta)}(x) = 0$ (i.e., $0 < x_j < 1$ for $j = 1,2,\ldots, N$). The ordinates corresponding to these points are denoted as $y_1, y_2, y_3, \ldots, y_N$, and the polynomial of degree $N-1$, $y_{N-1}(x)$, passing through these N points $(x_1, y_1), (x_2, y_2), \ldots, (x_N, y_N)$, is given by the following Lagrangian formula

$$y_{N-1}(x) = \sum_{j=1}^N l_j(x) y_j$$

where

$$l_j(x) = \frac{p_N(x)}{(x - x_j) p_N'(x_j)}$$

where $p_N(x)$ is the scaled Jacobi polynomial defined as

$$p_N(x) = (x - x_1)(x - x_2)(x - x_3) \cdots (x - x_N) = \frac{J_N^{(\alpha,\beta)}(x)}{\gamma_{N,N}}$$

(a) Show that the integral of the function $y_{N-1}(x)$ with respect to the weighting function $W(x) = x^\beta(1-x)^\alpha$ from 0 to 1 is given by the following quadrature

$$\int_0^1 \left[x^\beta(1-x)^\alpha\right] y_{N-1}(x)\, dx = \sum_{j=1}^N w_j y_j$$

where

$$w_j = \int_0^1 \left[x^\beta(1-x)^\alpha\right] l_j(x)\, dx$$

(b) Use the results of Problems 8.2 and 8.3 to show that the quadrature weights of part (a) can be written as

$$w_j = \int_0^1 \left[x^\beta(1-x)^\alpha\right]\left[l_j(x)\right]^2 dx$$

(c) Starting from the equation in part (b) and the definition of the Lagrangian polynomial $l_j(x)$, show that

$$w_j = \frac{(2N + \alpha + \beta + 1)c_N^{(\alpha,\beta)}}{x_j(1 - x_j)\left[\dfrac{dp_N(x_j)}{dx}\right]^2}$$

where

$$c_N^{(\alpha,\beta)} = \int_0^1 \left[x^\beta(1 - x)^\alpha\right]p_N^2(x)\,dx > 0$$

8.5₃. Let $x_1, x_2, x_3, \ldots, x_N$ be N interpolation points chosen as roots of the Jacobi polynomial $J_N^{(\alpha,\beta)}(x) = 0$ (i.e., $0 < x_j < 1$ for $j = 1, 2, \ldots, N$) and the $(N + 1)$th interpolation point is $x_{N+1} = 1$. The ordinates corresponding to these $(N + 1)$ points are denoted as $y_1, y_2, y_3, \ldots, y_N$, and y_{N+1}, and the polynomial of degree N, $y_N(x)$, passing through these $N + 1$ points $(x_1, y_1), (x_2, y_2), \ldots, (x_N, y_N), (x_{N+1}, y_{N+1})$ is given by the Lagrangian formula

$$y_N(x) = \sum_{j=1}^{N+1} l_j(x)y_j$$

where

$$l_j(x) = \frac{p_{N+1}(x)}{(x - x_j)p'_{N+1}(x_j)}$$

with $p_{N+1}(x)$ being the node polynomial defined as

$$p_{N+1}(x) = (x - x_1)(x - x_2)(x - x_3) \cdots (x - x_N)(x - x_{N+1})$$

$$= p_N^{(\alpha,\beta)}(x) \cdot (x - 1)$$

where $p_N^{(\alpha,\beta)}(x)$ is the scaled Jacobi polynomial

$$p_N^{(\alpha,\beta)}(x) = \frac{J_N^{(\alpha,\beta)}(x)}{\gamma_{N,N}}$$

(a) Show that the integral of the function $y_N(x)$ with respect to the weighting function $W(x) = x^\beta(1 - x)^\alpha$ from 0 to 1 is given by the quadrature

$$\int_0^1 \left[x^\beta(1 - x)^\alpha\right]y_N(x)\,dx = \sum_{j=1}^{N+1} w_j y_j$$

where

$$w_j = \int_0^1 \left[x^\beta (1 - x)^\alpha \right] l_j(x) \, dx$$

(b) Show that w_j (for $j = 1, 2, \ldots, N$) are identical to the quadrature weights obtained in Problem 8.4, where N interpolation points are used.

(c) Prove that $w_{N+1} = 0$, which implies that the extra interpolation point at x_{N+1} is not taken into account in the evaluation of the numerical quadrature.

8.6₃. Similar to Problem 8.5, let $x_1, x_2, x_3, \ldots, x_N$ be N interpolation points chosen as roots of the Jacobi polynomial $J_N^{(\alpha, \beta)}(x) = 0$ (i.e., $0 < x_j < 1$ for $j = 1, 2, \ldots, N$) and $x_0 = 0$ is the additional interpolation point. The ordinates corresponding to these $(N + 1)$ points are $y_0, y_1, y_2, y_3, \ldots, y_N$ and the polynomial of degree N, $y_N(x)$, passing through these $N + 1$ points $(x_0, y_0), (x_1, y_1), (x_2, y_2), \ldots, (x_N, y_N)$ is given by the Lagrangian formula

$$y_N(x) = \sum_{j=0}^{N} l_j(x) y_j$$

where

$$l_j(x) = \frac{p_{N+1}(x)}{(x - x_j) p'_{N+1}(x_j)}$$

with $p_{N+1}(x)$ being the node polynomial defined as

$$p_{N+1}(x) = (x - x_0)(x - x_1)(x - x_2) \cdots (x - x_N) = x \cdot p_N^{(\alpha, \beta)}(x)$$

where $p_N^{(\alpha, \beta)}(x)$ is the scaled Jacobi polynomial

$$p_N^{(\alpha, \beta)}(x) = \frac{J_N^{(\alpha, \beta)}(x)}{\gamma_{N, N}}$$

(a) Show that the integral of the function $y_N(x)$ with respect to the weighting function $W(x) = x^\beta (1 - x)^\alpha$ from 0 to 1 is given by the quadrature

$$\int_0^1 \left[x^\beta (1 - x)^\alpha \right] y_N(x) \, dx = \sum_{j=0}^{N} w_j y_j$$

where

$$w_j = \int_0^1 \left[x^\beta (1 - x)^\alpha \right] l_j(x) \, dx$$

(b) Show that w_j (for $j = 1, 2, \ldots, N$) are identical to the quadrature weights obtained in Problem 8.4, where N interpolation points are used, and $w_0 = 0$. This, like Problem 8.5, means that adding the extra interpolation point either at $x = 1$ or $x = 0$ does not help to improve the accuracy of the quadrature. Even when N extra interpolation points are added in addition to the N collocation points as roots of $J_N^{(\alpha, \beta)}(x) = 0$, there is no net effect of this addition. This is because the N ordinates y_j at the zeros of $J_N^{(\alpha, \beta)}(x) = 0$ are sufficient to integrate a polynomial of degree $2N - 1$ exactly.

8.7*. It is shown in Problems 8.5 and 8.6 that adding an extra interpolation point or even N interpolation points to the N interpolation points, which are zeros of $J_N^{(\alpha, \beta)}(x) = 0$, does not help to improve the evaluation of the integral by quadrature

$$\int_0^1 \left[x^\beta (1 - x)^\alpha \right] y(x)\, dx$$

Now reconsider Problem 8.5 where $(N + 1)$ interpolation points are used. The Nth degree polynomial passing through $N + 1$ points $(x_1, y_1), (x_2, y_2), \ldots, (x_N, y_N), (x_{N+1}, y_{N+1})$ is given by

$$y_N(x) = \sum_{j=1}^{N+1} l_j(x) y_j$$

where $l_j(x)$ is given as in Problem 8.5.

(a) Construct a $2N$th degree polynomial as follows

$$y_{2N}(x) = y_N(x) + G_{N-1} \cdot (1 - x) \cdot p_N^{(\alpha+1, \beta)}(x)$$

where G_{N-1} is any $(N - 1)$th degree polynomial and $p_N^{(\alpha, \beta)}(x)$ is the scaled Jacobi polynomial

$$p_N^{(\alpha+1, \beta)}(x) = \frac{J_N^{(\alpha+1, \beta)}(x)}{\gamma_{N, N}}$$

Evaluate the following integral

$$\int_0^1 \left[x^\beta (1 - x)^\alpha \right] y_{2N}(x)\, dx$$

by the quadrature method and show that it is equal to

$$\int_0^1 \left[x^\beta (1 - x)^\alpha \right] y_{2N}(x)\, dx = \sum_{j=1}^{N+1} w_j y_j$$

where

$$w_j = \int_0^1 \left[x^\beta (1 - x)^\alpha \right] l_j(x) \, dx$$

if the $N + 1$ interpolation points are chosen such that the first N interpolation points are roots of the Jacobi polynomial $J_N^{(\alpha+1, \beta)}(x) = 0$ and the $(N + 1)$th point is $x_{N+1} = 1$. This quadrature formula is called the Radau quadrature. It can integrate any polynomial of degree $2N$ exactly.

8.8₃. There are a number of ways to assume the form of the approximate solution. One way, which is used often in the text, is the power law expression

$$y_N = \sum_{j=0}^N a_j x^j$$

where the subscript N means that the approximate polynomial has the degree N. Another way, also equally useful, is the use of the Jacobi polynomials as the expansion terms, given as

$$y_N = \sum_{j=0}^N b_j J_j^{(\alpha, \beta)}(x)$$

where $J_j^{(\alpha, \beta)}(x)$ is the Jacobi polynomial of degree j.
(a) Make use of the following orthogonality properties of the Jacobi polynomial

$$\int_0^1 \left[x^\beta (1 - x)^\alpha \right] J_j^{(\alpha, \beta)}(x) J_k^{(\alpha, \beta)}(x) \, dx = 0 \qquad \text{for } k \neq j$$

to show how the coefficients b_j are determined in terms of the coefficients a_j.

8.9*. Modelling of a cylindrical catalyst with an nth order chemical reaction under the isothermal conditions gives rise to the following equation

$$D_e \frac{1}{r} \frac{d}{dr} \left(r \frac{dC}{dr} \right) - \rho_p k C^n = 0$$

where kC^n is the chemical reaction rate per unit mass of the catalyst, and D_e is the effective diffusivity. Assuming that the fluid surrounding the catalyst is vigorously stirred, the following boundary conditions can be taken

$$r = 0; \quad \frac{dC}{dr} = 0 \quad \text{and} \quad r = R; \quad C = C_0$$

where C_0 is the constant bulk concentration, and R is the particle radius.

(a) Show that the above dimensional equation can be cast into the following nondimensional form

$$\frac{1}{x}\frac{d}{dx}\left(x\frac{dy}{dx}\right) - \phi^2 y^n = 0$$

subject to

$$x = 0; \quad \frac{dy}{dx} = 0 \quad \text{and} \quad x = 1; \quad y = 1$$

What are the definitions of y, x, and ϕ that yield the above form of nondimensional equations?

(b) Because of the symmetry of the problem around the point $x = 0$, it is convenient to use the following symmetry transformation $u = x^2$. Show that the new equations written in terms of the new variable u are

$$4u\frac{d^2 y}{du^2} + 4\frac{dy}{du} - \phi^2 y^n = 0$$

$$u = 1; \quad y = 1$$

Note that the center boundary condition is not needed because of the symmetry transformation.

(c) To solve the equations in part (b) using the method of weighted residual, assume that the approximation solution has the form

$$y_a = 1 + a_1(1 - u)$$

where a_1 is the unknown coefficient to be found. Note that this assumed form is made to satisfy the boundary condition at $u = 1$. Use this approximate solution in the governing equation of part (b) to show that the residual is

$$R(a_1, u) = -4a_1 - \phi^2\left[1 + a_1(1 - u)\right]^n$$

The residual is a function of both the independent variable u and the unknown coefficient a_1.

(d) Apply the method of collocation with a test function $\delta(u - u_1)$ to show that the equation for a_1 is

$$\int_0^1 R(a_1, u)\delta(u - u_1)\, du = -4a_1 - \phi^2\left[1 + a_1(1 - u_1)\right]^n = 0$$

where u_1 is some arbitrary collocation point in the domain $(0, 1)$. The above equation is a nonlinear algebraic equation, which can be solved for a_1 once the collocation point u_1 is chosen.

(e) Apply the Galerkin method with the test function $(1 - u)$ to show that the equation for a_1 is expressed in the following form of the integral

$$\int_0^1 R(a_1, u)(1 - u)\, du$$

$$= \int_0^1 \left\{ -4a_1 - \phi^2 [1 + a_1(1 - u)]^n \right\}(1 - u)\, dx = 0$$

For integer n, the above integral can be analytically integrated, while for noninteger n, it must be evaluated numerically or using some form of numerical quadrature.

(f) Now try the numerical quadrature[3] approach to approximate the integral of part (e) and show that the approximation is

$$\int_0^1 R(a_1, u)(1 - u)\, du \approx \sum_{j=1}^M w_j R(a_1, u_j)$$

where u_j are quadrature points, w_j are the quadrature weights, and M is the number of such points. Show that the optimal choice of these M quadrature points are roots of $J_M^{(1,0)}(u)$.

Now take only one quadrature point, and this point will be the root of the following Jacobi polynomial of degree 1, $J_1^{(1,0)}(x)$, which is found in Problem 8.14 as

$$J_1^{(1,0)}(x) = 3u - 1 = 0 \qquad \text{i.e.,} \qquad u_1 = \frac{1}{3}$$

Show that the resulting equation for a_1 for the Galerkin method is identical to the equation for a_1 obtained by the collocation method. This means that collocation and Galerkin methods yield the same

---

[3]The approximation of the integral

$$\int_0^1 W(x) f(x)\, dx \qquad \text{with} \qquad W(x) = x^\beta (1 - x)^\alpha$$

by the quadrature method is

$$\int_0^1 W(x) f(x)\, dx \approx \sum_{k=1}^M w_k f(x_k)$$

where there exists an optimal choice of quadrature points, and those points are roots of the Mth degree Jacobi polynomial $J_M^{(\alpha,\beta)}(x) = 0$.

answer for a_1, if the collocation point is chosen as a root of the proper Jacobi polynomial, and the quadrature approximation of the integral is exact.[4]

(g) If the chemical reaction is first order ($n = 1$), show that the quadrature approximation of the integral obtained for the Galerkin method

$$\int_0^1 R(a_1, u)(1 - u) \, du = w_1 R(a_1, u_1)$$

is exact. This means that the Galerkin and the collocation methods will yield identical approximate solutions if the collocation point is chosen as root of the Jacobi polynomial $J_1^{(1,0)} = 0$ (i.e., $u_1 = 1/3$). For such a situation, the collocation method is called the orthogonal collocation method because the Jacobi polynomial belongs to a class of orthogonal functions.

(h) Now consider the second order chemical reaction ($n = 2$), show that the number of quadrature points required to yield exact evaluation of the integral

$$\int_0^1 R(a_1, u)(1 - u) \, du$$

by the method of quadrature is 2, with the two quadrature points being roots of $J_2^{(1,0)}(x) = 0$. Obtain the expression for this Jacobi polynomial of degree 2 and hence derive the solutions for these two quadrature points.

(i) The previous parts (a to h) consider only the one-term approximate solution. Now consider the following N-terms trial solution

$$y_a = 1 + \sum_{j=1}^{N} a_j (1 - u) u^{j-1}$$

where a_j ($j = 1, 2, \ldots, N$) are N unknown coefficients to be found. Show that the residual generated by this choice of approximate solution is

$$R_N(a_1, a_2, \ldots, a_N; u)$$

$$= 4 \sum_{j=1}^{N} a_j \left[(j - 1)^2 u^{j-2} - j^2 u^{j-1} \right] - \phi^2 \left[1 + (1 - u) \sum_{j=1}^{N} a_j u^{j-1} \right]^n$$

(j) Now apply the collocation method with the N test functions $\delta(u - u_k)$ (for $k = 1, 2, \ldots, N$), where u_k are collocation points to show that

---

[4] Using M quadrature points (which are roots of the proper Jacobi polynomial) in the numerical quadrature, the quadrature approximation will be exact if the function $R(u)$ is a polynomial of degree less than or equal to $2M - 1$.

the N nonlinear algebraic equations to be solved for a_j's are

$$R_N(a_1, a_2, \ldots, a_N; u_k)$$

$$= 4 \sum_{j=1}^{N} a_j \left[(j-1)^2 u_k^{j-2} - j^2 u_k^{j-1} \right] - \phi^2 \left[1 + (1 - u_k) \sum_{j=1}^{N} a_j u_k^{j-1} \right]^n$$

$$= 0$$

for $k = 1, 2, 3, \ldots, N$.

(k) Now apply the Galerkin method with the N test functions $(1 - u)u^{k-1}$ (for $k = 1, 2, \ldots, N$) to show that the N equations for a_j's written in the form of integral are

$$\int_0^1 R_N(a_1, a_2, \ldots, a_N; u) \left[(1-u)u^{k-1} \right] du = 0$$

for $k = 1, 2, \ldots, N$,

where R_N is given in part (i). These integrals must be evaluated numerically or approximated by the following quadrature using M quadrature points

$$\int_0^1 \left[R_N(a_1, a_2, \ldots, a_N; u)u^{k-1} \right](1 - u) \, du$$

$$\approx \sum_{j=1}^{M} w_j \left[R_N(a_1, a_2, \ldots, a_N; u_j)u_j^{k-1} \right] =$$

for $k = 1, 2, \ldots, N$

The M quadrature points are roots of the Jacobi polynomial of degree M, $J_M^{(1,0)}(x) = 0$. Show that the above quadrature approximation is the exact representation of the integral if

$$M \geq \frac{N(n+1)}{2}$$

(l) Show that one trivial way to satisfy the nonlinear algebraic equations in part (k) is to set

$$R_N(a_1, a_2, \ldots, a_N; u_j) = 0 \qquad \text{for} \qquad j = 1, 2, \ldots, M$$

and hence show that a_j's can be determined if the number of quadrature points are chosen the same as the number of coefficient, N, in the trial solution. When this is the case, the Galerkin is "best" approximated by the collocation method if the collocation points are chosen as roots of the Nth degree Jacobi polynomial $J_N^{(1,0)}(x) = 0$.

(m) Prove that if the chemical reaction is first order, the collocation method with N collocation points chosen as roots of $J_N^{(1,0)}(x) = 0$ is identical to the Galerkin method.

This example illustrates how collocation points should be optimally chosen so that they can closely match the Galerkin method.

8.10$_3$. The Jacobi polynomial can be expressed conveniently as

$$J_N^{(\alpha, \beta)}(x) = \sum_{j=0}^{N} (-1)^{N-j} \gamma_j x^j \qquad \text{with} \qquad \gamma_0 = 1$$

The N coefficients, γ_j ($j = 1, 2, \ldots, N$), are determined from the following N orthogonality condition equations

$$\int_0^1 \left[x^\beta (1-x)^\alpha \right] J_k^{(\alpha, \beta)}(x) \cdot J_N^{(\alpha, \beta)}(x) \, dx = 0$$

$$\text{for } k = 0, 1, 2, \ldots, N-1$$

(a) Show that the above orthogonality condition equations are equivalent to

$$\int_0^1 \left[x^\beta (1-x)^\alpha \right] x^k J_N^{(\alpha, \beta)}(x) \, dx = 0 \qquad \text{for} \qquad k = 0, 1, 2, \ldots, N-1$$

(b) Use the equations of part (a) to then show that the linear equations for solving for γ_j's are

$$\mathbf{A}\boldsymbol{\gamma} = \mathbf{b}$$

where $\mathbf{A}$ is a coefficient matrix of size $N \times N$ and $\mathbf{b}$ is a constant vector, taking the form

$$a_{ij} = \frac{\Gamma(\beta + i + j)\Gamma(\alpha + 1)(-1)^{N-j}}{\Gamma(\beta + \alpha + i + j + 1)}$$

$$b_i = -(-1)^N \frac{\Gamma(\beta + \alpha + i)\Gamma(\alpha + 1)}{\Gamma(\beta + \alpha + i + 1)}$$

for $i, j = 1, 2, \ldots, N$.

8.11$_3$. Show that

$$C_N^{(\alpha, \beta)} = \int_0^1 \left[x^\beta (1-x)^\alpha \right] \left[J_N^{(\alpha, \beta)}(x) \right]^2 dx$$

$$= \frac{\left[\Gamma(\beta + 1) \right]^2 \Gamma(N + \alpha + 1)(N!)}{\Gamma(N + \beta + 1)\Gamma(N + \alpha + \beta + 1)(2N + \alpha + \beta + 1)}$$

Hint: Use the Rodrigues formula and apply integration by parts N times.

8.12₂. Start with the definition of the Jacobi polynomial to prove that

$$\frac{dJ_N^{(\alpha,\beta)}(x)}{dx} = \frac{N(N+\alpha+\beta+1)}{\beta+1} J_{N-1}^{(\alpha+1,\beta+1)}(x)$$

8.13₃. Use the orthogonality condition defining Jacobi polynomial to prove that all N zeros of the Jacobi polynomial $J_N^{(\alpha,\beta)}(x) = 0$ are real and that they lie between 0 and 1.

8.14₁. (a) Use the Newton formula (Eq. 8.87) to determine roots of the following Jacobi polynomials

$$J_N^{(0,0)}(x), \quad J_N^{(1,1)}(x), \quad J_N^{(0,1)}(x), \quad J_N^{(1,0)}(x)$$

for $N = 1, 2, 3$, and 5.

Explain the shift of the zeros in the domain $[0, 1]$ as α and β change.

(b) Write the differential equation satisfied by each polynomial.

8.15₁. Repeat problem 8.14 using the Jacobi polynomial

$$J_N^{(1,-1/2)}(x), \quad J_N^{(1,0)}(x) \quad \text{and} \quad J_N^{(1,1/2)}(x)$$

for $N = 1, 2, 3$, and 5. These roots are used as interpolation points in the orthogonal collocation analysis of a slab, a cylinder, and a spherical particle, respectively.

8.16₁. Calculate the matrices **A** and **B** for two collocation points chosen as roots of $J_2^{(0,-1/2)}$ using the formula (8.103) and show that the sum of all rows is identically zero.

Hint: Use the definition of the matrices **A** and **B** in Eq. 8.102.

8.17₂. Use the five different methods of weighted residual to obtain approximate solutions for the equation

$$\frac{d}{dx}\left[(1+y)\frac{dy}{dx}\right] = 10y$$

subject to the following conditions

$$x = 0; \qquad \frac{dy}{dx} = 0$$

and

$$x = 1; \qquad y = 1$$

(a) First, try the approximate solution

$$y_a = 1 + a_1(1 - x^2)$$

which satisfies the boundary conditions. Substitute this approximate solution to the equation to form a residue, then use the test function appropriate for each method of weighted residual to obtain a solution for a_1, hence y_a. Compare the approximate solution with the exact solution obtained in part (c).

(b) To improve the accuracy of the approximate solution, use the following trial solution with two unknown coefficients

$$y_a = 1 + a_1(1 - x^2) + a_2 x^2 (1 - x^2)$$

Substitute this approximate solution with yet to be determined coefficients into the governing equation to form a residue, which is then forced to zero in some average sense. For each of the methods of weighted residual, use two test functions to determine the two coefficients, then compare this approximate solution with that in part (a).

(c) To obtain the exact solution of the governing equation, put

$$p = \frac{dy}{dx}$$

then show that the equation will take the form

$$p \frac{d}{dy}[(1 + y)p] = 10y$$

Next, set $u = (1 + y)$ to show the new equation is

$$p \frac{d(up)}{du} = 10(u - 1)$$

To put this equation in separable form, multiply both sides of the equation by u and show that the separable form is

$$(up)\, d(up) = 10u(u - 1)\, du$$

Integrate this new separable form with the condition at $x = 0$

$$x = 0; \qquad y = y_0; \qquad p = \frac{dy}{dx} = 0$$

to obtain the equation

$$p = \frac{dy}{dx} = \frac{\sqrt{20}}{1 + y}\left[\frac{(1 + y)^3 - (1 + y_0)^3}{3} - \frac{(1 + y)^2 - (1 + y_0)^2}{2}\right]^{1/2}$$

where y_0 is the value of y at $x = 0$, which is yet to be determined at this stage.

Put the new equation in a separable form and integrate it from x to 1 to show that the solution for y is simply

$$\int_y^1 \frac{(1+s)\,ds}{\left[\dfrac{(1+s)^3 - (1+y_0)^3}{3} - \dfrac{(1+s)^2 - (1+y_0)^2}{2}\right]^{1/2}} = \sqrt{20}\,(1-x)$$

To find y_0, simply put $x = 0$ into the above equation and show an implicit equation for y_0 is

$$\int_{y_0}^1 \frac{(1+s)\,ds}{\left[\dfrac{(1+s)^3 - (1+y_0)^3}{3} - \dfrac{(1+s)^2 - (1+y_0)^2}{2}\right]^{1/2}} = \sqrt{20}$$

where s is the dummy integration variable.

8.18₃. Transport of solute through membrane is often limited by the ability of the solute to move (diffuse) through the membrane. If diffusion through the membrane is the rate controlling step, the usual relation to describe such transport is Fick's law (see Problem 6.11). Usually the diffusion coefficient increases with concentration. Set up a material balance of a solute within a thin shell in the membrane to show that the governing equation will take the form at steady state

$$\frac{d}{dr}\left[D(C)\frac{dC}{dr}\right] = 0$$

where r is the coordinate, taking the origin at the feed side of the membrane. If on the collection side, the solute is swept away quickly with a carrier fluid, then the solute concentrations at both sides of the membrane are C_0 and 0, respectively.

(a) By setting $y = C/C_0$, $x = r/L$ and $f(y) = D(C)/D(C_0)$, where L is the membrane thickness, show that the mass balance equation will take the following dimensionless form

$$\frac{d}{dx}\left[f(y)\frac{dy}{dx}\right] = 0$$

The boundary conditions at two sides of the membrane in nondimensional form are

$$x = 0; \qquad y = 1$$
$$x = 1; \qquad y = 0$$

For the purpose of computation in this problem, take the following two forms for $f(y)$

$$f(y) = \begin{cases} 1 + \sigma(y - 1) \\ \exp[\sigma(y - 1)] \end{cases}$$

(a) Taking note of the asymmetric boundary conditions, use the following equation as a trial solution

$$y_a = 1 - x + a_1(x - x^2)$$

Apply the methods of collocation and Galerkin to obtain the approximate solutions.

(b) Repeat part (a) with the following trial solution having two coefficients

$$y_a = 1 - x + a_1(x - x^2) + a_2(x - x^3)$$

Compare the solutions with those obtained in part (a), and suggest an objective basis for assessing improvement between the two.

Use the following exact solution with which to compare the two approximate solutions

$$\frac{\int_y^1 f(s)\, ds}{\int_0^1 f(s)\, ds} = 1 - x$$

8.19₁. Problems 8.17 and 8.18 deal with simple diffusion problems with symmetry and asymmetry boundary conditions. The methods of weighted residual can also be applied to cases where a source term appears in the equation and such a source term can be a discontinuous function within the spatial domain, such as the following problem of diffusion of material in a slab with a mass production source. The governing equations are

$$\frac{d^2y}{dx^2} + f(x) = 0$$

where

$$f(x) = \begin{cases} 1 & 0 < x < \dfrac{1}{2} \\ 0 & \dfrac{1}{2} < x < 1 \end{cases}$$

The boundary conditions at two sides of the slab are taken to be

$$x = 0; \quad y = 0$$
$$x = 1; \quad y = 0$$

This set of equations can also describe the heat conduction in a slab solid object with a heat source within the slab.

(a) Polynomial trial solutions have been used in Problems 8.17 and 8.18. This time try the following trial solution of trigonometric form

$$y_a = a_1 \sin(\pi x)$$

and use the collocation method to find the coefficient a_1. Choose $x_1 = 1/2$ (take $f(1/2) = 1/2$).

(b) Improve the trial solution by having two terms in the solution; that is,

$$y_a = a_1 \sin(\pi x) + a_2 \sin(2\pi x)$$

This problem shows that the trial solutions need not be in the polynomial form used in the text as well as in Problems 8.17 and 8.18, and it also shows that as the number of terms used in the trial solution increases the analysis involving the polynomial is somewhat simpler than that using functions such as trigonometric functions.

8.20*. The mass and heat balance equations in a catalyst for a first order reaction are

$$\frac{d^2 y}{dx^2} - \phi^2 \exp\left(\frac{\gamma\theta}{1+\theta}\right) y = 0$$

$$\frac{d^2\theta}{dx^2} + \beta\phi^2 \exp\left(\frac{\gamma\theta}{1+\theta}\right) y = 0$$

where ϕ is the Thiele modulus, β is the dimensionless heat of reaction, and γ is the dimensionless activation energy.

Assuming the fluid medium surrounding the catalyst particle is very well stirred, the boundary conditions are

$$x = 0; \qquad \frac{dy}{dx} = \frac{d\theta}{dx} = 0$$

$$x = 1; \qquad y = 1, \qquad \theta = 0$$

The quantity of interest is the effectiveness factor, which indicates how well a catalyst is utilized by the reactant. It is defined as

$$\eta = \frac{1}{\phi^2} \frac{dy(1)}{dx}$$

(a) Apply the one-point orthogonal collocation method to solve the above coupled equations. Multiple solutions are possible in this case. Determine the range of ϕ such that more than one steady state is possible.

(b) Another way of solving this problem is to eliminate one variable by first relating it to the other variable. Multiply the equation for y by β

and add this to the θ equation to show that

$$\beta \frac{d^2 y}{dx^2} + \frac{d^2 \theta}{dx^2} = 0$$

(c) Integrate the above equation once to show that

$$\beta \frac{dy}{dx} + \frac{d\theta}{dx} = 0$$

(d) Integrate again and obtain

$$\beta y + \theta = \beta$$

This shows temperature and composition are linearly related.

(e) Use the result in part (d) to eliminate the temperature from the differential equation for y, and show that the final equation is

$$\frac{d^2 y}{dx^2} - \phi^2 \exp\left(\frac{\gamma \beta (1 - y)}{1 + \beta (1 - y)} \right) y = 0$$

Now there is only one equation and one unknown, y.

(f) Apply the one-point orthogonal collocation to solve the equation in part (e), and compare with the result in part (a).

8.21*. Consider the problem of diffusion and reaction in a spherical catalyst particle. The chemical reaction is assumed to follow the Langmuir–Hinshelwood kinetics

$$R_{rxn} = \frac{kC}{1 + KC + K_I C^2}$$

The film mass transfer resistance is taken as negligible compared to the internal diffusion resistance.

(a) Derive the mass balance equation for the reactant in the catalyst particle, put it in the nondimensional format, to give the result

$$\frac{1}{x^2} \frac{d}{dx}\left(x^2 \frac{dy}{dx} \right) - \phi^2 \frac{y}{1 + \beta y + \gamma y^2} = 0$$

(b) Use the one-point collocation to investigate the behavior of the system.

(c) Choose the parameters such that only one steady-state is possible, and obtain a better approximate solution to the problem using the orthogonal collocation method.

(d) Choose the parameters where multiple steady states occur and solve for all steady-state concentration distributions inside the particle.

Chapter 9

Introduction to Complex Variables and Laplace Transforms

9.1 INTRODUCTION

A principal engineering application of the theory of functions of complex variables is to effect the inversion of the so-called Laplace transform. Because the subjects are inextricably linked, we treat them together. The Laplace transform is an integral operator defined as:

$$F(s) = \int_0^\infty f(t)e^{-st}\,dt \tag{9.1}$$

where, in general, s is a complex variable defined here as:

$$s = \sigma + i\omega \tag{9.2}$$

The inverse process (that is given $F(s)$, find $f(t)$) is obtained through the Fourier–Mellin complex integral

$$f(t) = \frac{1}{2\pi i}\lim_{\omega \to \infty}\int_{\sigma_0 - i\omega}^{\sigma_0 + i\omega} e^{st}F(s)\,ds \tag{9.3}$$

henceforth called the Inversion theorem. Such complex integrals can be viewed as *contour integrals*, since they follow the two-dimensional path traced out by the curve of $s(\sigma, \omega)$. As we show presently, it is possible to invert Laplace transformed problems without recourse to complex integration, by laying out a few common sense building blocks, which are widely tabulated. However, for unusual problems the analyst must refer back to the fundamental Inversion theorem, Eq. 9.3. The derivation of Eq. 9.3 is based on the Fourier series representation of any $f(t)$, as detailed in Appendix C.

331

9.2 ELEMENTS OF COMPLEX VARIABLES

The complex number $\beta = a + ib$ is made up of two parts: the real part is a

$$Re(\beta) = a \qquad (9.4)$$

and the imaginary part is b

$$Im(\beta) = b \qquad (9.5)$$

In the same way, we may define a complex variable, s

$$s = \sigma + i\omega \qquad (9.6)$$

where it is clear that σ and ω are variables; hence, they may take many changing values. It is convenient to represent a complex variable on a rectangular coordinate system such that the abscissa represents real parts (σ) while the ordinate reflects values of the imaginary variable (ω), as shown in Fig. 9.1. Thus, any complex number or variable can be thought of as a vector quantity as illustrated in Fig. 9.1. We define the magnitude of s (line length) as the modulus

$$|s| = |\sigma + i\omega| = \sqrt{\sigma^2 + \omega^2} \qquad (9.7)$$

Henceforth, bars around a variable shall denote magnitude. The angle that s makes with the real axis is called the argument or simply the angle $\angle s$ (the angle corresponding to the complex variable s)

$$\theta = \angle s = \tan^{-1}(\omega/\sigma) \qquad (9.8)$$

When $\omega = 0$, s becomes identical to σ and can be treated as a real variable. When this happens, the processes of integration, differentiation, and so forth,

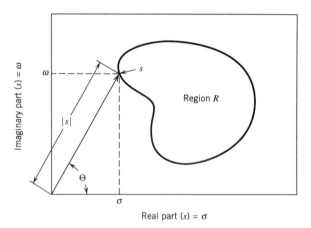

Figure 9.1 Representation of s in the complex plane.

then follow the usual rules of calculus. However, for the case of the complete complex variable, $s = \sigma + i\omega$, we must reexamine the usual rules of functions, especially the property of continuity (i.e., continuous functions, with continuous derivatives). We shall uncover these rules of continuity, called *analyticity*, in the form of the famous Cauchy–Riemann conditions. First, we shall need to know the elementary rules for multiplication, division, and so forth of complex variables.

The complex conjugate of s is simply

$$\bar{s} = \sigma - i\omega \tag{9.9}$$

and it is clear that since $i^2 = -1$

$$s \cdot \bar{s} = \sigma^2 + \omega^2 = |s|^2 \tag{9.10}$$

The product of two complex variables can always be reduced to a new complex variable, containing a real plus imaginary part

$$s_1 \cdot s_2 = (\sigma_1 + i\omega_1)(\sigma_2 + i\omega_2) = (\sigma_1\sigma_2 - \omega_1\omega_2) + i(\sigma_1\omega_2 + \sigma_2\omega_1) \tag{9.11}$$

Division by complex numbers calls upon using the complex conjugate definition as

$$\frac{s_1}{s_2} = \frac{(\sigma_1 + i\omega_1)}{(\sigma_2 + i\omega_2)} = \frac{(\sigma_1 + i\omega_1)}{(\sigma_2 + i\omega_2)} \cdot \frac{\sigma_2 - i\omega_2}{\sigma_2 - i\omega_2} = \frac{(\sigma_1 + i\omega_1)(\sigma_2 - i\omega_2)}{\sigma_2^2 + \omega_2^2}$$

$$\tag{9.12}$$

Carrying out the multiplication then yields a new complex variable

$$\frac{s_1}{s_2} = \left(\frac{\sigma_1\sigma_2 + \omega_1\omega_2}{\sigma_2^2 + \omega_2^2} \right) + i\left(\frac{\sigma_2\omega_1 - \sigma_1\omega_2}{\sigma_2^2 + \omega_2^2} \right) \tag{9.13}$$

Thus, any array of multiplication and division will always yield finally a single real plus imaginary part.

Both of these operations ($s_1 \cdot s_2$ or s_1/s_2) can be performed more easily (shown in Section 9.3) using a polar form for s. Now, since s has both magnitude $|s|$ and angle θ, we can write by inspection of Fig. 9.1

$$\sigma = |s|\cos\theta, \qquad \omega = |s|\sin\theta \tag{9.14}$$

hence, we see the polar representation of s is simply

$$s = |s|(\cos\theta + i\sin\theta) \tag{9.15}$$

Note, for a given complex number, the angle θ can take an infinity of values by simply adding multiples of 2π to the smallest possible angle. Thus, if the smallest angle is $\theta_0 = \tan^{-1}\omega/\sigma$, then θ can take values: $\theta_0 + 2\pi, \theta_0 + 4\pi$, etc. So, in general: $\theta = \theta_0 + k2\pi$; $k = 0, \pm 1, \pm 2, \pm 3\ldots$, and θ_0 is called the principal value of θ with $0 \le \theta_0 \le 2\pi$.

9.3 ELEMENTARY FUNCTIONS OF COMPLEX VARIABLES

In the course of analysis, elementary functions of the complex variable s arise, such as: $\exp(-s)$, $\sin\sqrt{s}$, $\log(s)$, and so forth. We must carefully define these operations so that in the limit $\omega \to 0$, these operations reduce to conventional operations on real variables. We begin our study on the most elementary function, the power law relationship

$$f(s) = s^n \tag{9.16}$$

where n is taken to be a positive integer or zero. Replacing $s = \sigma + i\omega$

$$f(s) = (\sigma + i\omega)^n \tag{9.17}$$

we can insert the polar form to see

$$f(s) = |s|^n (\cos\theta + i\sin\theta)^n \tag{9.18}$$

We next combine this result with properties of the $\exp(s)$, which can be proved to have the usual power series representation

$$\exp(s) = 1 + s + \frac{s^2}{2!} + \frac{s^3}{3!} + \cdots \tag{9.19}$$

which is convergent for all values of the complex variable s.

It is also true that $\exp(s_1) \cdot \exp(s_2) = \exp(s_1 + s_2)$. Moreover, for the integer n

$$(e^s)^n = e^{ns} \quad (n = 1, 2, 3 \dots) \tag{9.20}$$

The trigonometric functions can be defined in terms of the exponential as

$$\sin(s) = \frac{1}{2i}(\exp(is) - \exp(-is)) \tag{9.21}$$

and

$$\cos(s) = \frac{1}{2}(\exp(is) + \exp(-is)) \tag{9.22}$$

This can be seen by inserting the power series (9.19) to get

$$\sin(s) = s - \frac{s^3}{3!} + \frac{s^5}{5!} + \cdots \tag{9.23}$$

$$\cos(s) = 1 - \frac{s^2}{2!} + \frac{s^4}{4!} + \cdots \tag{9.24}$$

which of course reduces to the usual series expansion when s is real. The useful

Euler formula arises by combining the trigonometric functions

$$\cos(s) + i\sin(s) = \exp(is) \tag{9.25}$$

which can also be applied when s is real, say $s = \theta$

$$\exp(i\theta) = \cos(\theta) + i\sin(\theta) \tag{9.26}$$

If we compare Eq. 9.18 with Eq. 9.26, we deduce De Moivre's theorem

$$s^n = |s|^n(\cos\theta + i\sin\theta)^n = |s|^n \cdot \exp(in\theta) \tag{9.27}$$

from which we see

$$(\cos\theta + i\sin\theta)^n = \cos(n\theta) + i\sin(n\theta) \tag{9.28}$$

hence,

$$s^n = |s|^n(\cos(n\theta) + i\sin(n\theta)) \tag{9.29}$$

Thus, it is convenient on multiplication or division to use Euler's formula, since we have shown in Eq. 9.15 that $s = |s|(\cos\theta + i\sin\theta)$, hence from Eq. 9.26 $s = |s|\exp(i\theta)$, so

$$\frac{s_1}{s_2} = \frac{|s_1|}{|s_2|}\exp[i(\theta_1 - \theta_2)] \tag{9.30}$$

We see it is much easier to add and subtract angles than direct multiplication of complex numbers. These can be reduced to real and imaginary parts by reversing the process using Eq. 9.26

$$s_1 \cdot s_2 = |s_1||s_2|(\cos(\theta_1 + \theta_2) + i\sin(\theta_1 + \theta_2)) \tag{9.31}$$

and

$$\frac{s_1}{s_2} = \frac{|s_1|}{|s_2|}(\cos(\theta_1 - \theta_2) + i\sin(\theta_1 - \theta_2)) \tag{9.32}$$

9.4   MULTIVALUED FUNCTIONS

A peculiar behavior pattern that arises in complex variables must be recognized very early, and that is the multivalued behavior exhibited by certain functions. We next consider the most elementary such case, the square root function

$$f(s) = \sqrt{s} \tag{9.33}$$

In analysis of real variables, no particular problems arise when taking square roots. However, considerable circumspection must be given when s is complex.

To see this, let

$$s = |s| \exp(i\theta) = |s|(\cos\theta + i\sin\theta) \tag{9.34}$$

where

$$\theta = \theta_0 + 2\pi \cdot k; \quad k = 0, \pm 1, \pm 2, \pm 3, \text{ etc.} \tag{9.35}$$

Inserting into 9.33 gives

$$f = \sqrt{|s|} \, \exp\left(i\left[\frac{\theta_0}{2} + \pi k\right]\right) \tag{9.36}$$

where as before $\theta_0 = \tan^{-1}\omega/\sigma$. If we wrote the function f as its real plus imaginary part, two distinct functions arise

$$f_0 = \sqrt{|s|} \left[\cos\frac{\theta_0}{2} + i\sin\frac{\theta_0}{2}\right] \tag{9.37}$$

and

$$f_1 = \sqrt{|s|} \left[\cos\left(\frac{\theta_0}{2} + \pi\right) + i\sin\left(\frac{\theta_0}{2} + \pi\right)\right] \tag{9.38}$$

All other values of k reproduce one or the other of these two.

Thus, any selected value of σ and ω will yield a value of the complex variable $s = \sigma + i\omega$, and this will produce two functions; namely,

$$f_0 = (\sigma^2 + \omega^2)^{1/4}\left[\cos\left(\frac{\tan^{-1}(\omega/\sigma)}{2}\right) + i\sin\left(\frac{\tan^{-1}(\omega/\sigma)}{2}\right)\right] \tag{9.39}$$

$$f_1 = (\sigma^2 + \omega^2)^{1/4}\left[\cos\left(\frac{\tan^{-1}(\omega/\sigma)}{2} + \pi\right) + i\sin\left(\frac{\tan^{-1}(\omega/\sigma)}{2} + \pi\right)\right]$$

$$\tag{9.40}$$

For example, suppose $\sigma = 0$, $\omega = 1$, so that

$$\theta_0 = \tan^{-1}(1/0) = \frac{\pi}{2} \, \text{rad}(90°)$$

Hence

$$f_0 = 1 \cdot \left(\frac{\sqrt{2}}{2} + i\frac{\sqrt{2}}{2}\right) \tag{9.41}$$

$$f_1 = 1 \cdot \left(\frac{-\sqrt{2}}{2} - i\frac{\sqrt{2}}{2}\right) \tag{9.42}$$

Thus, two branches of the function f are formed, differing by an angle of π radians. Thus, one value of the complex variable s leads to two possible complex values for the function f. To make any progress on the applied level, we shall need a mechanism to restrict such functions to single-valued behavior. We shall illustrate methods for doing this later.

We see that we cannot take the usual properties of uniqueness and continuity for granted in dealing with complex variables. To make progress, then, we shall need to formalize the properties of continuity and single valuedness. In analysis of complex variables, these properties are called analytic. We discuss them next.

9.5   CONTINUITY PROPERTIES FOR COMPLEX VARIABLES: ANALYTICITY

As last illustrated, some peculiar behavior patterns arise with complex variables, so care must be taken to insure that functions are well behaved in some rational sense. This property is called analyticity, so any function $w = f(s)$ is called analytic within some two-dimensional region R if at all arbitrary points, say s_0, in the region it satisfies the conditions:

1. It is single-valued in region R,
2. It has a unique, finite value in R,
3. It has a unique, finite derivative at s_0, which satisfies the Cauchy–Riemann conditions.

The Cauchy–Riemann conditions are the essential properties for continuity of derivatives, quite apart from those encountered in real variables. To see these, write the general complex function

$$w = f(s) \tag{9.43}$$

to be a continuous function with $s = \sigma + i\omega$, and suppose the real and imaginary parts are such that

$$w = u + iv \tag{9.44}$$

The partial derivatives can be obtained two ways

$$\frac{\partial w}{\partial \sigma} = \frac{\partial u}{\partial \sigma} + i\frac{\partial v}{\partial \sigma} \tag{9.45}$$

and

$$\frac{\partial w}{\partial \sigma} = \frac{df}{ds}\frac{\partial s}{\partial \sigma} = \frac{df}{ds} \tag{9.46}$$

Equating the two yields

$$\frac{df}{ds} = \frac{\partial u}{\partial \sigma} + i\frac{\partial v}{\partial \sigma} \tag{9.47}$$

In a similar way, partials with respect to ω give

$$\frac{\partial w}{\partial \omega} = \frac{\partial u}{\partial \omega} + i\frac{\partial v}{\partial \omega} \tag{9.48}$$

$$\frac{\partial w}{\partial \omega} = \frac{df}{ds}\frac{\partial s}{d\omega} = i\frac{df}{ds} \tag{9.49}$$

hence,

$$i\frac{df}{ds} = \frac{\partial u}{\partial \omega} + i\frac{\partial v}{\partial \omega} \tag{9.50}$$

Obviously, the total derivative of df/ds must be the same for the two cases, so multiplying Eq. 9.50 by $(-i)$ gives

$$\frac{df}{ds} = -i\frac{\partial u}{\partial \omega} + \frac{\partial v}{\partial \omega} \tag{9.51}$$

Now, equating real and imaginary parts of Eqs. 9.47 and 9.51 gives the sought after continuity conditions

$$\frac{\partial u}{\partial \sigma} = \frac{\partial v}{\partial \omega}$$

$$\frac{\partial v}{\partial \sigma} = -\frac{\partial u}{\partial \omega}$$

These are the Cauchy–Riemann conditions, and when they are satisfied, the derivative dw/ds becomes a unique single-valued function, which can be used in the solution of applied mathematical problems. Thus, the continuity property of a complex variable derivative has two parts, rather than the one customary in real variables. Analytic behavior at a point is called "regular," to distinguish from nonanalytic behavior, which is called "singular". Thus, points wherein analyticity breaks down are referred to as *singularities*. Singularities are not necessarily bad, and in fact their occurrence will be exploited in order to effect a positive outcome (e.g., the inversion of the Laplace transform!).

Many of the important lessons regarding analytic behavior can best be understood by way of a series of examples, to follow.

EXAMPLE 9.1

If $w(s) = s^2$, prove that the function satisfies the Cauchy–Riemann conditions and find the region where the function is always analytic (i.e., regular behavior).

These are several ways to represent the function, either in terms of σ, ω, or in polar form ($|s|$ and θ). We introduce $s = \sigma + i\omega$ and see

$$w = s^2 = (\sigma + i\omega)^2 = (\sigma^2 - \omega^2) + 2i\sigma\omega$$

It is clear if this must equal $w = u + iv$, then

$$u = \sigma^2 - \omega^2$$

$$v = 2\sigma\omega$$

We operate on these using Cauchy–Riemann rules

$$\frac{\partial u}{\partial \sigma} = 2\sigma; \qquad \frac{\partial u}{\partial \omega} = -2\omega$$

$$\frac{\partial v}{\partial \sigma} = 2\omega; \qquad \frac{\partial v}{\partial \omega} = 2\sigma$$

hence, we see it is true that

$$\frac{\partial u}{\partial \sigma} = \frac{\partial v}{\partial \omega} \qquad \text{and} \qquad \frac{\partial v}{\partial \sigma} = -\frac{\partial u}{\partial \omega}$$

Also, we see for all finite values of s that $w(s)$ is also finite, and that w is single valued in any region R, which is finite in size.

EXAMPLE 9.2

Consider $w(s) = 1/s$ and determine if the function satisfies the Cauchy–Riemann conditions and find the region for analytic behavior.

It is clear at the outset that behavior in the region $s \to 0$ is singular, and we are alerted to possible irregular behavior there. First, we form the real and imaginary parts as before

$$w(s) = \frac{1}{\sigma + i\omega} = \frac{1}{\sigma + i\omega} \frac{\sigma - i\omega}{\sigma - i\omega} = \frac{\sigma - i\omega}{\sigma^2 + \omega^2}$$

so that

$$u = \frac{\sigma}{\sigma^2 + \omega^2} \qquad \text{and} \qquad v = -\frac{\omega}{\sigma^2 + \omega^2}$$

We see that

$$\frac{\partial u}{\partial \sigma} = \frac{\partial v}{\partial \omega} = \frac{\omega^2 - \sigma^2}{(\sigma^2 + \omega^2)^2}$$

$$\frac{\partial v}{\partial \sigma} = -\frac{\partial u}{\partial \omega} = \frac{-2\sigma\omega}{(\sigma^2 + \omega^2)^2}$$

so that at a general point in space, the Cauchy–Riemann conditions appear to be satisfied. However, consider the behavior near the origin, along a line where

$\omega = 0$ so that

$$u = \frac{1}{\sigma} \qquad \therefore \frac{\partial u}{\partial \sigma} = -\frac{1}{\sigma^2}$$

This partial derivative tends to $-\infty$ as $\sigma \to 0$. In a similar way, we can inspect v along a line where $\sigma = 0$ to see

$$v = -\frac{1}{\omega} \qquad \therefore \frac{\partial v}{\partial \omega} = +\frac{1}{\omega^2}$$

which shows that this partial derivative tends to $+\infty$ as $\omega \to 0$. Thus, we obtain different derivatives depending on how we approach the origin, and the Cauchy–Riemann equality breaks down at the origin! Thus, the function is everywhere analytic, except at the origin, where a singularity exists (we shall call this a pole-type singularity, since the function takes on very steep, pole-like behavior as we approach $s = 0$). If the function took the form

$$w(s) = \frac{1}{(s - a)}$$

then misbehavior would have occurred at the point $s = a$, so we would declare a pole existed at $s = a$.

We have thus encountered two important types of singularities raising non-analytic behavior:

1. Multivalued function singularity: $w(s) = \sqrt{s}$
2. Pole singularities: $w(s) = 1/s$

A third type, called "essential singularity," arises infrequently, but should be recognized. A classic case of this type is the function

$$w(s) = \exp\left(\frac{1}{s}\right)$$

It can be studied by reference to a pole-type singularity. One method of testing implicit functions for pole behavior is to try to "remove" the pole. Thus, suppose the function $w = f(s)$ becomes infinite at the point $s = a$, and we are suspicious that a pole exists at that point. If we can define a new function

$$g(s) = (s - a) \cdot f(s) \tag{9.52}$$

that becomes analytic at $s = a$, then the pole has been removed (cancelled out). If not, then we would try higher powers, up to say n,

$$g(s) = (s - a)^n f(s) \tag{9.53}$$

and if analytic behavior for $g(s)$ finally occurs, then we have removed an nth order pole. This simply means that $f(s)$ must have contained a term

$$\frac{1}{(s-a)^n}$$

The order of the pole is thus determined by the minimum value of n necessary to remove it (cancel it).

For an "essential" singularity, it is not possible to remove the infinite discontinuity. To see this for the classic example mentioned, expand $\exp(1/s)$ in series.

$$w(s) = \exp\left(\frac{1}{s}\right) = 1 + \frac{1}{s} + \frac{1}{2!s^2} + \frac{1}{3!s^3} + \cdots + \frac{1}{n!}\frac{1}{s^n} + \cdots \quad (9.54)$$

If we tried to remove the singularity at the origin, by multiplying first by s, then s^2, then s^3, and so forth, we would still obtain an infinite spike at $s = 0$. Since the singularity *cannot* be removed, it is called an *essential* singularity.

9.5.1 Exploiting Singularities

At the outset, we have aimed our study toward the implementation of the Inversion theorem given in Eq. 9.3. It is clear that this requires integration in the complex domain. In fact, we can show this integration becomes exceedingly simple if the region of interest contains elementary pole singularities, leading to an elementary summation of the remnants of such poles (called *residues*). Thus, the occurrence of singular behavior allows easy exploitation to effect a positive mathematical result.

To implement this, we shall need to know something about integration in the complex domain.

9.6 INTEGRATION: CAUCHY'S THEOREM

Integration in the complex domain is necessarily two-dimensional, since variations in real and imaginary variables occur together. Thus, if we wish to find the integral of some arbitrary complex function $f(s)$, then we must stipulate the values of s such as those traced out by the curve C in Fig. 9.2.

To represent the integral corresponding to the path C, we may write, by noting $f = u + iv$ and $s = \sigma + i\omega$

$$\int_C f(s)\,ds = \int_C (u\,d\sigma - v\,d\omega) + i\int_C (v\,d\sigma + u\,d\omega) \quad (9.55)$$

Each integral on the right-hand side is now a real integral and the limits are from σ_1 to σ_2 and ω_1 to ω_2, corresponding to the points terminating the curve C at points A and B (positions s_1 and s_2). We note that if both $f(s)$ and s were

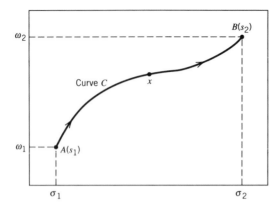

Figure 9.2 Curve in complex plane.

real (i.e., $v = \omega = 0$), then the line integral would be simply

$$\int_C u\, d\sigma$$

The values of s represented by the path in Fig. 9.2 could be integrated in two parts, if we wished, and we would denote this as

$$\int_{AXB} f(s)\, ds = \int_{AX} f(s)\, ds + \int_{XB} f(s)\, ds \tag{9.56}$$

and there should be no confusion on the meaning of the paths AX and XB.

Under certain conditions, the line integral in Eq. 9.55 is independent of path. Suppose both integrals on the right-hand side are exact, so for example, we could write

$$u\, d\sigma - v\, d\omega = dF(\sigma, \omega) \tag{9.57}$$

This implies also

$$dF = \frac{\partial F}{\partial \sigma}\, d\sigma + \frac{\partial F}{\partial \omega}\, d\omega \tag{9.58}$$

Comparing the last two equations gives the requirement

$$u = \frac{\partial F}{\partial \sigma} \qquad \text{and} \qquad -v = \frac{\partial F}{\partial \omega} \tag{9.59}$$

Now, since the order of differentiation is immaterial

$$\frac{\partial^2 F}{\partial \omega\, \partial \sigma} = \frac{\partial^2 F}{\partial \sigma\, \partial \omega} \tag{9.60}$$

then it is easy to see that

$$\frac{\partial u}{\partial \omega} = -\frac{\partial v}{\partial \sigma} \tag{9.61}$$

which of course is one of the Cauchy–Riemann conditions. Exactness for the second integral on the RHS of Eq. 9.55 similarly implies

$$v \, d\sigma + u \, d\omega = dG(\sigma, \omega) \tag{9.62}$$

and following the same procedure, this finally yields the second Cauchy–Riemann condition

$$\frac{\partial u}{\partial \sigma} = \frac{\partial v}{\partial \omega} \tag{9.63}$$

We have now seen that the requirement that the two integrals be exact differentials is exactly the requirement that the Cauchy–Riemann conditions be satisfied. This means of course that the line integral

$$\int_C f(s) \, ds$$

is independent of the path C joining the end points at s_1 and s_2, provided of course that the curve C lies within a region R wherein $f(s)$ is analytic. In such cases, the curve C need not be prescribed and we may indicate the integral by denoting it with its limit only

$$\int_{s_1}^{s_2} f(s) \, ds$$

and since it is exact, we could write

$$f(s) \, ds = dg(s) \tag{9.64}$$

and the integral can be evaluated in the usual way

$$\int_{s_1}^{s_2} f(s) \, ds = g(s_2) - g(s_1) \tag{9.65}$$

where of course $g(s)$ is a function whose derivative is $f(s)$. Many such paths between A and B can be drawn, and the same integral results as long as analyticity is maintained for the region where the curves are drawn.

Further, if we allow the end points to coincide by drawing a closed curve as shown in Fig. 9.3, and the region enclosed by the curve contains no singularities (i.e., is analytic), then the important *First Integral theorem* of Cauchy arises

$$\oint f(s) \, ds = 0 \tag{9.66}$$

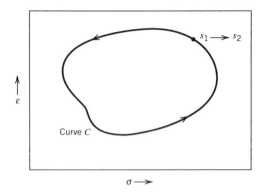

Figure 9.3 Closed contour, $s_1 \rightarrow s_2$.

The new symbol $\oint$ denotes integration around a simply connected (nonintersecting) curve C as shown in Fig. 9.3. For this curve, Eq. 9.65 shows $g(s_2) = g(s_1)$, since $s_1 \rightarrow s_2$.

It is important to stress that this curve encloses no singularities. In fact, it is the act of enclosing a singularity (or many singularities) that leads to a very simple, classical result for solving the Laplace Inversion theorem.

Now, in other cases (for example, curves that enclose singularities) the integral $f(s)$ around a closed contour of s values may or may not vanish. We shall define a positive curve direction such that a moving observer would keep the enclosed area to the left.

The exploitation of Cauchy's First theorem requires us to test the theorem for exceptional behavior. This allows, as we shall see, direct applications to the Laplace Inversion theorem.

We inspect the simplest single-valued function given by

$$f(s) = \frac{1}{s} \tag{9.67}$$

The derivative $df/ds = -1/s^2$ exists at all points except $s = 0$. This means that $f(s)$ is analytic in any region R which *does not* include the origin. Thus, any closed curve (of s values) not enclosing the origin will satisfy Cauchy's First theorem, hence, for any such curve

$$\oint_C \frac{ds}{s} = 0 \tag{9.68}$$

However, if the closed curve C encloses the origin, the integral need not vanish. It is convenient to inspect this curve using polar coordinates; to simplify the notation, we shall denote the magnitude of s as: $|s| = r$. Moreover, if we stipulate that the curve C is a unit circle around the origin, then $r = 1$ and we may write

$$s = e^{i\theta}, \; ds = ie^{i\theta} \, d\theta \qquad (9.69)$$

Using these, we can write the closed curve using definite limits

$$\oint_C \frac{ds}{s} = \int_0^{2\pi} e^{-i\theta}(ie^{i\theta} \, d\theta) = \int_0^{2\pi} i \, d\theta = 2\pi i$$

This is one of the fundamental results of contour integration and will find widespread applications; the point here being that the enclosure of a simple pole at the origin always yields $2\pi i$.

Suppose we perform the same test for a higher order pole, say $1/s^2$, again with $r = 1$

$$\oint_C \frac{ds}{s^2} = \int_0^{2\pi} e^{-2i\theta}(ie^{i\theta} \, d\theta) = \int_0^{2\pi} ie^{-i\theta} \, d\theta = i\left[\frac{e^{-i\theta}}{-i}\right]_0^{2\pi} = 0 \qquad (9.70)$$

In fact, if we perform the same test with the power relationship $f(s) = s^n$, where n is integer

$$\oint s^n \, ds = \int_0^{2\pi} e^{ni\theta}(ie^{i\theta} \, d\theta) = i\int_0^{2\pi} e^{(n+1)i\theta} \, d\theta = 0; \qquad \text{provided } n \neq -1 \qquad (9.71)$$

Now, if n is a positive integer, or zero, the above is obviously in accordance with Cauchy's First theorem, since then $f(s) = s^n$ is analytic for all finite values of s. However, if n becomes a negative integer, the s^n is clearly not analytic at the point $s = 0$. Nonetheless, the previous result indicates the closed integral vanishes even in this case, provided only that $n \neq -1$. Thus, only the simple pole at the origin produces a finite result, when the origin is enclosed by a closed contour.

The same result occurs for a point displaced from the origin

$$\oint_C (s - a)^n \, ds = 0 \qquad (n \neq -1) \qquad (9.72)$$

$$\oint_C \frac{1}{(s - a)} \, ds = 2\pi i \qquad (9.73)$$

provided the point at position a is enclosed by the closed curve C. We shall use this as a basis for constructing a method of inverting Laplace transforms when simple or multiple poles exist.

9.7 CAUCHY'S THEORY OF RESIDUES

The final elementary component of complex analysis necessary to effect closure of the Inversion theorem (Eq. 9.3) is Residue theory.

We have seen in Section 9.5 that simple poles or nth order poles at the origin are removable type singularities, so that if $f(s)$ contains a singularity at the origin, say a pole of order N, then it can be removed and the new function so generated will be analytic, even at the origin

$$g(s) = s^N f(s) \tag{9.74}$$

so that a closed contour C (of s values), encircling the origin, will cause $g(s)$ to be analytic inside and on the boundaries of C. If such is the case, the function $g(s)$ can be expanded in an ascending power series in s. The power series for $f(s)$ must include s^{-N}, so it could be expressed as

$$f(s) = \frac{B_N}{s^N} + \frac{B_{N-1}}{s^{N-1}} + \cdots + \frac{B_1}{s} + \sum_{n=0}^{\infty} C_n s^n \tag{9.75}$$

This series is often called the Laurent expansion. If this were multiplied by s^N, it is clear that $g(s)$ would in fact be an ascending series in s. The actual series for $f(s)$ need not be known, but the above clearly represents its behavior if it is known that an Nth order pole exists.

Now, from the previous lesson, the term-by-term integration of the series $f(s)$ for a contour C enclosing the origin will yield zero for each term, except the term B_1/s. Thus, we can write without further formalities:

$$\int_C f(s) \, ds = 2\pi i B_1 \tag{9.76}$$

Hence, the value of the contour integral of $f(s)$ is simply $2\pi i$ times the coefficient of s^{-1} in the Laurent expansion. The coefficient B_1 is called the "residue" of the function.

It doesn't matter if the singularity is not at the origin, since the same expansion is valid if it is taken around a singularity at the point $s = a$; in this case the series of powers are expressed as $(s - a)$, and one must find the coefficient of the term $(s - a)^{-1}$ in the Laurent expansion. On a practical level, it is easy to transfer the origin from zero to another point, as we illustrate in the next example.

EXAMPLE 9.3

Evaluate

$$\oint_C \frac{e^{st} \, ds}{(s - a)^2}$$

around a circle centered at the origin. It is clear if the radius of the circle is such that $r < a$, then the function is analytic within the center and we can write immediately

$$\oint \frac{e^{st}\, ds}{(s-a)^2} = 0$$

by Cauchy's theorem.

However, if we stipulate that the singularity is enclosed by the circle, so that $r > a$, then there is a pole of order 2 at $s = a$ within the contour. If we transfer the origin to $s = a$ by putting $p = s - a$, then

$$\oint_C \frac{e^{st}\, ds}{(s-a)^2} = \oint_C \frac{e^{(p+a)t}\, dp}{p^2} = e^{at}\oint_C \frac{e^{pt}\, dp}{p^2}$$

Next, we can expand $e^{pt} = 1 + pt + (pt)^2/2! + \cdots$

$$e^{at}\oint_C \frac{e^{pt}\, dp}{p^2} = e^{at}\oint_C \left[\frac{1}{p^2} + \frac{t}{p} + \frac{t^2}{2!} + \frac{t^3}{3!}p + \cdots\right] dp$$

All circular integrals are zero, according to the previous lesson, except the term containing t/p, which gives the residue te^{at}

$$\oint_C \frac{e^{(p+a)t}}{p^2}\, dp = \oint \frac{e^{st}\, ds}{(s-a)^2} = 2\pi i(te^{at})$$

9.7.1 Practical Evaluation of Residues

The Laurent expansion is often not obvious in many practical applications, so additional procedures are needed. Often, the complex function appears as a ratio of polynomials

$$f(s) = \frac{F(s)}{g(s)} \tag{9.77}$$

Now, if a simple pole exists at $s = a$, then obviously $(s - a)$ must be a factor in $g(s)$, so we could express the denominator as $g(s) = (s - a)G(s)$, provided $G(s)$ contains no other singularities at $s = a$. Clearly, the Laurent expansion must exist, even though it may not be immediately apparent, and so we can always write a hypothetical representation of the type given in Eq. 9.75:

$$f(s) = \frac{B_1}{s-a} + C_0 + C_1(s-a) + C_2(s-a)^2 + \cdots + C_n(s-a)^n + \cdots$$

since it is known that only a simple pole exists. Multiplying both sides by $(s - a)$, and then setting $s = a$ gives, when we replace $f(s)$ with

$$\frac{F(s)}{(s-a)G(s)}$$

that is

$$B_1 = (s - a)f(s)|_{s=a} = \frac{F(a)}{G(a)} \tag{9.78}$$

which is a handy way to find the residue for a simple pole.

Moreover, if $g(s)$ contained a number of distinct poles, such as

$$f(s) = \frac{F(s)}{(s - a)(s - b)(s - c)\ldots} \tag{9.79}$$

then $f(s)$ will have a residue at each pole, and these can be evaluated independently, one at a time, by the method illustrated. For a contour enclosing all such poles, then the contour integral around $f(s)$ will be simply the *sum* of *all residues*, multiplied by $2\pi i$ as before.

The procedure just outlined can be followed even if the pole cannot be factored out of $g(s)$; in such a case, Eq. 9.78 would become

$$B_1 = (s - a)f(s)|_{s=a} = \frac{(s - a)F(s)}{g(s)}\bigg|_{s=a} = \frac{0}{0} \tag{9.80}$$

which is indeterminate. However, if we apply L'Hopital's rule, then we see

$$B_1 = \lim_{s \to a} \frac{\dfrac{d}{ds}[(s - a)F(s)]}{\dfrac{dg(s)}{ds}} \tag{9.81}$$

which gives

$$B_1 = \lim_{s \to a} \left[\frac{F(s) + (s - a)F'(s)}{g'(s)} \right] = \frac{F(a)}{g'(a)} \tag{9.82}$$

EXAMPLE 9.4

Find the integral

$$\oint_C \frac{e^{st} \cosh(s)\, ds}{\sinh(s - a)}$$

around a circle with center at the origin, such that $r > |a|$.

The residue for the pole at $s = a$ is obtained from Eq. 9.82

$$B_1 = \frac{e^{at} \cosh(a)}{\cosh(0)} = e^{at} \cosh(a)$$

Hence the integral is

$$\oint_C \frac{e^{st} \cosh(s)\, ds}{\sinh(s-a)} = 2\pi i e^{at} \cosh(a)$$

9.7.2 Residues at Multiple Poles

If $f(s)$ contains a pole of order m at $s = a$ (and no other singularities), then it can be expressed as

$$f(s) = \frac{F(s)}{(s-a)^m} \tag{9.83}$$

where m is integer, and $F(s)$ is analytic at $s = a$. We can always expand an analytic function in a Taylor series, so that $F(s)$ can be expanded around the point $s = a$

$$F(s) = F(a) + (s-a)F'(a) + \frac{(s-a)^2}{2!}F''(a) + \cdots$$
$$+ \frac{(s-a)^{m-1}}{(m-1)!}F^{m-1}(a) + \frac{(s-a)^m}{m!}F^m(a) + \cdots \tag{9.84}$$

On division by $(s-a)^m$, the coefficient of $1/(s-a)$ becomes apparent and is the residue of $f(s)$

$$f(s) = \frac{F(a)}{(s-a)^m} + \frac{F'(a)}{(s-a)^{m-1}} + \cdots + \frac{F^{m-1}(a)}{(m-1)!(s-a)} + \cdots \tag{9.85}$$

hence,

$$B_1 = \frac{1}{(m-1)!}\frac{d^{m-1}}{ds^{m-1}}F(s)\Big|_{s=a} = \frac{1}{(m-1)!}\frac{d^{m-1}}{ds^{m-1}}(s-a)^m f(s)\Big|_{s=a} \tag{9.86}$$

EXAMPLE 9.5

Redo Example 9.3 using the results from Eq. 9.86.
  The required integral is

$$\oint_C \frac{e^{st}\, ds}{(s-a)^2}$$

which contains a second order pole at $s = a$. The residue is computed directly from Eq. 9.86.

$$B_1 = \frac{1}{1!}\frac{d}{ds}e^{st}\Big|_{s=a} = te^{at} \qquad \therefore \oint f(s)\, ds = 2\pi i t e^{at}$$

9.8   INVERSION OF LAPLACE TRANSFORMS BY CONTOUR INTEGRATION

At the beginning of this chapter, we quoted the Mellin–Fourier Inversion theorem for Laplace transforms, worth repeating here

$$F(s) = \int_0^\infty e^{-st} f(t) \, dt \tag{9.87}$$

$$f(t) = \frac{1}{2\pi i} \lim_{\omega \to \infty} \int_{\sigma_0 - i\omega}^{\sigma_0 + i\omega} e^{st} F(s) \, ds \tag{9.88}$$

It may now be clear why the factor $1/2\pi i$ appears in the denominator of the Inversion theorem. It should also be clear that

$$\int_{\sigma - i\omega}^{\sigma + i\omega} e^{st} F(s) \, ds$$

is a *line integral* of the complex function $e^{st} F(s)$, t being treated as an elementary parameter, and s denoting the complex variable, $s = \sigma + i\omega$. We will have closed the loop on this journey when we can show that the line integral is formally equivalent to the contour integral, enclosing some specialized region C; that is, we wish to show

$$\int_{\sigma_0 - i\infty}^{\sigma_0 + i\infty} e^{st} F(s) \, ds = \oint_C e^{st} F(s) \, ds \tag{9.89}$$

When this is done, the reader can see that Laplace inversion is formally equivalent to contour integration in the complex plane. We shall see that exceptional behavior arises occasionally (singularities owing to multivaluedness, for example) and these special cases will be treated in the sections to follow. Our primary efforts will be directed toward the usual case, that is, pole and multiple pole singularities occurring in the Laplace transform function $F(s)$.

We shall first consider the Inversion theorem for pole singularities only. The complex function of interest will be $f(s) = e^{st} F(s)$. The contour curve, denoting selected values of s, is called the First Bromwich path and is shown in Fig. 9.4.

The real constant σ_0 is selected (symbolically) to be greater than the real part of any pole existing in the denominator of $F(s)$. Thus, all poles of $F(s)$ are to the left of the line labelled AB (Fig. 9.4). It is clear that the semicircle $BCDEA$ can become arbitrarily large in the limit as $R \to \infty$, thereby enclosing all possible poles within the region to the left of line AB.

In order to prove the line integral (in the limit as $\omega \to \infty$, which corresponds to the limit $R \to \infty$) becomes identical to the contour integral, we shall need to break the contour into parts as we did earlier in Eq. 9.56

$$\int_{ABDA} e^{st} F(s) \, ds = \int_{\sigma_0 - i\omega}^{\sigma_0 + i\omega} e^{st} F(s) \, ds + \int_{BDA} e^{st} F(s) \, ds \tag{9.90}$$

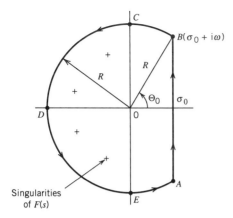

Figure 9.4 First Bromwich path for pole singularities.

It is clear by comparison with Eq. 9.89, where

$$\oint_C e^{st}F(s)\, ds = \int_{ABDA} e^{st}F(s)\, ds$$

that we must ensure that contour around the semicircle

$$\int_{BDA} e^{st}F(s)\, ds$$

becomes identically zero, as the region becomes arbitrarily large, that is as $R \to \infty$.

Along the arc $BCDEA$ (which we denote as curve C), we can set the magnitude $|s| = R$ so that $s = Re^{i\theta}$ and moreover, it will be assumed that $|F(s)| < \varepsilon R^{-k}$, where ε and k are positive constants. This is true, since all Laplace transforms of physically realizable processes are such that the denominator is of higher order in s than the numerator. With these stipulations in mind, we will now see the integral around the curve C tends to zero as $R \to \infty$.

To show this, first consider the semicircle CDE, to the left of the imaginary axis. On this part, the complex variable is $s = Re^{i\theta}$ and the angle varies between $\pi/2 < \theta < 3\pi/2$.

Hence, on this curve, we can write in polar form

$$\left| e^{st}F(s)\, ds \right| = \left| e^{Rt(\cos\theta + i\sin\theta)}F(Re^{i\theta})\, Re^{i\theta} i\, d\theta \right| \tag{9.91}$$

It was stipulated that the order of denominator is greater than numerator for $F(s)$, so $F(s)$ behaves as a negative power of R on the semicircle, and since $\cos\theta < 0$, then as $R \to \infty$ the magnitude in Eq. 9.91 tends to zero, since it includes a negative exponential in R multiplied by a decreasing power of R.

Next consider the arc BC in the first quadrant. The contribution to the total integral by this arc is bounded according to

$$\left| \int_{BC} e^{st}F(s)\, ds \right| < \int_{BC} |e^{st}F(s)\, ds| < \int_{\theta_0}^{\pi/2} e^{Rt\cos\theta} \frac{\varepsilon}{R^k} R\, d\theta \qquad (9.92)$$

where $\cos\theta_0 = \sigma_0/R$ as shown in Fig. 9.4. Now, since $\theta_0 \le \theta \le \pi/2$, it follows that $\cos\theta \le \cos\theta_0 = \sigma_0/R$, so we can write

$$\int_{\sigma_0}^{\pi/2} e^{Rt\cos\theta} \frac{\varepsilon}{R^k} R\, d\theta < \frac{\varepsilon}{R^{k-1}} \int_{\theta_0}^{\pi/2} e^{Rt\sigma_0/R}\, d\theta = \frac{\varepsilon}{R^{k-1}} e^{\sigma_0 t}\left(\frac{\pi}{2} - \theta_0\right) \qquad (9.93)$$

The quantity $\pi/2 - \theta_0$ equals $\pi/2 - \cos^{-1}(\sigma_0/R)$. For small arguments, the inverse cosine has expansion $\cos^{-1}(\sigma_0/R) \simeq \pi/2 - (\sigma_0/R)$ since for large R, then σ_0/R is small.

So finally, for large R, the upper bound behaves as

$$\frac{\varepsilon}{R^{k-1}} e^{\sigma_0 t}\left(\frac{\pi}{2} - \theta_0\right) \simeq \frac{\varepsilon e^{\sigma_0 t}}{R^{k-1}}\frac{\sigma_0}{R} = \frac{\varepsilon\sigma_0}{R^k} e^{\sigma_0 t} \qquad (9.94)$$

This final upper bound tends to zero as $R \to \infty$, since $k > 0$. Similarly, it is easy to show by the same arguments that the integral along the lower arc EA also tends to zero as $R \to \infty$.

Now we see the Cauchy Residue theorem gives the compact result

$$\oint_C e^{st}F(s)\, ds = \int_{BC} e^{st}F(s)\, ds + \int_{EA} e^{st}F(s)\, ds$$

$$+ \int_{CDE} e^{st}F(s)\, ds + \int_{AB} e^{st}F(s)\, ds \qquad (9.95)$$

$$= 2\pi i \sum \left[\text{Residues of } e^{st}F(s) \text{ inside } C\right]$$

Since the first three of these (BC, EA, and CDE) are proved nil, we can now formally write

$$f(t) = \frac{1}{2\pi i}\int_{\sigma_0 - i\infty}^{\sigma_0 + i\infty} e^{st}F(s)\, ds = \sum_{Br_1}\left[\text{Residues of } e^{st}F(s)\right] \qquad (9.96)$$

where of course the line integral

$$\int_{AB} e^{st}F(s)\, ds$$

is identical with

$$\int_{AB} e^{st}F(s)\, ds = \lim_{\omega \to \infty}\int_{\sigma_0 - i\omega}^{\sigma_0 + i\omega} e^{st}F(s)\, ds \qquad (9.97)$$

and the notation for curve C in the present case (for pole singularities) is called the first Bromwich path (Br_1). We shall need a slightly different integration path when multivalued functions arise (such as $\sqrt{s}$).

9.8.1 Summary of Inversion Theorem for Pole Singularities

The final simplified result for the Inversion theorem when pole singularities exist in $F(s)$ is recapitulated as

$$F(s) = \int_0^\infty f(t)e^{-st}\,dt \tag{9.98}$$

$$f(t) = \sum_{Br_1}(\text{residues of } F(s)e^{st}) \tag{9.99}$$

where Br_1 is a contour C, which embraces all possible pole (including higher order poles) singularities arising in the denominator of $F(s)$.

EXAMPLE 9.6

Use the Inversion theorem to find $f(t)$ corresponding to the Laplace transform

$$F(s) = \frac{1}{(s+1)(s+2)}$$

Two poles exist along the negative real axis: $s = -1$ and $s = -2$. We need to find the sum of the residues of $F(s)e^{st}$, arising from these two distinct poles. Since each pole can be easily factored out, computation of each residue can be accomplished using the procedure given in Eq. 9.78.

$$\text{res}(s = -1) = (s+1)F(s)e^{st}|_{s=-1} = e^{-t}$$

$$\text{res}(s = -2) = (s+2)F(s)e^{st}|_{s=-2} = -e^{-2t}$$

$$f(t) = \sum \text{residues of } F(s)e^{st} = e^{-t} - e^{-2t}$$

EXAMPLE 9.7

Find the corresponding $f(t)$ for the Laplace transform

$$F(s) = \frac{1}{s^2(s+1)^2}$$

This is a case where two second order poles exist: one at $s = 0$ and one at $s = -1$. The residues of $e^{st}F(s)$ can be found using the formula in Eq. 9.86, where $m = 2$ denotes the pole order

$$\text{res}(s = 0) = \frac{1}{1!} \frac{d}{ds} \left[\frac{e^{st}}{(s+1)^2} \right]_{s=0}$$

$$\text{res}(s = 0) = \frac{te^{st}}{(s+1)^2}\bigg|_{s=0} - \frac{2}{(s+1)^3} e^{st}\bigg|_{s=0} = t - 2$$

$$\text{res}(s = -1) = \frac{1}{1!} \frac{d}{ds} \left[\frac{e^{st}}{s^2} \right]_{s=-1} = \frac{te^{st}}{s^2}\bigg|_{s=-1} - 2\frac{e^{st}}{s^3}\bigg|_{s=-1}$$

$$\text{res}(s = -1) = te^{-t} + 2e^{-t} = (2 + t)e^{-t}$$

hence, $f(t)$ is the sum of the residues

$$f(t) = t - 2 + (2 + t)e^{-t}$$

N.B.: at $t = 0$, $f(0) = 0$.

9.9 LAPLACE TRANSFORMATIONS: BUILDING BLOCKS

The Laplace transform can be used to effect solutions for ordinary and partial differential equations. It is suited for initial-value problems, and is particularly useful for solving simultaneous equations. It has found great use in dealing with simulations using forcing functions, such as step and impulse type, which are cumbersome to handle using traditional methods. We set out in this section to lay a foundation composed of certain elementary building blocks, which have wide utility in solving practical problems. As we show, it is often possible to write the inverse of the Laplace transform by inspection, when certain elementary building blocks are recognized.

9.9.1 Taking the Transform

The first step in applying Laplace transform is to learn how to perform the elementary integration

$$\mathscr{L}f(t) = \int_0^\infty e^{-st} f(t) \, dt \tag{9.100}$$

Essentially, this integrates time out of the relationship and replaces it with a variable s (which we have already seen is in fact a complex variable). For ordinary differential equations, the operation will be seen to reduce the problem to algebraic manipulation.

The most elementary function is a simple constant K

$$\mathscr{L}[K] = \int_0^\infty Ke^{-st} \, dt = K\left(\frac{e^{-st}}{-s}\right)\bigg|_0^\infty = \frac{K}{s} \tag{9.101}$$

Thus, it is clear when a transform α/s appears, then the inversion, denoted as $\mathscr{L}^{-1}\alpha/s$, is simply α.

The next highest level is a linear function of time, say t

$$\mathscr{L}[t] = \int_0^\infty te^{-st}\,dt = \frac{e^{-st}}{s^2}(-st-1)\Big|_0^\infty = \frac{1}{s^2} \tag{9.102}$$

Again, it is clear when one sees α/s^2, then the inversion $\mathscr{L}^{-1}\alpha/s^2 = \alpha t$.

At a higher level, we can consider the power function t^n

$$\mathscr{L}[t^n] = \int_0^\infty t^n e^{-st}\,dt \tag{9.103}$$

To complete this integral, it is convenient to define a new variable

$$\beta = st; \qquad d\beta = s\,dt$$

hence,

$$\mathscr{L}[t^n] = s^{-1-n}\int_0^\infty \beta^n \exp(-\beta)\,d\beta$$

Comparing this with Eq. 4.14 in Chapter 4

$$\mathscr{L}[t^n] = \frac{\Gamma(n+1)}{s^{n+1}}$$

Moreover, if n is a positive integer, $\Gamma(n+1) = n!$, so

$$\mathscr{L}[t^n] = \frac{n!}{s^{n+1}}$$

The final building block is the frequently occurring exponential function

$$\mathscr{L}[e^{at}] = \int_0^\infty e^{at}e^{-st}\,dt = e^{-(s-a)t}\,dt = \frac{e^{-(s-a)t}}{-(s-a)}\Big|_0^\infty \tag{9.104}$$

If we insure that the real part of s is always greater than a (as we have seen in the previous section regarding the first Bromwich path), then the upper limit is zero and

$$\mathscr{L}[e^{at}] = \frac{1}{s-a}$$

Similarly,

$$\mathscr{L}[e^{-bt}] = \frac{1}{s+b}$$

Thus, it is clear when we see $1/(s + \alpha)$ the inversion is immediately

$$\mathscr{L}^{-1}\left(\frac{1}{s + \alpha}\right) = \exp(-\alpha t)$$

This is also verified by residue theory, which states for pole singularities that

$$f(t) = \mathscr{L}^{-1}F(s) = \Sigma \text{ residues of } e^{st}F(s) \qquad (9.105)$$

and the residue of $e^{st}/(s + \alpha)$ is simply $\exp(-\alpha t)$.

Many of the typical functions encountered have been integrated and their transforms are tabulated in Appendix D. For example,

$$\mathscr{L}[\sin \omega t] = \frac{\omega}{s^2 + \omega^2}$$

This transform can be accomplished by using the Euler formula

$$e^{i\omega t} = \cos \omega t + i \sin \omega t$$

to perform the complex integration of $e^{i\omega t}$.

$$\mathscr{L}[\sin \omega t] = \text{Im} \int_0^\infty e^{i\omega t} e^{-st} \, dt = \text{Im}\left[\frac{e^{-(s - i\omega)t}}{-(s - i\omega)}\right]_0^\infty$$

$$= \text{Im}\left[\frac{1}{s - i\omega}\right] = \text{Im}\left[\frac{s + i\omega}{s^2 + \omega^2}\right] = \frac{\omega}{s^2 + \omega^2}$$

Similarly, we could extract the real part and obtain $\mathscr{L}[\cos \omega t]$, which by inspection is

$$\mathscr{L}[\cos \omega t] = \frac{s}{s^2 + \omega^2}$$

The question of uniqueness in the inversion process has been answered (Hildebrand 1965) and is given expression in the understated Lerch's theorem: "if one function $f(t)$ corresponding to the known transform $F(s)$ can be found, it is the correct one." Not all functions of s are transforms, since continuity and other considerations must be taken into account. But, if $F(s) \to 0$ as $s \to \infty$ and $sF(s)$ is bounded as $s \to \infty$, then $F(s)$ is the transform of some function $f(t)$, which is at least piecewise continuous in some interval $0 \leq t \leq \tau$ and such function is of exponential order. When the initial value of $f(t)$ is desired and $F(s)$ is known, the following limit is useful

$$\lim_{s \to \infty} sF(s) = f(0)$$

provided that $f(t)$ and $f'(t)$ are at least piecewise continuous and of exponential order. By exponential order, we mean the product $\exp(-\sigma_0 t)|f(t)|$ is bounded for large values of t.

9.9.2   Transforms of Derivatives and Integrals

To apply Laplace transforms to practical problems of integro-differential equations, we must develop a formal procedure for operating on such functions. We consider the ordinary first derivative, leading to the general case of nth order derivatives. As we shall see, this procedure will require initial conditions on the function and its successive derivatives. This means of course that only initial value problems can be treated by Laplace transforms, a serious shortcoming of the method. It is also implicit that the derivative functions considered are valid only for continuous functions. If the function sustains a perturbation, such as a step change, within the range of independent variable considered, then the transform of the derivatives must be modified to account for this, as we will show later.

The Laplace transform of the ordinary first derivative is defined as

$$\mathscr{L}\left[\frac{df}{dt}\right] = \int_0^\infty \frac{df}{dt}e^{-st}\,dt = \int_0^\infty e^{-st}\,df \tag{9.106}$$

Using integration by parts, we see

$$\mathscr{L}\left[\frac{df}{dt}\right] = \left[f(t)e^{-st}\right]\Big|_0^\infty + s\int_0^\infty f(t)e^{-st}\,dt$$

But the new integral defines the Laplace transform of $f(t)$, which we denote as $F(s)$. Thus, since s can always be selected to insure that e^{-st} is damped faster than $f(t)$ can increase; hence,

$$\mathscr{L}\left[\frac{df}{dt}\right] = sF(s) - f(0)$$

where, as usual, $\mathscr{L}f(t) = F(s)$.

Similarly, we can use integration by parts again to find the Laplace transform of the second derivative

$$\mathscr{L}\left[\frac{d^2f}{dt^2}\right] = s^2F(s) - sf(0) - f'(0) \tag{9.107}$$

where we see that the initial condition $f(0)$ and the initial velocity $df(0)/dt$ must be known. This is what was meant when we stated earlier that the method was only suitable for initial value problems.

The procedure can be extended to nth order derivatives, requiring $(n-1)$ initial conditions.

$$\mathscr{L}\left[\frac{d^nf}{dt^n}\right] = s^nF(s) - \left[s^{n-1}f(0) + s^{n-2}f'(0) + \cdots + sf^{n-2}(0) + f^{n-1}(0)\right] \tag{9.108}$$

where initial derivatives up to $(n-1)$ are necessary.

There is some similarity to the Heaviside operator, in the sense that first, second, and so forth derivatives yield s, s^2 operators in a manner similar to D, D^2 in Heaviside operators; however, Laplace transforms also require additional knowledge of initial conditions. The transform of integrals is also similar in form to the Heaviside operator, as we see next.

Denoting as before that $\mathscr{L}f(t) = F(s)$, the transform of a continuous integral function is defined:

$$\mathscr{L}\left[\int_0^t f(t)\, dt\right] = \int_0^\infty \left[\int_0^t f(\tau)\, d\tau\right] e^{-st}\, dt \tag{9.109}$$

where we have used dummy variables for the interior integral. This can be integrated by parts if we let

$$u = \int_0^t f(\tau)\, d\tau$$

$$\frac{du}{dt}\, dt = f(t)\, dt$$

$$dv = e^{-st}\, dt \qquad \therefore v = -\frac{e^{-st}}{s}$$

Substituting these yields

$$\mathscr{L}\left[\int_0^t f(\tau)\, d\tau\right] = \left[-\frac{e^{-st}}{s} \cdot \int_0^t f(\tau)\, d\tau\right]_0^\infty + \frac{1}{s}\int_0^\infty f(t)e^{-st}\, dt \tag{9.110}$$

However, the last term defines $F(s)$, hence, if we ensure that e^{-st} damps faster than $\int_0^t f(\tau)\, d\tau$

$$\mathscr{L}\left[\int_0^t f(\tau)\, d\tau\right] = \frac{1}{s}F(s) \tag{9.111}$$

We see that s appears in the denominator, just as in the Heaviside operation.

Occasionally, it is useful to differentiate Laplace transforms with respect to the continuous variable s; this procedure becomes useful in the method of moments, as a parameter estimation tool. Thus, if we define the Laplace transform in the usual way

$$F(s) = \int_0^\infty f(t)e^{-st}\, dt \tag{9.112}$$

Then, if s is continuous, we can differentiate this

$$\frac{dF(s)}{ds} = \int_0^\infty -tf(t)e^{-st}\, dt \tag{9.113}$$

Thus, the derivative of $F(s)$ is in fact the (negative) integral of the moment of $f(t)$; that is, the product of $t \cdot f(t)$. Thus, for the usual (unweighted moment)

$$-\frac{dF(s)}{ds}\bigg|_{s=0} = \int_0^\infty tf(t)\, dt \tag{9.114}$$

Thus, if $F(s)$ is the transform of a model of a process (containing unknown parameters), and $f(t)$ is known experimentally, then a numerical integration of the RHS yields one equation to find one unknown. Suppose the experimental integral yields a number, μ_1, and the process has a transform

$$F(s) = \frac{1}{\tau s + 1} \tag{9.115}$$

hence,

$$-\frac{dF}{ds}\bigg|_{s=0} = \mu_1 = -\left(\frac{-\tau}{(s\tau + 1)^2}\right)_{s=0} = \tau$$

Thus, the first moment, μ_1, is exactly equal to the parameter τ.

Second and higher order moments can be defined by repeated differentiation.

$$\frac{d^2F(s)}{ds^2}\bigg|_{s=0} = \mu_2 = \int_0^\infty t^2 f(t)\, dt$$

$$(-1)^n \frac{d^n F(s)}{ds^n}\bigg|_{s=0} = \mu_n = \int_0^\infty t^n f(t)\, dt$$

The process of differentiating with respect to s suggests a technique for treating certain nonconstant coefficient ODE. Thus, we have previously seen that

$$\mathcal{L}[t^n f(t)] = \int_0^\infty t^n f(t) e^{-st}\, dt = (-1)^n \frac{d^n}{ds^n} F(s) \tag{9.116}$$

Thus, we can write

$$\mathcal{L}\left[t\frac{df}{dt}\right] = \int_0^\infty t\frac{df}{dt} e^{-st}\, dt = -\frac{d}{ds}\left\{\mathcal{L}\left[\frac{df}{dt}\right]\right\}$$

$$= -\frac{d}{ds}[sF(s) - f(0)] = -s\frac{dF(s)}{ds} - F(s) \tag{9.117}$$

This process can be carried forward for

$$t^2 \frac{df}{dt}, \quad t\frac{d^2f}{dt^2}, \quad t^2 \frac{d^2f}{dt^2}, \quad \text{etc.}$$

and the final forms for these nonconstant coefficients operating on differentials are tabulated in Table 9.1. This will allow certain classes of nonconstant coefficient ODE to be treated by Laplace transforms.

Table 9.1 Transforms of Differentials and Products

$f(t)$	$\mathscr{L}f(t) = F(s)$
$\dfrac{df}{dt}$	$sF(s) - f(0)$
$\dfrac{d^2 f}{dt^2}$	$s^2 F(s) - sf(0) - f'(0)$
$t\dfrac{df}{dt}$	$-s\dfrac{dF(s)}{ds} - F(s)$
$t^2\dfrac{df}{dt}$	$s\dfrac{d^2 F(s)}{ds^2} + 2\dfrac{dF(s)}{ds}$
$t\dfrac{d^2 f}{dt^2}$	$-s^2\dfrac{dF(s)}{ds} - 2sF(s) + f(0)$
$t^2\dfrac{d^2 f}{dt^2}$	$s^2\dfrac{d^2 F(s)}{ds^2} + 4s\dfrac{dF(s)}{ds} + 2F(s)$

9.9.3 The Shifting Theorem

Often, one of the elementary building blocks is recognized, except for an additive constant. Thus, suppose that s always appears added to a constant factor a, that is $F(s + a)$. Then it is easy to show that the original time function was multiplied by the exponential $\exp(-at)$ and

$$\mathscr{L}\left[e^{-at}f(t)\right] = F(s + a)$$

To prove this, write the transform integral explicitly

$$\mathscr{L}\left[e^{-at}f(t)\right] = \int_0^\infty e^{-at}e^{-st}f(t)\, dt \tag{9.118}$$

If we replace $p = s + a$, the integral is

$$\mathscr{L}\left[e^{-at}f(t)\right] = F(p) = F(s + a) \tag{9.119}$$

Similarly, for the product $e^{bt}f(t)$, we obtain

$$\mathscr{L}\left[e^{bt}f(t)\right] = F(s - b) \tag{9.120}$$

Thus, suppose we wish to invert the transform

$$G(s) = \frac{\omega}{(s - b)^2 + \omega^2} \tag{9.121}$$

and we recognize that $\mathscr{L}[\sin \omega t] = \omega/(s^2 + \omega^2)$, so it is clear that

$$\mathscr{L}^{-1}G(s) = e^{bt} \sin \omega t \tag{9.122}$$

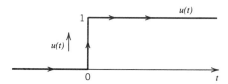

Figure 9.5 Unit step function at $t = 0$.

9.9.4   Transform of Distribution Functions

Certain operations in chemical engineering are modelled by elementary distribution functions. Such operations as the instantaneous closing of a valve can be modelled by the so-called step function. The quick injection of solute into a flowing stream, such as that done in gas chromatography, can be modelled by the so-called impulse function. Moreover, these distribution-type functions can be shifted in time.

Consider first the unit step function, illustrated schematically in Fig. 9.5. It is, of course, impossible to cause real physical systems to follow exactly this square-wave behavior, but it is a useful simulation of reality when the process is much slower than the action of closing a valve, for instance. The Laplace transform of $u(t)$ is identical to the transform of a constant

$$\mathscr{L}[u(t)] = \int_0^\infty 1e^{-st}\,dt = \frac{1}{s} \tag{9.123}$$

since $u(t)$ takes a value of unity when time is slightly greater than zero. If the step function is delayed as shown in Fig. 9.6, then the integral can be performed in two parts

$$\mathscr{L}[u(t - \tau)] = \int_0^\tau 0e^{-st}\,dt + \int_\tau^\infty 1e^{-st}\,dt = \frac{e^{-\tau s}}{s} \tag{9.124}$$

In fact, it can be proved that all delayed functions are multiplied by $\exp(-\tau s)$, if τ represents the delay time. For instance, suppose a ramp function, $r(t) = t$

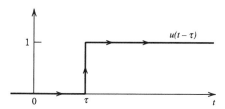

Figure 9.6 Delayed unit step function.

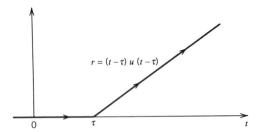

Figure 9.7 Delayed ramp function.

is delayed, as shown in Fig. 9.7. As before, we split the integral into two parts

$$\mathscr{L}[r(t - \tau)] = \int_0^\tau 0e^{-st}\, dt + \int_\tau^\infty (t - \tau)e^{-st}\, dt \qquad (9.125)$$

Recalling Eq. 9.102, which depicts $\mathscr{L}[t]$, we see

$$\mathscr{L}[r(t - \tau)] = \frac{e^{-\tau s}}{s^2} \qquad (9.126)$$

We note that as the time delay is reduced, so $\tau \to 0$, then the original step and ramp functions at time zero are recovered.

The unit impulse function, $\delta(t)$, is often called the Dirac delta function. It behaves in the manner shown in Fig. 9.8. The area under the curve is always unity, and as θ becomes small, the height of the pulse tends to infinity. We can define this distribution in terms of unit step functions

$$\delta(t) = \lim_{\theta \to 0} \left[\frac{u(t) - u(t - \theta)}{\theta} \right] \qquad (9.127)$$

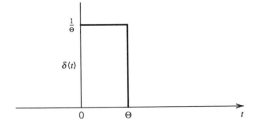

Figure 9.8 Simulation of $\delta(t)$.

Taking Laplace transforms of both sides gives

$$\mathcal{L}[\delta(t)] = \lim_{\theta \to 0} \left[\frac{1 - \exp(-\theta s)}{\theta s} \right] = \frac{0}{0} \tag{9.128}$$

To resolve the indeterminacy, we note for small arguments, the expansion for

$$e^{-\theta s} \simeq 1 - \theta s + \cdots$$

so in the limit

$$\mathcal{L}[\delta(t)] = \lim_{\theta \to 0} \left[\frac{1 - 1 + \theta s}{\theta s} \right] = 1 \tag{9.129}$$

Alternately, we could have used L'Hopital's rule to resolve the indeterminacy. The delayed impulse function can be easily shown to be

$$\mathcal{L}[\delta(t - \tau)] = 1 \cdot e^{-\tau s} \tag{9.130}$$

It is useful to note that the impulse function arises from the rate of change of a step function since

$$\mathcal{L}\left[\frac{du(t)}{dt} \right] = \int_0^\infty e^{-st} \frac{du}{dt} \, dt = u(t)e^{-st}\Big|_0^\infty + s \int_0^\infty u(t)e^{-st} \, dt \tag{9.131}$$

$$= s\mathcal{L}u(t) = s \cdot \frac{1}{s} = 1$$

where it is clear we have defined $u(0) = 0$. This implies at times infinitesimally larger than zero that u becomes unity; this is often denoted as $u(0^+) = 1$.

Thus, the derivative of the step function also defines the impulse function. This is seen to be also true for any response function; the time derivative of the step response produces the impulse response.

9.10   PRACTICAL INVERSION METHODS

We have derived the general Inversion theorem for pole singularities using Cauchy's Residue theory. This provides the fundamental basis (with a few exceptions, such as $\sqrt{s}$) for inverting Laplace transforms. However, the useful building blocks, along with a few practical observations, allow many functions to be inverted without undertaking the formality of the Residue theory. We shall discuss these practical, intuitive methods in the sections to follow. Two widely used practical approaches are: (1) partial fractions, and (2) convolution.

9.10.1   Partial Fractions

Often, the Laplace transform to be inverted appears in factored form. In such cases, the factors are of the form $1/(s - a)$, $1/(s - a)^2$, or generally $1/(s - a)^n$.

For such factors, the building blocks already enunciated would yield

$$\mathscr{L}^{-1} \frac{1}{s - a} = e^{at} \tag{9.132a}$$

$$\mathscr{L}^{-1} \frac{1}{(s - a)^2} = te^{at} \tag{9.132b}$$

$$\mathscr{L}^{-1} \frac{1}{(s - a)^n} = \frac{1}{(n - 1)!} t^{n-1} e^{at} \tag{9.133}$$

where the last two are obtainable directly from the multiple-pole Residue theory discussed in Section 9.7.2. The partial fraction expansion is best illustrated by a series of examples.

EXAMPLE 9.8

Find the corresponding $f(t)$ for the transform

$$F(s) = \frac{1}{(s + a)(s + b)}$$

This could be easily done using the Residue theory, but we could also expand the function into partial fractions as

$$F(s) = \frac{A}{s + a} + \frac{B}{s + b} = \frac{1}{(s + a)(s + b)}$$

Multiplying both sides by $(s + a)$, then setting $s = -a$ allows A to be found

$$A = \frac{1}{b - a}$$

Similarly, multiplying by $(s + b)$, then setting $s = -b$ gives the value for B

$$B = \frac{1}{a - b}$$

Now we can invert term by term

$$\mathscr{L}^{-1}F(s) = \mathscr{L}^{-1} \frac{1}{(b - a)} \cdot \frac{1}{s + a} + \mathscr{L}^{-1} \frac{1}{(a - b)} \cdot \frac{1}{s + b}$$

so finally,

$$f(t) = \frac{1}{(b - a)} e^{-at} + \frac{1}{(a - b)} e^{-bt}$$

EXAMPLE 9.9

Find the inversion for

$$F(s) = \frac{1}{s^2(s + 1)^2}$$

using partial fraction expansion.

This has already been accomplished using Residue theory in Example 9.7. We write the partial fraction expansion as

$$\frac{1}{s^2(s + 1)^2} = \frac{A}{s} + \frac{B}{s^2} + \frac{C}{(s + 1)} + \frac{D}{(s + 1)^2}$$

It is clear by multiplying by $s^2(s + 1)^2$ that

$$s(s + 1)^2 A + B(s + 1)^2 + Cs^2(s + 1) + Ds^2 = 1$$

Evaluate D by multiplying by $(s + 1)^2$ then set $s = -1$

$$D = 1$$

Evaluate B by multiplying by s^2 then set $s = 0$

$$B = 1$$

Next, equating coefficients of s^3 in the polynomial expansion requires

$$A = -C$$

and matching coefficients of s^2 requires that

$$C = 2$$

hence,

$$A = -2$$

So, inverting term by term yields

$$\mathscr{L}^{-1} \frac{1}{s^2(s + 1)^2} = \mathscr{L}^{-1}\left(-\frac{2}{s}\right) + \mathscr{L}^{-1}\left(\frac{1}{s^2}\right) + \mathscr{L}^{-1}\left(\frac{2}{s + 1}\right) + \mathscr{L}^{-1}\left(\frac{1}{(s + 1)^2}\right)$$

$$f(t) = -2 + t + 2e^{-t} + te^{-t}$$

As before, we see $f(0) = 0$.

EXAMPLE 9.10

Find the inverse of the function

$$F(s) = \frac{2s^2 + 3s - 4}{(s - 2)(s^2 + 2s + 2)}$$

using partial fractions.

First, write the partial expansion

$$F(s) = \frac{A}{s - 2} + \frac{Bs + C}{(s + 1)^2 + 1}$$

Find A by multiplying by $(s - 2)$, then set $s = 2$

$$A = 1$$

Next, require B and C to be such that the original numerator is recovered

$$s^2 + 2s + 2 + (Bs + C)(s - 2) = 2s^2 + 3s - 4$$

This shows that $B = 1$ and $C = 3$; hence,

$$F(s) = \frac{1}{s - 2} + \frac{s + 3}{(s + 1)^2 + 1} = \frac{1}{s - 2} + \frac{(s + 1)}{(s + 1)^2 + 1} + \frac{2}{(s + 1)^2 + 1}$$

Since we recognize that $\mathcal{L} \cos t = s/(s^2 + 1)$ and $\mathcal{L} \sin t = 1/(s^2 + 1)$, we can then invoke the shifting theorem and invert term by term to get

$$f(t) = e^{2t} + e^{-t} \cos t + 2e^{-t} \sin t$$

9.10.2 Convolution Theorem

Occasionally, products of factors occur wherein each factor has a known inversion, and it is desired to find the product of the two. The mechanics for doing this is accomplished by the method of "convolution," the derivation of which is given in standard operational mathematics textbooks (Churchill 1958). Thus, if the product occurs

$$F(s) = G(s)H(s) \tag{9.134}$$

and the inverse of each is known, that is $g(t)$ and $h(t)$ are recognizable, then the inversion of the product is given by the convolution integral

$$f(t) = \mathscr{L}^{-1}[G(s)H(s)] = \int_0^t g(\tau)h(t-\tau)\,d\tau = \int_0^t g(t-\tau)h(\tau)\,d\tau$$

$$\textbf{(9.135)}$$

The two forms shown suggest that the convolution integral is symmetrical. We illustrate the application of convolution in the next series of examples.

EXAMPLE 9.11

Find the inversion of the product

$$\beta(s) = \frac{1}{s}\frac{1}{s+a}$$

This could be performed directly, since it is recognized that $1/s$ implies integration. Thus, recalling Eq. 9.111, which stated

$$\mathscr{L}\left[\int_0^t f(\tau)\,d\tau\right] = \frac{1}{s}F(s)$$

we can see, since $F(s) = 1/(s+a)$ and $f(t) = e^{-at}$ in the present case, that

$$\beta(t) = \int_0^t e^{-a\tau}\,d\tau = \frac{e^{-aT}}{-a}\bigg|_0^t = \frac{1}{a}(1 - e^{-at})$$

Of course, we could also apply the convolution integral, taking $G(s) = 1/s$ and $H(s) = 1/(s+a)$, where $g(t) = u(t)$ and $h(t) = e^{-at}$, hence,

$$\beta(t) = \int_0^t u(\tau)e^{-a(t-\tau)}\,d\tau = \frac{e^{-at}e^{a\tau}}{a}\bigg|_0^t$$

$$\beta(t) = \frac{1}{a}(1 - e^{-at})$$

EXAMPLE 9.12

Use the convolution integral to find the inversion of

$$F(s) = \frac{1}{s^2}\frac{1}{(s+1)^2}$$

This has been already worked out in Example 9.7 by the Residue theory, and in Example 9.9 by partial fractions. We shall take $G(s) = 1/(s^2)$, $H(s) = 1/(s+1)^2$ so that $g(t) = t$ and according to Eq. 9.132, $h(t) =$

te^{-t}, hence, the convolution integral should be

$$f(t) = \int_0^t \tau(t - \tau)\exp(-(t - \tau))\, d\tau$$

which should yield the same result as the second form of convolution:

$$f(t) = \int_0^t (t - \tau)\tau \exp(-\tau)\, d\tau$$

Taking the first form of convolution

$$f(t) = t \exp(-t)\int_0^t \tau \exp(\tau)\, d\tau - \exp(-t)\int_0^t \tau^2 \exp(\tau)\, d\tau$$

$$f(t) = t \exp(-t)\left[\exp(\tau)(\tau - 1)\right]_0^t$$

$$- \exp(-t)\left[\tau^2 \exp(\tau)\big|_0^t - 2\exp(\tau)(\tau - 1)\big|_0^t\right]$$

Cancelling terms, yields finally,

$$f(t) = (t - 2) + (t + 2)\exp(-t)$$

as before. The reader should show that the second form of the convolution integral gives an identical result.

9.11   APPLICATIONS OF LAPLACE TRANSFORMS FOR SOLUTIONS OF ODE

The nature of the Laplace transform has now been sufficiently studied so that direct applications to solution to physicochemical problems are possible. Because the Laplace transform is a linear operator, it is not suitable for nonlinear problems. Moreover, it is a suitable technique only for initial-value problems. We have seen (Table 9.1) that certain classes of variable coefficient ODE can also be treated by Laplace transforms, so we are not constrained by the constant coefficient restriction required using Heaviside operators.

EXAMPLE 9.13

It is desired to find the transient response for an operating CSTR undergoing forcing by time variation in the inlet composition. Assume isothermal behavior and linear rate of consumption of species A according to

$$A \xrightarrow{k} \text{products}$$

Constant volume and flowrate can be assumed. Find the response when the inlet takes (1) a step change and (2) impulse disturbance.

The dynamic material balance for constant volumetric flow is

$$V\frac{dc_A(t)}{dt} = qc_{A_0}(t) - qc_A(t) - kVc_A(t) \tag{9.136}$$

Before the disturbance enters the system, the reactor is operating at steady state, so initially

$$0 = q\bar{c}_{A_0} - q\bar{c}_A - kV\bar{c}_A \tag{9.137}$$

where the overbar denotes the initially steady-state condition; this of course implies

$$c_A(0) = \bar{c}_A$$

$$c_{A_0}(0) = \bar{c}_{A_0}$$

It is convenient to rearrange the equations to a standard form, which makes the system time constant explicit. Thus, we wish to rearrange so that the appearance is like

$$\tau\frac{dy}{dt} + y = f(t)$$

This can be done by rewriting the equations

$$V\frac{dc_A}{dt} + (kV + q)c_A = qc_{A_0}$$

Next, divide by $kV + q$ and define the system time constant as

$$\tau = \frac{V}{kV + q}$$

hence, we find

$$\tau\frac{dc_A}{dt} + c_A = \varepsilon c_{A_0} \tag{9.138}$$

where the dimensionless fraction $\varepsilon = q/(q + kV)$ appears.

If we apply the Laplace transform at this point, we shall need to include the initial condition, owing to the appearance of the derivative dc_A/dt; thus,

$$\mathscr{L}\left[\frac{dc_A}{dt}\right] = sC_A(s) - c_A(0)$$

It is possible to eliminate such excess baggage by writing the stimulus (c_{A_0}) and response (c_A) variables in terms of deviation from initial steady state, so write

$$c_A(t) = \hat{c}_A(t) + \bar{c}_A$$

$$c_{A_0}(t) = \hat{c}_{A_0}(t) + \bar{c}_{A_0} \tag{9.139}$$

Thus, the hat variable represents excursions away from the initially steady condition. It is also clear that

$$c_A(0) = \bar{c}_A \qquad \therefore \hat{c}_A(0) = 0$$

So in terms of $\hat{c}_A(t)$, the dynamic initial condition is zero. Inserting Eqs. 9.139 into Eq. 9.138, and recalling the steady state, Eq. 9.137, we obtain

$$\tau\frac{d\hat{c}_A}{dt} + \hat{c}_A = \varepsilon\hat{c}_{A_0} \tag{9.140}$$

since the steady state requires $\bar{c}_A = \varepsilon\bar{c}_{A_0}$, hence these terms cancel.

Now, we must stipulate something about the magnitude of the disturbance entering the system. Suppose the inlet composition is suddenly doubled in size and operates like a step function. This means the inlet deviation $\hat{c}_{A0}$ behaves as

$$\hat{c}_{A_0}(t) = u(t)\bar{c}_{A_0} \tag{9.141}$$

as shown in Fig. 9.9.

In general, any positive or negative fraction of $\bar{c}_{A_0}$ could be used for the magnitude of the inlet step function. It is clear that the complete inlet composition is

$$c_{A_0}(t) = u(t)\bar{c}_{A_0} + \bar{c}_{A_0}; \qquad \hat{c}_{A_0} = u(t)\bar{c}_{A_0}$$

Taking the Laplace transform of Eq. 9.140, and noting $\hat{c}_A(0) = 0$, we obtain since $\mathscr{L}u(t) = 1/s$

$$\tau s\hat{C}_A(s) + \hat{C}_A(s) = \varepsilon\frac{\bar{c}_{A_0}}{s} \tag{9.142}$$

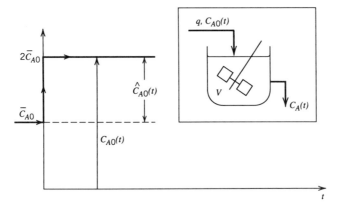

Figure 9.9 Inlet composition disturbance for Example 9.13.

hence,

$$\hat{C}_A(s) = \frac{\varepsilon \bar{c}_{A_0}}{s(\tau s + 1)} \tag{9.143}$$

It is convenient, with reference to our previous work, to write this as

$$\hat{C}_A(s) = \left(\frac{\varepsilon \bar{c}_{A_0}}{\tau} \right) \left[\frac{1}{s(s + 1/\tau)} \right] \tag{9.144}$$

Except for the multiplicative constant, this is identical to Example 9.11, so the inverse, noting the equivalence $a = 1/\tau$, is simply,

$$\hat{c}_A(t) = \varepsilon \bar{c}_{A_0}[1 - \exp(-t/\tau)] \tag{9.145}$$

The absolute response is obtained by adding this to $\bar{c}_A$ to get

$$c_A(t) = \bar{c}_A + \varepsilon \bar{c}_{A_0}[1 - \exp(-t/\tau)] \tag{9.146}$$

but since the steady state required

$$\bar{c}_A = \varepsilon \bar{c}_{A_0}$$

we can then write the results in dimensionless terms

$$\frac{\hat{c}_A(t)}{\bar{c}_A} = 1 - \exp(-t/\tau) \tag{9.147}$$

This shows the final value of $\hat{c}_A(\infty) = \bar{c}_A$ and of course $c_A(\infty) = 2\bar{c}_A$.

Another useful observation arises by inspecting the response as it passes through the point $t = \tau$; under these conditions we have

$$\frac{\hat{c}_A(\tau)}{\bar{c}_A} = 1 - e^{-1} = \frac{e - 1}{e} \simeq 0.632 \tag{9.148}$$

This means that the time constant τ can be deduced when the response for $\hat{c}_A(t)$ is around 63% of the distance to the final steady state. Figure 9.10 illustrates the step response.

The impulse response can be found in similar manner, by taking $\hat{c}_{A_0}(t)$ to be

$$\hat{c}_{A_0}(t) = c_0 \delta(t) \tag{9.149}$$

The weighting factor c_0 accounts for how much solute A was injected, relative to the molar inflow of $q\bar{c}_{A_0}$.

Thus, suppose the CSTR flow rate is, $q = 0.1$ liter/sec and $\bar{c}_{A_0} = 1$ mole/liter. Now suppose we inject 100 cc of solute A of composition 1 mole/liter $(= \bar{c}_{A_0})$ into the flowing inlet stream. We shall model this injection as an "impulse," since it was done quickly, relative to the system time constant τ. The net molar

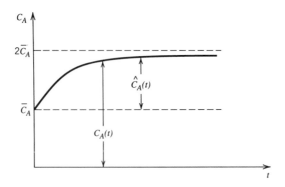

Figure 9.10 Response of $c_A(t)$ to a step change in $c_{A_0}(t)$.

inflow is then

$$qc_{A_0}(t) = q\bar{c}_{A_0} + 0.1 \text{ liter} \times \bar{c}_{A_0} \times \delta(t)$$

where it is clear that $\delta(t)$ has units of reciprocal time, $\sec^{-1}$. We now see

$$c_{A_0}(t) = \bar{c}_{A_0} + \frac{0.1\delta(t)}{q}\bar{c}_{A_0}$$

hence,

$$\hat{c}_{A_0} = \left(\frac{0.1}{q}\right)\delta(t)\bar{c}_{A_0}$$

by comparing this with Eq. 9.149 we see that the weighting factor

$$c_0 = 0.1\frac{\bar{c}_{A_0}}{q}$$

with units moles $A/(\text{lit}/\text{sec})$. Referring to Eq. 9.140, we have for an impulse stimulus

$$\tau\frac{d\hat{c}_A}{dt} + \hat{c}_A = \varepsilon c_0\delta(t) \qquad (9.150)$$

Taking Laplace transforms, noting $\hat{c}_A(0) = 0$ as before, we find

$$\hat{C}_A(s) = \frac{\varepsilon c_0}{\tau s + 1} = \frac{\varepsilon c_0}{\tau}\frac{1}{s + 1/\tau} \qquad (9.151)$$

This is the simple exponential building block; hence,

$$\hat{c}_A(t) = \frac{\varepsilon c_0}{\tau} \cdot \exp(-t/\tau) \qquad (9.152)$$

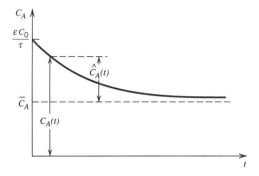

Figure 9.11 Impulse response of CSTR.

Except for the weighting factor c_0, this could have been obtained directly by differentiating the step response. The response (not to scale) is illustrated in Fig. 9.11.

It is seen that $\hat{c}_A (\infty) \to 0$, and the system returns to the original steady state. This is a clear advantage for "impulse" experiments, since the reactor is disturbed for only brief periods of time. Note, since c_0 and ε are known, the intercept at $t = 0$ allows the estimate of system time constant τ. Again, it is also clear that the impulse response is simply the time derivative of the step response, since

$$s\left[\frac{1}{s}\,\frac{1}{(\tau s + 1)}\right] = \frac{1}{\tau s + 1}$$

Considerable simplifications arise using deviation variables (sometimes referred to as perturbation variables, implying small magnitudes), not only because of the elimination of finite initial conditions. In fact, the structure evolving from such variables substitution allows nonlinear systems to be easily "linearized," provided deviations are not too large. We consider such circumstances in the next example, wherein the "stimulus" arises from a disturbance in the volumetric flow rate, so that q is a function of time, $q(t)$.

EXAMPLE 9.14

The CSTR in the previous example is subjected to a step change in the system flowrate, while maintaining a constant volume for the reactor. Find the exit composition response, $c_A(t)$, to such disturbances.

The deviation variables in the present case will be

$$q(t) = \bar{q} + \hat{q}(t) \qquad (9.153a)$$

$$c_A(t) = \bar{c}_A + \hat{c}_A(t) \qquad (9.153b)$$

and $c_{A_0}(t) = \bar{c}_{A_0}$ is maintained at a constant value. Inserting these into the material balance for component A yields

$$V\frac{d\hat{c}_A}{dt} = \left[\bar{q}\bar{c}_{A_0} - \bar{q}\bar{c}_A - kV\bar{c}_A\right] + \left[\hat{q}(t)\hat{c}_A(t)\right]$$
$$+ \hat{q}(t)\bar{c}_{A_0} - \hat{q}(t)\bar{c}_A - \bar{q}\hat{c}_A(t) - kV\hat{c}_A(t)$$

$$(9.154)$$

The first bracketed group of terms are identically zero, since this group defines the original steady-state condition, which in simplified form is the same as the last example,

$$\bar{c}_A = \varepsilon\bar{c}_{A_0}$$

The second bracketed term represents the nonlinearity, which is quite small if the deviations are small, hence take

$$\left[\hat{q}(t) \cdot \hat{c}_A(t)\right] \cong 0$$

The remaining items can be arranged into the usual format

$$V\frac{d\hat{c}_A}{dt} + [\bar{q} + kV]\hat{c}_A = \hat{q}(t)\bar{c}_{A_0} - \hat{q}(t)\bar{c}_A \qquad (9.155)$$

Now, since the steady state requires $\bar{c}_A = \varepsilon\bar{c}_{A_0}$, we see the RHS becomes simply $\hat{q}(t)(1 - \varepsilon)\bar{c}_{A_0}$. If we define the time constant as before: $\tau = V/(\bar{q} + kV)$, we finally get

$$\tau\frac{d\hat{c}_A}{dt} + \hat{c}_A = \frac{(1 - \varepsilon)\bar{c}_{A_0}}{\bar{q} + kV} \cdot \hat{q}(t) \qquad (9.156)$$

Suppose the flow rate takes a step change with magnitude equal to 50% of the original flow; this would be expressed as

$$\hat{q}(t) = 0.5\bar{q} \cdot u(t)$$

Any fraction could be used, but we must be alert to the constraint that $\hat{q}(t)\hat{c}_A(t) \approx 0$, which depends strongly on the reaction rate constant k. Taking Laplace transforms as before, noting

$$\hat{c}_A(0) = 0$$

yields

$$\hat{C}_A(s) = \left[\frac{(1 - \varepsilon)\bar{c}_{A_0}}{\bar{q} + kV}\right](0.5\bar{q})\frac{1}{s(s\tau + 1)} \qquad (9.157)$$

We recall that $\varepsilon = \bar{q}/(\bar{q} + kV)$ for the present case, so that the simplified transform is now

$$\hat{C}_A(s) = \varepsilon(1 - \varepsilon)0.5\bar{c}_{A_0}\frac{1}{s(s\tau + 1)} \qquad (9.158)$$

The transform is the same as the last example, except for multiplicative constants; hence, replacing $\bar{c}_A = \varepsilon\bar{c}_{A_0}$

$$\frac{\hat{c}_A(t)}{\bar{c}_A} = (1 - \varepsilon)0.5[1 - \exp(-t/\tau)]$$

The fraction of steady flow, which is the magnitude of the step change in $\hat{q}(t)$, appears as a direct multiplier of the final result. The largest value of $\hat{c}_A(t)$ occurs at $t = \infty$, so we can check to see if the nonlinear term, $\hat{c}_A\hat{q}$, was indeed small relative to the steady case

$$\frac{\hat{c}_A(\infty)\hat{q}}{\bar{c}_A\bar{q}} = \frac{(1 - \varepsilon)0.5\bar{c}_A \cdot 0.5\bar{q}}{\bar{c}_A\bar{q}} = \frac{1}{4}(1 - \varepsilon)$$

Now, since $\varepsilon < 1$, and $(1 - \varepsilon) < 1$, then $(1 - \varepsilon)/4$ is relatively small. We would have even better assurance had we selected a magnitude of 10% step change, since then we would have had 0.01 $(1 - \varepsilon)$, and this is very close to zero.

Nonlinearities can always be "linearized" in some sense, and the structure of deviation variables makes this quite easy. Thus, suppose we wish to linearize a function $f(c_A)$ around the steady value $\bar{c}_A$; expanding the function in Taylor series yields

$$f(c_A) = f(\bar{c}_A) + \left.\frac{\partial f}{\partial c_A}\right|_{\bar{c}_A}(c_A - \bar{c}_A) + \cdots \qquad (9.159)$$

Now, since $c_A - \bar{c}_A = \hat{c}_A$, which defines a small deviation, then higher order terms can be ignored and the linearized function becomes simply

$$f(c_A) \simeq f(\bar{c}_A) + \left.\frac{\partial f}{\partial c_A}\right|_{\bar{c}_A}\hat{c}_A \qquad (9.160)$$

Suppose in the CSTR examples that the reaction was of order n (where $n \neq 1$). The material balance would then be nonlinear

$$V\frac{dc_A}{dt} = qc_{A0} - qc_A - kVc_A^n \qquad (9.161)$$

If we introduce deviation variables, say for the case whereby time variation such as $c_{A_0}(t)$ and $c_A(t)$ exist, then we must remove the nonlinearity arising from the kinetics term. To do this, define $f(c_A) = c_A^n$ and use the above Taylor expansion to see

$$c_A^n \simeq \bar{c}_A^n + n\bar{c}_A^{n-1}\hat{c}_A(t) \qquad (9.162)$$

Now, $\bar{c}_A^n$ and $n\bar{c}_A^{n-1}$ are constants, independent of time variation. Inserting the above and noting the steady state must satisfy

$$0 = q\bar{c}_{A_0} - q\bar{c}_A - kV\bar{c}_A^n \tag{9.163}$$

hence the equation for deviation from steady state becomes

$$V\frac{d\hat{c}_A}{dt} = q\hat{c}_{A_0} - q\hat{c}_A - kV\left(n\bar{c}_A^{n-1}\right)\hat{c}_A \tag{9.164}$$

which is now linear in $\hat{c}_A(t)$. This can be handled exactly as in the previous examples, except the rate constant is modified by the factor $n\bar{c}_A^{n-1}$ (Caution: the steady state must be solved first, since $\bar{c}_A$ appears in the above dynamic equation).

EXAMPLE 9.15

We studied reaction and diffusion within cylindrically shaped catalyst pellets in Example 3.3, where it was shown

$$r\frac{d^2C_A}{dr^2} + \frac{dC_A}{dr} - r\left(\frac{k}{D_A}\right)C_A = 0 \tag{9.165}$$

subject to $C_A = C_{As}$ at $r = R$ and $dC_A/dr = 0$ at $r = 0$. Find the solution $C_A(r)$ using Laplace transforms.

First, arrange variables to dimensionless form

$$y = \frac{C_A}{C_{As}} \quad \text{and} \quad x = r\sqrt{\frac{k}{D_A}}$$

giving

$$x\frac{d^2y}{dx^2} + \frac{dy}{dx} - xy = 0 \tag{9.166}$$

Take Laplace transforms one at a time, noting from Table 9.1

$$\mathscr{L}\left[x\frac{d^2y}{dx^2}\right] = -s^2\frac{dY}{ds} - 2sY(s) + y(0)$$

$$\mathscr{L}\left[\frac{dy}{dx}\right] = sY(s) - y(0)$$

and finally, the transform $\mathscr{L}[xy]$ is simply the first moment given by Eq. 9.113

$$\mathscr{L}[xy] = -\frac{dY(s)}{ds}$$

Inserting these gives finally,

$$(s^2 - 1)\frac{dY(s)}{ds} + sY(s) = 0 \tag{9.167}$$

We have thus reduced the second order ODE to a separable first order problem

$$\frac{dY}{Y} = -\frac{s}{s^2 - 1}\, ds = -\frac{1}{2}\frac{ds^2}{s^2 - 1} = -\frac{1}{2}\frac{d(s^2 - 1)}{(s^2 - 1)}$$

This can be integrated directly to yield

$$Y(s) = \frac{A}{\sqrt{s^2 - 1}}$$

We have not discussed inversion of singularities of the multi-valued type, but we shall proceed anyway, calling on Appendix D for the tabulation of $f(t)$ versus $F(s)$ to effect the final solution. To evaluate the arbitrary constant A, we recall the condition

$$y\left(R\sqrt{k/D_A}\right) = 1$$

We shall first invert

$$\frac{1}{\sqrt{s^2 - 1}}$$

and then evaluate A. Thus, Appendix D shows for

$$\mathscr{L}^{-1}\frac{1}{\sqrt{s^2 - 1}} = \mathscr{L}^{-1}\frac{1}{\sqrt{(s + 1)(s - 1)}} = I_0(x)$$

where we note $x = r\sqrt{k/D_A}$, so we now have

$$y(x) = \frac{C_A(r)}{C_{As}} = AI_0\left(r\sqrt{\frac{k}{D_A}}\right)$$

The boundary condition at $r = R$ yields A, which is

$$A = \frac{1}{I_0\left(R\sqrt{k/D_A}\right)}$$

hence,

$$\frac{C_A(r)}{C_{As}} = \frac{I_0\left(r\sqrt{k/D_A}\right)}{I_0\left(R\sqrt{k/D_A}\right)}$$

which is identical to the result in Example 3.3. We shall discuss inversion of singularities of the multivalued type ($1/\sqrt{s}$, etc.) in the next section. Contour integration in the complex plane is required around a special contour C called the second Bromwich path.

9.12 INVERSION THEORY FOR MULTIVALUED FUNCTIONS: THE SECOND BROMWICH PATH

In Section 9.4, we discussed the peculiar behavior arising from functions such as $\sqrt{s}$. We noted, for example, that $f(s) = \sqrt{s}$ yielded two possible values for f, which we denoted in Eqs. 9.37 and 9.38 as f_0 and f_1. However, such behavior is not analytic, since it was stipulated in Section 9.5 that functions must be single valued (in some defined region of analyticity, which we called R). In order to manipulate such functions so that uniqueness is assured, we must stipulate conditions that cause single-valued behavior. The easiest way to do this, considering the specific case of $f(s) = \sqrt{s}$, is to restrict the angle θ so that $0 \leq \theta \leq 2\pi$. Now, since we defined θ in terms of the smallest angle θ_0 as $\theta = \theta_0 + 2\pi k$ in Eq. 9.35, where θ_0 is called the principal value of θ, then the only allowable value of f is f_0, and so multivalued behavior is thus eliminated by taking $k = 0$, which ensures $\theta = \theta_0$ and $0 < \theta \leq 2\pi$. We illustrate the consequences of this restriction in Fig. 9.12 by comparing allowed values of θ in the s-plane for the case when $|s| = r = $ constant, with the corresponding values of $f(s)$ in the f-plane.

We could also have restricted the angle to the region $-\pi < \theta < \pi$, to yield the same result. The fundamental reason causing the misbehavior in the first instance was the encirclement of $s = 0$, which is the zero value of $f(s) = \sqrt{s}$. Had the origin not been encircled in this case, then multivalued behavior would never have arisen. For example, consider the selections of s giving rise to the contour shown in Fig. 9.13.

Thus, by constructing the barrier along the negative real axis, so that $-\pi \leq \theta \leq \pi$, and by requiring that this barrier never be crossed, we have thus assured

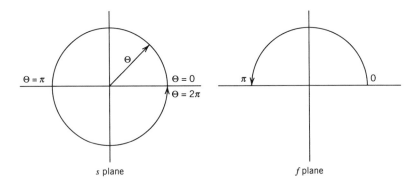

Figure 9.12 Eliminating multivalued behavior by restricting θ to: $0 \leq \theta \leq 2\pi$.

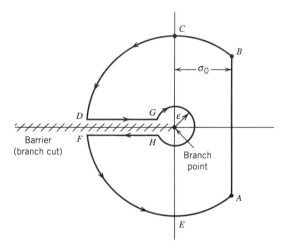

Figure 9.13 Contour C_2 (second Bromwich path),
which does not enclose branch point at $s = 0$.

that the angle s corresponding to these selections of s, will always be such that
$-\pi \leq \theta \leq \pi$. This is so because positive angles are taken as anticlockwise,
while negative angles are clockwise. Except for the line DG, circle ε and line
HF, the contour is identical to Fig. 9.4, which we called the first Bromwich path.
This ensures that s is no longer infinitely circular (i.e., θ does not encircle the
origin). The barrier could have been drawn anywhere in the plane, as illustrated
in Fig. 9.14, and its sole purpose is to prevent encirclement of the point $s = 0$.

The "branch point" is the particular value of s where the function (in this
example, $f(s) = \sqrt{s}$) becomes zero or infinite. So the branch point for $f(s) = \sqrt{s}$
is $s = 0$ and the branch point for $f(s) = 1/\sqrt{s}$ is also $s = 0$.

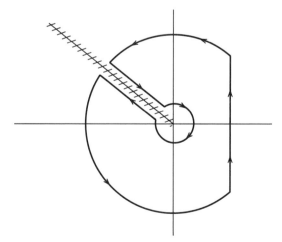

Figure 9.14 Alternative Br_2 path.

Thus, if any contour is drawn so that the branch point is encircled, then multivalued behavior arises. The principle to ensure analyticity is simple: branch points cannot be encircled. There is considerable range and scope for choosing branch cuts to ensure analytic behavior. Now, if no other singularities exist in the contour selected, then Cauchy's First Integral theorem is valid, and we denote the new contour C_2 as the second Bromwich path, Br_2; hence,

$$\oint_{Br_2} f(s) \, ds = 0 \tag{9.168}$$

We next consider how to cope with the inversion of transforms containing multivalued functions. To illustrate this, we consider inverting

$$F(s) = \frac{1}{\sqrt{s}} \tag{9.169}$$

We shall use the second Bromwich path drawn in Fig. 9.13, where the branch point is located at $s = 0$ and the branch cut extends to infinity. We write first that, since no other singularities exist

$$\oint_{Br_2} \frac{e^{st}}{\sqrt{s}} \, ds = 0 \tag{9.170}$$

At the outset, we note the inversion can be easily accomplished using elementary building blocks using the substitution $\beta = \sqrt{t}$ and noting that $\mathrm{erf}(\infty) = 1$; hence,

$$\mathscr{L}\frac{1}{\sqrt{t}} = \int_0^\infty \frac{1}{\sqrt{t}} e^{-st} \, dt = \int_0^\infty \frac{1}{\beta} e^{-s\beta^2} 2\beta d\beta = 2\int_0^\infty e^{-s\beta^2} \, d\beta = \sqrt{\frac{\pi}{s}} \tag{9.171}$$

Hence, we conclude $\mathscr{L}^{-1}(1/\sqrt{s}) = 1/\sqrt{\pi t}$. However, we wish to develop a general approach, applicable to any (or many) multivalued functions. To do this, we decompose the integral in Eq. 9.170 into its component parts (see Fig. 9.13)

$$\oint_{Br_2} \frac{e^{st}}{\sqrt{s}} \, ds = \int_{BCD} \frac{e^{st}}{\sqrt{s}} \, ds + \int_{DG} \frac{e^{st}}{\sqrt{s}} \, ds + \oint_\varepsilon \frac{e^{st}}{\sqrt{s}} \, ds$$
$$+ \int_{HF} \frac{e^{st}}{\sqrt{s}} \, ds + \int_{FEA} \frac{e^{st}}{\sqrt{s}} \, ds + \int_{\sigma_0 - i\omega}^{\sigma_0 + i\omega} \frac{e^{st}}{\sqrt{s}} \, ds = 0 \tag{9.172}$$

We take careful note that the last term in the summation defines the Laplace inversion, according to the Fourier–Mellin theorem in Eq. 9.3, since in the limit as $\omega \to \infty$ (or $R \to \infty$)

$$f(t) = \frac{1}{2\pi i} \int_{\sigma_0 - i\infty}^{\sigma_0 + i\infty} e^{st} F(s) \, ds \tag{9.173}$$

Thus, we wish to solve for:

$$\int_{\sigma_0 - i\omega}^{\sigma_0 + i\omega} \frac{e^{st}}{\sqrt{s}} \, ds$$

in terms of the remaining integrals, in the limit as $\omega \to \infty$, which is the same as taking $R \to \infty$.

The arguments put forth earlier regarding integrals around the first Bromwich path are still applicable to parts of the Br_2 curve (e.g., Eqs. 9.92–9.94), so when $R \to \infty$, we can write immediately

$$\int_{BCD} \frac{e^{st}}{\sqrt{s}} \, ds = \int_{FEA} \frac{e^{st}}{\sqrt{s}} \, ds = 0 \qquad \text{(9.174)}$$

The remaining three terms can be treated as follows; first, consider the small circular path in the limit as $\varepsilon \to 0$, using polar coordinates $s = \varepsilon \exp(i\theta)$

$$\oint_\varepsilon \frac{e^{st}}{\sqrt{s}} \, ds = \int_\pi^{-\pi} \frac{\exp(\varepsilon t[\cos\theta + i\sin\theta]) i\varepsilon \exp(i\theta) \, d\theta}{\sqrt{\varepsilon} \, \exp\left(i\dfrac{\theta}{2}\right)}$$

$$\text{(9.175)}$$

$$= \lim_{\varepsilon \to 0} i\sqrt{\varepsilon} \int_\pi^{-\pi} \exp\left[\varepsilon t(\cos\theta + i\sin\theta) + \frac{1}{2}i\theta\right] d\theta = 0$$

This leaves only the line integrals DG and HF. Consider DG first where we define $s = re^{i\pi} = -r$, hence $\sqrt{s} = \sqrt{r}\, e^{i\pi/2} = i\sqrt{r}$

$$\int_{DG} \frac{e^{st}}{\sqrt{s}} \, ds = \int_{-\infty}^{0} \frac{e^{st}}{\sqrt{s}} \, ds = \int_{\infty}^{0} \frac{e^{-rt}}{i\sqrt{r}}(-dr) = +i\int_{\infty}^{0} \frac{e^{-rt}}{\sqrt{r}} \, dr = -i\int_{0}^{\infty} \frac{e^{-rt}}{\sqrt{r}} \, dr$$

$$\text{(9.176)}$$

Now, let us perform the same type of integral along HF; let $s = re^{-i\pi} = -r$, so that $\sqrt{s} = \sqrt{r}\, e^{-i\frac{\pi}{2}} = -i\sqrt{r}$; hence,

$$\int_{HF} \frac{e^{st}}{\sqrt{s}} \, ds = \int_{0}^{-\infty} \frac{s^{st}}{\sqrt{s}} \, ds = \int_{0}^{\infty} \frac{e^{-rt}}{-i\sqrt{r}}(-dr) = -i\int_{0}^{\infty} \frac{e^{-rt}}{\sqrt{r}} \, dr \quad \text{(9.177)}$$

Thus, solving Eq. 9.172 for

$$\int_{\sigma_0 - i\omega}^{\sigma_0 + i\omega} \frac{e^{st}}{\sqrt{s}} \, ds$$

in the limit

$$\lim_{\omega \to \infty} \int_{\sigma_0 - i\omega}^{\sigma_0 + i\omega} \frac{e^{st}}{\sqrt{s}} \, ds = +2i\int_{0}^{\infty} \frac{e^{-rt}}{\sqrt{r}} \, dr \qquad \text{(9.178)}$$

This is simply the integral along the lines FHGD, which is the negative of DGHF.

Dividing this by $2\pi i$ yields the Fourier–Mellin Inversion theorem, which defines the inverse transform of $\sqrt{s}$

$$f(t) = \frac{1}{\pi} \int_0^\infty \frac{e^{-rt}}{\sqrt{r}} \, dr = \mathscr{L}^{-1} \frac{1}{\sqrt{s}} \qquad (9.179)$$

Taking $rt = \beta^2$, we find

$$f(t) = \frac{1}{\pi} \int_0^\infty \frac{e^{-\beta^2} 2\beta}{\dfrac{\beta}{\sqrt{t}} t} \, d\beta = \frac{2}{\pi\sqrt{t}} \int_0^\infty e^{-\beta^2} \, d\beta \qquad (9.180)$$

But, the error function is

$$\operatorname{erf}(x) = \frac{2}{\sqrt{\pi}} \int_0^x \exp(-\beta^2) \, d\beta \qquad \text{and} \qquad \operatorname{erf}(\infty) = 1$$

hence,

$$f(t) = \frac{1}{\sqrt{\pi t}} \qquad (9.181)$$

9.12.1   Inversion When Poles and Branch Points Exist

When pole singularities exist, along with branch points, the Cauchy theorem is modified to account for the (possibly infinite) finite poles that exist within the Bromwich Contour; hence, $\Sigma \operatorname{res}(F(s)e^{st})$ must be added to the line integrals taken along the branch cut, so

$$f(t) = \frac{1}{2\pi i} \oint_{Br_2} e^{st} F(s) \, ds = \frac{1}{2\pi i} \int_{FHGD} e^{st} F(s) \, ds + \sum \left[\text{residues of } e^{st} F(s) \right]$$

$$(9.182)$$

If no branch points exist, this reduces to the formula given in Eq. 9.99. Often, elementary building blocks can be used to invert awkward transforms. For example, we have seen that $\mathscr{L}^{-1}(1/\sqrt{s}) = 1/\sqrt{\pi t}$ and moreover, since $1/s$ denotes integration, then it is straightforward to find the inversion of

$$F(s) = \frac{1}{s} \frac{1}{\sqrt{s}} \qquad (9.183)$$

as simply

$$\mathscr{L}^{-1}F(s) = \int_0^t \frac{1}{\sqrt{\pi\beta}}\, d\beta = \frac{2}{\sqrt{\pi}}\,\sqrt{\beta}\,\Big|_0^t = 2\sqrt{\frac{t}{\pi}} \qquad (9.184)$$

Thus, any transform can, in theory, be inverted provided the behavior in the complex plane is such that $|F(s)| \to 0$ as $|s| \to \infty$. All physical processes must satisfy this condition, so in principle an inversion always exists, although the mathematics to find it can often be intractable. For this reason, numerical techniques have been developed, which have wide applicability for practical problems. We discuss these approximate, numerical techniques in the next section.

9.13   NUMERICAL INVERSION TECHNIQUES

Previous sections dealt with the analytical development of Laplace transform and the inversion process. The method of residues is popular in the inversion of Laplace transforms for many applications in chemical engineering. However, there are cases where the Laplace transform functions are very complicated and for these cases the inversion of Laplace transforms can be more effectively done via a numerical procedure. This section will deal with two numerical methods of inverting Laplace transforms. One was developed by Zakian (1969), and the other method uses a Fourier series approximation (Crump 1976). Interested readers may also wish to perform transforms using a symbolic algebra language such as Maple (Heck 1993).

9.13.1   The Zakian Method

The Laplace transform of $f(t)$ is denoted as $F(s)$, given in Eq. 9.1, and is rewritten again here for convenience

$$F(s) = \int_0^{\infty} f(t)e^{-st}\, dt \qquad (9.185)$$

Here, we assume that $f(t)$ is integrable (piecewise continuous) and is of exponential order σ_0; that is, $|f(t)| < M\exp(\sigma_0 t)$.

Recalling the Dirac's delta function, defined in Eq. 8.10, and if we replace x_k of Eq. 8.10 by unity, we obtain

$$\int_{1^-}^{1^+} \delta(x - 1)\, dx = 1 \qquad (9.186)$$

If we now make the following substitution

$$x = \frac{y}{t} \qquad (9.187)$$

then the Dirac delta function in Eq. 9.186 will become

$$\int_{t^-}^{t^+} \delta\left(\frac{y}{t} - 1\right) dy = t \tag{9.188}$$

Thus, if t lies between 0 and T, we can replace the above integration range to $[0, T]$ without changing the value of the integral; that is,

$$\int_0^T \delta\left(\frac{y}{t} - 1\right) dy = t \tag{9.189}$$

Using the property of the Dirac delta function, we can write

$$f(t) = \frac{1}{t}\int_0^T f(y) \delta\left(\frac{y}{t} - 1\right) dy \tag{9.190}$$

valid for any values of t between 0 and T.

Following Zakian (1969), the delta function, $\delta(y/t - 1)$ can be expressed in terms of a series of exponential functions

$$\delta\left(\frac{y}{t} - 1\right) = \sum_{i=1}^{\infty} K_i \exp\left(-\alpha_i \frac{y}{t}\right) \tag{9.191}$$

where the parameters α_i and K_i satisfy the following criteria

(a) α and K are either real or they occur in complex-conjugate pairs; for example, $\alpha_1 = \alpha_2^*$ and $K_1 = K_2^*$
(b) α_i and K_i depend on N, where N is the number of terms kept in the partial series (Eq. 9.192)
(c) $\lim_{N\to\infty}\text{Real}(\alpha_i) = \infty$, $\lim_{N\to\infty}|K_i| = \infty$
(d) $\text{Real}(\alpha_i) > 0$
(e) α_i are distinct; that is, $\alpha_i = \alpha_j$ if and only if $i = j$

If we denote the sequence δ_N as the partial series of δ as

$$\delta_N\left(\frac{y}{t} - 1\right) = \sum_{i=1}^{N} K_i \exp\left(-\alpha_i \frac{y}{t}\right) \tag{9.192}$$

and the function sequence $f_N(t)$ as

$$f_N(t) = \frac{1}{t}\int_0^T f(y) \delta_N\left(\frac{y}{t} - 1\right) dy \tag{9.193}$$

then for every continuity point t of f, we have

$$f(t) = \lim_{N\to\infty} f_N(t) \tag{9.194}$$

Substitution of δ_N of Eq. 9.192 into Eq. 9.193 yields

$$f_N(t) = \frac{1}{t}\int_0^T f(y) \sum_{i=1}^{N} K_i \exp\left(-\alpha_i \frac{y}{t}\right) dy \tag{9.195}$$

If we interchange the summation and integral signs, we obtain

$$f_N(t) = \frac{1}{t}\sum_{i=1}^{N} K_i \int_0^T f(y) \exp\left(-\alpha_i \frac{y}{t}\right) dy \tag{9.196}$$

Now, letting $T \to \infty$ and using the definition of Laplace transform (Eq. 9.1 or 9.185) gives

$$f_N(t) = \frac{1}{t}\sum_{i=1}^{N} K_i F\left(\frac{\alpha_i}{t}\right) \tag{9.197}$$

for $0 < t < t_c$, where

$$t_c = \min_{i=1,2,\ldots,N}\left[\mathrm{Re}\left(\frac{\alpha_i}{\sigma_0}\right)\right] \tag{9.198}$$

Using the property (c) above, $N \to \infty$, $\mathrm{Re}(\alpha_i) \to \infty$, and hence, $t_c \to \infty$, we have the following explicit inversion formula

$$f(t) = \lim_{N\to\infty}\left[\frac{1}{t}\sum_{i=1}^{N} K_i F\left(\frac{\alpha_i}{t}\right)\right] \tag{9.199a}$$

Since the constants α and K appear in complex conjugate (see property a), the explicit inversion formula can be written as follows

$$f(t) = \lim_{N\to\infty}\left[\frac{2}{t}\sum_{i=1}^{N/2}\mathrm{Real}\left\{K_i F\left(\frac{\alpha_i}{t}\right)\right\}\right] \tag{9.199b}$$

Table 9.2 gives a set of five constants for α and K ($N/2 = 5$) (Zakian 1970). For most applications, the above set of constants yields good numerical inverse as we shall see in the following example.

Table 9.2 Set of Five Constants for α and K for the Zakian Method

i	α_i	K_i
1	$1.283767675e + 01 + i\ 1.666063445$	$-3.69020821e + 04 + i\ 1.96990426e + 05$
2	$1.222613209e + 01 + i\ 5.012718792e + 00$	$+6.12770252e + 04 - i\ 9.54086255e + 04$
3	$1.09343031e + 01 + i\ 8.40967312e + 00$	$-2.89165629e + 04 + i\ 1.81691853e + 04$
4	$8.77643472e + 00 + i\ 1.19218539e + 01$	$+4.65536114e + 03 - i\ 1.90152864e + 00$
5	$5.22545336e + 00 + i\ 1.57295290e + 01$	$-1.18741401e + 02 - i\ +1.41303691e + 02$

Table 9.3 Comparison Between the Numerical Inverse Obtained by the Zakian Method and the Analytical Solution for Example 9.16

t	Zakian Numerical Inverse	Analytical Inverse	Relative Error $\times$ 100
0.5	0.01629	0.01633	0.233
1.5	0.2803	0.2810	0.228
2.5	0.8675	0.8694	0.221
3.5	1.662	1.666	0.217
4.5	2.567	2.572	0.216
5.5	3.523	3.531	0.212
6.5	4.503	4.513	0.215
7.5	5.494	5.505	0.211
8.5	6.489	6.502	0.208
9.5	7.501	7.501	0.211

EXAMPLE 9.16

To show an example of numerical inversion by the Zakian method, we take the function treated in Example 9.12

$$F(s) = \frac{1}{s^2} \frac{1}{(s + 1)^2}$$

which has the following analytical inverse

$$f(t) = (t - 2) + (t + 2) \exp(-t)$$

Table 9.3 shows the excellent agreement between the analytical and numerical inverse. A maximum percentage relative error of less than 3% is observed.

EXAMPLE 9.17

To illustrate problems for transcendental functions in the numerical inversion method of Zakian, we take the function

$$F(s) = \frac{1}{(s^2 + 1)}$$

which has the inverse

$$f(t) = \sin(t)$$

Table 9.4 shows the comparison between the analytical and numerical inverse. The error increases with an increase in time, suggesting that the Zakian method does not work well with oscillating functions.

Table 9.4 Comparison Between the Numerical Inverse Obtained by the Zakian Method and the Analytical Solution for Example 9.17

Time	Zakian Numerical Inverse	Analytical Inverse	Relative Error × 100
0.5	0.4785	0.4794	0.1894
2.5	0.5982	0.5985	0.04848
4.5	− 0.9750	− 0.9775	0.2606
6.5	0.2146	0.2151	0.2443
8.5	0.8004	0.7985	0.2377
10.5	− 0.8752	− 0.8797	0.5090
12.5	0.06947	− 0.06632	4.744
14.5	0.9239	0.9349	1.181
16.5	− 0.5281	− 0.7118	25.81

EXAMPLE 9.18

Finally we test the Zakian technique with the exponentially increasing function treated in Example 9.10. The Laplace transform is

$$F(s) = \frac{2s^2 + 3s - 4}{(s - 2)(s^2 + 2s + 2)}$$

which the analytical inverse is

$$f(t) = e^{2t} + e^{-t}\cos(t) + 2e^{-t}\sin(t)$$

Table 9.5 shows the numerical inverse obtained by the Zakian method, and the performance of the routine is even worse than the previous example.

In conclusion, the Zakian numerical technique yields good results only when the inverse does not exhibit oscillating behavior or when it increases exponentially with time. The Fourier approximation method presented in the next section will remedy the disadvantages encountered by the Zakian method.

Table 9.5 Comparison Between the Numerical Inverse Obtained by the Zakian Method and the Analytical Solution for Example 9.18

t	Zakian Numerical Inverse	Analytical Inverse	Percentage Rel. Error
0.5	$0.3829e + 01$	$0.3832e + 01$	0.073
1.5	$0.2050e + 02$	$0.2055e + 02$	0.224
4.5	$0.8082e + 04$	$0.8103e + 04$	0.261
9.5	$0.7401e + 04$	$0.1785e + 09$	100

9.13.2 The Fourier Series Approximation

Another method of getting the numerical inverse is by the Fourier series approximation. The Laplace transform pairs are given as in Eqs. 9.1 and 9.3, written again for convenience

$$F(s) = \int_0^\infty e^{-st} f(t) \, dt \qquad (9.200a)$$

$$f(t) = \frac{1}{2\pi i} \int_{\sigma - i\omega}^{\sigma + i\omega} e^{st} F(s) \, ds \qquad (9.200b)$$

We assume that $f(t)$ is piecewise continuous and of exponential order σ_0, that is, $|f(t)| < M \exp(\sigma_0 t)$.

The inverse equation (Eq. 9.200b) can be written in terms of the integral with respect to ω as

$$f(t) = \frac{e^{\sigma t}}{\pi} \int_0^\infty \{ \text{Re}[F(s)] \cos(\omega t) - \text{Im}[F(s)] \sin(\omega t) \} \, d\omega \qquad (9.201)$$

where σ can be any number greater than σ_0.

Following the procedure of Crump (1976), the approximation to the inverse is given by

$$f(t) \approx \frac{e^{\sigma t}}{T} \left\{ \frac{1}{2} F(\sigma) + \sum_{k=1}^{\infty} \text{Re}\left[F\left(\sigma + \frac{k\pi i}{T} \right) \right] \cos\left(\frac{k\pi t}{T} \right) \right.$$
$$\left. - \text{Im}\left[F\left(\sigma + \frac{k\pi i}{T} \right) \right] \sin\left(\frac{k\pi t}{T} \right) \right\} \qquad (9.202)$$

where $T > t$, and σ is any number greater than the exponential order of the function $f(t)$. This equation is simply the trapezoidal rule of the inverse equation 9.201.

Numerically, the computation is done as follows (Crump 1976). Suppose the numerical value of $f(t)$ is needed for a range of t up to $t_{\max}$, and the relative error is to be no greater than E. First T is chosen such that

$$2T > t_{\max} \qquad (9.203)$$

and the value of σ is chosen as

$$\sigma = \sigma_0 - \frac{\ln(E)}{2T} \qquad (9.204)$$

The parameter σ_0, which is the exponential order of the function $f(t)$, can be computed from the transform $F(s)$. We simply take σ_0 to be a number slightly larger than

$$\max\{ \text{Re}(P) \}$$

Table 9.6 Comparison Between the Numerical Inverse Obtained by the
Fourier Method and the Analytical Solution for Example 9.16

Time	Fourier Numerical Inverse of Example 9.16	Analytical Inverse	Relative Error $\times$ 100
0.5	0.01633	0.01633	0.045
1.5	0.28094	0.28096	0.006
2.5	0.86938	0.86938	0.000
3.5	1.66607	1.66609	0.001
4.5	2.57227	2.57221	0.002
5.5	3.53071	3.53065	0.002
6.5	4.51285	4.51278	0.002
7.5	5.50534	5.50525	0.001
8.5	6.50223	6.50214	0.001
9.5	7.50097	7.50086	0.001

Table 9.7 Comparison Between the Numerical Inverse Obtained by the
Fourier Method and the Analytical Solution for Example 9.17

Time	Fourier Numerical Inverse of Example 9.17	Analytical Inverse	Relative Error $\times$ 100
0.5	0.47943	0.47943	0.000
2.5	0.59847	0.59847	0.000
4.5	-0.97752	-0.97753	0.001
6.5	0.21518	0.21512	0.028
8.5	0.79856	0.79849	0.010
10.5	-0.87969	-0.87970	0.001
12.5	0.06632	-0.06632	0.003
14.5	0.93489	0.93490	0.000
16.5	-0.71176	-0.71179	0.004

Table 9.8 Comparison Between the Numerical Inverse Obtained by the
Fourier Method and the Analytical Solution for Example 9.18

t	Fourier Numerical Inverse of Example 9.18	Analytical Inverse	Percentage Rel. Error
0.5	$0.38319e + 01$	$0.38321e + 01$	0.005
1.5	$0.20546e + 02$	$0.20546e + 02$	0.001
4.5	$0.81030e + 04$	$0.81031e + 04$	0.000
7.5	$0.32700e + 07$	$0.32690e + 07$	0.028
9.5	$0.17857e + 09$	$0.17848e + 09$	0.051

where P denotes a pole of $F(s)$.

Taking the Examples 9.16, 9.17, and 9.18, we compute the numerical inverse and show the results in the Tables 9.6, 9.7, and 9.8, where it is seen that the Fourier series approximation is a better approximation than the Zakian method.

The Fourier series approximation is a better method to handle oscillating functions, but it requires more computation time than the Zakian method. With the advent of high speed personal computers, this is not regarded as a serious disadvantage.

9.14 REFERENCES

1. Churchill, R. V. *Operational Mathematics*, second edition. McGraw Hill, New York (1958).
2. Crump, K. S. "Numerical Inversion of Laplace Transforms Using a Fourier Series Approximation," *J. Assoc. Comput. Machinery*, 23, 89–96 (1976).
3. Heck, A. *Introduction to Maple*. Springer-Verlag, New York (1993).
4. Hildebrand, F. B. *Advanced Calculus for Applications*. Prentice Hall, Englewood Cliffs, New Jersey (1962).
5. Zakian, V. "Numerical Inversion of Laplace Transform," Electronics Letters, 5, 120–121 (1969).
6. Zakian, V. "Rational Approximation to Transfer-Function Matrix of Distributed System," Electronics Letters, 6, 474–476 (1970).

9.15 PROBLEMS

9.1₂. Prove that the function

$$f = \sigma - i\omega = \bar{s}$$

is not analytic anywhere.

9.2₂. Introduce polar coordinates to Cauchy's Integral theorem $\oint f(s)\, ds = 0$ in the form $f(s) = u + iv$ and $s = re^{i\theta}$ and show that the property of exactness, discussed in Section 9.6, leads to a polar form of the Cauchy–Riemann conditions

$$\frac{\partial u}{\partial r} = \frac{1}{r}\frac{\partial v}{\partial \theta}; \qquad \frac{1}{r}\frac{\partial u}{\partial \theta} = -\frac{\partial v}{\partial r}$$

9.3₃. The properties of the logarithm of complex variables are related to the exponential function e^s introduced in Eq. 9.19 by writing

$$s = \exp(f(s))$$

then

$$f(s) = \mathrm{Ln}(s) = \mathrm{Ln}\big[re^{i(\theta_0 + 2k\pi)}\big]$$

where k is zero or integer. This shows that $f(s)$ is in fact a multivalued function. We shall use capital Ln to represent logs of complex variables.
(a) If $f(s) = u + iv$, show that $u = \ln(r)$ and $v = \theta = \theta_0 + 2k\pi$
(b) If $s = i$, show that

$$Ln(i) = \left(\frac{4k+1}{2}\right)\pi i \ (k = 0, \pm 1, \pm 2, \ldots)$$

The value corresponding to $k = 0$ is often called the "principal value of the logarithm."
(c) If $s = x$ (x is a positive, real number), show that

$$Ln(x) = \ln x + 2k\pi i$$

This shows that the complex logarithm of a positive real number may differ from the usual real logarithm by an arbitrary multiple of 2π. The two become identical only when $k = 0$, which corresponds to the principal value; that is, when $\theta = \theta_0$.

9.4₃. Determine all possible values of the function

$$f(s) = \sin^{-1}(s)$$

when $s = 2$. Hint: treat the general case first by writing

$$s = \sin f = \frac{e^{if} - e^{-if}}{2i} = \frac{e^{2if} - 1}{2ie^{if}}$$

where we have used Eq. 9.21

$$\text{Ans: } f = \frac{\pi}{2} + 2k\pi \pm \ln(2 + \sqrt{3})$$

9.5₂. Prove that $\int_c 1/(s - a)\, ds = 0$ for a closed curve encircling the origin with $|s| < a$, but for $|s| > a$, $\int 1/(s - a)\, ds = 2\pi i$.

9.6*. Suppose we have a function of the form

$$g(s) = f(s)/(s - s_0)$$

where $f(s)$ is taken to be analytic within a region R and s_0 lies within R. If $f(s_0) \neq 0$, then $g(s)$ is not analytic at $s = s_0$. Prove that

$$f(s_0) = \frac{1}{2\pi i} \oint_C \frac{f(s)}{s - s_0}\, ds$$

where C is a closed curve that lies in R and includes the point s_0 (this is called Cauchy's Second Integral theorem). Hint: Isolate the point $s = s_0$ by drawing an additional circle of radius C_1 around the point, then allow s to traverse both curves through a cut between the closed curves so that analytic behavior is guaranteed in the region separating C and C_1 so that

$$\oint_C \frac{f(s)}{s-a} ds - \oint_{C_1} \frac{f(s)}{s-a} ds = 0$$

9.7$_3$. Find the residues of the function $f(s) = 1/(1 - e^s)$, at all poles. Hint: The zeros of $\exp(s) = 1$ can be obtained using the Logarithm discussed in Problem 9.3$_3$ since

$$\text{Ln}(1) = s = \ln(1) + 2k\pi i = 2k\pi i$$

Ans: -1

9.8$_3$. Use the Laplace transform to find the value of the sine integral

$$\int_0^\infty \frac{\sin(t\beta)}{\beta} d\beta = \frac{\pi}{2} \text{sgn}(t)$$

where

$$\text{sgn}(t) = \begin{cases} 1, & t > 0 \\ 0, & t = 0 \\ -1, & t < 0 \end{cases}$$

Hint: Define the Laplace transform of the general integral as

$$\mathscr{L} \int_0^\infty f(t, \beta) \, d\beta = \int_0^\infty \mathscr{L} f(t, \beta) \, d\beta$$

where

$$f(t, \beta) = \sin(t\beta)/\beta.$$

9.9$_3$. Prove the Laplace transform of the complementary error function is given by

$$\mathscr{L} \, \text{erfc}\left(a/\sqrt{4t}\right) = e^{-a\sqrt{s}}/s$$

where $a > 0$ is a real constant.

9.10$_2$. (a) Use the convolutions integral to solve for $Y(s)$ in the Volterra integral

$$y(t) = f(t) + \int_0^t y(u) g(t - u) \, du$$

and show that

$$Y(s) = \frac{F(s)}{1 - G(s)}$$

(b) Apply the results from (a) and show that

$$y(t) = \sin t + \int_0^t \sin[2(t - u)] y(u) \, du$$

has the explicit solution

$$y(t) = 3 \sin t - \sqrt{2} \, \sin(\sqrt{2} t)$$

9.11$_2$. Find the function $f(t)$ corresponding to the Laplace transforms

(a) $\dfrac{1}{s(s + 1)^3}$ (b) $\dfrac{s}{(s^2 + 1)(s^2 + 4)}$

Ans: (a) $f(t) = 1 - e^t(t + 1) - \dfrac{1}{2} t^2 e^{-t}$

(b) $f(t) = \dfrac{1}{3}[\cos t - \cos 2t]$

9.12$_2$. (a) Show that a particular solution to Kummer's equation (see Example 3.4)

$$t \frac{d^2 y}{dt^2} + (1 - t) \frac{dy}{dt} + \beta y = 0$$

can be obtained by the inversion of the Laplace transform

$$y(t) = A \mathcal{L}^{-1} \left[\frac{(1 - s)^\beta)}{s^{1+\beta}} \right]$$

where A is an arbitrary constant. [Hint: use Table 9.1.]
(b) Show for the case $\beta = 1$ that

$$y(t) = A(t - 1)$$

and for the case $\beta = 2$

$$y(t) = A \left(1 - 2t + \frac{t^2}{2!} \right)$$

9.13$_3$. Use Laplace transforms to solve the equation

$$t \frac{d^2 y}{dt^2} + (t - 1) \frac{dy}{dt} - y = 0$$

for the condition $y(0) = 0$ and show that

$$y(t) = A[(t - 1) + e^{-t}]$$

where A is an arbitrary constant. What is the value of $y'(0)$?

9.14$_2$. The dynamic equation for a damped, spring-inertia system was derived in Problem 1.6$_2$. If the initial position and velocity are zero, the Laplace transform of the second order equation yields

$$(\tau^2 s^2 + 2\tau\zeta s + 1)X(s) = F(s)$$

Hence, to find the dynamics for an arbitrary forcing function $F(s)$, we shall need to invert

$$X(t) = \mathscr{L}^{-1}\left[\frac{F(s)}{\tau^2 s^2 + 2\zeta\tau s + 1}\right]$$

For the specific case of forcing by a unit step function ($F(s) = 1/s$), we shall consider three possible behavior patterns

$\zeta > 1$ (overdamped)
$\zeta = 1$ (critically damped)
$\zeta < 1$ (underdamped)

(a) Perform the inversion when $\zeta > 1$ by first showing the transform can be rearranged as

$$X(s) = \frac{1}{\tau^2} \cdot \frac{1}{s} \frac{1}{(s+a)^2 - b^2}$$

where

$$a = \frac{\zeta}{\tau} \quad \text{and} \quad b = \frac{1}{\tau}\sqrt{\zeta^2 - 1}$$

(b) Since $1/s$ implies integration, use this along with the shifting theorem to show from Transform 15

$$X(t) = \frac{1}{\tau^2}\int_0^t e^{-at}\frac{1}{b}\sinh(bt)\, dt$$

hence

$$X(t) = 1 - e^{-at}\left[\cosh(bt) + \frac{a}{b}\sinh(bt)\right]$$

so finally, replacing a and b

$$X(t) = 1 - \exp(-\zeta t/\tau)\left[\cosh\left(\frac{\sqrt{\zeta^2 - 1}}{\tau}t\right) + \frac{\zeta}{\sqrt{\zeta^2 - 1}}\sinh\left(\frac{\sqrt{\zeta^2 - 1}}{\tau}t\right)\right]$$

(c) For the case $\zeta = 1$, show that the integral inversion formula in part (b) becomes

$$X(t) = \frac{1}{\tau^2}\int_0^t te^{-at}\, dt$$

hence,

$$X(t) = 1 - (1 + t/\tau)\exp(-t/\tau)$$

(d) Consider the underdamped case ($\zeta < 1$) and repeat the procedure in part (b), noting since b becomes purely imaginary, then use cosh $(ix) = \cos x$ and sinh $(ix) = i \sin x$, so write immediately

$$X(t) = 1 - \exp(-\zeta t/\tau)\left[\cos\left(\frac{\sqrt{1-\zeta^2}}{\tau}\right)t \right.$$

$$\left. + \frac{\zeta}{\sqrt{1-\zeta^2}} \sin\left(\frac{\sqrt{1-\zeta^2}}{\tau}\right)t\right]$$

(e) Use these results to compute response curves for the three cases. Why is critical damping so important for the design of systems such as shown in Problem 1.6?

9.15₃. A cascade of equivolume CSTRs are connected in series and under linear kinetics to deplete species A in $A \overset{k}{\to}$ products.

Denoting $C_n(t)$ as the composition of A leaving the nth reactor, the material balance is

$$V\frac{dC_n}{dt} = qC_{n-1} - qC_n - kVC_n$$

(a) Introduce perturbation variables of the type

$$C_n(t) = \overline{C}_n + \hat{C}_n(t)$$

and show that

$$\tau\frac{d\hat{C}_n}{dt} + \hat{C}_n = \varepsilon\hat{C}_{n-1}$$

where

$$\varepsilon = \frac{q}{kV + q}; \qquad \tau = \frac{V}{kV + q}$$

(b) Take Laplace transforms and show that

$$\hat{C}_n(s) = \frac{\varepsilon\hat{C}_{n-1}(s)}{\tau s + 1}$$

(c) Solve the linear finite difference equation by assuming a solution of the form

$$\hat{C}_n(s) = A(r)^n$$

and show that

$$\hat{C}_n(s) = \hat{C}_0(s)\left(\frac{\varepsilon}{\tau s + 1}\right)^n$$

where $\hat{C}_0(s)$ denotes the transform of the composition perturbation entering the first reactor.

(d) Show that a step change of magnitude $\alpha\bar{C}_0(0 < \alpha < 1)$ entering the first reactor causes the following response for the nth reactor

$$\hat{C}_n(t) = \varepsilon^n \alpha\bar{C}_0\left[1 + \frac{1}{\tau^n}\frac{1}{(n-1)!}\frac{d^{n-1}}{ds^{n-1}}\left(\frac{e^{st}}{s}\right)\bigg|_{s=-1/\tau}\right]$$

(e) For the case of two reactors, show that the exit perturbation is

$$\hat{C}_2(t) = \varepsilon^2 \alpha\bar{C}_0\left[1 - \frac{t}{\tau}\exp(-t/\tau) - \exp(-t/\tau)\right]$$

(f) Using results from part (e), deduce the impulse response, $\hat{C}_0(t) = W_0\delta(t)$, where the weighting W_0 replaces $\alpha\bar{C}_0$, and obtain

$$\hat{C}_2(t) = \varepsilon^2 W_0\frac{t}{\tau^2}\exp(-t/\tau)$$

and then show the composition maximum occurs when $t = \tau$.

Chapter 10

Solution Techniques for Models Producing PDEs

10.1 INTRODUCTION

In Chapter 1, we showed how conservation laws often lead to situations wherein more than one independent variable is needed. This gave rise to so-called partial derivatives, defined for example in the distance–time domain as

$$\lim_{\Delta x \to 0} \frac{f(x + \Delta x, t) - f(x, t)}{\Delta x} = \frac{\partial f}{\partial x} \tag{10.1}$$

$$\lim_{\Delta t \to 0} \frac{f(x, t + \Delta t) - f(x, t)}{\Delta t} = \frac{\partial f}{\partial t} \tag{10.2}$$

The first implies holding time constant while differentiating with respect to x, while the second implies holding x constant while differentiating with respect to t. These implicit properties must be kept in mind when integrating partial derivatives. Thus, for ordinary derivatives, the integral of $dy/dx = 0$ yields $y(x) = $ constant. However, for the partial derivatives, we must account for the implied property

$$\frac{\partial f}{\partial x} = 0 \qquad \therefore f = g(t) \tag{10.3}$$

Thus, instead of adding an arbitrary constant, we must in general add an arbitrary function of the independent variable, which was held constant during integration.

The general function of three variables can be written

$$u = f(x, y, z) \tag{10.4}$$

The total differential of u can be written using the chain rule

$$du = \frac{\partial f}{\partial x} dx + \frac{\partial f}{\partial y} dy + \frac{\partial f}{\partial z} dz \qquad (10.5)$$

For continuously differentiable functions, the order of differentiation is immaterial

$$\frac{\partial}{\partial x}\left(\frac{\partial f}{\partial y}\right) = \frac{\partial}{\partial y}\left(\frac{\partial f}{\partial x}\right) = \frac{\partial^2 f}{\partial x \partial y} = \frac{\partial^2 f}{\partial y \partial x} \qquad (10.6)$$

For convenience, we often abbreviate partial derivatives as

$$\frac{\partial^2 f}{\partial x^2} = f_{xx} \quad ; \quad \frac{\partial^2 f}{\partial y^2} = f_{yy} \quad ; \quad \frac{\partial^2 f}{\partial x \partial y} = f_{xy} \qquad (10.7)$$

Occasionally, implicit functions arise

$$g(x, y) = 0 \qquad (10.8)$$

and we wish to find

$$y = f(x) \qquad (10.9)$$

To do this, we apply the chain rule using Eq. 10.5 on Eq. 10.8 to yield

$$dg = \frac{\partial g}{\partial x} dx + \frac{\partial g}{\partial y} dy = 0 \qquad (10.10)$$

Solving for dy/dx gives

$$\frac{dy}{dx} = -\frac{\left(\dfrac{\partial g}{\partial x}\right)}{\left(\dfrac{\partial g}{\partial y}\right)} = \frac{df}{dx} \qquad (10.11)$$

Because of the implicit properties of partial derivatives, ∂g cannot be cancelled as would be the case for ordinary derivatives.

EXAMPLE 10.1

If:

$$x^2 + y^2 = \lambda x, \qquad \text{find } \frac{dy}{dx}.$$

We can accomplish this exercise in two ways. The first way simply requires solution for y (taking the positive root):

$$y = \sqrt{\lambda x - x^2}$$

Hence,

$$\frac{dy}{dx} = \frac{1}{2} \frac{1}{\sqrt{\lambda x - x^2}} (\lambda - 2x) = \frac{\left(\frac{\lambda}{2} - x\right)}{\sqrt{\lambda x - x^2}}$$

On the other hand, we could have defined

$$g(x, y) = 0 = x^2 + y^2 - \lambda x$$

The partial derivatives are

$$\frac{\partial g}{\partial x} = 2x - \lambda; \qquad \frac{\partial g}{\partial y} = 2y$$

hence, using Eq. 10.11

$$\frac{dy}{dx} = - \frac{\left(\frac{\partial g}{\partial x}\right)}{\left(\frac{\partial g}{\partial y}\right)} = - \frac{(2x - \lambda)}{2y}$$

Eliminating y gives

$$\frac{dy}{dx} = \frac{\left(\frac{\lambda}{2} - x\right)}{\sqrt{\lambda x - x^2}}$$

as before.

This procedure can be easily extended to three variables, given the implicit form

$$F(x, y, z) = 0 \qquad\qquad \textbf{(10.12)}$$

and we wish to find

$$z = g(x, y) \qquad\qquad \textbf{(10.13)}$$

One of the procedures found useful in solving partial differential equations is the so-called combination of variables method, or similarity transform. The strategy here is to reduce a partial differential equation to an ordinary one by judicious combination of independent variables. Considerable care must be given to changing independent variables.

Consider the partial differential equation

$$\frac{\partial u}{\partial y} = \frac{\partial^2 u}{\partial x^2} \qquad\qquad \textbf{(10.14)}$$

Suppose there exists a combined variable such that

$$u(x, y) = f(\eta) \qquad \text{and} \qquad \eta = g(x, y) \tag{10.15}$$

We first write the total differential

$$du = \frac{df}{d\eta} d\eta \tag{10.16}$$

But η is a function of two variables, so write

$$d\eta = \frac{\partial \eta}{\partial x} dx + \frac{\partial \eta}{\partial y} dy \tag{10.17}$$

hence,

$$du = f'(\eta) \left[\frac{\partial \eta}{\partial x} dx + \frac{\partial \eta}{\partial y} dy \right] \tag{10.18}$$

But since $u(x, y)$, we can also write

$$du = \frac{\partial u}{\partial x} dx + \frac{\partial u}{\partial y} dy \tag{10.19}$$

Now, on comparing Eq. 10.18 and 10.19, and equating like coefficients of dx and dy, we find

$$\frac{\partial u}{\partial x} = f'(\eta) \frac{\partial \eta}{\partial x} \tag{10.20}$$

$$\frac{\partial u}{\partial y} = f'(\eta) \frac{\partial \eta}{\partial y} \tag{10.21}$$

To obtain second derivatives, we follow the same procedure. Thus, to find $\partial^2 u / \partial x^2$, we write in general

$$\frac{\partial u}{\partial x} = \psi(\eta, x, y) = f'(\eta) \frac{\partial \eta}{\partial x} \tag{10.22}$$

The total differentials are

$$d\left(\frac{\partial u}{\partial x} \right) = \frac{\partial^2 u}{\partial x^2} dx + \frac{\partial^2 u}{\partial y \partial x} dy \tag{10.23}$$

$$d\psi = \frac{\partial \psi}{\partial \eta} d\eta + \frac{\partial \psi}{\partial x} dx + \frac{\partial \psi}{\partial y} dy \tag{10.24}$$

Replacing $d\eta$ from Eq. 10.17 yields

$$d\psi = \left[\frac{\partial \psi}{\partial \eta} \cdot \frac{\partial \eta}{\partial x} + \frac{\partial \psi}{\partial x} \right] dx + \left[\frac{\partial \psi}{\partial \eta} \cdot \frac{\partial \eta}{\partial y} + \frac{\partial \psi}{\partial y} \right] dy \tag{10.25}$$

Equating like coefficients of dx and dy between Eqs. 10.23 and 10.25 yields

$$\frac{\partial^2 u}{\partial x^2} = \frac{\partial \psi}{\partial \eta} \cdot \frac{\partial \eta}{\partial x} + \frac{\partial \psi}{\partial x} \qquad (10.26)$$

$$\frac{\partial^2 u}{\partial y \partial x} = \frac{\partial \psi}{\partial \eta} \cdot \frac{\partial \eta}{\partial y} + \frac{\partial \psi}{\partial y} \qquad (10.27)$$

Suppose we had selected for η

$$\eta = \frac{x}{\sqrt{4y}} \qquad (10.28)$$

This would yield from Eqs. 10.20 and 10.21

$$\frac{\partial u}{\partial x} = f'(\eta) \frac{1}{\sqrt{4y}} \qquad (10.29)$$

$$\frac{\partial u}{\partial y} = f'(\eta) \frac{\left(-\frac{1}{2}\eta\right)}{y} \qquad (10.30)$$

From Eq. 10.22, this implies the dependence $\psi(\eta, y)$ instead of $\psi(\eta, x, y)$, hence Eq. 10.26 becomes simply

$$\frac{\partial^2 u}{\partial x^2} = \frac{\partial \psi}{\partial \eta} \cdot \frac{\partial \eta}{\partial x} \qquad (10.31)$$

It is clear that the partial $\partial \psi / \partial \eta$ is required. There is a persistent mistake in the literature by writing this as an ordinary derivative. Now from Eq. 10.22

$$\frac{\partial^2 u}{\partial x^2} = \frac{\partial}{\partial \eta} \left[f'(\eta) \frac{\partial \eta}{\partial x} \right] \frac{\partial \eta}{\partial x} \qquad (10.32)$$

hence finally we arrive at the unambiguous result

$$\frac{\partial^2 u}{\partial x^2} = f''(\eta) \left(\frac{\partial \eta}{\partial x} \right)^2 = f''(\eta) \left(\frac{1}{\sqrt{4y}} \right)^2 \qquad (10.33)$$

Inserting this and Eq. 10.30 into the original equation (Eq. 10.14) yields

$$f''(\eta) \frac{1}{4y} = f'(\eta) \frac{\left(-\frac{1}{2}\eta\right)}{y} \qquad (10.34)$$

Cancelling y shows that a single independent variable results so that Eq. 10.28 was in fact a proper choice to reduce this particular PDE to an ODE, viz.

$$\frac{d^2f}{d\eta^2} + 2\eta\frac{df}{d\eta} = 0 \tag{10.35}$$

The importance of variables transformation cannot be overstated, for certain boundary conditions, since very useful solutions can be obtained to otherwise intractable problems. The above particular solution is suitable for the case of dynamic diffusion into an unbounded domain, subject to a step change at the origin (taking y to be the time variable and x to be the unbounded length variable).

10.1.1   Classification and Characteristics of Linear Equations

The general linear equation of second order can be expressed

$$P\frac{\partial^2 z}{\partial x^2} + Q\frac{\partial^2 z}{\partial x\partial y} + R\frac{\partial^2 z}{\partial y^2} = S \tag{10.36}$$

where the coefficients P, Q, R depend only on x, y, whereas S depends on $x, y, z, \partial z/\partial x, \partial z/\partial y$. The terms involving second derivatives are of special importance, since they provide the basis for classification of type of PDE. By analogy with the nomenclature used to describe conic sections (planes passed through a cone) written as

$$ax^2 + 2bxy + cy^2 = d \tag{10.37}$$

then we can classify Eq. 10.36 for constant coefficients when P, Q, R take values a, b, c, respectively. The discriminant for the case of constant coefficients a, b, c is defined

$$\Delta = b^2 - 4ac \tag{10.38}$$

so that when

$$\Delta < 0: \quad \text{Elliptic equation}$$
$$\Delta = 0: \quad \text{Parabolic equation} \tag{10.39}$$
$$\Delta > 0: \quad \text{Hyperbolic equation}$$

Typical examples occurring in chemical engineering are, Fick's second law of diffusion

$$\frac{\partial C}{\partial t} = D\frac{\partial^2 C}{\partial x^2} \tag{10.40}$$

which is parabolic, since $a = D$, $b = 0$, and $c = 0$, and Newton's law for wave motion

$$\frac{\partial^2 u}{\partial t^2} = \rho \frac{\partial^2 u}{\partial y^2} \qquad (10.41)$$

which is hyperbolic, since $a = 1$, $b = 0$, and $c = -\rho$, and finally Laplace's equation for heat conduction

$$\frac{\partial^2 T}{\partial x^2} + \frac{\partial^2 T}{\partial y^2} = 0 \qquad (10.42)$$

which is elliptic, since $a = 1$, $b = 0$, and $c = 1$.

The *homogeneous* linear equation of second order occurs frequently. It is the special form when $S = 0$. In particular, when P, Q, R are constants, denoted by a, b, c, respectively, we write

$$a \frac{\partial^2 z}{\partial x^2} + b \frac{\partial^2 z}{\partial x \partial y} + c \frac{\partial^2 z}{\partial y^2} = 0 \qquad (10.43)$$

which contains only second derivatives and is called a homogeneous equation. The discriminant is $(b^2 - 4ac)$. A solution to the linear homogeneous equation can be proposed as

$$z = f(y + \lambda x) \qquad (10.44)$$

This implies the variable change

$$\eta = y + \lambda x \qquad (10.45)$$

so from the previous lessons, we see

$$\frac{\partial^2 z}{\partial x^2} = \lambda^2 f''(\eta); \quad \frac{\partial^2 z}{\partial x \partial y} = \lambda f''(\eta); \quad \frac{\partial^2 z}{\partial y^2} = f''(\eta) \qquad (10.46)$$

Inserting these into Eq. 10.43 shows

$$a\lambda^2 + b\lambda + c = 0 \qquad (10.47)$$

which yields two characteristic values λ_1, λ_2. Now, since the original equation was linear, then superposition requires (as in ODEs) the general solution

$$z = f(y + \lambda_1 x) + g(y + \lambda_2 x) \qquad (10.48)$$

In fact, we see the discriminant $(b^2 - 4ac)$ appears within the quadratic equation for λ

$$\lambda = \frac{-b \pm \sqrt{b^2 - 4ac}}{2a} \qquad (10.49)$$

Whereas the functions f and g are unspecified and quite arbitrary at this point, nonetheless, much can be learned by observation of the arguments, now established to be $(y + \lambda x)$. Now, suppose $a = 1$, $b = -3$, and $c = 2$, hence $(b^2 - 4ac) = 1 > 0$, so the equation is hyperbolic with roots $\lambda_1 = 1, \lambda_2 = 2$; hence, in general we can write

$$z = f(y + x) + g(y + 2x) \tag{10.50}$$

Moreover, if $a = 0$ and $b \neq 0$, only one root obtains

$$z = f\left(y - \frac{b}{c}x\right) + g(x) \tag{10.51}$$

and the second solution, g, depends only on x, since in general $g(x)$ also satisfies the original equation. Finally, if both $a = b = 0$, then it is clear by successive integrations of $\partial^2 z / \partial y^2$ that

$$z = f(x) + yg(x) \tag{10.52}$$

which is linear with respect to y. To see this, first integrate partially with respect to y

$$\frac{\partial z}{\partial y} = g(x)$$

Integrate again to find $z = f(x) + yg(x)$.

Another exceptional case occurs when repeated roots arise. This can be treated in a manner analogous to the case for repeated roots in the Method of Frobenius (CASE II). Thus, the second solution is obtained by taking the limit

$$\frac{\partial}{\partial \lambda} f(y + \lambda x) \Big|_{\lambda = \lambda_1} = xf'(y + \lambda_1 x) = xg(x + \lambda_1 x) \tag{10.53}$$

so for repeated roots $\lambda = \lambda_1$ we have

$$z = f(y + \lambda_1 x) + xg(x + \lambda_1 x) \tag{10.54}$$

and it is clear that the two solutions are linearly independent.

Now, for elliptic equations such that $(b^2 - 4ac) < 0$, complex conjugate roots occur, so for this case, taking Laplace's equation as example

$$\frac{\partial^2 z}{\partial x^2} + \frac{\partial^2 z}{\partial y^2} = 0 \tag{10.55}$$

we have

$$z = f(y + ix) + g(y - ix) \tag{10.56}$$

And for the hyperbolic equation, such that $(b^2 - 4ac) > 0$, we write the wave equation as an example

$$\frac{\partial^2 u}{\partial t^2} - \rho \frac{\partial^2 u}{\partial y^2} = 0 \tag{10.57}$$

with general solution

$$u = f\left(y + \sqrt{\rho}\, t\right) + g\left(y - \sqrt{\rho}\, t\right) \tag{10.58}$$

An example of parabolic equations with $(b^2 - ac) = 0$ is

$$\frac{\partial^2 z}{\partial x^2} - 2\frac{\partial^2 z}{\partial x \partial y} + \frac{\partial^2 z}{\partial y^2} = 0 \tag{10.59}$$

so that a repeated root $\lambda = 1$ occurs. The general solution is

$$z = f(x + y) + xg(x + y) \tag{10.60}$$

which is made up of linearly independent solutions.

Although we have derived general solutions, nothing has been said about the specific functional form taken by f and g, since these are completely arbitrary, other than the requirement that they must be continuous, with continuous derivatives. For example, considering the parabolic equation given in (10.59), the specific function $f(x + y) = c \sinh(x + y)$ is seen, by substitution, to satisfy the equation. The discussion in this section has determined only the arguments, not specific functions.

10.2   PARTICULAR SOLUTIONS FOR PDEs

In the previous sections, we learned that certain general, nonspecific solutions evolved for linear, partial differential equations (PDEs), but these were of little value on a computational, applied engineering level. However, we did learn that a subtle change of variables could often lead to significant simplification. In fact, the overall strategy in solving a PDE is usually to reduce the PDE to an ODE, or sets of ODEs. This strategy can only be accomplished for certain particular boundary conditions. Thus, the solution methods discussed next are termed particular solutions, or collections of particular solutions. They are suited only for very specific *particular* boundary (or initial) conditions, and lead to well-known, specific functions in mathematical physics. This is certainly more satisfying to an engineer than writing a solution as $f(x + 2y)$.

Thus, the remainder of this chapter deals with solving linear, homogeneous equations, using the following three approaches to reduce PDEs to ODEs.

1. *Combination of Variables* Sometimes called a similarity transform, this technique seeks to combine all *independent* variables into one variable, which may then produce a single ODE. It is applicable only to cases where

variables are unbounded e.g., $0 < t < \infty, 0 < x < \infty$. It is suitable for only one type of boundary or initial condition and produces a single particular solution.

2. *Separation of Variables* This is the most widely used method in applied mathematics, and its strategy is to break the *dependent* variable into component parts, each depending (usually) on a single independent variable; invariably, it leads to a multiple of particular solutions.

3. *Laplace Transforms* The Laplace transform is an integral technique, which basically integrates time out of the equation, thereby reducing a PDE to an ODE.

There are other methods to treat single and simultaneous PDE, most especially the Finite Transform methods, which are treated generally in Chapter 11. The Laplace transform is especially useful in this regard, as we saw in Chapter 9, since it easily copes with simultaneous equations.

It is clear from these remarks that no general solution is possible, and the methods that have evolved are closely tied to specific boundary conditions. We discuss boundary and initial conditions next.

10.2.1   Boundary and Initial Conditions

On considering boundary conditions, it must be borne in mind that at least two independent variables exist in attempting to solve partial differential equations. We reflect on the derivations for Model II in Chapter 1 and note that principally three types of boundary or initial conditions arise.

1. Function specified at boundary; e.g., for $T(r, z)$

$$T = f(r) \text{ at } z = 0; \quad T = g(z) \quad \text{at} \quad r = R$$

2. Derivative of function specified at boundary

$$\frac{\partial T}{\partial r} = 0 \quad \text{at} \quad r = 0; \quad -k\frac{\partial T}{\partial r} = Q \quad \text{at} \quad r = R$$

3. Mixed functions at boundary

$$-k\frac{\partial T}{\partial r} = h(T - T_W) \quad \text{at} \quad r = R, \quad \text{for all} \quad z$$

The first type includes initial conditions, which may be written for $T(t, x)$

$$T(0, x) = f(x) \tag{10.61}$$

This means at time zero, temperature T is distributed according to the function $f(x)$, which does not exclude a simple constant such as $f(x) = T_i$. At some boundary, time variation may also occur, so

$$T(t, 0) = g(t) \tag{10.62}$$

Neither of these type (1) conditions are homogeneous. However, if the boundary value is a fixed constant, such as

$$T(t,0) = T_0$$

then we could define a new variable $\theta = T - T_0$ and see

$$\theta(t,0) = 0 \qquad\qquad (\mathbf{10.63})$$

which is indeed homogeneous at the boundary.

For type (2) conditions, the classical symmetry condition nearly always arises in cylindrical and spherical coordinate systems, along the centerline

$$\frac{\partial T}{\partial r} = 0 \qquad \text{at} \qquad r = R \qquad\qquad (\mathbf{10.64})$$

For such coordinate systems, r is always positive, so to insure symmetrical profiles of T, we must invoke Eq. 10.64. This condition can also arise at a finite position, for example, at an insulated tube wall ($r = R$) where conduction flux tends to zero, so that

$$-k\frac{\partial T}{\partial r} = 0 \qquad \text{at} \qquad r = R \qquad\qquad (\mathbf{10.65})$$

For electrically heated tube walls, the heat entering is uniform and constant, so that we would write at a tube boundary

$$-k\frac{\partial T}{\partial r} = Q \qquad \text{at} \qquad r = R \qquad\qquad (\mathbf{10.66})$$

Since flux is a vector quantity, it is clear that $\partial T/\partial r > 0$ near the wall, hence Q is a negative quantity, as it should be since it moves in the anti-r direction. It is not possible to convert the electrically heated wall condition to a homogeneous boundary condition for T. However, the insulated wall (Eq. 10.65) is homogeneous without further modification.

The mixed type (3) includes both a function and its derivative (or its integral). For example, we showed in Chapter 1 the homogeneous type (3) condition

$$-k\frac{\partial T}{\partial r} = h(T - T_W) \qquad \text{at} \qquad r = R \qquad\qquad (\mathbf{10.67})$$

which simply says the conduction flux is exactly balanced by heat transfer at the wall. By replacing $\theta = T - T_w$ (provided T_w is constant), we see this condition is also homogeneous

$$-k\frac{\partial \theta}{\partial r} = h\theta \qquad \text{at} \qquad r = R \qquad\qquad (\mathbf{10.68})$$

This is a powerful boundary condition, since it actually contains both type (1) and type (2) as subsets. Thus, for large values of h (high velocity) then $h \to \infty$, and Eq. 10.68 reduces to $\theta(z, R) = 0$, which is the homogeneous type (1) condition. On the other hand, if h gets very small, then in the limit when h approaches 0, we obtain type (2), specifically Eq. 10.65, which is an insulated boundary.

Occasionally, a mixed condition arises as an integro-differential balance. Suppose a stirred pot of solvent is used to extract oil from (assumed) spherically shaped, porous seeds. The mass rate from the boundary of the many seeds (m in number) is computed from Fick's law, where $c(r, t)$ denotes composition within the seeds:

$$N = -m(4\pi R^2)D\frac{\partial c(R, t)}{\partial r}$$

and the accumulation in the solvent is $V dC/dt$, so that the oil conservation balance for the solvent phase is:

$$V\frac{dC}{dt} = -m(4\pi R^2)D\frac{\partial c(R, t)}{\partial r}$$

where $C(t)$ is the oil composition in the stirred solvent. If the initial solvent contains no oil, then integration yields the integro boundary condition:

$$VC(t) = -m\int_0^t 4\pi R^2 D\frac{\partial c(R, t)}{\partial r}\, dt \tag{10.69}$$

Such boundary conditions can be handled most easily using a Laplace transform solution method, as we demonstrate later in this chapter.

The number of boundary or initial conditions required to solve an ordinary differential equation, as we saw in Chapter 2, corresponded to the number of arbitrary constants of integration generated in the course of analysis. Thus, we showed that nth order equations generated n arbitrary constants; this also implies the application of boundary or initial conditions totalling n conditions. In partial differential equations, at least two independent variables exist, so, for example, an equation describing transient temperature distribution in a long cylindrical rod of metal; that is,

$$\rho C_p\frac{\partial T}{\partial t} = k\frac{1}{r}\frac{\partial}{\partial r}\left(r\frac{\partial T}{\partial r}\right) = k\left(\frac{\partial^2 T}{\partial r^2} + \frac{1}{r}\frac{\partial T}{\partial r}\right) \tag{10.70}$$

will require at least one condition for time (usually an initial condition) and two for fixed spatial positions (say, $r = 0, r = R$). As a rule, then, we usually need one fixed condition for each order of each partial derivative. This is not always the case; for example, an initial condition is not necessary if the time-periodic solution is sought; for example,

$$T(t, r) = f(r)e^{i\omega t} \tag{10.71}$$

For such cases, only conditions at the boundaries ($r = 0, r = R$) are needed. Similar arguments apply when dealing with the angular position in cylindrical or spherical coordinates, namely, that solution must be periodic in angle so that it repeats itself every 2π radians.

The range of variables can be open (unbounded) or closed (bounded). If all the independent variables are closed, we classify such equations as "boundary value problems." If only initial values are necessary, with no bounds specified then we speak of "initial value problems." In the next section, we develop the method of *combination of variables*, which is strictly applicable only to initial value problems.

10.3   COMBINATION OF VARIABLES METHOD

We introduced the idea of change of variables in Section 10.1. The coupling of variables transformation and suitable initial conditions often lead to useful particular solutions. Consider the case of an unbounded solid material with initially constant temperature T_0 in the whole domain $0 < x < \infty$. At the face of the solid, the temperature is suddenly raised to T_s (a constant). This so-called step change at the position $x = 0$ causes heat to diffuse into the solid in a wavelike fashion. For an element of this solid of cross-sectional area A, density ρ, heat capacity C_p, and conductivity k, the transient heat balance for an element Δx thick is

$$(q_x A)|_x - (q_x A)|_{x+\Delta x} = A\Delta x \rho C_p \frac{\partial T}{\partial t} \tag{10.72}$$

Dividing by $A\Delta x$, and taking limits gives

$$-\frac{\partial q_x}{\partial x} = \rho C_p \frac{\partial T}{\partial t} \tag{10.73}$$

Introducing Fourier's law as $q_x = -k\,\partial T/\partial x$ then yields

$$\alpha \frac{\partial^2 T}{\partial x^2} = \frac{\partial T}{\partial t} \tag{10.74}$$

where $\alpha = k/\rho C_p$ is thermal diffusivity (cm^2/s). It is clear that both independent variables are unbounded, that is, $0 < x < \infty$, $0 < t < \infty$. Next, we carefully list the initial conditions, and any other behavioral patterns we expect

$$T = T_0 \quad \text{at} \quad t = 0, \quad \text{all } x \tag{10.75}$$

$$T = T_s \quad \text{at} \quad x = 0, \quad \text{all } t \tag{10.76}$$

For some position within the solid matrix (say x_1), we certainly expect the temperature to rise to T_s, eventually, that is, when t approaches infinity. Moreover, for some finite time (say t_1) we certainly expect, far from the face at $x = 0$, that temperature is unchanged; that is, we expect at t_1 that $T = T_0$ as x approaches infinity. We can write these expectations as follows

$$T \rightarrow T_s, \quad t \rightarrow \infty, \quad x > 0 \tag{10.77}$$

$$T \rightarrow T_0, \quad x \rightarrow \infty, \quad t > 0 \tag{10.78}$$

There appears to be some symmetry for conditions at zero and infinity, and this is the first prerequisite for a combination of variables approach to be applicable. Now, the combination of variables method is often called a "similarity transform." This nomenclature arises from the manner we use to select a combined variable. Let us next write a rough approximation to Eq. 10.74, to an order of magnitude

$$\alpha \frac{\Delta T}{x^2} \sim \frac{\Delta T}{t} \tag{10.79}$$

This suggests roughly $\alpha t \sim x^2$, which we interpret to mean the change in αt is similar in size to a change in x^2. Or, equivalently, a change in $(\alpha t)^{1/2}$ is similar to a change in x. On viewing the symmetry of our boundary conditions at zero and infinity, these qualitative arguments suggest the combined variable

$$\eta_0 = \frac{x}{\sqrt{\alpha t}} \tag{10.80}$$

which is dimensionless. Comparing this with conditions listed in Eqs. 10.75 to 10.78 we see

$$T = T_0, \quad \text{when either} \quad t = 0 \quad \text{or} \quad x \to \infty, \quad \text{so} \quad \eta_0 \to \infty \tag{10.81}$$

$$T = T_s, \quad \text{when either} \quad t \to \infty, \quad \text{or} \quad x = 0, \quad \text{so} \quad \eta_0 \to 0 \tag{10.82}$$

An alternative way to set the combined variable is to write η in terms of penetration distance $\delta(t)$, which represents the distance from the surface when the first temperature rise (of arbitrary size) occurs.

$$\eta = \frac{x}{\delta(t)} \tag{10.83}$$

Now, to use this variable change, we write the hypothesis

$$T(x, t) = f(\eta) \tag{10.84}$$

The hypothesis fails if any variable other than η appears in the final equation. We next transform variables, using the procedure outlined in Section 10.1. It is worth repeating here; Eq. 10.84 implies equality of total differentials

$$dT(x, t) = df(\eta) \tag{10.85}$$

We now apply the chain rule to both sides

$$\frac{\partial T}{\partial x} dx + \frac{\partial T}{\partial t} dt = \frac{df}{d\eta} d\eta \tag{10.86}$$

Now taking the total differentiation of $\eta(x, t)$, defined in Eq. 10.83, we get

$$d\eta = \frac{\partial \eta}{\partial x} dx + \frac{\partial \eta}{\partial t} dt \tag{10.87}$$

Inserting this into Eq. 10.86 gives the necessary comparative information

$$\frac{\partial T}{\partial x} dx + \frac{\partial T}{\partial t} dt = f'(\eta) \left[\frac{\partial \eta}{\partial x} dx + \frac{\partial \eta}{\partial t} dt \right] \tag{10.88}$$

Equating like coefficients from both sides gives

$$\frac{\partial T}{\partial x} = f'(\eta) \frac{\partial \eta}{\partial x} = f'(\eta) \frac{1}{\delta(t)} \tag{10.89}$$

$$\frac{\partial T}{\partial t} = f'(\eta) \frac{\partial \eta}{\partial t} = f'(\eta) \left(-\frac{\eta}{\delta} \frac{d\delta}{dt} \right) \tag{10.90}$$

We now need to find $\partial^2 T/\partial x^2$. To do this, we represent the RHS of Eq. 10.89 as $\psi(\eta, t)$ and the LHS as $F(x, t)$; thus, we have taken

$$F(x, t) = \frac{\partial T}{\partial x} \tag{10.91}$$

$$\psi(\eta, t) = f'(\eta) \frac{1}{\delta(t)} \tag{10.92}$$

Now, equating total differentials of F and ψ as the next stage

$$dF(x, t) = d\psi(\eta, t) \tag{10.93}$$

we again apply the chain rule

$$\frac{\partial F}{\partial x} dx + \frac{\partial F}{\partial t} dt = \frac{\partial \psi}{\partial \eta} d\eta + \frac{\partial \psi}{\partial t} dt \tag{10.94}$$

Here, we take *careful note* that the first term on the RHS is the partial with respect to η! Many serious mistakes are made by not observing this fact.

The sought-after term can be obtained again by replacing $d\eta$ from Eq. 10.87 and equating like coefficients of dx and dt

$$\frac{\partial F}{\partial x} = \frac{\partial \psi}{\partial \eta} \frac{\partial \eta}{\partial x} \tag{10.95}$$

As an aside, we also obtained the (redundant) information

$$\frac{\partial F}{\partial t} = \frac{\partial \psi}{\partial \eta} \frac{\partial \eta}{\partial t} + \frac{\partial \psi}{\partial t} \tag{10.96}$$

Replacing F and ψ from Eqs. 10.91 and 10.92 gives

$$\frac{\partial^2 T}{\partial x^2} = f''(\eta)\frac{1}{\delta(t)}\frac{\partial \eta}{\partial x} = f''(\eta)\frac{1}{\delta^2(t)} \tag{10.97}$$

Inserting the results from Eq. 10.97 and 10.90 into the original equation gives finally the condition

$$\alpha\frac{f''}{\delta^2} = f'\left(-\eta\frac{1}{\delta}\frac{d\delta}{dt}\right) \tag{10.98a}$$

As we show next, it is convenient to set

$$\frac{\delta}{\alpha}\frac{d\delta}{dt} = 2 \tag{10.98b}$$

and since $\delta(0) = 0$, we find

$$\delta(t) = \sqrt{4\alpha t} \tag{10.98c}$$

which thereby justifies the original hypothesis; that is $T(x, t) = f(\eta)$ and now $\eta = x/(4\alpha t)^{1/2}$. This expression can be integrated twice, by first letting $p = df/d\eta$

$$p' + 2\eta p = 0 \tag{10.99}$$

yielding

$$p = A\exp(-\eta^2) = \frac{df}{d\eta} \tag{10.100}$$

The reason for selecting Eq. 10.98b is now obvious. A perfectly acceptable solution is obtained by integrating again

$$f(\eta) = A\int \exp(-\eta^2)d\eta + B \tag{10.101}$$

However, it would be useful to express our result in terms of known tabulated functions. In Chapter 4, we showed that the definite integral, the error function, takes a form similar to the above

$$\text{erf}(\eta) = \frac{2}{\sqrt{\pi}}\int_0^\eta \exp(-\beta^2)d\beta \tag{10.102}$$

Now, since the constants A, B are completely arbitrary, we can write $f(\eta)$ as

$$f(\eta) = C\,\text{erf}(\eta) + B \tag{10.103}$$

The conditions required to find the arbitrary constants C, B are simply

$$f(0) = T_s \text{ and } f(\infty) = T_0 \qquad (\textbf{10.104})$$

hence $B = T_s$ and $C = (T_0 - T_s)$, since $\text{erf}(\infty) = 1$, by definition (that is, the error function has been normalized using the factor $2/\sqrt{\pi}$ to insure $\text{erf}(\infty) = 1$). Written in terms of T and (x, t) we finally have

$$T(x, t) = (T_0 - T_s)\text{erf}\left(\frac{x}{\sqrt{4\alpha t}}\right) + T_s \qquad (\textbf{10.105})$$

and the reader can easily verify that the four original conditions (Eqs. 10.75–10.78) are satisfied. A plot of $(T - T_s)/(T_0 - T_s)$ versus $x/\sqrt{(4\alpha t)}$ is shown in Fig. 10.1.

If we change the original boundary conditions, for example, by requiring a constant heat flux at the boundary $x = 0$, the above solution is invalid. However, if we differentiated Eq. 10.74 with respect to x, multiplied by $-k$, then introduce the vector $q_x = -k\partial T/\partial x$ we see

$$\alpha\frac{\partial^2}{\partial x^2}(q_x) = \frac{\partial}{\partial t}(q_x) \qquad (\textbf{10.106})$$

This yields an identical equation to the previous one, but the unknown

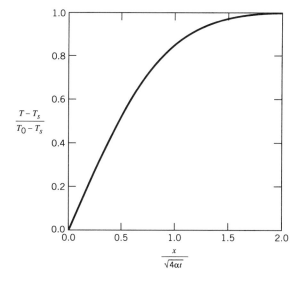

Figure **10.1** Plot of $(T - T_s)/(T_0 - T_s)$ versus $x/\sqrt{4\alpha t}$.

dependent variable is q_x instead of T. Now, for the new boundary conditions, we may write

$$q_x = 0 \quad \text{at} \quad t = 0, \quad x > 0 \tag{10.107}$$

$$q_x = q_s \quad \text{at} \quad x = 0, \quad t > 0 \tag{10.108}$$

$$q_z \to q_s, \quad t \to \infty, \quad x > 0 \tag{10.109}$$

$$q_x \to 0, \quad x \to \infty, \quad t > 0 \tag{10.110}$$

We see these conditions are identical to Eqs. 10.75–10.78 except the symbols for variables are changed, so we can write immediately

$$q_x(x,t) = q_s - q_s \, \text{erf}\left(\frac{x}{\sqrt{4\alpha t}}\right) \tag{10.111}$$

This is written more conveniently in terms of the complementary error function, $\text{erfc}(\eta) = 1 - \text{erf}(\eta)$; hence,

$$q_x(x,t) = q_s \, \text{erfc}\left(\frac{x}{\sqrt{4\alpha t}}\right) \tag{10.112}$$

To find temperature profiles for the constant flux case, simply replace $q_x = -k\partial T/\partial x$ and integrate partially with respect to x. To do this, we need to call upon material in Chapter 4. If the temperature far from the heated surface is denoted by $T_0 = T(\infty, t)$, then partial integration yields

$$T(x,t) - T_0 = \left(-\frac{q_s}{k}\right)\int_\infty^x \text{erfc}\left(\frac{x}{\sqrt{4\alpha t}}\right) dx \tag{10.113}$$

Since the partial integration implies holding t constant, we can rewrite this integral in terms of η

$$T(x,t) - T_0 = \left(-\frac{q_s}{k}\right)\sqrt{4\alpha t}\int_\infty^{\frac{x}{\sqrt{4\alpha t}}} \text{erfc}(\eta)\, d\eta \tag{10.114}$$

The integral

$$\int \text{erf}(\eta)\, d\eta$$

is given in Chapter 4, so we have

$$T(x,t) - T_0 = \left(-\frac{q_s}{k}\right)\sqrt{4\alpha t}\left[\eta(1 - \text{erf}(\eta)) - \frac{1}{\sqrt{\pi}}\exp(-\eta^2)\right]_\infty^{\frac{x}{\sqrt{4\alpha t}}} \tag{10.115}$$

The lower limit on the first term is troublesome, but can be shown to zero, since for large arguments (Chapter 4)

$$\operatorname{erf}(x) \sim 1 - \frac{a_1}{1 + px} \exp(-x^2) \qquad (10.116)$$

where $a_1, p > 0$. Thus, in the limit

$$\lim x(1 - \operatorname{erf}(x)) = \lim \frac{a_1 x}{1 + px} \exp(-x^2) \to 0 \qquad (10.117)$$

Hence, the temperature distribution for constant flux at $x = 0$ becomes

$$T(x, t) = T_0 + \frac{q_s}{k} \sqrt{\frac{4\alpha t}{\pi}} \exp\left(-\frac{x^2}{4\alpha t}\right) - \frac{q_s}{k} x \left[1 - \operatorname{erf}\left(\frac{x}{\sqrt{4\alpha t}}\right)\right] \quad (10.118)$$

In particular, the temperature rise at the interface $x = 0$ is important

$$T(0, t) = T_0 + \frac{q_s}{k} \sqrt{\frac{4\alpha t}{\pi}} \qquad (10.119)$$

The particular solution $(C \operatorname{erf}(\eta) + B)$ derived here has many applications, and arises in innumerable transport and physics problems.

EXAMPLE 10.2   *LAMINAR FLOW CVD REACTOR*

Chemical vapor deposition (CVD) has become an important technique to grow electronically active layers for all kinds of solid-state devices. Thus, the active metal organic (MO) vapor is swept into a two-dimensional slit reactor by a carrier gas, and deposition occurs at the hot top and bottom plates as shown in Fig. 10.2.

The reaction at the plate surfaces can be written

$$MO \to M + O$$

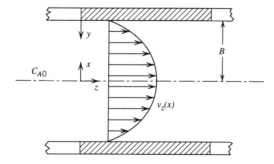

Figure 10.2 Schematic diagram of CVD reactor.

Develop an expression to compute rate of loss of MO for diffusion limited, laminar flow conditions.

Under laminar flow conditions, the velocity profile can be taken to be fully developed

$$v_z = v_{max}\left[1 - \left(\frac{x}{B}\right)^2\right] \tag{10.120}$$

where

$$v_{max} = \frac{\Delta pB^2}{2\mu L} = \frac{3}{2}v_0$$

and

L = reactor length
B = half-width of reactor
μ = gas viscosity
Δp = pressure drop
v_0 = average velocity

Denoting the metal organic as species A, the material balance, ignoring axial diffusion, can be shown to be

$$v_{max}\left[1 - \left(\frac{x}{B}\right)^2\right]\frac{\partial C_A}{\partial z} = D_A\frac{\partial^2 C_A}{\partial x^2} \tag{10.121}$$

where D_A denotes molecular diffusion of species A.

Suitable boundary conditions are

$$\frac{\partial C_A}{\partial x} = 0 \quad \text{at} \quad x = 0 \quad \text{(symmetry)} \tag{10.122}$$

$$-D_A\frac{\partial C_A}{\partial x} = kC_A \quad \text{at} \quad x = \pm B \tag{10.123}$$

$$C_A = C_A^0 \quad \text{at} \quad z = 0 \tag{10.124}$$

The second boundary condition indicates diffusive flux is exactly balanced by rate of decomposition at the wall. Under the stipulated conditions of "diffusion-limited," the implication is that $k \to \infty$, so that for the present analysis

$$\lim_{k \to \infty}\left[-\frac{D_A}{k}\frac{\partial C_A}{\partial x}\right] = 0 = C_A \quad \text{at} \quad x = \pm B \tag{10.125}$$

If such reactors are quite short, so that contact time is small, most of the depletion of A occurs very near the wall, whereas the core composition hardly changes. Thus, the principal impact of the velocity profile occurs near the wall. This suggests we introduce a coordinate system (y) emanating from the wall and since, for half the symmetrical reactor, we have:

$$x + y = B \qquad (10.126)$$

then velocity in terms of y is

$$v_z = v_{max}\left[2\left(\frac{y}{B}\right) - \left(\frac{y}{B}\right)^2\right] \qquad (10.127)$$

This can be approximated to a linear form near the wall since in that vicinity $y/B \ll 1$; hence,

$$v_z \simeq 2v_{max}\left(\frac{y}{B}\right) \qquad (10.128)$$

Thus, replacing x with y from Eq. 10.126 into the diffusion term, we have the approximate expression (after Leveque 1928):

$$2v_{max}\left(\frac{y}{B}\right)\frac{\partial C_A}{\partial z} = D_A \frac{\partial^2 C_A}{\partial y^2} \qquad (10.129)$$

subject to the diffusion-limited wall condition

$$C_A = 0 \quad \text{at} \quad y = 0, \quad z > 0 \qquad (10.130)$$

and the entrance condition

$$C_A = C_{A0} \quad \text{at} \quad z = 0, \quad y > 0 \qquad (10.131)$$

For very thin layers near the wall, the bulk flow appears far away, so we write for short penetration

$$C_A \to C_{A0}, y \to \infty \qquad (10.132)$$

On the other hand, if the reactors were very long ($z \to \infty$), we would expect all of the metal organic component to be depleted, so that

$$C_A \to 0, z \to \infty \qquad (10.133)$$

The symmetry of the boundary conditions suggests a combination of variables approach. It is clear that similarity must exist as before, since

$$2v_{max} y^3 \sim BD_A z$$

so that our first guess for a combined (dimensionless) variable is

$$\eta_0 = \frac{y}{\left(\dfrac{BD_A z}{2v_{max}}\right)^{1/3}} \qquad (10.134)$$

The exact combined variable can be found as before by writing $\eta = y/\delta(z)$, which leads to the convenient form

$$\eta = \frac{y}{\left(\dfrac{9BD_A z}{2v_{max}}\right)^{1/3}} \tag{10.135}$$

It is easy to see that only two values of η will satisfy the four conditions, Eq. 10.130–10.133; thus, taking $C_A = f(\eta)$

$$f = C_{A0} \quad \text{at} \quad \eta = \infty \qquad (z = 0 \quad \text{or} \quad y = \infty) \tag{10.136}$$

$$f = 0 \quad \text{at} \quad \eta = 0 \qquad (z = \infty \quad \text{or} \quad y = 0) \tag{10.137}$$

We have thus satisfied the conditions necessary to undertake a combined variable solution. Following the same procedure as in the thermal diffusion problem, we see

$$\frac{\partial C_A}{\partial z} = \frac{\partial f}{\partial \eta}\frac{\partial \eta}{\partial z} = f'(\eta)\left[-\frac{1}{3}\eta\frac{1}{z}\right] \tag{10.138}$$

$$\frac{\partial C_A}{\partial y} = \frac{\partial f}{\partial \eta}\frac{\partial \eta}{\partial y} = f'(\eta)\left[\frac{1}{\left(\dfrac{9BD_A z}{2v_{max}}\right)^{1/3}}\right] \tag{10.139}$$

$$\frac{\partial^2 C_A}{\partial y^2} = \frac{\partial}{\partial \eta}\left[f'(\eta)\frac{1}{\left(\dfrac{9BD_A z}{2v_{max}}\right)^{1/3}}\right]\frac{\partial \eta}{\partial y} \tag{10.140}$$

$$\frac{\partial^2 C_A}{\partial y^2} = f''(\eta)\frac{1}{\left(\dfrac{9BD_A z}{2v_{max}}\right)^{2/3}}$$

Inserting these into the defining equation shows

$$2v_{max}\left(\frac{y}{B}\right)\left[-f'(\eta)\frac{1}{3}\eta\frac{1}{z}\right] = D_A f''(\eta)\frac{1}{\left(\dfrac{9BD_A z}{2v_{max}}\right)^{2/3}} \tag{10.141}$$

Collecting and combining terms, we find

$$f''(\eta) + 3\eta^2 f'(\eta) = 0 \tag{10.142}$$

Integrating once by replacing $p = f'$

$$f'(\eta) = A\exp(-\eta^3) = \frac{df}{d\eta} \tag{10.143}$$

Integrating again gives

$$f(\eta) = A \int \exp(-\eta^3)\, d\eta + B \qquad (10.144)$$

It is convenient to introduce definite limits

$$f(\eta) = D \int_0^\eta \exp(-\beta^3)\, d\beta + B \qquad (10.145)$$

Evaluating D, B using Eqs. 10.136 and 10.137 shows

$$B = 0,\, D = \frac{C_{A0}}{\displaystyle\int_0^\infty \exp(-\beta^3)\, d\beta} \qquad (10.146)$$

As we showed in Chapter 4, the integral can be found in terms of the tabulated Gamma function

$$\Gamma(x) = \int_0^\infty t^{x-1} e^{-t}\, dt \qquad (10.147)$$

by letting $\beta^3 = t, 3\beta^2 d\beta = dt$; hence,

$$\int_0^\infty \exp(-\beta^3)\, d\beta = \frac{1}{3}\Gamma\left(\frac{1}{3}\right) = \Gamma\left(\frac{4}{3}\right) \qquad (10.148)$$

This allows our final result to be written

$$\frac{C_A}{C_{A0}} = \frac{\displaystyle\int_0^\eta \exp(-\beta^3)\, d\beta}{\Gamma\left(\dfrac{4}{3}\right)} \qquad (10.149)$$

This will be valid only for short contact conditions up to $\eta \sim 1$.

The mass flux at the wall is the important quantity, and this can be computed from

$$N_0(z) = D_A\left[\frac{\partial C_A}{\partial y}\right]_{y=0} = D_A\left[\frac{df}{d\eta}\right]_{\eta=0}\frac{\partial \eta}{\partial y} \qquad (10.150)$$

so the local flux is

$$N_0(z) = \frac{D_A C_{A0}}{\Gamma\left(\dfrac{4}{3}\right)}\left[\frac{2v_{max}}{9BD_A z}\right]^{1/3} \qquad (10.151)$$

The average flux rate can be computed from

$$\overline{N_0} = \frac{1}{L}\int_0^L N_0(z)\,dz \qquad (10.152)$$

which yields finally, for a single plate

$$\overline{N_0} = \frac{\frac{3}{2}D_A}{\Gamma\left(\frac{4}{3}\right)}C_{A0}\left[\frac{2v_{max}}{9BD_A L}\right]^{1/3} \qquad (10.153)$$

For two active plates, the total surface area for width W is $2(WL)$ so the net loss rate (moles/sec) of metal organic (A) is

$$R = 2(WL)\overline{N_0}$$

hence,

$$R = \frac{3}{\Gamma\left(\frac{4}{3}\right)}WC_{A0}\left[\frac{2}{9}\frac{v_{max}L^2 D_A^2}{B}\right]^{1/3} \qquad (10.154)$$

It is usual to express rates in terms of average velocity, which is $v_0 = 2v_{max}/3$. Combining the numerical factors gives in terms of v_0,

$$R = \frac{3^{2/3}}{\Gamma\left(\frac{4}{3}\right)}WC_{A0}\left[\frac{v_0 L^2 D_A^2}{B}\right]^{1/3} \qquad (10.155)$$

where $3^{2/3}/\Gamma(4/3) = 2.33$. This type of analysis was used by Van De Ven et al. (1986) to predict CVD behavior.

10.4 SEPARATION OF VARIABLES METHOD

We have seen in the previous section that the combination of variables (similarity transform) is applicable only to initial value problems, with unbounded independent variables. Nonetheless, we gave an example where such an approach could be applied under conditions of short contact time (which is the same as short penetration for the length variable). Thus, we pretended that a boundary was infinite in extent, when in fact it was physically finite. For large contact time (deep penetration) such an approach breaks down, and the solution becomes invalid.

To cope with finite boundaries, we next introduce the separation of variables method. This method is thus capable of attacking boundary value problems and leads invariably to sets of ODE (rather than a single one). Since only linear equations are amenable to this method, we can obviously invoke the principle of superposition. This means that solutions of sets of ODE can be summed to produce a complete solution. These sets of equations will be shown to be countably infinite, so that the problem of finding an infinite set of arbitrary constants of integration arises. This problem will finally be resolved, for homogeneous boundary conditions, by invoking a condition called *orthogonality*. The properties of orthogonal functions will allow the arbitrary constants to be computed one at a time, and is the singularly most important property to finally close up the complete solution. Thus, we reserve a separate section to study orthogonality and the famous Sturm–Liouville conditions.

We illustrate the method by a practical chemical engineering example in the next section. We shall proceed as far along on this problem as possible until new information is needed to fully resolve a complete solution.

10.4.1 Coated Wall Reactor

Consider a hollow, tubular reactor (Fig. 10.3), the inside walls of which are coated with a catalyst (or the tube itself could be constructed of a catalyst metal).

A feed stock composed of species A entering with an inert fluid passes through the reactor. At the tube wall, the irreversible catalytic reaction takes place

$$A \rightarrow \text{Products}$$

If the reactor is not too long, the velocity profile is not fully developed, so that it is taken to be plug shaped and the heterogeneous reaction at the wall ($r = R$) is just balanced by diffusion of A from the flowing phase

$$-D_A \frac{\partial C_A}{\partial r}\bigg|_R = k C_A(R, z) \tag{10.156}$$

At steady state, the balance between convection and diffusion for an annular element (ignoring axial diffusion) is

$$v_0 2\pi r \Delta r C_A|_z - v_0 2\pi r \Delta r C_A|_{z+\Delta z} + \Delta z (2\pi r J_A)|_r - \Delta z (2\pi r J_A)|_{r+\Delta r} = 0$$

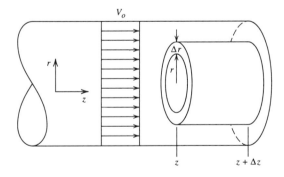

Figure 10.3 Schematic diagram of a coated wall reactor.

Dividing by $2\pi\Delta r\Delta z$, and taking limits, yields

$$-v_0 r\frac{\partial C_A}{\partial z} - \frac{\partial}{\partial r}(rJ_A) = 0$$

Inserting Fick's law, $J_A = -D_A\partial C_A/\partial r$ finally gives

$$\frac{D_A}{r}\frac{\partial}{\partial r}\left(r\frac{\partial C_A}{\partial r}\right) = v_0\frac{\partial C_A}{\partial z} \tag{10.157}$$

It is convenient to combine parameters to produce dimensionless, independent variables. This eliminates lots of clutter and reduces algebraic repetition. First, define dimensionless radius as $\xi = r/R$. When this is inserted into Eq. 10.157, the second dimensionless group becomes evident

$$\frac{D_A}{R^2}\frac{1}{\xi}\frac{\partial}{\partial \xi}\left(\xi\frac{\partial C_A}{\partial \xi}\right) = v_0\frac{\partial C_A}{\partial z}$$

Now, if we combine all remaining parameters with z, the results must be dimensionless, so let

$$\zeta = \frac{z}{v_0}\frac{D_A}{R^2} \tag{10.158}$$

This is actually the ratio of local residence time (z/v_0) divided by diffusion time, R^2/D_A, so that it is clearly dimensionless. Now the PDE is free of parameters

$$\frac{1}{\xi}\frac{\partial}{\partial \xi}\left(\xi\frac{\partial C_A}{\partial \xi}\right) = \frac{\partial C_A}{\partial \zeta} \tag{10.159}$$

Moreover, we see the variables range as, taking $C_A(r,0) = C_{A0}$ (the inlet composition)

$$0 \le \xi < 1, \quad 0 \le \zeta \le \infty, \quad 0 \le C_A \le C_{A0} \tag{10.160}$$

Our strategy is to reduce the PDE to ODEs. The method of separation of variables requires that two functions describe the solution $\varphi(\xi)$ and $Z(\zeta)$. The φ depends only on radial position ξ, and $Z(\zeta)$ depends only on axial location ζ. There are several possibilities of combining these to construct $C_A(\xi, \zeta)$. It is usual to start with a product of functions

$$C_A(\xi, \zeta) = \varphi(\xi)Z(\zeta) \tag{10.161}$$

This must satisfy Eq. 10.159 and any associated boundary/initial conditions, so insert the above into Eq. 10.159 to find

$$Z(\zeta)\frac{1}{\xi}\frac{d}{d\xi}\left(\xi\frac{d\varphi}{d\xi}\right) = \varphi(\xi)\frac{dZ}{d\zeta} \tag{10.162}$$

We can see we have already accomplished one of our original purposes, that is, we now have only ordinary differential relations. We next separate variables by dividing throughout by $Z(\zeta)\varphi(\xi)$

$$\frac{\frac{1}{\xi}\frac{d}{d\xi}\left(\xi\frac{d\varphi}{d\xi}\right)}{\varphi(\xi)} = \frac{\frac{dZ}{d\zeta}}{Z} \tag{10.163}$$

Henceforth, it is convenient to use primes to denote differentiation, since it is clear that only one independent variable belongs to each function, thus let

$$Z' = \frac{dZ}{d\zeta}, \qquad \varphi' = \frac{d\varphi}{d\xi} \tag{10.164}$$

Although the variables are now separated, we have not yet decoupled the equations so that they can be solved separately. Since one side of Eq. 10.163 cannot vary without the other's consent, it is clear that they must both equal the same (arbitrary) constant, which we denote as $-\lambda^2$. Hence,

$$\frac{\left(\varphi'' + \frac{1}{\xi}\varphi'\right)}{\varphi} = \frac{Z'}{Z} = -\lambda^2 \tag{10.165}$$

We have selected a negative constant (or zero, since $\lambda = 0$ is still a possibility) on physical grounds, since it is clear that the first order solution

$$Z = K_0 \exp(\pm\lambda^2\zeta)$$

obeys physical expectations only when $-\lambda^2$ is proposed as the separation constant; that is, species A disappears for an infinitely long reactor, so that $C_A(r, \infty) \rightarrow 0$. We now have two separated equations

$$\varphi'' + \frac{1}{\xi}\varphi' + \lambda^2\varphi = 0 \tag{10.166}$$

$$Z' + \lambda^2 Z = 0 \tag{10.167}$$

By comparison with Bessel's equation in Section 3.5, we see that the solution to Eq. 10.166 is

$$\varphi(\xi) = AJ_0(\lambda\xi) + BY_0(\lambda\xi) \tag{10.168}$$

and of course, the Z solution is

$$Z(\zeta) = K_0 \exp(-\lambda^2\zeta) \tag{10.169}$$

We need to place the boundary/initial conditions in terms of φ and Z. Now, along the center line of the tube, we must have symmetry, so the homogeneous condition there is

$$\frac{\partial C_A}{\partial r} = \frac{\partial C_A}{\partial \xi} = 0 \quad \text{at} \quad \xi = 0 \qquad (10.170)$$

Inserting the proposed solution

$$C_A = \varphi(\xi) Z(\zeta)$$

we have

$$\left[Z(\zeta) \frac{d\varphi}{d\xi} \right]_{\xi=0} = 0 \quad \therefore \varphi'(0) = 0 \qquad (10.171)$$

since at least $Z(\zeta)$ must be finite. At the tube wall, we have from Eq. 10.156

$$-D_A \frac{Z(\zeta)}{R} \frac{d\varphi}{d\xi} = kZ(\zeta)\varphi \quad \text{at} \quad \xi = 1 \qquad (10.172)$$

Cancelling $Z(\zeta)$ and combining the parameters to form a Hatta number = Ha = kR/D_A, we have the homogeneous wall condition

$$\varphi'(1) + \text{Ha} \, \varphi(1) = 0 \qquad (10.173)$$

Thus, the two boundary conditions for $\varphi(\xi)$ (Eqs. 10.171 and 10.173) are seen to be homogeneous, which we shall presently see is of paramount importance in our quest to find an analytical solution. We also note that the centerline condition, $\varphi'(0) = 0$, is formally equivalent to the requirement that $\varphi(0)$ is at least finite. This latter condition is often easier to apply in practice than the formal application of the zero derivative condition.

To cover all possibilities (other than $+\lambda^2$, which we have excluded on physical grounds), we must also inspect the case when $\lambda = 0$. The solutions to Eqs. 10.166 and 10.167 for this case are obtained by direct integration of the exact differentials

$$d\left(\xi \frac{d\varphi}{d\xi} \right) = 0 \qquad (10.174)$$

$$dZ = 0 \qquad (10.175)$$

yielding the functions

$$\varphi = A_0 \ln \xi + B_0, \quad Z = K_1 \qquad (10.176)$$

Combining all the solutions obtained thus far gives (where products of constants are given new symbols)

$$C_A = \exp(-\lambda^2 \zeta)\left[CJ_0(\lambda \xi) + DY_0(\lambda \xi) \right] + E \ln \xi + F \qquad (10.177)$$

We learned in Section (3.5) that $Y_0(0) \to -\infty$, and we know that $\ln(0) \to -\infty$, hence finiteness at $\xi = 0$ requires that we immediately set $D = E = 0$. Moreover, the physical state of the system, as argued earlier, is such that we require $C_A \to 0$ as $\zeta \to \infty$, hence we must also set $F = 0$. We finally have the compact form

$$C_A(\xi, \zeta) = C \exp(-\lambda^2 \zeta) J_0(\lambda \xi) \qquad (10.178)$$

where, exclusive of the arbitrary constant C, $Z(\zeta) = \exp(-\lambda^2 \zeta)$ and $\varphi(\xi) = J_0(\lambda \xi)$. The reader can easily verify that $\varphi'(0) = 0$.

We have two conditions remaining that have not been used

$$C_A = C_{A0}, \quad \zeta = 0; \quad \varphi'(1) + \text{Ha } \varphi(1) = 0 \qquad (10.179)$$

and we have two unknown constants, C and λ. Apply next the wall condition to see, using Eq. 3.197

$$\lambda J_1(\lambda) = \text{Ha } J_0(\lambda) \qquad (10.180)$$

There are many, countably infinite, positive values of λ that satisfy this transcendental relation, a few of which are tabulated in Table 3.3. This obviously means there are many functions

$$\varphi_n(\xi) = J_0(\lambda_n \xi)$$

which satisfy the original Eq. 10.166. The sum of all solutions is the most general result, by superposition, so we must rewrite Eq. 10.178 to reflect this

$$C_A(\xi, \zeta) = \sum_{n=1}^{\infty} C_n \exp(-\lambda_n^2 \zeta) J_0(\lambda_n \xi) \qquad (10.181)$$

where λ_1 denotes the smallest (excluding zero, which has already been counted) root to the *eigenvalue equation*

$$\lambda_n J_1(\lambda_n) = \text{Ha } J_0(\lambda_n) \qquad (10.182)$$

This relation gives the countably infinite characteristic values (eigenvalues) for the system of equations.

It remains only to find the infinite number of values for C_n, using the final condition

$$C_A(\xi, 0) = C_{A0} = \sum_{n=1}^{\infty} C_n J_0(\lambda_n \xi) \qquad (10.183)$$

This appears to be a formidable task. To accomplish it, we shall need some new tools. Under certain conditions, it may be possible to compute C_n without a trial-and-error basis. To do this, we shall need to study a class of ODE with homogeneous boundary conditions called the Sturm–Liouville equation. We shall return to the coated-wall reactor after we garner knowledge of the properties of orthogonal functions.

10.5 ORTHOGONAL FUNCTIONS AND STURM–LIOUVILLE CONDITIONS

The new tool we needed in the previous coated-wall reactor problem can be stated very simply as

$$\int_a^b r(x)\varphi_n(x)\varphi_m(x)\,dx = 0 \quad \text{if} \quad m \neq n \qquad (10.184)$$

Thus, two functions

$$\varphi_n(x) \quad \text{and} \quad \varphi_m(x), \quad \text{where } n \neq m$$

are said to be orthogonal over the interval a, b if the integral of the product of the functions with respect to a weighting function $r(x)$ over the range a, b is identically zero. This is a powerful result, but it only arises for a certain class of ODEs, called Sturm–Liouville equations. By comparing a specific equation with the general Sturm–Liouville form, the needed weighting function $r(x)$ can be deduced.

10.5.1 The Sturm–Liouville Equation

Any equation that can be put into the general form

$$\frac{d}{dx}\left[p(x)\frac{dy}{dx}\right] + [q(x) + \beta r(x)]y(x) = 0 \qquad (10.185)$$

where β is a constant, and p, q, r are continuous functions of x, is said to be of Sturm–Liouville type. For example, if we compare Eq. 10.166, rewritten as

$$\frac{d}{d\xi}\left(\xi\frac{d\varphi}{d\xi}\right) + \xi\lambda^2\varphi = 0 \qquad (10.186)$$

we would conclude that it is of the Sturm–Liouville type if we take

$$p(\xi) = \xi, \quad q(\xi) = 0, \quad r(\xi) = \xi, \quad \text{and} \quad \beta = \lambda^2$$

Thus, for any discrete set of values β_n, then corresponding solutions $y = \varphi_n$ will be obtained. Suppose for two distinct values, say β_n and β_m, we have solutions φ_n and φ_m. Each of these must satisfy Eq. 10.185, so write

$$\frac{d}{dx}\left[p(x)\frac{d\varphi_n}{dx}\right] + [q(x) + \beta_n r(x)]\varphi_n = 0 \qquad (10.187)$$

$$\frac{d}{dx}\left[p(x)\frac{d\varphi_m}{dx}\right] + [q(x) + \beta_m r(x)]\varphi_m = 0 \qquad (10.188)$$

Our aim is to derive the orthogonality condition (Eq. 10.184) and suitable boundary conditions. To do this, multiply the first of the above equations by φ_m

and the second by φ_n, then subtract the two equations to find $q(x)$ eliminated

$$\varphi_m \frac{d}{dx}\left[p(x)\frac{d\varphi_n}{dx}\right] - \varphi_n \frac{d}{dx}\left[p(x)\frac{d\varphi_m}{dx}\right] + (\beta_n - \beta_m)r(x)\varphi_n\varphi_m = 0$$

$$(\mathbf{10.189})$$

The range of independent variables is given as a physical condition, so integrate the above between arbitrary points a and b

$$(\beta_n - \beta_m)\int_a^b r(x)\varphi_n\varphi_m \, dx = \int_a^b \varphi_n \frac{d}{dx}\left[p(x)\frac{d\varphi_m}{dx}\right] dx$$

$$- \int_a^b \varphi_m \frac{d}{dx}\left[p(x)\frac{d\varphi_n}{dx}\right] dx \quad (\mathbf{10.190})$$

It is clear that the LHS is almost the same as Eq. 10.184. We inspect ways to cause the RHS to become zero (when n is different from m). Each of the integrals on the RHS can be integrated by parts, for example, the first integral can be decomposed as

$$\int_a^b \varphi_n \, d\left[p(x)\frac{d\varphi_m}{dx}\right] = \left[\varphi_n p(x)\frac{d\varphi_m}{dx}\right]_a^b - \int_a^b p(x)\frac{d\varphi_m}{dx}\frac{d\varphi_n}{dx} \, dx \quad (\mathbf{10.191})$$

The second integral in Eq. 10.190 yields

$$\int_a^b \varphi_m \, d\left[p(x)\frac{d\varphi_n}{dx}\right] = \left[\varphi_m p(x)\frac{d\varphi_n}{dx}\right]_a^b - \int_a^b p(x)\frac{d\varphi_n}{dx}\frac{d\varphi_m}{dx} \, dx \quad (\mathbf{10.192})$$

When these are subtracted, the integrals cancel, so that we finally have from Eq. 10.190

$$(\beta_n - \beta_m)\int_a^b r(x)\varphi_n\varphi_m \, dx = p(x)\left[\varphi_n\frac{d\varphi_m}{dx} - \varphi_m\frac{d\varphi_n}{dx}\right]_a^b \quad (\mathbf{10.193})$$

We see that RHS will vanish if any two of the three *homogeneous boundary conditions* are imposed

(i) $x = a$ or b, $y = 0$ $\therefore \varphi_n = \varphi_m = 0$

(ii) $x = a$ or b, $\dfrac{dy}{dx} = 0$ $\therefore \varphi_n' = \varphi_m' = 0$

(iii) $x = a$ or b, $\dfrac{dy}{dx} = Ny$ $\therefore \varphi_n' = N\varphi_n, \quad \varphi_m' = N\varphi_m$

We now see the singular importance of *homogeneous boundary conditions*, which we have reiterated time and again. For example, referring back to the coated-wall reactor problem in the previous section, we had the conditions

$$\varphi'_n(0) = 0 \quad \text{and} \quad \varphi'_n(1) = -\text{Ha } \varphi_n(1) \qquad (10.194)$$

By comparison, we identify $a = 0$, $b = 1$, $x = \xi$, and $N = -\text{Ha}$. Thus, it is easy to see

$$\varphi_n(1)\varphi'_m(1) - \varphi_m(1)\varphi'_n(1) = 0 \qquad (10.195)$$

since for n or m,

$$\varphi'(1) = -\text{Ha } \varphi(1) \qquad (10.196)$$

then

$$\varphi_n(1)[-\text{Ha } \varphi_m(1)] - \varphi_m(1)[-\text{Ha } \varphi_n(1)] = 0 \qquad (10.197)$$

Also, at the lower limit

$$\varphi_n(0)\varphi'_m(0) - \varphi_m(0)\varphi'_n(0) = 0 \qquad (10.198)$$

which must be true for the reactor problem since $\varphi'_n(0) = 0$ for n or m.

We can now finish the reactor problem using the fact that the cylindrical equation was of Sturm–Liouville type with homogeneous boundary conditions. When this is the case, the solutions φ_n, φ_m are said to be *orthogonal functions* with respect to the weighting function $r(x)$. Since we identified the weighting function for the reactor as $r(\xi) = \xi$, we can write when n is different from m

$$\int_0^1 \xi \varphi_n(\xi)\varphi_m(\xi)\, d\xi = 0 \qquad (10.199)$$

where

$$\varphi_n(\xi) = J_0(\lambda_n\xi), \qquad \varphi_m(\xi) = J_0(\lambda_m\xi)$$

We apply this condition to the final stage of the reactor problem represented by Eq. 10.183

$$C_{A0} = \sum_{n=1}^{\infty} C_n J_0(\lambda_n\xi) \qquad (10.200)$$

Multiply both sides by $\xi J_0(\lambda_m\xi)$ and then integrate between 0 and 1

$$C_{A0}\int_0^1 \xi J_0(\lambda_m\xi)\, d\xi = \sum_{n=1}^{\infty} C_n \int_0^1 \xi J_0(\lambda_n\xi) J_0(\lambda_m\xi)\, d\xi \qquad (10.201)$$

But all terms within the summation are identically zero (owing to the orthogonality condition, Eq. 10.199), except for the case $n = m$. Thus, among the infinite integrations, only one remains, and we can solve directly for C_n by requiring $n = m$

$$C_n = \frac{C_{A0} \int_0^1 \xi J_0(\lambda_n \xi) \, d\xi}{\int_0^1 \xi J_0^2(\lambda_n \xi) \, d\xi} \tag{10.202}$$

We call upon Eq. 3.204 for the integral in the numerator, and Eq. 3.207 for the denominator, respectively

$$\int_0^1 \xi J_0(\lambda_n \xi) \, d\xi = \frac{J_1(\lambda_n)}{\lambda_n} \tag{10.203}$$

and

$$\int_0^1 \xi J_0^2(\lambda_n \xi) \, d\xi = \frac{1}{2} \left[J_0^2(\lambda_n) + J_1^2(\lambda_n) \right] \tag{10.204}$$

The eigenvalue expression, Eq. 10.182, could be used to eliminate $J_1(\lambda_n)$, since

$$J_1(\lambda_n) = \text{Ha} \frac{J_0(\lambda_n)}{\lambda_n} \tag{10.205}$$

Inserting this and solving for C_n finally gives

$$C_n = C_{A0} \frac{1}{J_0(\lambda_n)} \frac{2\,\text{Ha}}{\left(\lambda_n^2 + \text{Ha}^2 \right)} \tag{10.206}$$

Inserting this into Eq. 10.181 gives the analytical solution to compute $C_A(r, z)$

$$\frac{C_A(r, z)}{C_{A0}} = \sum_{n=1}^{\infty} \frac{2\,\text{Ha}}{\left(\lambda_n^2 + \text{Ha}^2 \right)} \frac{J_0(\lambda_n \xi)}{J_0(\lambda_n)} \exp\left(-\lambda_n^2 \zeta \right) \tag{10.207}$$

where eigenvalues are determined from Eq. 10.205.

Table 3.3 shows, if Ha = 10, the first three eigenvalues are 2.1795, 5.0332, and 7.9569.

The flux at the wall is simply

$$N_0 = -\left[D_A \frac{\partial C_A}{\partial r} \right]_{r=R} = k C_A(R, z) \tag{10.208}$$

Far downstream, such that

$$\zeta = z D_A / \left(R^2 v_0 \right) > 1$$

the first term in the series dominates, so that approximately

$$N_0(z) \simeq kC_{A0} \frac{2\,\text{Ha}}{[\lambda_1^2 + \text{Ha}^2]} \exp\left(-\lambda_1^2 \frac{zD_A}{R^2 v_0}\right) \tag{10.209}$$

We can also compute the average composition downstream, and since v_0 is constant, this is written

$$\overline{C_A} = \frac{2\pi \int_0^R C_A(r,z)r\,dr}{2\pi \int_0^R r\,dr} = \frac{2}{R^2}\int_0^R C_A(r,z)r\,dr \tag{10.210}$$

In terms of ξ, this is

$$\overline{C_A} = 2\int_0^1 C_A(\xi,\zeta)\xi\,d\xi \tag{10.211}$$

Using the Bessel integral relationship, Eq. 3.204, we find

$$\overline{C_A} = 4\,\text{Ha}\,C_{A0}\sum_{n=1}^\infty \frac{J_1(\lambda_n)\exp(-\lambda_n^2\zeta)}{\lambda_n J_0(\lambda_n)[\lambda_n^2 + \text{Ha}^2]} \tag{10.212}$$

But since the eigenvalue expression is $J_1(\lambda_n) = \text{Ha}\,J_0(\lambda)/\lambda_n$, we can eliminate $J_1(\lambda_n)$ again to get finally,

$$\frac{\overline{C_A}}{C_{A0}} = 4\,\text{Ha}^2\sum_{n=1}^\infty \frac{\exp(-\lambda_n^2\zeta)}{\lambda_n^2[\lambda_n^2 + \text{Ha}^2]} \tag{10.213}$$

The plot of the reduced concentration (Eq. 10.213) versus ζ is shown in Fig. 10.4 for various values of Ha number.

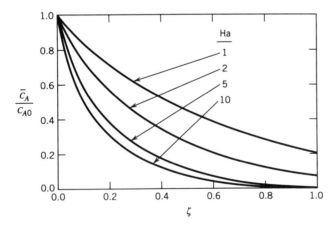

Figure 10.4 Plot of Eq. 10.213 versus ζ.

Far downstream, where $\zeta \gg 1$, we can truncate the series to a single term, so that approximately

$$\frac{\overline{C_A}}{C_{A0}} \simeq \frac{4\,\mathrm{Ha}^2}{\lambda_1^2\left[\lambda_1^2 + \mathrm{Ha}^2\right]}\,\exp\!\left(-\lambda_1^2\,\frac{zD_A}{R^2 v_0}\right) \qquad (10.214)$$

Thus, the $\ln(\overline{C_A}/C_{A0})$ varies linearly with position z. This suggests studying experimental reactors of different length (fixed v_0) or varying v_0 (fixed z). Baron et al. (1952) used this model to study oxidation of sulfur dioxide on a vanadium pentoxide catalyst.

EXAMPLE 10.3 *COOLING FLUIDS IN LAMINAR PIPE FLOW*

In Chapter 1, we considered the problem of heat transfer in laminar pipe flow. If it is permissible to ignore axial conduction, the transport equation we derived there was given by Eq. 1.31 with the term $k\partial^2 T/\partial z^2 \sim 0$

$$2v_0\left[1 - \left(\frac{r}{R}\right)^2\right]\frac{\partial T}{\partial z} = \alpha\frac{1}{r}\frac{\partial}{\partial r}\left(r\frac{\partial T}{\partial r}\right) \qquad (10.215)$$

where $\alpha = k/\rho C_p$ defines thermal diffusivity. The constant entrance temperature is denoted T_0 and the constant wall temperature is T_w. Apply the method of separation of variables to find $T(r, z)$.

To solve this linear problem, it is convenient, as before, to introduce dimensionless independent variables; thus,

$$\xi = \frac{r}{R}; \qquad \zeta = \frac{z\alpha}{R^2 v_{\max}}; \qquad v_{\max} = 2v_0 \qquad (10.216)$$

so that the uncluttered equation becomes

$$(1 - \xi^2)\frac{\partial T}{\partial \zeta} = \frac{1}{\xi}\frac{\partial}{\partial \xi}\left(\xi\frac{\partial T}{\partial \xi}\right) \qquad (10.217)$$

The boundary/initial conditions are

$$T = T_w \qquad \text{at} \qquad \xi = 1 \qquad (10.218)$$

$$\frac{\partial T}{\partial \xi} = 0 \qquad \text{at} \qquad \xi = 0 \qquad (10.219)$$

$$T = T_0 \qquad \text{at} \qquad \zeta = 0 \qquad (10.220)$$

It is clear by inspection that conditions must be homogeneous in the ξ domain, if we are to effect an analytical solution. To insure this condition at the outset, we follow the advice of Chapter 1 and define

$$\theta = T - T_w \qquad (10.221)$$

The new conditions on θ are now

$$\theta = 0, \quad \xi = 1; \quad \frac{\partial \theta}{\partial \xi} = 0, \quad \xi = 0; \quad \theta = T_0 - T_w, \quad \zeta = 0 \quad \textbf{(10.222)}$$

the first two of which are clearly homogeneous. Next replace T with θ in Eq. 10.217, and propose

$$\theta(\xi, \zeta) = \varphi(\xi)Z(\zeta) \quad \textbf{(10.223)}$$

Inserting this into Eq. 10.217, and rearranging to separate variables, gives

$$\frac{\frac{1}{\xi}\frac{d}{d\xi}\left(\xi\frac{d\varphi}{d\xi}\right)}{\left[(1-\xi^2)\varphi\right]} = \frac{Z'}{Z} = -\lambda^2 \quad \textbf{(10.224)}$$

The negative separation constant is implied to insure that $(T - T_w) = \theta$ diminishes as $\zeta \to \infty$. The two ODEs resulting are

$$\frac{d}{d\xi}\left(\xi\frac{d\varphi}{d\xi}\right) + \lambda^2\xi(1-\xi^2)\varphi = 0 \quad \textbf{(10.225)}$$

$$\frac{dZ}{d\zeta} + \lambda^2 Z = 0 \quad \textbf{(10.226)}$$

It is clear that Eq. 10.225 is a Sturm–Liouville equation with

$$p(\xi) = \xi, \quad q(\xi) = 0, \quad r(\xi) = \xi(1-\xi^2) \quad \text{and} \quad \beta = \lambda^2$$

It is also clear that the homogeneous boundary conditions in Eq. 10.222 require that

$$\varphi'(0) = 0, \qquad \varphi(1) = 0 \quad \textbf{(10.227)}$$

which are also of the Sturm–Liouville type. These statements insure that the solutions φ_n and φ_m are in fact orthogonal with respect to $r(\xi) = \xi(1-\xi^2)$. The polynomials generated by the solution to Eq. 10.225 are called *Graetz polynomials*. The only Graetz function, which is finite at the centerline, can be found by applying the method of Frobenius to get (Problem 3.4)

$$\varphi(\xi) = Gz(\xi, \lambda) \quad \textbf{(10.228)}$$

$$Gz(\xi, \lambda) = 1 - \left(\frac{\lambda^2}{4}\right)\xi^2 + \left(\frac{\lambda^2}{16}\right)\left(1 + \frac{\lambda^2}{4}\right)\xi^4 + \cdots \quad \textbf{(10.229)}$$

Combining the solutions for $Z(\zeta)$ and φ gives for $\lambda \neq 0$

$$\theta = AGz(\xi, \lambda)\exp(-\lambda^2\zeta) \quad \textbf{(10.230)}$$

As in the previous reactor problem, the solution when $\lambda = 0$ is simply: $B\ln\xi + C$. But since $\ln\xi$ is not finite at the centerline, set $B = 0$. The remaining solutions are summed as before

$$\theta = AGz(\xi, \lambda)\exp(-\lambda^2\zeta) + C \tag{10.231}$$

It is clear for an infinitely long heat exchanger, that $T \to T_w$ as $z \to \infty$, hence $\theta \to 0$ as $\zeta \to \infty$, so that we set $C = 0$. We now have the compact result

$$\theta = AGz(\xi, \lambda)\exp(-\lambda^2\zeta) \tag{10.232}$$

If we apply the wall condition, $\theta = 0$ at $\xi = 1$, then we obtain, since $\exp(-\lambda^2\zeta)$ is at least finite

$$Gz(1, \lambda) = 0, \qquad \xi = 1 \tag{10.233}$$

There exists countably infinite roots to satisfy Eq. 10.229 when $\xi = 1$

$$0 = 1 - \left(\frac{\lambda^2}{4}\right) + \left(\frac{\lambda^2}{16}\right)\left(1 + \frac{\lambda^2}{4}\right) + \cdots \tag{10.234}$$

the smallest positive root takes a value of 2.704, and the next largest is 6.66 (Jacob 1949).

Since there are many solutions, the general result by superposition is

$$\theta(\xi, \zeta) = \sum_{n=1}^{\infty} A_n Gz_n(\xi)\exp(-\lambda_n^2\zeta) \tag{10.235}$$

Henceforth, we shall denote $Gz_n(\xi)$ as the eigenfunction $Gz_n(\xi, \lambda_n)$. The condition at the inlet can now be applied

$$(T_0 - T_w) = \sum_{n=1}^{\infty} A_n Gz_n(\xi) \tag{10.236}$$

Multiplying both sides by the weighting function $r(\xi) = \xi(1 - \xi^2)$ and by $Gz_m(\xi)$, where $n \neq m$, then integrating between 0 to 1, gives

$$(T_0 - T_w)\int_0^1 Gz_m(\xi)\xi(1 - \xi^2)\,d\xi$$

$$= \sum_{n=1}^{\infty} A_n \int_0^1 Gz_n(\xi)Gz_m(\xi)\xi(1 - \xi^2)\,d\xi \tag{10.237}$$

Invoking the orthogonality conditions eliminates all terms save one (when $n = m$) within the summation, so that we can solve for A_n

$$A_n = \frac{(T_0 - T_w) \int_0^1 Gz_n(\xi)\xi(1 - \xi^2)\,d\xi}{\int_0^1 Gz_n^2(\xi)\xi(1 - \xi^2)\,d\xi} \tag{10.238}$$

The Graetz eigenfunctions and eigenvalues are not widely tabulated. However, Brown (1960) has computed a large number of eigenvalues suitable for most problems. The final result can be written in dimensionless form as

$$\frac{T - T_w}{T_0 - T_w} = \sum_{n=1}^{\infty} A_n Gz_n(\xi)\exp\left(-\lambda_n^2 \zeta\right) \tag{10.239}$$

where the eigenvalues are found from $Gz(1, \lambda_n) = 0$. The above series converges very slowly (as does the polynomial represented by Eq. 10.229), hence a large number of terms must be retained, especially for the entrance region. In this regard, Schechter (1967) has worked out an approximate solution based on variational calculus, which needs only two terms to find an accurate prediction of mean fluid temperature. The average (mixing-cup) temperature for the present case can be found by integration using Eq. 10.239

$$\bar{T}(\zeta) = \frac{\int_0^1 T(\xi, \zeta)(1 - \xi^2)\xi\,d\xi}{\int_0^1 (1 - \xi^2)\xi\,d\xi} \tag{10.240}$$

10.6 INHOMOGENEOUS EQUATIONS

We have used elementary change of variables as a method to convert certain inhomogeneous boundary conditions to homogeneous form. This was a clear imperative in order to use, in the final steps, the properties of orthogonality. Without this property, series coefficients cannot be computed individually.

In certain cases, the boundary inhomogeneity cannot be removed by elementary substitution (e.g., constant boundary flux). In other cases, the defining equation itself is not homogeneous. Both sets of circumstances lead to inhomogeneous equations.

A fairly general way of coping with inhomogeneous PDE is to apply methods already introduced in Chapter 9, specifically, the concept of deviation variables (similar, but not to be confused with perturbation variables). This technique is best illustrated by way of examples.

EXAMPLE 10.4 *TRANSIENT COOLING OF NUCLEAR FUEL PELLETS*

Spherical pellets of uranium undergo self-generated heating at a rate per unit volume Q, while being cooled at the boundary through a heat transfer coefficient h, so for a single pellet on start-up ($\alpha = k/\rho C_p$)

$$\frac{\partial T}{\partial t} = \alpha \frac{1}{r^2} \frac{\partial}{\partial r}\left(r^2 \frac{\partial T}{\partial r}\right) + \frac{Q}{\rho C_p} \qquad (10.241)$$

subject to boundary and initial condition

$$r = R; \qquad -k \frac{\partial T}{\partial r} = h(T - T_f) \qquad (10.242a)$$

$$t = 0; \qquad T(0, r) = T_0 \qquad (10.242b)$$

where T_f is the flowing coolant fluid temperature (constant). Find the relationship to predict the transient response of the pellet, $T(r, t)$.

It is clear that the heat balance produces an inhomogeneous equation, owing to the (assumed) constant heat source term Q. This means of course that a separation of variables approach will not lead to a Sturm–Liouville equation. However, the following expedient will produce a result that is homogeneous; let temperature be decomposed into two parts: a steady state (the future steady state) and a deviation from steady state

$$T(r, t) = \overline{T}(r) + y(r, t) \qquad (10.243)$$

Of course, the steady state must be represented by

$$\alpha \frac{1}{r^2} \frac{d}{dr}\left(r^2 \frac{d\overline{T}}{dr}\right) + \frac{Q}{\rho C_p} = 0 \qquad (10.244)$$

which can be integrated directly to yield

$$d\left(r^2 \frac{d\overline{T}}{dr}\right) = -\frac{Q}{k} r^2 \, dr$$

hence,

$$r^2 \frac{d\overline{T}}{dr} = -\frac{Q}{k} \frac{r^3}{3} + C_1$$

$$\frac{d\overline{T}}{dr} = -\frac{Q}{k} \frac{r}{3} + \frac{C_1}{r^2}$$

But symmetry must be maintained at the centerline, hence take $C_1 = 0$, and integrate again

$$\overline{T}(r) = -\frac{Q}{k}\frac{r^2}{6} + C_2$$

Now, cooling at the boundary requires Eq. 10.242a to be obeyed at all times, so that

$$Q\frac{R}{3} = h\left(-\frac{Q}{k}\frac{R^2}{6} + C_2 - T_f\right)$$

hence, solve for C_2 to get

$$C_2 = \frac{QR}{3h} + \frac{QR^2}{6k} + T_f$$

Hence, the steady-state temperature profile (as $t \to \infty$) is

$$\overline{T}(r) = T_f + \frac{QR}{3h} + \frac{QR^2}{6k}\left[1 - \left(\frac{r}{R}\right)^2\right] \qquad (10.245)$$

Thus, when the decomposed form of temperature is substituted into the defining equation and boundary conditions, we get

$$\frac{\partial y}{\partial t} = \alpha \frac{1}{r^2}\frac{\partial}{\partial r}\left(r^2\frac{\partial y}{\partial r}\right) \qquad (10.246a)$$

$$r = 0; \qquad \frac{\partial y}{\partial r} = 0 \qquad (10.246b)$$

$$r = R; \qquad -k\frac{\partial y}{\partial r} = hy \qquad (10.246c)$$

$$t = 0; \qquad y(r,0) = T_0 - \overline{T}(r) \qquad (10.246d)$$

where the steady part removes Q according to Eq. 10.244.

It is clear in the limit as $t \to \infty$, that y must vanish. In terms of dimensionless independent variables, we use

$$\xi = \frac{r}{R}; \qquad \tau = \frac{\alpha t}{R^2}$$

hence, Eq. 10.246a becomes

$$\frac{\partial y}{\partial \tau} = \frac{1}{\xi^2}\frac{\partial}{\partial \xi}\left(\xi^2\frac{\partial y}{\partial \xi}\right) \qquad (10.247a)$$

$$\xi = 0; \qquad \frac{\partial y}{\partial \xi} = 0 \qquad (10.247b)$$

$$\xi = 1; \qquad -\frac{\partial y}{\partial \xi} = \text{Bi}\, y \qquad (10.247c)$$

where

$$\mathrm{Bi} = \frac{hR}{k}$$

The equation for y and boundary conditions are fully homogeneous. Apply separation of variables in the usual way by proposing

$$y(\xi, \tau) = \varphi(\xi)\theta(\tau) \tag{10.248}$$

we obtain

$$\frac{\dfrac{1}{\xi^2}\dfrac{d}{d\xi}\left(\xi^2\dfrac{d\varphi}{d\xi}\right)}{\varphi} = \frac{\dfrac{d\theta}{d\tau}}{\theta} = -\lambda^2 \tag{10.249}$$

So, the required solutions are obtained from

$$\frac{1}{\xi^2}\frac{d}{d\xi}\left(\xi^2\frac{d\varphi}{d\xi}\right) + \lambda^2\varphi = 0 \tag{10.250}$$

$$\frac{d\theta}{d\tau} + \lambda^2\theta = 0 \tag{10.251}$$

The equation for φ is of the Sturm–Liouville type if we stipulate $p(\xi) = \xi^2, q(\xi) = 0$, and the weighting function must be $r(\xi) = \xi^2$. The equation for φ is easier to solve by defining $\varphi = u(\xi)/\xi$; hence, obtain

$$\frac{d^2u}{d\xi^2} + \lambda^2 u = 0 \tag{10.252}$$

which has solution

$$u(\xi) = A_0 \sin(\lambda\xi) + B_0 \cos(\lambda\xi) \tag{10.253}$$

so that

$$\varphi(\xi) = A_0\frac{\sin(\lambda\xi)}{\xi} + B_0\frac{\cos(\lambda\xi)}{\xi} \tag{10.254}$$

The solution for θ is

$$\theta = K\exp(-\lambda^2\tau) \tag{10.255}$$

When $\lambda = 0$, the solution can be seen to be simply $C/\xi + D$, so that the combined solution is

$$y = \left[A\frac{\sin(\lambda\xi)}{\xi} + B\frac{\cos(\lambda\xi)}{\xi}\right]\exp(-\lambda^2\tau) + \frac{C}{\xi} + D \tag{10.256}$$

Finiteness at the centerline requires $B = C = 0$, and since $y \to 0$ as $\tau \to \infty$, then $D = 0$, so the compact form results

$$y(\xi, \tau) = A \frac{\sin(\lambda\xi)}{\xi} \exp(-\lambda^2\tau) \tag{10.257}$$

which vanishes as $\tau \to \infty$. If we apply the boundary condition at $\xi = 1$, there results

$$\sin(\lambda) - \lambda \cos(\lambda) = \text{Bi} \sin(\lambda) \tag{10.258}$$

There are countably infinite values of λ, which satisfy this; hence, we use λ_n to count these (noting $\lambda = 0$ has already been counted), and write the eigenvalue expression

$$\lambda_n \cot(\lambda_n) - 1 = -\text{Bi} \tag{10.259}$$

For large Bi (high velocity coolant), this reduces to

$$\sin(\lambda_n) = 0$$

hence, when Bi $\to \infty$, $\lambda_n = n\pi (n = 1, 2, 3, \dots)$. Even for finite Bi, the spacing for successive eigenvalues is very close to π.

The general solution now must be written as a superposition of all possible solutions

$$y(\xi, \tau) = \sum_{n=1}^{\infty} A_n \frac{\sin(\lambda_n\xi)}{\xi} \exp(-\lambda_n^2\tau) \tag{10.260}$$

It remains only to find A_n using the orthogonality condition together with the initial condition (Eq. 10.246d)

$$(T_0 - T_f) - QR\left(\frac{1}{3h} + \frac{R}{6k}\right) + \frac{QR^2}{6k}\xi^2 = \sum_{n=1}^{\infty} A_n \frac{\sin(\lambda_n\xi)}{\xi} \tag{10.261}$$

If we multiply throughout by the eigenfunction $\sin(\lambda_n\xi)/\xi$ and the weighting function $r(\xi) = \xi^2$, then integrate over the range of orthogonality, all terms disappear on the RHS except for the case when $\lambda_n = \lambda_m$

$$A_n \int_0^1 \sin^2(\lambda_n\xi) \, d\xi = \int_0^1 \xi \sin(\lambda_n\xi)$$

$$\times \left[(T_0 - T_f) - \frac{QR}{3}\left(\frac{1}{h} + \frac{R}{2k}\right) + \frac{QR^2}{6k}\xi^2\right] d\xi$$

Performing the integrals, we obtain

$$A_n \left[\frac{1}{2} - \frac{1}{2} \frac{\sin(\lambda_n)\cos(\lambda_n)}{\lambda_n} \right]$$

$$= \left[(T_0 - T_f) - \frac{QR}{3} \left(\frac{1}{h} + \frac{R}{2k} \right) \right] \left(\frac{\sin(\lambda_n) - \lambda_n \cos(\lambda_n)}{\lambda_n^2} \right)$$

$$+ \frac{QR^2}{6k} \left[\frac{(3\lambda_n^2 - 6)\sin(\lambda_n) - (\lambda_n^3 - 6\lambda_n)\cos(\lambda_n)}{\lambda_n^4} \right]$$

Collecting terms, using the eigenvalue expression to simplify, we finally obtain

$$\frac{y(\xi,\tau)}{T_0 - T_f} = 2\,\mathrm{Bi} \sum_{n=1}^{\infty} \frac{\left[\dfrac{N_k}{\lambda_n^2} - 1 \right]}{\left[\mathrm{Bi} - 1 + \cos^2 \lambda_n \right]} \frac{\cos \lambda_n}{\lambda_n} \frac{\sin(\lambda_n \xi)}{\xi} \exp\left(-\lambda_n^2 \tau \right) \quad \textbf{(10.262)}$$

where

$$N_k = \frac{QR^2}{k(T_0 - T_f)}; \qquad \mathrm{Bi} = \frac{hR}{k}$$

The parameter Bi is the ratio of film to conduction transfer, and N_k is the ratio of the generation to conduction transfer.

The response at the center is illustrated in Fig. 10.5.

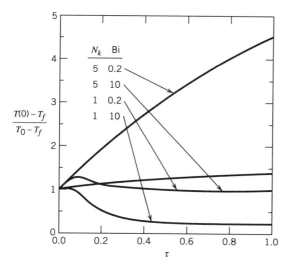

Figure 10.5 Temperature deviation at pellet center.

EXAMPLE 10.5 *CONDUCTION IN FINITE SLABS*

Consider a homogeneous slab, initially at a temperature T_0, when suddenly the face at position $x = 0$ is raised to T_s ($> T_0$), while the face at $x = L$ is held at $T = T_0$

$$T(0, t) = T_s \qquad (10.263a)$$

$$T(L, t) = T_0 \qquad (10.263b)$$

$$T(x, 0) = T_0 \qquad (10.263c)$$

The transient conduction equation in rectangular coordinates is

$$\frac{\partial T}{\partial t} = \alpha \frac{\partial^2 T}{\partial x^2} \qquad (10.264)$$

Find the temperature response $T(x, t)$.

At the outset, it is clear that the boundary conditions are inhomogeneous, for if we defined

$$\theta = T - T_0$$

then it is true that the boundary condition at $x = L$ would become homogeneous, but the one at $x = 0$ is still inhomogeneous.

Again, we propose a structure made up of steady plus unsteady parts,

$$T(x, t) = \bar{T}(x) + y(x, t) \qquad (10.265)$$

The (future) steady-state part obviously satisfies

$$\frac{d^2 \bar{T}}{dx^2} = 0; \qquad \bar{T}(0) = T_s; \qquad \bar{T}(L) = T_0 \qquad (10.266)$$

hence,

$$\bar{T}(x) = (T_0 - T_s)\left(\frac{x}{L}\right) + T_s \qquad (10.267)$$

Inserting Eq. 10.265 into Eq. 10.264 gives

$$\frac{\partial y}{\partial t} = \alpha \frac{\partial^2 y}{\partial x^2} \qquad (10.268)$$

since

$$\alpha \frac{d^2 \bar{T}}{dx^2} = 0$$

More importantly, the boundary conditions (10.263) show

$$y(0, t) = y(L, t) = 0 \qquad \qquad (\mathbf{10.269})$$

which are clearly homogeneous. The price we pay for this simplification occurs in the initial condition for y

$$t = 0; \quad y(x, 0) = T(x, 0) - \bar{T}(x) = T_0 - T_s - (T_0 - T_s)\left(\frac{x}{L}\right) \quad (\mathbf{10.270})$$

It is now convenient to use dimensionless independent variables, so let

$$\xi = \frac{x}{L}; \qquad \tau = \frac{\alpha t}{L^2}$$

yielding

$$\frac{\partial y}{\partial \tau} = \frac{\partial^2 y}{\partial \xi^2}$$

subject to

$$y(0, \tau) = y(1, \tau) = 0$$
$$y(\xi, 0) = (T_0 - T_s)(1 - \xi)$$

Separation of variables proceeds as before by letting

$$y = \varphi(\xi)\theta(\tau)$$

giving

$$\frac{\dfrac{d^2\varphi}{d\xi^2}}{\varphi} = \frac{\dfrac{d\theta}{d\tau}}{\theta} = -\lambda^2$$

hence,

$$\frac{d^2\varphi}{d\xi^2} + \lambda^2\varphi = 0 \qquad \qquad (\mathbf{10.271}a)$$

$$\frac{d\theta}{d\tau} + \lambda^2\theta = 0 \qquad \qquad (\mathbf{10.271}b)$$

The solutions of the ODEs are by now familiar

$$\varphi(\xi) = A\sin(\lambda\xi) + B\cos(\lambda\xi) \qquad \qquad (\mathbf{10.272}a)$$

$$\theta = K\exp(-\lambda^2\tau) \qquad \qquad (\mathbf{10.272}b)$$

The solution for the case when $\lambda = 0$ is easily seen to be $C\xi + D$, so that the superposition of the two cases yields

$$y = [A \sin(\lambda \xi) + B \cos(\lambda \xi)] \exp(-\lambda^2 \tau) + C\xi + D \qquad (10.273)$$

Now, since $y(0, \tau) = 0$, then $D = 0$ and $B = 0$. Moreover, since $y(1, \tau) = 0$, we must also have

$$0 = A \sin(\lambda) \exp(-\lambda^2 \tau) + C$$

The only nontrivial way this can be satisfied is to set

$$C = 0; \qquad \sin(\lambda) = 0$$

But there exists countably infinite λ to satisfy $\sin(\lambda) = 0$, hence, write the eigenvalue result

$$\lambda_n = n\pi \qquad (n = 1, 2, 3, \dots) \qquad (10.274)$$

This suggests a superposition of all solutions as

$$y = \sum_{n=1}^{\infty} A_n \sin(n\pi\xi) \exp(-n^2\pi^2\tau) \qquad (10.275)$$

Inspection of Eq. 10.271 shows this to be of the Sturm–Liouville type if $p(x) = 1$, $q = 0$, and $r(x) = 1$. Moreover, if we inspect the original form, $y = \varphi(\xi)\theta(\tau)$, then clearly

$$y(0, \tau) = 0 = \varphi(0)\theta(\tau) \qquad \therefore \varphi(0) = 0$$

and

$$y(1, \tau) = 0 = \varphi(1)\theta(\tau) \qquad \therefore \varphi(1) = 0$$

This means that all solutions $\varphi_n(\xi)$ are such that

$$\varphi_n(0) = \varphi_n(1) = 0 \qquad \text{for} \qquad n = 1, 2, 3, \dots \qquad (10.276)$$

So since the boundary values for all eigenfunctions are homogeneous and of the Sturm–Liouville type, we can write the orthogonality condition immediately

$$\int_0^1 \varphi_n(x)\varphi_m(x) \, dx = 0 \qquad \text{for} \qquad n \neq m \qquad (10.277)$$

We will use this in the final stage to find the coefficients A_n. The last remaining unused condition is the initial value

$$y(\xi, 0) = (T_0 - T_s)(1 - \xi) = \sum_{n=1}^{\infty} A_n \sin(n\pi\xi) \qquad (10.278)$$

Multiplying both sides by

$$\varphi_m(\xi)\, d\xi = \sin(m\pi\xi)\, d\xi$$

integrating and invoking the orthogonality condition, Eq. 10.278 yields for the case $n = m$

$$(T_0 - T_s)\int_0^1 (1 - \xi)\sin(n\pi\xi)\, d\xi = A_n\int_0^1 \sin^2(n\pi\xi)\, d\xi \quad \textbf{(10.279)}$$

hence,

$$A_n = \frac{2(T_0 - T_s)}{\lambda_n} = \frac{2(T_0 - T_s)}{n\pi} \tag{10.280}$$

and the final expression for $y(\xi, \tau)$ is

$$\frac{y(\xi, \tau)}{(T_0 - T_s)} = 2\sum_{n=1}^{\infty} \frac{\sin(n\pi\xi)}{n\pi}\exp(-n^2\pi^2\tau) \tag{10.281}$$

10.7 APPLICATIONS OF LAPLACE TRANSFORMS FOR SOLUTIONS OF PDEs

The Laplace transform is not limited to ordinary derivatives, but it can also be applied to functions of two independent variables $f(x, t)$. Usually the transform is taken with respect to the time variable, but it can also be applied to spatial variables x, as long as $0 < x < \infty$. The appropriate variable is usually selected based on the suitability of initial conditions, since only initial value problems can be treated by Laplace transforms. Henceforth, we shall imply transforms with respect to time, unless otherwise stated.

The first derivative is treated just as before, except the initial condition must now in general depend on the second independent variable; thus,

$$\mathscr{L}\left[\frac{\partial f(x, t)}{\partial t}\right] = \int_0^\infty e^{-st}\frac{\partial f}{\partial t}\, dt = \left[f(x, t)e^{-st}\right]_0^\infty + s\int_0^\infty f(x, t)e^{-st}\, dt \quad \textbf{(10.282)}$$

The second term on the RHS defines the Laplace transform, and the upper bound in the first term can be made to be dominated by appropriate selection of s; hence,

$$\mathscr{L}\left[\frac{\partial f}{\partial t}\right] = sF(x, s) - f(x, 0) \tag{10.283}$$

where it is emphasized that F depends on x.

Similarly, the second time derivative can be obtained as

$$\mathscr{L}\left[\frac{\partial^2 f(x,t)}{\partial t^2}\right] = \int_0^\infty e^{-st}\frac{\partial^2 f}{\partial t^2}\,dt = s^2 F(x,s) - sf(x,0) - \left.\frac{\partial f(x,t)}{\partial t}\right|_{t=0}$$

(**10.284**)

Transforms of spatial derivatives are also easily obtained as

$$\mathscr{L}\left[\frac{\partial f}{\partial x}\right] = \int_0^\infty e^{-st}\frac{\partial f}{\partial x}\,dt \qquad (\mathbf{10.285})$$

The partial implies holding time constant, and, moreover, since time is completely integrated out of the equation, thus we can take the ordinary derivative with respect to x and write

$$\int_0^\infty e^{-st}\frac{\partial f}{\partial x}\,dt = \frac{d}{dx}\int_0^\infty e^{-st}f(x,t)\,dt = \frac{d}{dx}F(x,s) \qquad (\mathbf{10.286})$$

So we finally arrive at

$$\mathscr{L}\left[\frac{\partial f}{\partial x}\right] = \frac{d}{dx}F(x,s) \qquad (\mathbf{10.287})$$

where s is now a parameter. Similarly,

$$\mathscr{L}\left[\frac{\partial^2 f}{\partial x^2}\right] = \frac{d^2 F(x,s)}{dx^2} \qquad (\mathbf{10.288})$$

and for mixed partials

$$\mathscr{L}\left[\frac{\partial^2 f}{\partial x\,\partial t}\right] = \frac{d}{dx}\int_0^\infty e^{-st}\frac{\partial f}{\partial t}\,dt = \frac{d}{dx}\left[sF(x,s) - f(x,0)\right] \quad (\mathbf{10.289})$$

We now have all the necessary tools to apply the Laplace transform method to *linear* partial differential equations, of the initial value type. The power of the Laplace transform in PDE applications is the ease with which it can cope with simultaneous equations. Few analytical methods have this facility.

EXAMPLE 10.6

We saw in Chapter 1 that packed bed adsorbers can be described by the simultaneous, coupled PDEs

$$V_0\frac{\partial C}{\partial z} + \varepsilon\frac{\partial C}{\partial t} + (1-\varepsilon)K\frac{\partial C^*}{\partial t} = 0 \qquad (\mathbf{10.290})$$

$$(1-\varepsilon)K\frac{\partial C^*}{\partial t} = k_c a(C - C^*) \qquad (\mathbf{10.291})$$

where $C(z, t)$ is the flowing solute composition and C^* is the value that would be in equilibrium with the solid phase, such that $q = KC^*$, where K is the linear partition coefficient. In such packed beds, ε denotes voidage, V_0 depicts superficial fluid velocity, and $k_c a$ is the volumetric mass transfer coefficient. If the initial conditions are such that

$$C(z, 0) = 0 \quad \text{and} \quad C^*(z, 0) = 0 \qquad (10.292)$$

and at the bed entrance

$$C(0, t) = C_0 \quad \text{(constant inlet composition)} \qquad (10.293)$$

find the transient behavior of the exit composition C using Laplace transforms.

It is possible to apply the Laplace transform directly, but this leads to some intractable problems later. The equations are easier to treat if their form is changed at the outset. The thrust of the following variable transformation is to eliminate one of the derivatives, as we shall see.

We shall first need to express velocity as an interstitial value, which is the actual linear velocity moving through the interstices of the bed

$$V = \frac{V_0}{\varepsilon} \qquad (10.294)$$

so we rewrite the overall bed balance using V

$$V\frac{\partial C}{\partial z} + \frac{\partial C}{\partial t} + \frac{(1-\varepsilon)K}{\varepsilon}\frac{\partial C^*}{\partial t} = 0 \qquad (10.295)$$

Next, since a time response cannot occur until the local residence time is exceeded, define the relative time scale as

$$\theta = t - z/V \qquad (10.296)$$

which is the difference between real time and local fluid residence time. We effect the change of variables

$$C(z, t) = C(z, \theta)$$
$$C^*(z, t) = C^*(z, \theta) \qquad (10.297)$$

Considering the transformation of $C(z, t)$ to $C(z, \theta)$, we first write the total differential

$$dC(z, t) = dC(z, \theta)$$

Next, expand this by the chain rule

$$\left.\frac{\partial C}{\partial z}\right|_t dz + \left.\frac{\partial C}{\partial t}\right|_z dt = \left.\frac{\partial C}{\partial z}\right|_\theta dz + \left.\frac{\partial C}{\partial \theta}\right|_z d\theta$$

In order to form an identity, we shall equate multipliers of dz on left- and right-hand sides, and the same for dt; to do this, we need to find $d\theta$, which is from Eq. 10.296

$$d\theta = dt - \frac{dz}{V}$$

Now, equating coefficients of dz and dt shows

$$\left.\frac{\partial C}{\partial z}\right|_t = \left.\frac{\partial C}{\partial z}\right|_\theta - \frac{1}{V}\left.\frac{\partial C}{\partial \theta}\right|_z$$

$$\left.\frac{\partial C}{\partial t}\right|_z = \left.\frac{\partial C}{\partial \theta}\right|_z$$

Similarly, we can easily see

$$\left.\frac{\partial C^*}{\partial t}\right|_z = \left.\frac{\partial C^*}{\partial \theta}\right|_z$$

Inserting these into Eqs. 10.295 and 10.291 gives

$$V\left.\frac{\partial C}{\partial z}\right|_\theta = -\left(\frac{k_c a}{\varepsilon}\right)(C - C^*) \tag{10.298}$$

$$(1 - \varepsilon)K\frac{\partial C^*}{\partial \theta} = k_c a(C - C^*) \tag{10.299}$$

To make this system of coupled equations even more compact, combine the remaining constant parameters with the independent variables thusly:

$$\zeta = \left(\frac{k_c a}{\varepsilon}\right) \cdot \frac{z}{V} \cdots \text{dimensionless distance}$$

$$\tag{10.300}$$

$$\tau = \left(\frac{k_c a}{K(1 - \varepsilon)}\right) \cdot \theta \cdots \text{dimensionless relative time}$$

When this is done, we have the very clean form of equations

$$\frac{\partial C}{\partial \zeta} = -(C - C^*) \tag{10.301}$$

$$\frac{\partial C^*}{\partial \tau} = (C - C^*) \tag{10.302}$$

The relative time scale is such that at any position z, the time required for a slug of fluid to reach this point is exactly $t = z/V$, which corresponds to $\theta = 0$ or in dimensionless terms, $\tau = 0$. Thus, at this point in time, the portion of the bed in front of the slug is completely clean and is identical with the initial condition, so

$$C(\zeta, 0) = 0 \tag{10.303}$$

$$C^*(\zeta, 0) = 0 \tag{10.304}$$

Moreover, for all $\tau > 0$ at the entrance where $\zeta = 0$, the concentration is fixed

$$C(0, \tau) = C_0 \qquad (10.305)$$

Thus, even with a transformation of variables, the initial and boundary conditions are unchanged.

Taking Laplace transforms with respect to τ, we obtain

$$\frac{dC(\zeta, s)}{d\zeta} = -(C(\zeta, s) - C^*(\zeta, s))$$

$$sC^*(\zeta, s) = (C(\zeta, s) - C^*(\zeta, s))$$

Solving the second equation for C^* yields

$$C^* = \frac{C}{s + 1}$$

hence the first ODE becomes

$$\frac{dC}{d\zeta} = -C + \frac{C}{s + 1} = -C\frac{s}{s + 1}$$

The integral of this is simply

$$C(\zeta, s) = A(s)\exp\left(-\frac{s}{s + 1}\zeta\right) = A\exp(-\zeta)\exp\left(\frac{\zeta}{s + 1}\right) \quad (10.306)$$

where $A(s)$ is the arbitrary constant of integration. The transform of the step change at the bed entrance is

$$\mathscr{L}C(0, \tau) = \mathscr{L}C_0 = C_0/s$$

so the arbitrary constant becomes $A = C_0/s$ and we are left to invert the function

$$C(\zeta, s) = C_0\frac{1}{s}\exp(-\zeta)\exp\left(\frac{\zeta}{s + 1}\right) \qquad (10.307)$$

The shifting theorem could be used to good effect except for the term $1/s$. The transforms in Appendix D shows that $\mathscr{L}^{-1}(1/s)e^{-k/s}$ equals $J_0(2\sqrt{kt})$, which is directly applicable to the present problem if we can replace s with $(s + 1)$ in the multiplier. To do this, we inspect the process of integration with respect to ζ

$$\int_0^\zeta \exp(-\beta)\exp\left(\frac{\beta}{s + 1}\right) d\beta = \left(\frac{s + 1}{s}\right)\left[1 - \exp(-\zeta)\exp\left(\frac{\zeta}{s + 1}\right)\right] \quad (10.308)$$

This shows that the exponential can be expressed in integral form, which also allows direct inversion of the transform; hence,

$$C(\zeta, s) = C_0 \left[\frac{1}{s} - \int_0^\zeta \frac{\exp(-\beta)\exp\left(\dfrac{\beta}{s+1}\right)}{s+1} \, d\beta \right]$$

Using the shifting theorem and noting that

$$J_0(2\sqrt{-kt}) = J_0(2i\sqrt{kt}) = I_0(2\sqrt{kt})$$

according to Equations 3.158 and 3.159, we obtain finally,

$$C(\tau, \zeta) = \left[1 - \int_0^\zeta \exp(-\beta)\exp(-\tau) I_0(2\sqrt{\beta\tau}) \, d\beta \right] C_0 u(\tau) \quad \textbf{(10.309)}$$

where $u(\tau)$ is inserted to remind that a response occurs only when $\tau > 0$, and $u(0) = 0$.

Since the impulse response is simply the time derivative of the step response, we can also obtain

$$C(\tau, \zeta)_I = C_0 \sqrt{\frac{\zeta}{\tau}} \exp(-\zeta - \tau) I_1(2\sqrt{\zeta\tau}) \quad \textbf{(10.310)}$$

where C_0 now denotes the weighting factor for the impulse input, that is, $C(0, \tau) = C_0 \delta(\tau)$. In arriving at Eq. 10.310, we have used the J function, tabulated in Perry (1973), defined here as

$$J(\zeta, \tau) = 1 - \int_0^\zeta \exp(-\beta - \tau) I_0(2\sqrt{\beta\tau}) \, d\beta$$

which has the useful property

$$J(\zeta, \tau) + J(\tau, \zeta) = 1 + \exp(-\tau - \zeta) I_0(2\sqrt{\tau\zeta})$$

The step and impulse responses are illustrated in Fig. 10.6.

As a lead into treating the case of multivalued singularities, we shall reconsider the thermal diffusion problem discussed in Section 10.3, where a combination of variables approach was used. Again, we consider a semiinfinite slab, initially at a temperature T_0 throughout, when suddenly the face at $x = 0$ is raised to T_s, where $T_s > T_0$. The partial differential equation describing the dynamics of heat diffusion was derived to be

$$\alpha \frac{\partial^2 T}{\partial x^2} = \frac{\partial T}{\partial t}$$

We can apply Laplace transforms directly, but this of course will carry with it the initial condition on T. We can cause this initial condition to be zero by

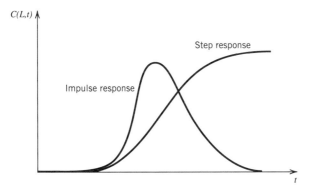

Figure 10.6 Response curves at exit of packed bed adsorber, $L = 100$ cm, $\varepsilon = 0.4$, $K = 2$, $k_c a = 0.1$ sec^{-1}, $V_0 = 4$ cm/sec.

defining the new variable $\theta = T - T_0$, so we now have

$$\alpha \frac{\partial^2 \theta}{\partial x^2} = \frac{\partial \theta}{\partial t} \qquad (10.311)$$

Taking Laplace transforms yields

$$\alpha \frac{d^2 \theta(x, s)}{dx^2} = s\theta(x, s) \qquad (10.312)$$

This is a second order ODE with characteristic roots $\pm \sqrt{s/\alpha}$, so the general solution is:

$$\theta(x, s) = A(s)\exp\left(-\sqrt{\frac{s}{\alpha}}\, x\right) + B(s)\exp\left(\sqrt{\frac{s}{\alpha}}\, x\right) \qquad (10.313)$$

For any real or complex value of s it is clear the second solution is inadmissible since it is required that $\theta \to 0$ as $x \to \infty$, hence, we take $B(s) = 0$. To find $A(s)$, we use the condition at $x = 0$, the transform of which is

$$\mathscr{L}\theta(0, t) = \mathscr{L}(T_s - T_0) = (T_s - T_0)/s \qquad (10.314)$$

so that $A(s) = (T_s - T_0)/s$ and we now have

$$\theta(x, s) = \frac{(T_s - T_0)}{s}\exp\left(-\sqrt{\frac{s}{\alpha}}\, x\right) \qquad (10.315)$$

The table in Appendix D shows that the inverse Laplace transform $\mathscr{L}^{-1}(1/s)\exp(-k\sqrt{s}) = \mathrm{erfc}(k/2\sqrt{t}\,)$, which we can use to complete the above solution. We note that the equivalence $k = x/\sqrt{\alpha}$; hence,

$$\theta(x, t) = T(x, t) - T_0 = (T_s - T_0)\text{erfc}\left(\frac{x}{\sqrt{4\alpha t}}\right) \tag{10.316}$$

which is identical to the result obtained in Section 10.3.

EXAMPLE 10.7

The Nusselt problem is similar to the Graetz problem studied in Example 10.3, except the velocity profile is plug shaped. Starting with the transport equation

$$V_0 \frac{\partial T}{\partial z} = \alpha \frac{1}{r} \frac{\partial}{\partial r}\left(r \frac{\partial T}{\partial r}\right)$$

subject to

$$T = T_w \quad \text{at} \quad r = R$$

$$T = T_0 \quad \text{at} \quad x = 0$$

$$\frac{\partial T}{\partial r} = 0 \quad \text{at} \quad r = 0$$

apply Laplace transforms with respect to distance z to find the relation to predict $T(z, r)$.

This model is suitable for the case of heating liquids in laminar flow, since the wall layer sustains lower viscosity than central core fluid, hence the overall velocity profile becomes more plug shaped.

The first stage in the solution is to rectify the defining equation as

$$\tau = \frac{z}{V_0} \frac{\alpha}{R^2}, \qquad \text{dimensionless local residence time}$$

$$\xi = \frac{r}{R}, \qquad \text{dimensionless radial coordinate}$$

$$\theta = \frac{T - T_0}{T_w - T_0}, \qquad \begin{array}{l} \text{dimensionless temperature,} \\ \text{which assures zero initial condition} \end{array}$$

Introducing the new variables reposes the problem without excess baggage

$$\frac{\partial \theta}{\partial \tau} = \frac{1}{\xi} \frac{\partial}{\partial \xi}\left(\xi \frac{\partial \theta}{\partial \xi}\right)$$

subject to

$$\theta(0, \xi) = 0$$

$$\theta(\tau, 1) = \theta_w = 1$$

$$\frac{\partial \theta}{\partial \xi} = 0, \qquad \xi = 0$$

We now take Laplace transforms with respect to τ, which is timelike

$$\theta(s,\xi) = \int_0^\infty e^{-s\tau} \theta(\tau,\xi)\, d\tau$$

to obtain

$$s\theta(s,\xi) = \frac{1}{\xi}\frac{d}{d\xi}\left(\xi \frac{d\theta(s,\xi)}{d\xi}\right)$$

Performing the differentiation and rearranging yields Bessel's equation

$$\xi^2 \frac{d^2\theta(s,\xi)}{d\xi^2} + \xi \frac{d\theta(s,\xi)}{d\xi} - \xi^2 s\theta(s,\xi) = 0$$

We can express solutions in terms of $I_0(\sqrt{s}\,\xi)$ or $J_0(i\sqrt{s}\,\xi)$; we select the latter because of its known transcendental properties (i.e., see Table 3.2, which gives eigenvalues for $J_0(\lambda_n) = 0$).
   The general solution can now be written as

$$\theta(s,\xi) = A(s)J_0(i\sqrt{s}\,\xi) + B(s)Y_0(i\sqrt{s}\,\xi)$$

However, at the centerline, the symmetry condition requires $\theta(\tau,\xi)$ (and as corollary, $\theta(s,\xi)$) to be at least finite, hence, the function $Y_0(i\sqrt{s}\,\xi)$ is inadmissible, so set $B(s) = 0$. This gives the uncluttered result

$$\theta(s,\xi) = A(s)J_0(i\sqrt{s}\,\xi)$$

The arbitrary constant $A(s)$ is found by transforming the condition at $\xi = 1$, so that

$$\mathcal{L}\theta(s,1) = \mathcal{L}(1) = 1/s$$

so that $A(s)$ becomes

$$A(s) = \frac{1}{s}\frac{1}{J_0(i\sqrt{s})}$$

hence, we finally obtain

$$\theta(s,\xi) = \frac{1}{s}\frac{J_0(i\sqrt{s}\,\xi)}{J_0(i\sqrt{s})}$$

At first glance, this appears quite troublesome, since it appears to contain simple poles (e.g., at $s = 0$) and also multivalued complex functions within the arguments of transcendental functions (i.e., $J_0(i\sqrt{s}\,)$). To show that the branch point disappears, expand the Bessel functions in series according to Eq. 3.150

$$J_0(i\sqrt{s}) = 1 - \frac{\left(\frac{1}{2}i\sqrt{s}\right)^2}{(1!)^2} + \frac{\left(\frac{1}{2}i\sqrt{s}\right)^4}{(2!)^2} - \cdots$$

which is an even series, thereby eliminating $\sqrt{s}$ as a factor, and it also eliminates i

$$\frac{1}{s}\frac{J_0(i\sqrt{s}\,\xi)}{J_0(i\sqrt{s})} = \frac{1}{s}\left[\frac{1 + \frac{1}{4}s\xi + \frac{1}{64}s^2\xi^4 + \cdots}{1 + \frac{1}{4}s + \frac{1}{64}s^2 + \cdots}\right]$$

From this, it is clear that branch points do not exist, and the denominator contains only simple poles, which are the countably infinite roots of the polynomial

$$s_n\left[1 + \frac{1}{4}s_n + \frac{1}{64}s_n^2 + \cdots\right] = 0$$

The bracketed term obviously contains no poles at $s_n = 0$, since at $s_n = 0$ it became unity. It is easy to find the poles (s_n) by defining

$$i\sqrt{s_n} = \lambda_n$$

$$-s_n = \lambda_n^2$$

hence, the countably infinite poles are obtained from

$$J_0(\lambda_n) = 0$$

which from Table 3.2 gives, for example, the first two

$$s_1 = -(2.4048)^2 = -\lambda_1^2$$

$$s_2 = -(5.5201)^2 = -\lambda_2^2$$

There also exists a simple pole at $s = 0$, arising from the term $1/s$. The inversion of $\theta(s, \xi)$ can now be written formally as

$$\theta(\tau, \xi) = \sum_{s_n = 0}^{\infty} \text{Residues}\left\{e^{s_n\tau}\frac{J_0(i\sqrt{s_n}\,\xi)}{sJ_0(i\sqrt{s_n})}\right\}$$

We calculate the residue at $s = 0$ using Eq. 9.78

$$\text{Res}\left[e^{s\tau}F(s); \quad s = 0\right] = \left.\frac{e^{s\tau}J_0\left(i\sqrt{s}\,\xi\right)}{J_0\left(i\sqrt{s}\,\right)}\right|_{s=0} = 1$$

The remaining residues can be obtained using the method illustrated by Eq. 9.82

$$\text{Res}\left[e^{s\tau}F(s); s_n = -\lambda_n^2\right] = \left.\frac{e^{s\tau}J_0\left(i\sqrt{s}\,\xi\right)}{\dfrac{d}{ds}\left[sJ_0\left(i\sqrt{s}\,\right)\right]}\right|_{s_n = -\lambda_n^2}$$

Now, the derivative is

$$\frac{d}{ds}\left[sJ_0\left(i\sqrt{s}\,\right)\right] = J_0\left(i\sqrt{s}\,\right) + s\frac{dJ_0(u)}{du}\frac{di\sqrt{s}}{ds}$$

where we have defined $u = i\sqrt{s}$. The derivative $dJ_0(u)/du$ can be obtained from Eq. 3.197

$$\frac{dJ_0(u)}{du} = -J_1(u)$$

hence we have

$$\frac{d}{ds}\left[sJ_0\left(i\sqrt{s}\,\right)\right] = J_0\left(i\sqrt{s}\,\right) - \frac{i}{2}\sqrt{s}\,J_1\left(i\sqrt{s}\,\right)$$

So, remembering $i\sqrt{s_n} = \lambda_n$, the residues are

$$\text{Res}\left[e^{s\tau}F(s); s_n = -\lambda_n^2\right] = \sum_{n=1}^{\infty}e^{-\lambda_n^2\tau}\frac{J_0(\lambda_n\xi)}{\dfrac{-\lambda_n}{2}J_1(\lambda_n)}$$

Summing up all residues, yields finally

$$\theta(\tau, \xi) = 1 - 2\sum_{n=1}^{\infty}\frac{J_0(\lambda_n\xi)}{\lambda_nJ_1(\lambda_n)}\exp\left(-\lambda_n^2\tau\right)$$

where $J_0(\lambda_n) = 0$; $n = 1, 2, 3 \ldots$. The corresponding values of $J_1(\lambda_n)$ can be interpolated using Table 3.1, or calculated directly using the expansion given in Eq. 3.149 where $\Gamma(n + 2) = (n + 1)!$

The average for θ can be obtained from

$$\bar{\theta}(\tau) = 2\int_0^1\theta(\tau, \xi)\xi\,d\xi = 1 - 4\sum_{n=1}^{\infty}\exp\left(-\lambda_n^2\tau\right)/\lambda_n^2$$

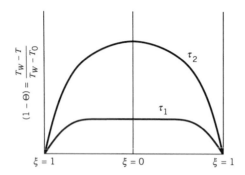

Figure 10.7 Dimensionless temperature profiles for plug flow heat exchanger.

since the integral

$$\int_0^1 J_0(\lambda_n \xi) \xi \, d\xi = \frac{1}{\lambda_n} \xi J_1(\lambda_n \xi) \Big|_0^1 = J_1(\lambda_n)/\lambda_n$$

is obtained from the general formula in Eq. 3.204. For conditions where $\tau \gg 1$, the series can be truncated since only the first exponential time function is significant, hence, approximately

$$\bar{\theta}(\tau) \simeq 1 - 4\exp(-\lambda_1^2 \tau)/\lambda_1^2$$

It is also clear in the limit as $\tau \to 0$, the summation of terms in Eq. 9.233 must be such that

$$2\sum_{n=1}^{\infty} \frac{J_0(\lambda_n \xi)}{\lambda_n J_1(\lambda_n)} = 1$$

The behavior of $\theta(\tau, \xi)$ is illustrated in Fig. 10.7.

10.8 REFERENCES

1. Baron, T., W. R. Manning, and H. F. Johnstone, "Reaction Kinetics in a Tubular Reactor," *Chem. Eng. Prog.*, 48, 125 (1952).
2. Brown, G. M., "Heat and Mass Transfer in a Fluid in Laminar Flow in a Circular or Flat Conduit," *AIChE Journal*, 10, 179–183 (1960).
3. Crank, J., *The Mathematics of Diffusion*, Clarendon Press, Oxford (1975).
4. Do, D. D., and R. G. Rice, "Validity of the Parabolic Profile Assumption in Adsorption Studies," *AIChE Journal*, 32, 149–154 (1986).
5. Jacob, M., *Heat Transfer*, Vol. 1, John Wiley & Sons, New York (1949).
6. Leveque, J., *Ann. Mines*, 13, pp. 201, 305, 381 (1928).
7. Masamune, S., and J. M. Smith, "Adsorption of Ethyl Alcohol on Silica Gel," *AIChE Journal*, 11, 41-45 (1965).

8. Mickley, A. S., T. K. Sherwood, and C. E. Reid, *Applied Mathematics in Chemical Engineering*, McGraw-Hill, New York (1957).
9. Perry, R. H., and C. H. Chilton, *Chemical Engineer's Handbook*, Fifth Edition, McGraw Hill, New York (1973).
10. Rice, R. G., and M. A. Goodner, "Transient Flow Diffusometer," *Ind. Eng. Chem. Fundam.*, 25, 300 (1986).
11. Schechter, R. S., *The Variational Method in Engineering*, McGraw Hill, New York (1967).
12. Van de Ven, J., G. M. J. Rutten, M. J. Raaijmakers, and L. J. Giling, "Gas Phase Depletion and Flow Dynamics in Horizontal MOCVD Reactors," *Journal Crystal Growth*, 76, 352–372 (1986).

10.9 PROBLEMS

10.1₂ A very thin thermistor probe is placed into the center of a spherically shaped, unknown metal for the purposes of deducing thermal diffusivity. By weighing the metal ball, and measuring its diameter ($2R$), the density was calculated (ρ). The ball is held in an oven overnight and reaches an initially uniform temperature T_s. It is then placed in the middle of a stream of very fast flowing, cool liquid of temperature T_f. The ball begins to cool and the thermistor records the centerline temperature.
(a) Show that an elemental heat balance on the sphere yields

$$\frac{\partial T}{\partial t} = \alpha \frac{1}{r^2} \frac{\partial}{\partial r}\left(r^2 \frac{\partial T}{\partial r}\right)$$

where solid thermal diffusivity ($\alpha = k/\rho C_p$) can be treated as constant for small temperature excursions.
(b) Introduce dimensionless independent variables by letting

$$\xi = \frac{r}{R}; \qquad \tau = \frac{\alpha t}{R^2}$$

and apply the method of separation of variables of the form

$$T(\xi, \tau) = \varphi(\xi)\theta(\tau)$$

and give arguments that one must obtain

$$\frac{1}{\xi^2} \frac{d}{d\xi}\left(\xi^2 \frac{d\varphi}{d\xi}\right) + \lambda^2 \varphi = 0$$

$$\frac{d\theta}{d\tau} + \lambda^2 \theta = 0$$

(c) Equation for φ can be expressed as

$$\xi^2 \frac{d^2\varphi}{d\xi^2} + 2\xi \frac{d\varphi}{d\xi} + \lambda^2 \xi^2 \varphi = 0$$

which has a structure that is matched by the *Generalized Bessel equation*. In particular, if we take

$$a = 2; \quad b = 0; \quad c = 0; \quad d = \lambda^2; \quad s = 1$$

then show that

$$p = \frac{1}{2}; \quad \frac{\sqrt{d}}{s} = \lambda \quad \text{(real)}$$

$$\varphi(\xi) = \frac{1}{\sqrt{\xi}} \left[A_0 J_{1/2}(\lambda\xi) + B_0 J_{-1/2}(\lambda\xi) \right]$$

Show that this can be written in simpler form as

$$\varphi(\xi) = A_1 \frac{\sin(\lambda\xi)}{\xi} + B_1 \frac{\cos(\lambda\xi)}{\xi}$$

(d) Next, show that the solution for the case when $\lambda = 0$ yields

$$T = \frac{C}{\xi} + D$$

and the complete solution must be

$$T(\xi, \tau) = \exp(-\lambda^2\tau) \left[A \frac{\sin(\lambda\xi)}{\xi} + B \frac{\cos(\lambda\xi)}{\xi} \right] + \frac{C}{\xi} + D$$

(e) What values should B and C take to insure admissibility? Give arguments to support your assertions.

(f) What value must the constant D take to insure physical sense?

(g) Apply the boundary condition at $r = R$ or $\xi = 1$ and thereby deduce system eigenvalues.

(h) What is the specific function representing $\varphi(\xi)$? What are the boundary values for the function $\varphi(\xi)$? Are these homogeneous? Is the ODE describing $\varphi(\xi)$, along with its boundary conditions, of the Sturm–Liouville type? If so, what is the proper weighting function?

(i) Apply the initial condition and any other appropriate conditions to complete the analytical solution.

(j) Suppose the sphere's radius was $R = 1$ cm, and the initial temperature was measured to be $30°$ C. The flowing coolant temperature was constant at a temperature of $20°$ C. After 6.05 sec, the centerline temperature measured $20.86°$ C. What is the thermal diffusivity of the solid? Centerline temperature behavior is illustrated in Fig. 10.8 as a function of $\alpha t / R^2$.

Ans: $\alpha = 1/19$ cm^2/sec

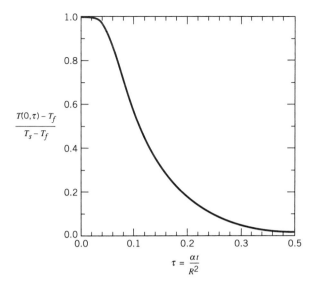

Figure 10.8 Centerline temperature behavior as a function of $\alpha t / R^2$.

10.2₃. We have seen in Example 10.2 (CVD Reactor) that conditions of short contact time (or short penetration) can lead to considerable simplifications, to produce practical results. Consider the case of heat transfer for laminar tube flow, under conditions of short contact time. Ignoring axial conduction, the steady-state heat balance was shown for parabolic velocity profile

$$2v_0\left[1 - \left(\frac{r}{R}\right)^2\right]\frac{\partial T}{\partial z} = \alpha\frac{1}{r}\frac{\partial}{\partial r}\left(r\frac{\partial T}{\partial r}\right)$$

where thermal diffusivity is $\alpha = k/\rho C_p$. For short penetration from a wall of temperature T_w, we introduce the wall coordinate $y = R - r$, so that the conduction term close to the wall is approximately

$$\frac{1}{r}\frac{\partial}{\partial r}\left(r\frac{\partial T}{\partial r}\right) \simeq \frac{\partial^2 T}{\partial y^2}$$

and the velocity close to the wall is

$$v_z \simeq 4v_0\left(\frac{y}{R}\right)$$

(a) By analogy with the CVD-reactor problem (Example 10.2), use the combined variable

$$\eta = \frac{y}{\left[\dfrac{9R\alpha z}{4v_0}\right]^{1/3}}$$

$$(T - T_w) = f(\eta)$$

and show that, for a fluid entering at T_0

$$\frac{T - T_w}{T_0 - T_w} = \frac{1}{\Gamma\left(\dfrac{4}{3}\right)} \int_0^\eta \exp(-\beta^3)\, d\beta$$

(b) By defining the flux at the wall as

$$q_0 = k\left[\frac{\partial T}{\partial y}\right]_{y=0}$$

we can define a local heat transfer coefficient based on the inlet temperature difference, so let

$$q_0 = h_0(z)(T_0 - T_w)$$

hence, show that

$$h_0(z) = \frac{\left(\dfrac{4}{9}\right)^{1/3}}{\Gamma\left(\dfrac{4}{3}\right)} \left[\frac{v_0 \rho C_p k^2}{Rz}\right]^{1/3}$$

(c) Define the average heat transfer coefficient using

$$\overline{h_0} = \frac{1}{L} \cdot \int_0^L h_0(z)\, dz$$

and thus show the dimensionless result

$$Nu = 1.62\left[Re\, Pr\frac{D}{L}\right]^{1/3}; \qquad 1.62 \simeq \frac{1.5\left(\dfrac{8}{9}\right)^{1/3}}{\Gamma\left(\dfrac{4}{3}\right)}$$

where

$$Nu = \frac{\overline{h_0}D}{k}, \quad Re = \frac{Dv_0\rho}{\mu}, \quad Pr = \frac{\mu C_p}{k}, \quad D = 2R$$

This result, after Leveque (1928), compares reasonably well with experiments. The multiplicative constant has been adjusted based on experimental data to give a value 1.86.

10.3₃. The Loschmidt diffusion cell is used to experimentally determine binary gas diffusivity. For ideal gases, the transport equation, for the hollow cylindrical cells, is Fick's second law of diffusion

$$\frac{\partial x_A}{\partial t} = D_{AB}\frac{\partial^2 x_A}{\partial z^2}$$

where x_A denotes mole fraction A, z is distance and t is time. Thus, two equivolume cylindrical cells, each of length L are joined by a thin, removable membrane. One cell is loaded with pure component A, the second is loaded with pure B. The thin membrane is suddenly removed, and interdiffusion commences, according to the above transport equation. The initial and boundary conditions can be written

$$\frac{\partial x_A}{\partial z} = 0, \quad \text{at} \quad z = \pm L$$

$$x_A = 1, \quad -L \le z \le 0, \quad t = 0$$

$$x_A = 0, \quad 0 \le z \le L, \quad t = 0$$

(a) Apply the method of separation of variables and show
   (i) the expression to compute the system eigenvalues is:

$$\lambda_n = \left(\frac{2n+1}{2}\right)\pi; \quad n = 0, 1, 2, \cdots$$

   (ii) the expression for x_A takes the form

$$x_A = \frac{1}{2} + \sum_{n=0}^{\infty} A_n \sin\left(\lambda_n \frac{z}{L}\right)\exp\left(-\lambda_n^2 \frac{tD_{AB}}{L^2}\right)$$

(b) Show that $A_n = -1/\lambda_n$, using the initial condition.

(c) The average (well-mixed) composition of each chamber after an exposure time denoted as τ, is represented by the integrals

$$\left(\overline{x_A}\right)_{\text{BOT}} = \frac{\int_{-L}^{0} x_A \, dz}{\int_{-L}^{0} dz}$$

$$\left(\overline{x_A}\right)_{\text{TOP}} = \frac{\int_{0}^{L} x_A \, dz}{\int_{0}^{L} dz}$$

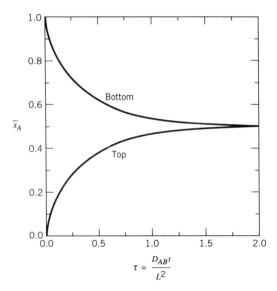

$$\tau = \frac{D_{AB}t}{L^2}$$

Figure 10.9

Find expressions to compute the average compositions; truncate the series for large time. Typical behavior for average compositions are illustrated in Fig. 10.9 as a function of $\tau = D_{AB}t/L^2$.

(d) For long contact time, deduce an approximate expression to calculate D_{AB} directly from average composition.

$$\text{Ans: (c)} \quad \bar{x}_A \approx \frac{1}{2} \pm \frac{1}{\lambda_1^2}\exp\left(-\lambda_1^2\frac{tD_{AB}}{L^2}\right); \quad \begin{array}{l} + \text{for bottom} \\ - \text{for top} \end{array}$$

10.4₃. Modern blood dialysis modules are made up of around 1 million microsized, hollow fibers bound to a tube sheet in a fashion similar to a traditional countercurrent heat exchanger. Thus, blood containing unwanted solute (such as uric acid) is forced into a header, which then distributes blood flow into each of the 1 million hollow fibers (tubes). A header at the exit recollects the cleansed blood, which is then pumped back to the patient. A dialysis solution (free of solutes, such as uric acid) passes in large volumes across the outer surface of the many hollow fiber tubes. The walls of the hollow fibers are semipermeable and allow certain substances, such as uric acid, to diffuse through the polymeric material and thence into the fast flowing dialysis solution.

It is important to predict from first principles the time on-line to reduce blood solute content to certain acceptable levels (much as a functioning human kidney should do).

Because the solute flux is small relative to total blood flow, we can model a single hollow fiber as a straight pipe with walls permeable to certain solutes, so at quasi-steady state in the flowing blood phase, the solute balance for species A is taken as

$$V_0 \frac{\partial C_A}{\partial z} = D_A \frac{1}{r} \frac{\partial}{\partial r} \left(r \frac{\partial C_A}{\partial r} \right)$$

where v_0 is the constant, uniform (plug) velocity profile, D_A is solute diffusivity in blood, and C_A is the solute molar composition. To account for mass transfer resistances through the wall and across the outer tube film resistance, we define an overall transport coefficient

$$\frac{1}{K_{0L}} = \frac{t_w}{D_w} + \frac{1}{k_0}$$

where t_w is wall thickness, D_w is diffusivity of species A through polymeric wall, and k_0 accounts for film resistance on the dialysis solution side. From this, we can write the flux at the blood–wall interface as

$$-D_A \left[\frac{\partial C_A}{\partial r} \right]_{r=R} = K_{0L} [C_A - C_D]_{r=R}$$

where C_D is concentration of solute A in dialysis solution, usually taken as zero.

(a) We shall denote the blood inlet solute composition as C_0 at the axial position $z = 0$. For a single pass system, show that the average composition C_A obeys the analytical solution at position $z = L$

$$\frac{\overline{C_A}}{C_0} = 4 \sum_{n=1}^{\infty} \frac{\exp\left[-\lambda_n^2 \frac{LD_A}{v_0 R^2} \right]}{\lambda_n^2 \left(1 + \frac{\lambda_n^2}{\text{Bi}^2} \right)}$$

where $\text{Bi} = K_{0L} R / D_A$ and the eigenvalues are obtained from

$$\lambda_n J_1(\lambda_n) = \text{Bi} \, J_0(\lambda_n)$$

Figure 10.10 illustrates behavior of the exit composition as Biot number changes.

(b) For normal kidney function, a uric acid level for men is 3.5–8.5 mg/dl, and for women it is 2.5–7.5. Suppose the blood from a patient sustains a level 20 mg/dl, which as a one-pass design goal we wish to reduce to 7.5. Furthermore, take the hollow-fiber length to be 22.5 cm, with an inner radius of 0.25 mm, and wall thickness of 0.1 mm. Diffusivity of uric acid through the polymeric wall is around 10^{-7} cm^2/sec, while diffusivity of uric acid in blood is taken as 2.5×10^{-6} cm^2/sec. You may correctly assume k_0 is quite large, so film resistance on the dialysis solution side is negligible. What is the required single-tube blood velocity to meet the design goal?

Ans: 0.18 mm/sec

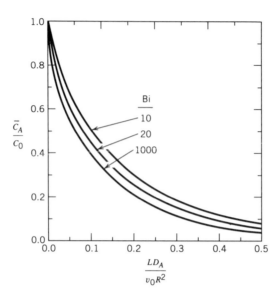

Figure 10.10

10.5_3. Adsorption of low-volatility substances on silica gel is controlled by surface diffusion (Masamune and Smith 1965) so that a model for uptake of solutes such as ethyl alcohol on spherical pellets suggests that equilibrium adsorption at the particle outer surface is followed by surface diffusion into the solid particle interior. Assuming the particle size is unchanged in time, the appropriate material balance is:

$$\frac{\partial q}{\partial t} = D\frac{1}{r^2}\frac{\partial}{\partial r}\left(r^2\frac{\partial q}{\partial r}\right)$$

subject to conditions

$$q = 0, \quad t = 0, \quad 0 < r < R$$

$$q = q^*, \quad t > 0, \quad r = R$$

$$\frac{\partial q}{\partial r} = 0, \quad t > 0, \quad r = 0$$

The composition of the solid phase $q(r, t)$ (moles solute/gram solid) can be determined by separation of variables. The case of spherical geometry can be mitigated somewhat by a change of variables: $q = u(r, t)/r$, hence, yielding the elementary result

$$\frac{\partial u}{\partial t} = D\frac{\partial^2 u}{\partial r^2}$$

(a) Find the average particle composition defined as

$$\bar{q} = \left(\frac{3}{R^3}\right)\int_0^R q(r,t)r^2\,dr$$

and show that the final result agrees with Crank (1975)

$$\frac{\bar{q}}{q^*} = 1 - \frac{6}{\pi^2}\sum_{n=1}^{\infty}\frac{1}{n^2}\exp\left[\frac{-n^2\pi^2 Dt}{R^2}\right]$$

(b) Do and Rice (1986) suggest a linear driving force (LDF) result can be obtained by assuming the existence of a parabolic profile within the particle

$$q = a_0(t) + a_2(t)r^2$$

Use this and the transport equation to show

$$\frac{\partial\bar{q}}{\partial t} = \frac{15D}{R^2}(q^* - \bar{q})$$

This LDF approximation has become widely used to model complicated processes, such as Pressure–Swing–Adsorption.

10.6*. When a thin membrane is stretched over a cylindrical hoop, the drumhead surface movement $w(r,t)$ obeys the approximate force balance

$$\sigma\frac{1}{r}\frac{\partial}{\partial r}\left(r\frac{\partial w}{\partial r}\right) = \rho\frac{\partial^2 w}{\partial t^2}$$

where ρ is membrane density grams/cm^2, and σ is the applied tension, dynes per centimeter. Suppose the initial condition is such that the membrane shape is

$$w(r,0) = w_0\left[1 - \left(\frac{r}{R}\right)^2\right]$$

where R denotes hoop radius.

(a) Show that the membrane vibrates according to the relation

$$\frac{w(r,t)}{w_0} = 4\sum_{n=1}^{\infty}\frac{J_2(\lambda_n)}{\lambda_n^2 J_1^2(\lambda_n)}J_0\left(\lambda_n\frac{r}{R}\right)\cos\left[t\sqrt{\frac{\sigma}{\rho}}\frac{\lambda_n}{R}\right]$$

where

$$J_0(\lambda_n) = 0$$

Note, recurrence relations show

$$J_2(\lambda_n) = \frac{2J_1(\lambda_n)}{\lambda_n}$$

so

$$A_n = \frac{8}{\lambda_n^3 J_1(\lambda_n)}$$

(b) Find an expression to compute the lowest possible frequency; this is related to the method of tuning drums, violins, and so on.

$$\text{Ans: } \omega_1 = \frac{2.405}{R}\sqrt{\frac{\sigma}{\rho}}$$

10.7₃. One method to assess axial dispersion in packed beds is to inject a small amount of solute upstream, N_0, and measure its distribution downstream $C_A(x, t)$. Thus, for plug flow with interstitial velocity v, the material balance can be shown to be

$$v\frac{\partial C_A}{\partial x} + \frac{\partial C_A}{\partial t} = D_a \frac{\partial^2 C_A}{\partial x^2}$$

where D_a denotes axial dispersion coefficient. A simpler equation results by introducing a coordinate moving with the fluid: $z = x - vt$.

(a) Show that the change of coordinate gives a form of Fick's second law

$$\frac{\partial C_A}{\partial t} = D_a \frac{\partial^2 C_A}{\partial z^2}$$

(b) A new particular solution (to some unspecified problem) can be obtained by differentiating the error function solution for step input; show that this gives

$$C_A = \frac{K}{\sqrt{4D_a t}} \exp\left(-\frac{z^2}{4D_a t}\right)$$

We shall use this solution to describe response to a pulse of solute.

(c) Find the constant K by noting that N_0 must be conserved at any time, t > 0 to get[1]

$$C_A(z, t) = \left(\frac{N_0}{A_f}\right)\frac{1}{\sqrt{4\pi D_a t}} \exp\left[-\frac{z^2}{4D_a t}\right]$$

where A_f is flow area, defined as total area times porosity.

---

[1] The behavior at $z = 0$ is similar to an impulse, in the sense that $C_A \to \infty$ as $t \to 0$.

(d) Use this result to suggest methods for finding D_a if experimental response curves at some position $x = L$ are known. Note, the residence time for injected solute is L/v, yet the maximum of C_A arrives earlier by an amount D_a/v^2, when $D_a/v^2 \ll L/v$. Prove this assertion.

10.8*. An incompressible fluid, initially at rest in a circular tube of length L, is subjected to the application of a step change in pressure gradient, so that for laminar conditions, the local velocity along the tube axis obeys

$$\rho \frac{\partial v}{\partial t} = \mu \frac{1}{r} \frac{\partial}{\partial r} \left(r \frac{\partial v}{\partial r} \right) + \frac{\Delta p}{L}$$

At the final steady state, the velocity profile obeys

$$\mu \frac{1}{r} \frac{\partial}{\partial r} \left(r \frac{\partial \bar{v}}{\partial r} \right) + \frac{\Delta p}{L} = 0$$

hence, since

$$r = R, \quad \bar{v} = 0 \quad \text{and} \quad r = 0, \quad \frac{d\bar{v}}{dr} = 0$$

then

$$\bar{v}(r) = \left(\frac{R^2 \Delta p}{4\mu L} \right) \left[1 - \left(\frac{r}{R} \right)^2 \right]$$

As it stands, the equation describing transient velocity is inhomogeneous, and a separation of variables approach will fail. This can be remedied by the following technique. Define velocity as being made up of two parts, a steady part plus a deviation from steady state

$$v(r, t) = \bar{v}(r) + y(r, t)$$

When this is inserted above, the steady part causes cancellation of $\Delta p/L$; hence,

$$\rho \frac{\partial y}{\partial t} = \mu \frac{1}{r} \frac{\partial}{\partial r} \left(r \frac{\partial y}{\partial r} \right)$$

(a) Show that the deviation velocity $y(r, t)$ must obey the initial condition

$$y(r, 0) = -2v_0 \left[1 - \left(\frac{r}{R} \right)^2 \right]$$

where

$$2v_0 = \frac{R^2 \Delta p}{4\mu L}$$

and v_0 is the average tube velocity at steady state; the no-slip and symmetry conditions are also obeyed; hence,

$$y(R, t) = 0 \quad \text{and} \quad \frac{\partial y(0, t)}{\partial r} = 0$$

(b) The equation and boundary conditions are now homogeneous; apply the separation variables method and show

$$y(\xi, \tau) = \sum_{n=1}^{\infty} A_n J_0(\lambda_n \xi) \exp(-\lambda_n^2 \tau)$$

where

$$\xi = \frac{r}{R}; \quad \tau = \frac{\mu t}{\rho R^2}; \quad J_0(\lambda_n) = 0$$

(c) Evaluate A_n and obtain the analytical prediction

$$\frac{y(\xi, \tau)}{v_0} = -\sum_{n=1}^{\infty} \frac{16}{\lambda_n^3 J_1(\lambda_n)} J_0(\lambda_n \xi) \exp(-\lambda_n^2 \tau)$$

The absolute velocity is calculated using

$$\frac{v(\xi, \tau)}{v_0} = 2(1 - \xi^2) - \sum_{n=1}^{\infty} \frac{16}{\lambda_n^3 J_1(\lambda_n)} J_0(\lambda_n \xi) \exp(-\lambda_n^2 \tau)$$

(d) Estimate the time required for the centerline velocity to reach 99% of the final steady value if $\mu/\rho = 1 \text{ cm}^2/\text{sec}$ and the tube radius is 1 cm. The development of centerline velocity is depicted in Fig. 10.11.

Ans.: $t_{SS} = 0.814$ sec

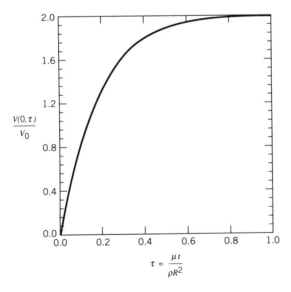

Figure 10.11

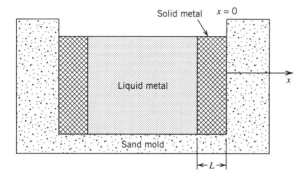

Figure 10.12

10.9*. The production of metals in industries involves solidification, and sand molds are used extensively (see Fig. 10.12).

(a) As a result of heat transfer during the solidification process, where is the dominant heat transfer resistance located? Explain your answer.

(b) To understand the heat flow in the sand mold, set up the coordinate x as shown in the figure. Explain why the coordinate system selected is a good one.

(c) If there is no heat resistance in the metal melt, what is the boundary condition at $x = 0$ (i.e., at metal/sand mold interface)?

(d) In practice, the temperature profile in the mold penetrates only a short distance. (Short penetration theory is applicable.) Considering this circumstance, suggest a suitable condition far from the metal–sand interface.

(e) The sand mold is initially at ambient temperature before the metal melt is poured into the mold. Write down the initial condition for the heat balance equation.

(f) Show that the appropriate transient heat balance is given by

$$\frac{\partial T}{\partial t} = \alpha \frac{\partial^2 T}{\partial x^2}$$

(g) Show that a particular solution to part (f) can be written in terms of the error function (Chapter 4)

$$\frac{T - T_M}{T_0 - T_M} = \text{erf}\left(\frac{x}{2\sqrt{\alpha t}}\right)$$

where T_M is the melt temperature, and T_0 is the initial sand mold temperature.

(h) Show the local heat flux at $x = 0$ is given by

$$q_x\big|_{x=0} = -k\frac{\partial T}{\partial x}\bigg|_{x=0} = \frac{k(T_M - T_0)}{\sqrt{\pi \alpha t}}$$

(i) Let L be the thickness of the solid metal (see Fig. 10.8), carry out the heat balance around the solid metal and show that the heat balance is

$$\rho_M H_M \frac{dL}{dt} = \frac{k(T_M - T_0)}{\sqrt{\pi \alpha t}}$$

What do ρ_M and H_M represent? What is the initial condition for this heat balance equation?

(j) Integrate the equation in part (i) to derive the solution for the metal thickness L as a function of time. Discuss this result and suggest ways to reduce the solidification time.

10.10*. In order to obtain the solubility of gas in a polymer membrane and the diffusivity of the dissolved gas within the membrane, an experimental system is set up as follows. A membrane is placed between two closed reservoirs—one is large and the other is small (see Fig. 10.13).

Initially, the membrane and the two reservoirs are thoroughly evacuated with a vacuum pump. The system is then isolated from the vacuum pump by closing in-line valves. Next, a dose of gas is introduced into the bottom large reservoir, such that its pressure is P_0. This gas then dissolves into the membrane and then diffuses to the top (small) reservoir, where its pressure is recorded with a highly sensitive pressure transducer.

(a) If composition varies only in the direction normal to the flat interface, perform a transient shell balance on a thin slice of membrane using Fick's law to describe diffusion flux of dissolved gas.

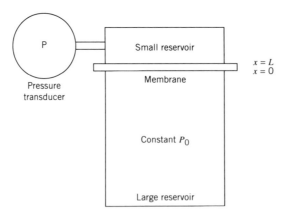

Figure 10.13

(b) Solubility can be described by Henry's law, so that at the lower interface (denoted as $x = 0$) take the boundary condition

$$C(x = 0; t) = HP_0$$

Pressure is so low that convective transport in the membrane by Darcy's law is negligible.

(c) Show that the solution to the mass balance equation is

$$\frac{C}{HP_0} = \left[1 - \frac{x}{L}\right] - \frac{2}{\pi} \sum_{n=1}^{\infty} \frac{\sin\left(n\pi \frac{x}{L}\right)}{n} \exp\left(-\frac{n^2 \pi^2 Dt}{L^2}\right)$$

(d) Note there are two distinct group of terms in this solution. What does the first (bracketed) term represent?

(e) If the volume of the small reservoir is V and the recorded pressure at the time t is P, write down the mass balance equation of this reservoir, and show that it has the form

$$\frac{V}{R_g T} \frac{dP}{dt} = -AD \frac{\partial C}{\partial x}\bigg|_{x=L}$$

where A denotes surface area.

(f) Substitute the solution for C in part (c) to the mass balance equation of part (e), and integrate to show that the solution for the pressure of the small reservoir is

$$\frac{P}{P_0} = \frac{AR_g T}{VL}\left\{HDt + \frac{2HL^2}{\pi^2} \sum_{n=1}^{\infty} \frac{\cos(n\pi)}{n^2}\left[1 - \exp\left(-\frac{n^2\pi^2 Dt}{L^2}\right)\right]\right\}$$

The response of the small reservoir pressure ratio is illustrated in Fig. 10.14.

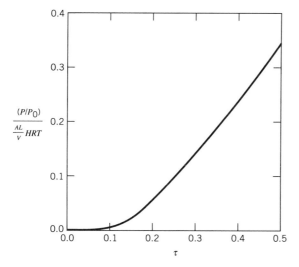

Figure 10.14

(g) At sufficiently large time, the solution of part (f) is reduced to

$$P = \frac{AR_g TP_0}{VL}\left(HDt - \frac{L^2 H}{6}\right)$$

Use this relation to show how the solubility and diffusivity are determined from experimental data.

10.11*. Dissolution of a solid particle in a finite volume (e.g., dissolution of a medicine tablet in an organ) may be considered as a kinetic process with the existence of a surface layer adjacent to the solid surface and the diffusion through this layer being the controlling step to the dissolution process. The dissolved material is then consumed by the different parts of the organ according to a first order chemical reaction.

(a) The thickness of the surface layer around the solid particle is taken as δ. Assuming the particle is spherical in shape, derive the mass balance equation for the dissolved species in this layer (see Fig. 10.15). Then show that under quasi-steady-state conditions the mass balance equation is

$$\frac{D}{r^2}\frac{\partial}{\partial r}\left(r^2 \frac{\partial C}{\partial r}\right) = 0$$

What does this quasi steady state imply? What do D and C represent?

(b) The boundary conditions to the mass balance in part (a) are:

$$r = R; \qquad C = C_0$$
$$r = R + \delta; \qquad C = C_b$$

What do R, C_0, and C_b represent?

(c) If the density of the particle is ρ and the molecular weight is M, show that the particle radius is governed by the following equation

$$\frac{\rho}{M}\frac{dR}{dt} = D\frac{\partial C}{\partial r}\Big|_{r=R}$$

(*Hint*: Carry out mass balance around the particle.)

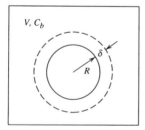

V, C_b

δ

R

Figure 10.15

(d) If the consumption of the dissolved species in the finite volume is very fast, that is, $C_b \approx 0$, solve the mass balance equation for the dissolved species in the surface layer in part (a) and the particle radius equation in part (c) to derive a solution for the particle radius as a function of time, and then determine the time it takes to dissolve completely the solid particle.

(e) If the consumption of the dissolved species in the finite volume is slow and the rate of consumption per unit volume is kC_b, derive a mass balance equation for the dissolved species in the finite volume. Solve this mass balance equation together with the mass balance in part (a) and the particle radius equation in part (d) to obtain solutions for the particle radius and the dissolved species concentrations as a function of time.

$$\text{Ans: } (d)\ t_f = \frac{\rho\delta\left(R_0 + \delta\ln\left(\dfrac{\delta}{R_0 + \delta}\right)\right)}{DMC_0}$$

10.12*. As a design engineer, you are asked by your boss to design a wetted wall tower to reduce a toxic gas in an air stream down to some acceptable level. At your disposal are two solvents, which you can use in the tower; one is nonreactive with the toxic gas but is cheap, whereas the other is reactive and quite expensive. In order to choose which solvent to use, you will need to analyze a model to describe the absorption of the toxic gas into the flowing solvent (see Fig. 10.16).

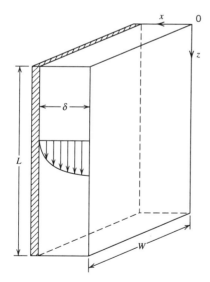

Figure 10.16

(a) For the nonreactive solvent, derive from first principles the mass balance equation for the absorbed toxic gas into the solvent and show

$$V_{max}\left[1 - \left(\frac{x}{\delta}\right)^2\right]\frac{\partial C}{\partial z} = D\frac{\partial^2 C}{\partial x^2}$$

(b) What are the boundary conditions for the mass balance equation obtained in part (a). Consider two cases. In Case 1, the gas stream is turbulent, whereas in Case 2, it is not.

(c) For the reactive solvent, derive from the first principles the mass balance for the absorbed toxic species and assume that the reaction between the absorbed toxic species and the solvent follows first order kinetics.

$$V_{max}\left[1 - \left(\frac{x}{\delta}\right)^2\right]\frac{\partial C}{\partial z} = D\frac{\partial^2 C}{\partial x^2} - kC$$

(d) Assuming that you have obtained the solution for the distribution of absorbed concentration of the toxic species, obtain a formula to calculate the mass flux (moles/area-time) into the falling film as a function of z.

$$N(z) = -D\frac{\partial C}{\partial x}\bigg|_{x=0}$$

Next, for a falling film of length L and width W, obtain a formula for the mass transfer (moles/time) into the film.

$$\text{Mass flux} = W\int_0^L N(z)\,dz = -WD\int_0^L \frac{\partial C}{\partial x}\bigg|_{x=0}dz$$

(e) For a given length and width and a given flow rate of liquid, which solvent do you expect to give higher absorption rate and why?

10.13₃. The heat exchanger described in Section 1.2 operates at steady state until an upset occurs in the inlet temperature.

(a) Show that the fluid temperature response obeys the PDE

$$\rho C_p\frac{\partial T}{\partial t} + V_0\rho C_p\frac{\partial T}{\partial z} + \left(\frac{2h}{R}\right)[T(z,t) - T_w] = 0$$

with the stipulation that the wall temperature (T_w) remains everywhere constant.

(b) Introduce deviation variables of the form

$$T(z,t) = \bar{T}(z) + \hat{T}(z,t)$$

$$T_0(t) = \bar{T}_0 + \hat{T}_0(t)$$

and show that the following equations result

$$\rho C_p \frac{\partial \hat{T}}{\partial t} + V_0 \rho C_p \frac{\partial \hat{T}}{\partial z} + \left(\frac{2h}{R}\right)\hat{T} = 0; \quad \hat{T}(z,0) = 0; \quad \hat{T}(0,t) = \hat{T}_0(t)$$

$$V_0 \rho C_p \frac{d\overline{T}}{dz} + \frac{2h}{R}\left[\overline{T} - \overline{T}_0\right] = 0$$

The solution to the steady-state equation has already been obtained and is given in Eq. 1.17 as

$$\overline{T}(z) = T_w + (\overline{T}_0 - T_w)\exp\left(-\frac{2h}{R}\frac{z}{V_0 \rho C_p}\right)$$

(c) If the inlet temperature sustains a step change of magnitude $\alpha \overline{T}_0$ $(0 < \alpha < 1)$, so that $\hat{T}_0(t) = \alpha \overline{T}_0 u(t)$, show that the Laplace transform $\mathscr{L}\hat{T}(z,t)$ is

$$\hat{T}(s,z) = \alpha \overline{T}_0 \frac{1}{s}\exp(-s\tau)\exp(-\tau/\tau_i)$$

where

$$\tau = \frac{z}{V_0} \quad \cdots \text{ local residence time}$$

$$\tau_i = \frac{2h}{R\rho C_p} \quad \cdots \text{ thermal time constant}$$

(d) Invert the Laplace transform and show that

$$\hat{T}(t,\tau) = \alpha \overline{T}_0 \exp\left(-\frac{\tau}{\tau_i}\right)u(t - \tau)$$

where

$$u(t - \tau) = \begin{cases} 0 & \text{when } t < \tau \\ 1 & \text{when } t > \tau \end{cases}$$

Thus, the disturbance does not appear at the heat exchanger exit until one full residence time has elapsed. The dynamics associated with the wall may have considerable significance, as illustrated in Problem 10.14*.

10.14*. Suppose the heat exchanger in Problem 10.13 is operated so that the coolant stream temperature outside the tubular wall is relatively constant, but the wall temperature is allowed to vary. Denote the inside and outside film coefficients as h_i and h_0, respectively, and if conduction along the wall axis can be ignored, it is easy to show the applicable thermal balances are

$$\rho C_p \frac{\partial T}{\partial t} + V_0 \rho C_p \frac{\partial T}{\partial z} + \frac{2h_i}{R}[T(z,t) - T_w(t)] = 0$$

$$A_w \rho_w C_{pw} \frac{\partial T_w}{\partial t} = h_i P_i (T - T_w) - h_0 P_0 (T_w - T_c)$$

where A_w denotes the wall cross-sectional area and T_c represents the (constant) coolant temperature. The inner and outer tube perimeters are approximately equal, so take $P_i \sim P_0 = P = 2\pi R$.

(a) Rearrange the dynamic equations into a more tractable form and show that

$$\frac{\partial T}{\partial t} + \frac{\partial T}{\partial \tau} + \frac{1}{\tau_i}(T - T_w) = 0$$

$$C\frac{\partial T_w}{\partial t} = \frac{1}{\tau_i}(T - T_w) - \frac{1}{\tau_0}(T_w - T_c)$$

where

$$\tau = z/V_0 \quad \cdots \text{ hot fluid residence time}$$

$$\tau_i = \frac{2h_i}{R\rho C_p} \quad \cdots \text{ inner thermal time constant}$$

$$\tau_0 = \frac{2h_0}{R\rho C_p} \quad \cdots \text{ outer thermal time constant}$$

$$C = \left[\left(\frac{R_0}{R}\right)^2 - 1\right]\frac{\rho_w C_{pw}}{\rho C_p} \quad \cdots \text{ thermal capacitance ratio}$$

$$R_0 = \text{ outer tube radius}$$

(b) Introduce perturbation variables of the form

$$T(t,\tau) = \bar{T}(\tau) + \hat{T}(t,\tau)$$

$$T_w(t,\tau) = T_w(\tau) + \hat{T}_w(t,\tau)$$

and show that the new dynamic equations are:

$$\frac{\partial \hat{T}}{\partial t} + \frac{\partial \hat{T}}{\partial \tau} + \frac{1}{\tau_i}(\hat{T} - \hat{T}_w) = 0$$

$$C\frac{\partial \hat{T}_w}{\partial t} = \frac{1}{\tau_i}(\hat{T} - \hat{T}_w) - \frac{1}{\tau}\hat{T}_w$$

where we note that T_c is constant (hence, $\hat{T}_c = 0$).

(c) Solve the steady-state equations and show that the overall heat transfer coefficient, $1/U = 1/h_i + 1/h_0$, arises naturally

$$\overline{T}(\tau) = T_c + (\overline{T}_0 - T_c)\exp\left(-\frac{2U}{R\rho C_p}\tau\right)$$

where $\overline{T}_0$ denotes the steady inlet temperature and U denotes overall heat transfer coefficient:

$$U = h_i h_0/(h_i + h_0)$$

the local wall temperature must obey

$$\overline{T}_w(\tau) = \frac{h_i}{h_i + h_0}\overline{T}(\tau) + \frac{h_0}{h_i + h_0}T_c$$

where in the limit as $h_0/h_w \to \infty$, we see that $\overline{T}_w \to \overline{T}_c$, which corresponds to the condition in Problem 10.13.

(d) To reduce algebra in subsequent manipulations, and since it is clear that exchangers do not respond to inlet disturbances until $t > \tau$, it is wise to introduce a change of variables by letting

$$\theta = t - \tau = t - z/V_0$$

Show that the dynamic equations become, since $\hat{T}(t, \tau) = \hat{T}(\theta, \tau)$

$$\frac{\partial \hat{T}}{\partial \tau} + \frac{1}{\tau_i}\left(\hat{T} - \hat{T}_w\right) = 0$$

$$C\frac{\partial \hat{T}_w}{\partial \theta} = \frac{1}{\tau_i}\left(\hat{T} - \hat{T}_w\right) - \frac{1}{\tau_0}\left(\hat{T}_w\right)$$

(e) Take Laplace transforms with respect to θ to obtain, for an inlet step of $\alpha\overline{T}_0 u(t) = \hat{T}(t, 0)$:

$$\hat{T}_w(s, \tau) = \frac{1/\tau_i}{Cs + 1/\tau_i + 1/\tau_0}\hat{T}(s, \tau)$$

and since $\hat{T}(t, 0) = \hat{T}(\theta, 0)$, because $\theta = t$ when $\tau = 0$; hence, $\mathscr{L}\hat{T}(\theta, 0) = \alpha\overline{T}_0/s$,

$$\hat{T}(s, \tau) = \alpha\overline{T}_0 \exp\left(-\frac{\tau}{\tau_i}\right)\frac{1}{s}\exp\left[\frac{\tau/\tau_i^2}{Cs + 1/\tau_i + 1/\tau_0}\right]$$

(f) Invert the Laplace transform, $\hat{T}(s, \tau)$, noting that

$$\mathscr{L}^{-1}\frac{1}{s}\exp\left(\frac{k}{s}\right) = I_0(2\sqrt{kt})$$

in Appendix D, and noting further that

$$\frac{1}{s}\exp\left(\frac{k}{s+a}\right) = \left(\frac{s+a}{s}\right)\left(\frac{1}{s+a}\right)\exp\left(\frac{k}{s+a}\right)$$

hence use the shifting-theorem to see

$$\hat{T}(\theta,\tau) = \alpha\overline{T}_0\exp\left(-\frac{\tau}{\tau_i}\right)$$

$$\times\left[u(\theta)e^{-a\theta}I_0(2\sqrt{k\theta}) + a\int_0^\theta e^{-a\beta}I_0(2\sqrt{k\beta})\,d\beta\right]$$

where $u(\theta)$ is inserted to remind that no response occurs unless $\theta > 0$, and the parameters are

$$a = \frac{1}{C}\left(\frac{1}{\tau_i} + \frac{1}{\tau_0}\right),\ \sec^{-1}$$

$$k = \frac{\tau}{C\tau_i^2},\ \sec^{-1}$$

$$\theta = t - \tau > 0,\ \sec$$

The response is similar to Problem 10.13 at small $\theta > 0$ (one residence time). Unlike the previous problem, dynamic behavior continues, owing to the thermal capacitance effect of the wall.

10.15₃. Show that a formally equivalent solution for Example 10.6 can be obtained by writing Eq. 10.307 as

$$C(\zeta,s) = \frac{C_0}{s}\exp(-\zeta)\exp\left(\frac{\zeta}{s+1}\right)$$

$$= C_0\left(\frac{s+1}{s}\right)\left(\frac{1}{s+1}\right)\exp(-\zeta)\exp\left(\frac{\zeta}{s+1}\right)$$

and hence, obtain

$$\frac{C(\tau,\zeta)}{C_0} = u(\tau)\exp(-\tau-\zeta)I_0(2\sqrt{\zeta\tau})$$

$$+ \int_0^\tau \exp(-\beta-\zeta)I_0(2\sqrt{\beta})\,d\beta$$

where $u(\tau)$ is inserted because no response occurs until $\tau > 0$. Compare this with Eq. 10.309 to prove the property for J functions

$$J(\zeta,\tau) + J(\tau,\zeta) = 1 + \exp(-\tau-\zeta)I_0(2\sqrt{\tau\zeta})$$

10.16*. Consider the stirred pot discussed in Section 10.2.1, whereby pure solvent is used to extract oil from spherically shaped (e.g., soy beans) seeds. Suppose m seeds of radius R containing oil composition C_0 are thrown into a well-stirred pot containing pure solvent. The material balances for transient extraction are seen to be, for the solvent solution

$$V\frac{dC}{dt} = -m(4\pi R^2)D\frac{\partial c(R,t)}{\partial r}$$

and for the porous seeds, we have

$$\frac{\partial c}{\partial t} = D\frac{1}{r^2}\frac{\partial}{\partial r}r^2\frac{\partial c}{\partial r}$$

(a) Express the equations using dimensionless independent variables and obtain the coupled integro-differential balances

$$C(\tau) = -3\alpha\int_0^\tau \frac{\partial c}{\partial \xi}\Big|_{\xi=1} d\tau$$

$$\frac{\partial c}{\partial \tau} = \frac{1}{\xi^2}\frac{\partial}{\partial \xi}\xi^2\frac{\partial c}{\partial \xi}$$

where

$$\xi = \frac{r}{R}; \quad \tau = \frac{tD}{R^2}; \quad \alpha = \frac{m}{V}\left(\frac{4}{3}\pi R^3\right) \cdots \text{(capacity ratio)}$$

(b) Apply Laplace transforms for the well-stirred condition (i.e., $C(\tau) = c(1,\tau)$) so that film resistance is negligible, and thus obtain the transient composition of oil in the solvent phase

$$\frac{C(\tau)}{C_0} = \frac{\alpha}{1+\alpha} - 6\alpha\sum_{n=1}^\infty \frac{\exp\left(-\lambda_n^2\tau\right)}{\left[\lambda_n^2 + 9\alpha + 9\alpha^2\right]}$$

where

$$\tan\lambda_n = \frac{3\alpha\lambda_n}{\lambda_n^2 + 3\alpha}; \quad \lambda_n > 0$$

Note: Use series expansions to show that a simple pole exists at $s = 0$.

Eigenvalues are tabulated in Table 10.1 for the values of $\alpha = 0.2$, 0.5, and 1, and Fig. 10.17 illustrates the response of solvent composition with time.

Table 10.1 Eigenvalues for $\tan \lambda_n = \dfrac{3\alpha\lambda_n}{\lambda_n^2 + 3\alpha}$

i	$\alpha = 0.2$	$\alpha = 0.5$	$\alpha = 1$
1	3.3117 1012	3.5058 8947	3.7263 8470
2	6.3756 5983	6.5023 8663	6.6814 3485
3	9.4875 1775	9.5776 6992	9.7155 6609
4	12.6137 2372	12.6830 1447	12.7927 1161
5	15.7459 5800	15.8020 4234	15.8923 9684
6	18.8812 6941	18.9283 0870	19.0048 4951
7	22.0183 5816	22.0588 3600	22.1251 0812
8	25.1565 6476	25.1920 7365	25.2504 4796
9	28.2955 1960	28.3271 3870	28.3792 6573
10	31.4349 9965	31.4634 9280	31.5105 6238
11	34.5748 6239	34.6007 8953	34.6436 8526
12	37.7150 1258	37.7387 9608	37.7781 9091
13	40.8553 8411	40.8773 5031	40.9137 6798
14	43.9959 2970	44.0163 3589	44.0501 9163
15	47.1366 1467	47.1556 6720	47.1872 9559

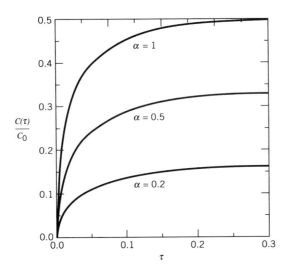

Figure 10.17

10.17*. The problem of desorption of bound solute from particles thrown into a well-stirred pot of solvent is formally equivalent to the leaching of oil from seeds, worked out in Problem 10.16. Thus, fluid-filled sorbent particles, initially containing solute on the solid phase, obey the linear partition relationship

$$q = Kc$$

where q denotes solid solute composition as moles solute per unit volume solid, and c is the solute in the fluid phase within the solid pore structure with composition moles per unit volume fluid.

The material balance for a single particle in the stirred pot is

$$\varepsilon \frac{\partial c}{\partial t} + (1 - \varepsilon) \frac{\partial q}{\partial t} = \varepsilon D_p \frac{1}{r^2} \frac{\partial}{\partial r} \left(r^2 \frac{\partial c}{\partial r} \right)$$

where D_p denotes pore diffusivity, ε is the particle void fraction (porosity) and $(1 - \varepsilon)$ represents solid volume fraction.

Inserting the linear partition coefficient and rearranging, we find the identical diffusion equation as in the oil seed problem

$$\frac{\partial c}{\partial t} = D \frac{1}{r^2} \frac{\partial}{\partial r} \left(r^2 \frac{\partial c}{\partial r} \right)$$

where the new diffusivity is seen to be

$$D = \frac{\varepsilon D_p}{\varepsilon + (1 - \varepsilon) K}$$

The mass balance for m particles exchanging solute with an initially clean solvent is also similar to the oil seed problem for constant solvent volume V

$$V \frac{dC}{dt} = -m(4\pi R^2) \varepsilon D_p \frac{\partial c(R, t)}{\partial r}$$

where the derivative is evaluated at the outer rim of the particle where $r = R$.

(a) Introduce dimensionless variables $\xi = r/R, \tau = tD/R^2$ and show that the process of physical desorption is formally equivalent to the leaching problem of 10.16 if the capacity ratio is defined as

$$\alpha = \frac{m \left(\frac{4}{3} \pi R^3 \right) [\varepsilon + (1 - \varepsilon) K]}{V}$$

(b) The reverse process of adsorption from concentrated solutions in well-stirred pots can be worked out using the above equations, except the initial conditions are reversed

$$C(0) = C_0, \qquad c(r, 0) = q(r, 0) = 0$$

Use Laplace transforms to show that the bulk, well-stirred solution varies according to the relation

$$\frac{C(\tau)}{C_0} = \frac{1}{1 + \alpha} + 6\alpha \sum_{n=1}^{\infty} \frac{\exp(-\lambda_n^2 t)}{(9\alpha + 9\alpha^2 + \lambda_n^2)}$$

where the eigenvalues are found from

$$\tan \lambda_n = \frac{3\alpha\lambda_n}{3\alpha + \lambda_n^2}$$

(c) As a final check on results, use an elementary, overall material balance (initial system solute = final system solute) to prove that, for desorption

$$C(\infty) = \frac{\alpha}{1 + \alpha} C_0$$

and for adsorption

$$C(\infty) = \frac{1}{1 + \alpha} C_0$$

The evaluation of the often-missed pole at $s = 0$ can be circumvented in this and Problem 10.16 by writing variables as a deviation from future steady state, that is,

$$c(r, t) = C(\infty) + \hat{c}(r, t)$$

It is easy to see that the deviation variable at $r = R$ is

$$\frac{\hat{c}}{C_0} = \mp 6\alpha \sum_{n=1}^{\infty} \frac{\exp(-\lambda_n^2 \tau)}{9\alpha + 9\alpha^2 + \lambda_n^2}$$

where the positive sign corresponds to adsorption and the negative to desorption.

10.18*. Heat regenerators operate in a manner similar to packed bed adsorbers. Thus, large beds of solids (ceramic or metal balls, river rocks, etc.) are heated using, for example, warm daytime air, and this stored heat is then recycled for home-heating at night. Thus, consider a packed bed of solid with voidage ε into which air of uniform temperature T_0 is injected. The bed sustains an initial temperature T_i. The air exiting the packed bed is thus heated, and the bed is cooled, hence, the heat exchange is a transient process much the same as adsorption or desorption of solute from granular beds.

(a) Perform a transient heat balance on the packed bed, assuming the superficial gas velocity is uniform and constant at a value U_0 and the walls are perfectly insulated, and thus show

$$\varepsilon \rho_f c_f \frac{\partial T_f}{\partial t} + (1 - \varepsilon) \rho_s c_s \frac{\partial T_s}{\partial t} + U_0 \rho_f c_f \frac{\partial T_f}{\partial x} = 0$$

where x denotes axial position from the bed inlet.

(b) Perform a transient heat balance on the solid phase, taking the volumetric gas film coefficient to be $h_G a$, where a denotes interfacial area of the solid phase per unit column volume and show

$$(1 - \varepsilon)\rho_s c_s \frac{\partial T_s}{\partial t} = -h_G a(T_s - T_f)$$

(c) Introduce interstitial velocity, $v = U_0/\varepsilon$ and the change of variables, which accounts for the reality that a thermal wavefront has a null response until the residence time (x/v) is exceeded

$$\theta = t - \frac{x}{v}$$

to show that

$$\rho_f c_f v \left(\frac{\partial T_f}{\partial x}\right)_\theta = \frac{h_G a}{\varepsilon}(T_s - T_f)$$

$$(1 - \varepsilon)\rho_s c_s \left(\frac{\partial T_s}{\partial \theta}\right)_x = -h_G a(T_s - T_f)$$

(d) Combine the remaining variables to remove excess baggage and obtain

$$\frac{\partial T_f}{\partial \zeta} = (T_s - T_f)$$

$$\frac{\partial T_s}{\partial \tau} = -(T_s - T_f)$$

where

$$\zeta = \frac{x h_G a}{\varepsilon \rho_f c_f v}; \qquad \tau = \frac{\theta h_G a}{(1 - \varepsilon)\rho_s c_s}$$

(e) Solve the remaining equation using Laplace transforms, noting the initial bed temperature is T_i and the inlet air temperature is T_0, and show[2]

$$\frac{T_i - T_f}{T_i - T_0} = \exp(-\zeta - \tau) I_0\left(2\sqrt{\tau \zeta}\right) \cdot u(\tau)$$

$$+ \exp(-\zeta) \int_0^\tau \exp(-\beta) I_0\left(2\sqrt{\beta \zeta}\right) d\beta$$

---

[2] Noting the symmetrical property of the J function

$$J(\zeta, \tau) = 1 - \int_0^\zeta \exp(-\tau - \beta) I_0\left(2\sqrt{\tau \beta}\right) d\beta$$

so that

$$J(\zeta, \tau) + J(\tau, \zeta) = 1 + \exp(-\tau - \beta) I_0\left(2\sqrt{\tau \zeta}\right)$$

hence, it can be shown that another form of solution is obtainable, which is strictly analogous to the results in Example 10.6.

(Note: $u(\tau)$ is added as a reminder that a response cannot occur until time exceeds local residence time, that is, when $\tau > 0$).

(f) Program the series expansion for the Bessel function (compare truncating the series expansion at 10, then 20 terms, and note when $x > 5$: $I_0(x) \approx e^x/\sqrt{2\pi x}$), and use the Trapezoidal or Simpsons rule to compute the integral and thus produce plots of the dimensionless exit temperature versus real time t using the following solid and air properties at $20°$ C

$$\text{packed height at exit} = 300 \text{ cm}$$

$$\text{voidage} = 0.4$$

$$\text{superficial gas velocity} = 30 \text{ cm/sec}$$

$$h_G a = 0.8 \text{ cal/cm}^3\text{-hr-°C}$$

$$\rho_f = 0.00144 \text{ g/cm}^3$$

$$\rho_s = 2 \text{ g/cm}^3$$

$$c_f = 0.2 \text{ cal/g-°C}$$

$$c_s = 0.237 \text{ cal/g-°C}$$

(g) Suppose the initial bed temperature (T_i) is $25°$ C, and the inlet temperature (T_0) is $20°$ C. Use the properties in part (f) to compute the time corresponding to air exiting the bed at $21°$ C.

10.19$_3$. A capillary-tube method to analyze binary liquid diffusivity was reported in Mickley et al. (1957): a straight, narrow bore capillary tube of known internal length has one end sealed and is filled with a binary solution of known composition. The capillary is maintained in a vertical position with the open end pointed upward. A slow stream of pure solvent is allowed to continuously sweep past the open mouth of the capillary tube. We shall designate species A as solute and species B as solvent. After an elapsed time, τ, the capillary is removed and the solution it contains is well mixed and then analyzed for composition, to find how much solute A was extracted. Generally, several such experiments are conducted at different values of τ.

If the diffusivity does not change too much with composition (actually, an average diffusivity is computed corresponding to the average composition between initial and final state), the transient form of Fick's law for equimolar counter diffusion and constant molar density is

$$\frac{\partial x_A}{\partial t} = D_{AB}\frac{\partial^2 x_A}{\partial z^2}$$

where z denotes position ($z = 0$ is the position of the open mouth and $z = L$ is the closed end). Suitable initial and boundary conditions are

$$t = 0, \qquad x_A = x_0 \qquad \text{(initial composition known)}$$

$$z = 0; \qquad x_A = 0 \qquad \text{(pure solvent at mouth)}$$

$$z = L; \quad D_{AB}\frac{\partial x_A}{\partial z} = 0 \qquad \text{(impermeable boundary)}$$

(a) Apply the method of separation of variables and show that suitable Sturm–Liouville conditions exist for the application of the orthogonality condition.

(b) Show that the appropriate eigenvalues for the stated conditions are given by

$$\lambda_n = \left(\frac{2n-1}{2}\right)\pi; \qquad n = 1, 2, 3, \cdots$$

(c) The fraction solute A remaining (after an experiment) is computed using

$$R = \frac{1}{L}\int_0^L \frac{x_A(z,t)}{x_0}\,dz$$

show that R obeys the analytical result

$$R = \frac{8}{\pi^2}\sum_{n=1}^\infty \frac{1}{(2n-1)^2}\exp\left[-\frac{(2n-1)^2\pi^2}{4}\left(\frac{D_{AB}t}{L^2}\right)\right]$$

(d) Indicate by a graphical representation how long time experiments could be used to find D_{AB} directly (long time implies $D_{AB}t/L^2 \gg 1$).

10.20*. The capillary-tube experiment described in Problem 10.19 was analyzed from the point of view of long time ($D_{AB}t/L^2 \gg 1$). We wish to reconsider this analysis in order to derive a suitable relationship for short-time conditions, such that $D_{AB}t/L^2 \ll 1$. For short-time conditions, the effect of concentration-dependent diffusivity is minimized (Rice and Goodner 1986). In the analysis to follow, use dimensionless independent variables

$$\theta = \frac{D_{AB}t}{L^2}, \qquad \zeta = \frac{z}{L}$$

(a) Apply Laplace transforms to the nondimensional transport equation and show that

$$\bar{R}(s) = \frac{1}{s} - \frac{1}{s^{3/2}}\left[\frac{1 - \exp(-2\sqrt{s})}{1 + \exp(-2\sqrt{s})}\right]$$

(b) The so-called *Initial-Value theorem* can be derived from the definition

$$\int_0^\infty f(t)e^{-st}\,dt = sF(s) - f(0)$$

The existence of the Laplace Transform depends on the condition that any exponential order for $f(t)$ is dominated by e^{-st}, so that in the limit $s \to \infty$

$$\lim_{s \to \infty} s \cdot F(s) = f(0)$$

So, for example, if we apply this all the way as $s \to \infty$ for the function $\bar{R}(s)$ we find

$$\lim_{s \to \infty} s\bar{R}(s) = 1 = R(0)$$

which correctly shows that the fraction remaining at time zero is unity (no extraction took place).

If we carry the limit for $\bar{R}(s)$ only partly, that is, let $s \to$ large, we find the compact transform

$$\bar{R}(s) \approx \frac{1}{s} - \frac{1}{s^{3/2}}$$

hence, for small time show that

$$R(t) \approx 1 - 2\left(\frac{D_{AB}t}{\pi L^2}\right)^{1/2}$$

(c) For moderate times $(D_{AB}t/L^2 \sim 1)$, the denominator of $\bar{R}(s)$ can be expanded using the Binomial theorem

$$\frac{1}{1 + \varepsilon} \approx 1 - \varepsilon + \varepsilon^2$$

hence, show that the additional terms are

$$R(t) \approx 1 - 2\left(\frac{D_{AB}t}{\pi L^2}\right)^{1/2} - \left(4\sqrt{\frac{D_{AB}t}{\pi L^2}}\right)\exp\left(-\frac{L^2}{D_{AB}t}\right)$$

$$+ 4\,\mathrm{erfc}\left(\sqrt{\frac{L^2}{D_{AB}t}}\right)$$

10.21₃. The analysis of heat transfer in Example 10.7, the so-called Nusselt problem, could have been inverted without resort to Residue theory, by a clever use of partial fraction expansion. If it is known that only distinct poles exist, as in this example, then $\theta(s, \xi)$ could be expanded as an infinity of partial fractions

$$\theta(s,\xi) = \frac{1}{s}\frac{J_0(i\sqrt{s}\,\xi)}{J_0(i\sqrt{s}\,)} = \frac{A_0}{s} + \sum_{n=1}^{\infty}\frac{A_n}{s - s_n}$$

(a) Show that the first coefficient is simply $A_0 = 1$.

(b) The remaining poles (eigenvalues) were found by setting $J_0(i\sqrt{s}\,) = 0$, hence, we obtained $s_n = -\lambda_n^2$ since $J_0(\lambda_n) = 0$. For any of the remaining coefficients, say A_j, we could use the usual partial-fraction algorithm to see

$$A_j = \lim_{s \to s_j} (s - s_j)\frac{f(s)}{g(s)}$$

where

$$f(s) = J_0(i\sqrt{s}\,\xi); \qquad g(s) = s \cdot J_0(i\sqrt{s}\,)$$

However, in doing so, we arrive at the indeterminancy $0/0$, since $g(s_j) = 0$. To resolve this, expand $g(s)$ around the point s_j using the Taylor series

$$g(s) = g(s_j) + \frac{\partial g}{\partial s}\bigg|_{s_j}(s - s_j) + \frac{1}{2!}\frac{\partial^2 g}{\partial s^2}\bigg|_{s_j}(s - s_j)^2 + \cdots$$

and show that

$$A_j = \frac{f(s_j)}{\dfrac{\partial g}{\partial s}\bigg|_{sj}} = -2\frac{J_0(\lambda_j\xi)}{\lambda_j J_1(\lambda_j)}$$

(c) Invert the partial fractions term by term and show that the results are identical to Example 10.7.

Chapter 11

Transform Methods for Linear PDEs

11.1 INTRODUCTION

In the previous chapters, partial differential equations with finite space domain were treated by the method of separation of variables. Certain conditions must be satisfied before this method could yield practical results. We can summarize these conditions as follows:

1. The partial differential equation must be linear.
2. One independent variable must have a finite domain.
3. The boundary conditions must be homogeneous for at least one independent variable.
4. The resulting ODEs must be solvable, preferably in analytical form.

With reference to item (4), quite often the ODEs generated by separation of variables do not produce easy analytical solutions. Under such conditions, it may be easier to solve the PDE by approximate or numerical methods, such as the orthogonal collocation technique, which is presented in Chapter 12. Also, the separation of variables technique does not easily cope with coupled PDE, or simultaneous equations in general. For such circumstances, transform methods have had great success, notably the Laplace transform. Other transform methods are possible, as we show in the present chapter.

The spatial domain for problems normally encountered in chemical engineering are usually composed of rectangular, cylindrical, or spherical coordinates. Linear problems having these types of domain usually result in ODEs (after the application of separation of variables) that are solvable. Solutions of these ODEs normally take the form of trigonometric, hyperbolic, Bessel, and so forth. Among special functions, these three are familiar to engineers because they arise so frequently. They are widely tabulated in handbooks, for example, the handbook by Abramowitz and Stegun (1964) provides an excellent resource on the properties of special functions.

If the boundary conditions arising in linear analysis are nonhomogeneous, they must be transformed to become homogeneous as taught in Chapters 1 and

10. This is normally done by transforming the dependent variable so that the new partial differential equation will have homogeneous boundary conditions.

Although the separation of variables method is easy to apply, nonetheless, considerable practice is required to use it successfully. In this section, a method called the Sturm–Liouville integral transform will be presented. This method is also known as the finite integral transform method. It has the distinct advantage of all operational mathematics, which is simplicity.

11.2   TRANSFORMS IN FINITE DOMAIN: STURM–LIOUVILLE TRANSFORMS

The strategy for using Sturm–Liouville transforms is, first, to carefully lay out the algebraic rules for this class of operator. Obviously, the defining equation and boundary conditions must be of the Sturm–Liouville type, as discussed in Chapter 10.

11.2.1   Development of Integral Transform Pairs

The method of finite integral transforms (FIT) involves an operator (similar to the Heaviside operator), which transforms the original equation into another, simpler domain. The solution in the new domain will be seen to be quite elementary. However, to be of practical value, it must be transformed back to the original space. The operation for this inverse transformation, together with the original operator, will form what we will call later the integral transform pair. Figure 11.1 illustrates the mapping of the integral transform pair. The Laplace transform and its inverse defined in Eqs. 9.1 and 9.3 form an integral transform pair.

We shall denote the transform pairs as L (forward operator) and L^{-1} (inverse operator), similar to the operators D and D^{-1} discussed in Chapter 2. Figure 11.1 outlines in graphic form the movement between domains, much the same as the Laplace transform moves from t to s domains, and back again. The solution methodology within the space A is very difficult and tortuous. The solution may be more easily obtained by transforming the original equation into space B, where the solution is more straightforward. Following this, the desired

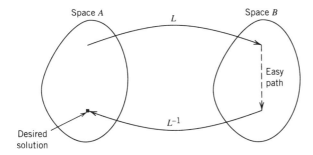

Figure 11.1 Schematic diagram of the transform pair.

solution can be obtained by the inversion process using the operator L^{-1}. The method becomes particularly attractive when solving certain types of simultaneous, coupled partial differential equations.

Because they depend on the Sturm–Liouville equation, the separation of variables method and the integral transform yield exactly the same solution, as you would expect. But the advantage of the integral transform is the simplicity of handling coupled PDEs, for which other methods are unwieldy. Moreover, in applying the finite integral transform, the boundary conditions need not be homogeneous (See Section 11.2.3).

EXAMPLE 11.1

To demonstrate the development of the integral transform pair in a practical way, consider the Fickian diffusion or Fourier heat conduction problem in a slab object (Fig. 11.2)

$$\frac{\partial y}{\partial t} = \frac{\partial^2 y}{\partial x^2} \qquad (11.1)$$

where y could represent a dimensionless concentration or temperature.

The partial differential Eq. 11.1 is subjected to conditions

$$x = 0; \qquad \frac{\partial y}{\partial x} = 0 \qquad (11.2a)$$

$$x = 1; \qquad y = 0 \qquad (11.2b)$$

$$t = 0; \qquad y = 1 \qquad (11.3)$$

We note that the boundary conditions (11.2) are homogeneous. We define the spatial domain as $(0, 1)$ but any general domain (a, b) can be readily converted to $(0, 1)$ by a simple linear transformation

$$x = \frac{(x' - a)}{(b - a)} \qquad (11.4)$$

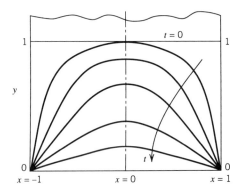

Figure 11.2 Temperature profile in a slab object.

where $x' \in (a, b)$ and $x \in (0, 1)$. This convenient symbolism simply means, x' is bounded by a and b, so $a \le x' \le b$ and $0 \le x \le 1$.

Now multiply the LHS and RHS of Eq. 11.1 by a continuous function $K_n(x)$ (called the kernel) such that the function $K_n(x)$ belongs to an infinite-dimensional space of twice differentiable functions in the domain $(0, 1)$. Eq. 11.1 then becomes

$$\frac{\partial y}{\partial t} K_n(x) = \frac{\partial^2 y}{\partial x^2} K_n(x) \tag{11.5}$$

We can see analogies with the Laplace transform, which has an unbounded domain and the kernel is

$$K(x) = e^{-sx} \tag{11.6}$$

If we integrate Eq. 11.5 with respect to x over the whole domain of interest, we find

$$\frac{d}{dt} \int_0^1 y(x, t) K_n(x) \, dx = \int_0^1 \frac{\partial^2 y}{\partial x^2} K_n(x) \, dx \tag{11.7a}$$

Since the function $K_n(x)$ is assumed to be twice differentiable, we can carry out integration by parts on the integral in the RHS of Eq. 11.7a twice to see

$$\frac{d}{dt} \int_0^1 y(x, t) K_n(x) \, dx = \left[\frac{\partial y}{\partial x} K_n(x) - y \frac{dK_n(x)}{dx} \right]_0^1 + \int_0^1 y \frac{d^2 K_n}{dx^2} \, dx \tag{11.7b}$$

Now, making use of the boundary conditions (11.2) to evaluate the square bracket term on the RHS of Eq. 11.7b, we have

$$\frac{d}{dt} \int_0^1 y(x, t) K_n(x) \, dx = \left[\frac{\partial y(1, t)}{\partial x} K_n(1) - y(0, t) \frac{dK_n(0)}{dx} \right] + \int_0^1 y \frac{d^2 K_n}{dx^2} \, dx \tag{11.8}$$

Up to this point, it looks as if the new Eq. 11.8 is just as complicated as the original Eq. 11.1. However, at this point we have not specified the specific form for the kernel function $K_n(x)$. It is Eq. 11.8 that provides the hint to simplify matters. It is our aim to solve for

$$\int_0^1 y(x, t) K_n(x) \, dx$$

so the RHS must also be known or written in terms of this integral product. Since the following variables are not known or specified

$$\frac{\partial y(1, t)}{\partial x} \qquad \text{and} \qquad y(0, t)$$

we have only one way to remove the bracketed terms (so-called unwelcome terms) in the RHS of Eq. 11.8 and that is by setting

$$x = 0; \qquad \frac{dK_n}{dx} = 0 \tag{11.9a}$$

$$x = 1; \qquad K_n = 0 \tag{11.9b}$$

Equations 11.9 specify boundary conditions for the function $K_n(x)$ (even though this function is still unknown at this stage), and it is noted that these boundary conditions are identical in form to the boundary conditions (11.2) for the function y.

Having defined the boundary conditions (11.9), Eq. 11.8 now takes the simple structure

$$\frac{d}{dt} \int_0^1 y(x, t) K_n(x) \, dx = \int_0^1 y \frac{d^2 K_n}{dx^2} \, dx \tag{11.10}$$

To proceed further, we shall need to specify $K_n(x)$. Equation 11.10 is now simpler, but cannot be solved because the LHS and RHS appear to involve two different functional forms. One way to resolve this difficulty is to define K_n by setting

$$\frac{d^2 K_n(x)}{dx^2} = -\xi_n^2 K_n(x) \tag{11.11}$$

from which the reader can see that the integral product yK_n exists on both sides of the equation.

We could have made the RHS of Eq. 11.11 a positive number rather than negative, but it can be proved that the negative number is the only option that will yield physically admissible solutions. This will be proved later in dealing with a general Sturm–Liouville system (see Section 11.2.2).

With Eq. 11.11, Eq. 11.10 becomes

$$\frac{d}{dt} \int_0^1 y(x, t) K_n(x) \, dx = -\xi_n^2 \int_0^1 y(x, t) K_n(x) \, dx \tag{11.12}$$

Now, Eq. 11.11 for $K_n(x)$ is subject to two boundary conditions (11.9) and is called the associated eigenproblem for Eq. 11.1. The function $K_n(x)$ is called the kernel or eigenfunction, whereas ξ_n is the corresponding eigenvalue. The solution for this particular eigenproblem subject to conditions (11.9) is

$$K_n(x) = \cos(\xi_n x) \tag{11.13}$$

where the countably infinite eigenvalues are

$$\xi_n = \left(n - \frac{1}{2} \right) \pi \qquad \text{for} \qquad n = 1, 2, 3, \dots, \infty \tag{11.14}$$

which arises since $K_n(1) = \cos(\xi_n) = 0$. The first eigenvalue is $\pi/2$, and successive eigenvalues differ by π.

In arriving at Eq. 11.13, the multiplicative constant of integration for $\cos(\xi_n x)$ has been arbitrarily set equal to unity. The actual values of the multiplicative constants are added later when the inversion process takes place.

It is cumbersome to carry the full integral representation all the way through the analysis, so we define $\langle y, K_n \rangle$ as an "image" of the function y, defined as

$$\langle y, K_n \rangle = \int_0^1 y K_n(x)\, dx \tag{11.15}$$

This is typical operator format; for example, we replace K_n with e^{-st}, we would have the Laplace transform with respect to time, written for an unbounded time domain as

$$\mathscr{L} y = \int_0^\infty y e^{-st}\, dt$$

Now Eq. 11.12 can be written compactly using this image as

$$\frac{d}{dt} \langle y, K_n \rangle = -\xi_n^2 \langle y, K_n \rangle \tag{11.16}$$

which is a simple, first order ODE in the variable $\langle y, K_n \rangle = f(t)$.

This result is clearly much simpler than the original partial differential equation 11.1. To solve this first order equation for $\langle y, K_n \rangle$ we need to specify an initial condition. This can be readily found by inspecting the "image" of the initial condition (11.3) where, since $y(x,0) = 1$, we have

$$t = 0; \quad \langle y, K_n \rangle = \langle 1, K_n \rangle \tag{11.17}$$

where $\langle 1, K_n \rangle$ by definition in Eq. 11.15 is simply

$$\langle 1, K_n \rangle = \int_0^1 K_n\, dx = \left. \frac{\sin(\xi_n x)}{\xi_n} \right|_0^1 = \frac{-(-1)^n}{\left(n - \frac{1}{2} \right) \pi}$$

We remind the reader that $K_n(x)$ is a function of x only.

The solution of the simple first order ODE in Eq. 11.16 subject to the initial condition (11.17) is quite simple, so the sought after $f(t)$ is

$$\langle y, K_n \rangle = \langle 1, K_n \rangle e^{-\xi_n^2 t} \tag{11.18}$$

where $\langle 1, K_n \rangle$ is a computable constant as shown in the previous step. Up to this point, one can easily recognize the similarity between the separation of variables and the integral transform approaches by noting the appearance of the exponential function on the RHS of Eq. 11.18 and by noting the analogous eigenproblem in Eq. 11.11.

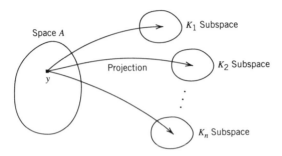

Figure 11.3 Mapping diagram.

The image of y by the forward integral transform is $\langle y, K_n \rangle$, and of course there are infinite eigenfunctions K_n so there are infinite solutions for $\langle y, K_n \rangle$ where $n = 1, 2, \ldots$. This is the basis for a special type of function space, called a Hilbert space. In fact, whether we like it or not, every time we deal with an infinity of series solutions, we are in a sense dealing with Hilbert space. In the context of the present problem, the Hilbert space is an infinite-dimensional space, and this space has infinite coordinates and each coordinate is represented by an eigenfunction. This arises because the eigenvalues given in Eq. 11.14 are countably infinite.

The mapping diagram (Fig. 11.3) shows that a function y in the space A is mathematically equivalent to the many images $\langle y, K_1 \rangle, \langle y, K_2 \rangle$, and so on. From elementary vector analysis in a three-dimensional Euclidean space, we may regard K_n as one of the coordinates and $\langle y, K_n \rangle$ is the projection of y onto the coordinate K_n.

We shall return to the subject of Hilbert space when we will deal with the generalized integral transform. For the present, let us return to the solution $\langle y, K_n \rangle$ in the space B. Of course, this solution is not what we desire. It is $y(x, t)$ that we are after. In analogy with Euclidean space, if we know the projection on the coordinate K_n as $\langle y, K_n \rangle$, the function y can be reconstructed in terms of a linear combination of all, countably infinite, coordinates; hence,

$$y = \sum_{m=1}^{\infty} a_m(t) K_m(x) \tag{11.19}$$

where a_m is some set of arbitrary functions of time.

The only task remaining is to find suitable values for $a_m(t)$, which must be clearly dependent on time, t. Since the function $K_n(x)$ is the eigenfunction, an orthogonality condition must be associated with it. This condition simply states that, under a proper definition of an *inner product* (to be defined shortly), the eigenfunctions are orthogonal to each other. As we have shown in Chapter 10, the orthogonality condition for the present problem (cf. Eq. 11.11 with

Sturm–Liouville equation) is obviously,

$$\int_0^1 K_n(x) K_m(x)\, dx = 0 \qquad \text{for} \qquad n \neq m \qquad (11.20)$$

The integral form of this orthogonality condition is identical to the definition of the integral transform (11.15) if we replace y with $K_m(x)$ and see

$$\langle K_n, K_m \rangle = \int_0^1 K_n(x) K_m(x)\, dx \qquad (11.21)$$

where obviously $\langle K_n, K_m \rangle = \langle K_m, K_n \rangle$.

We shall use this as the definition of the *inner product* for the present case. The eigenfunction $K_n(x)$ is orthogonal to all other eigenfunctions except to itself, since when $n = m$, the integral is finite; that is,

$$\int_0^1 K_n^2(x)\, dx \neq 0$$

By multiplying both sides of Eq. 11.19 by K_n, and integrating over the domain $[0, 1]$, we obtain the inner products on the RHS as

$$\langle y, K_n \rangle = a_n \langle K_n, K_n \rangle + \sum_{\substack{m=1 \\ m \neq n}}^{\infty} a_m \langle K_m, K_n \rangle \qquad (11.22)$$

Making use of the orthogonality condition (11.20), the summation of series terms in the RHS of Eq. 11.22 is identically zero. Thus, the coefficients a_n (or a_m) can be found directly

$$a_n = \frac{\langle y, K_n \rangle}{\langle K_n, K_n \rangle} = \frac{\langle 1, K_n \rangle e^{-\xi_n^2 t}}{\langle K_n, K_n \rangle} \qquad (11.23)$$

since $\langle y, K_n \rangle$ is known from Equation (11.18).

Substituting a_n from Eq. 11.23 into Eq. 11.19, we have, since $m = n$

$$y = \sum_{n=1}^{\infty} \frac{\langle y, K_n \rangle}{\langle K_n, K_n \rangle} K_n(x) \qquad (11.24)$$

Equations 11.15 and 11.24 define the integral transform pairs for the finite integral transform. The complete solution is now at hand, when the integrals $\langle 1, K_n \rangle$ and $\langle K_n, K_n \rangle$ are inserted along with $K_n = \cos(\xi_n x)$.

Since we now know the solutions to the eigenfunction and eigenvalue problems as Eqs. 11.13 and 11.14, then the solution for y is simply

$$y = 2 \sum_{n=1}^{\infty} \frac{\sin(\xi_n)}{\xi_n} \cos(\xi_n x) e^{-\xi_n^2 t} \qquad (11.25)$$

since it is easy to see

$$\langle 1, K_n \rangle = \int_0^1 \cos(\xi_n x)\, dx = \frac{\sin(\xi_n)}{\xi_n}$$

$$\langle K_n, K_n \rangle = \int_0^1 \cos^2(\xi_n x)\, dx = \left[\frac{x}{2} + \frac{\sin(2\xi_n x)}{4\xi_n}\right]_0^1 = \frac{1}{2}$$

with ξ_n defined in Eq. 11.14. This is exactly the same form obtainable by the method of separation of variables.

Equation 11.24 not only defines the *inversion process*, but also states that the function y can be expressed as a linear combination of all scaled projections

$$\frac{\langle y, K_n \rangle}{\left[\langle K_n, K_n \rangle\right]^{1/2}} \tag{11.26a}$$

with the unit vector in normalized form as

$$\phi_n(x) = \frac{K_n(x)}{\left[\langle K_n, K_n \rangle\right]^{1/2}} \tag{11.26b}$$

The normalized unit vector $\phi_n(x)$ means that $\langle \phi_n, \phi_n \rangle = 1$. Thus, the normalized form of Eq. 11.24 is usually written as

$$y = \sum_{n=1}^{\infty} \frac{\langle y, K_n \rangle}{\left[\langle K_n, K_n \rangle\right]^{1/2}} \cdot \frac{K_n(x)}{\left[\langle K_n, K_n \rangle\right]^{1/2}}$$

$$= \sum_{n=1}^{\infty} \frac{\langle 1, K_n \rangle}{\langle K_n, K_n \rangle} \cos(\xi_n x) \exp\left(-\xi_n^2 t\right) \tag{11.27}$$

which is a direct analogue of a three-dimensional Euclidean vector space. For example, a vector $\mathbf{a}$ has the representation in rectangular coordinates as

$$\mathbf{a} = a_1 \mathbf{i} + a_2 \mathbf{j} + a_3 \mathbf{k} \tag{11.28}$$

where $\mathbf{i}$, $\mathbf{j}$, and $\mathbf{k}$ are unit vectors of the three directional coordinates and a_1, a_2, a_3 are projections of the vector $\mathbf{a}$ onto these coordinates. Thus, it is clear that the finite integral transform is similar to the Euclidean vector space except that countably infinite coordinates exist.

Schematic plots of $y(x, t)$ as function of x and t are shown in Fig. 11.2.

11.2.2 The Eigenvalue Problem and the Orthogonality Condition

Using an elementary example, we have introduced the methodology and the conceptual framework for the finite integral transform, most notably the concept of the integral transform pairs. The function y is called the object and the product pair $\langle y, K_n \rangle$ is called the integral transform of y or the "image" of y.

The operator $\langle \; , \; \rangle$ is called the transform operator or the inner product. $K_n(x)$ is called the *kernel* of the transform and can be regarded as a coordinate in the infinite dimensional Hilbert space, within which an inner product is defined. The equation describing the kernel K_n is called the associated eigenproblem.

In this section, we will apply the finite integral transform to a general Sturm–Liouville system, and the integral transform is therefore called the Sturm–Liouville integral transform. Thus, all finite integral transforms are covered at once: Fourier, Hankel, and so forth.

It was clear in the previous example that cosine functions occurred in the natural course of analysis. In fact, the transformation we performed there is often called the finite Fourier transform. However, the broad category of such finite transforms are called Sturm–Liouville.

The eigenvalue problem of a general Sturm–Liouville system must be of the following general form, in analogy with the Sturm–Liouville relation given by Eq. 10.185

$$\frac{d}{dx}\left[p(x)\frac{dK(x)}{dx} \right] - q(x)K(x) + \xi r(x)K(x) = 0 \qquad (11.29)$$

where x lies between a and b and ξ is a constant (replacing β in Eq. 10.185). We have written $q(x)$ with a negative sign in the present case, and stipulate that $q(x) \geq 0$.

The boundary conditions for an eigenvalue problem must be homogeneous

$$x = a; \qquad A_1 K(a) + A_2 \frac{dK(a)}{dx} = 0 \qquad (11.30a)$$

$$x = b; \qquad B_1 K(b) + B_2 \frac{dK(b)}{dx} = 0 \qquad (11.30b)$$

where A_1, A_2 and B_1, B_2 are constants, which can take zero values.

These forms admit $dK/dx = 0$ or $K = 0$ at either $x = a$ or $x = b$. The conditions on the functions $p(x)$, $q(x)$, and $r(x)$ are not too restrictive. These are: $p(x)$, $q(x)$, $r(x)$, and dp/dx are continuous, and $p(x)$ and $r(x)$ are positive and $q(x)$ is nonnegative in the domain $[a, b]$.

Since generally there is more than one eigenvalue and its corresponding eigenfunction, we denote the nth eigenvalue and its eigenfunction as ξ_n and $K_n(x)$.

The Sturm–Liouville equations corresponding to eigenvalues ξ_n and ξ_m are

$$\frac{d}{dx}\left[p(x)\frac{dK_n(x)}{dx} \right] - q(x)K_n(x) + \xi_n r(x)K_n(x) = 0 \quad (11.31a)$$

$$\frac{d}{dx}\left[p(x)\frac{dK_m(x)}{dx} \right] - q(x)K_m(x) + \xi_m r(x)K_m(x) = 0 \quad (11.31b)$$

The corresponding general, homogeneous boundary conditions suitable for orthogonality are

$$x = a; \ A_1 K_n(a) + A_2 \frac{dK_n(a)}{dx} = A_1 K_m(a) + A_2 \frac{dK_m(a)}{dx} = 0 \quad \textbf{(11.32a)}$$

$$x = b; \ B_1 K_n(b) + B_2 \frac{dK_n(b)}{dx} = B_1 K_m(b) + B_2 \frac{dK_m(b)}{dx} = 0 \quad \textbf{(11.32b)}$$

Multiplying Eq. 11.31a by $K_m(x)$ and Eq. 11.31b by $K_n(x)$ and eliminating $q(x)$ between the two resulting equations, we obtain

$$- \frac{d}{dx}\left[p(x)\frac{dK_n(x)}{dx} \right]K_m(x) + \frac{d}{dx}\left[p(x)\frac{dK_m(x)}{dx} \right]K_n(x)$$

$$= (\xi_n - \xi_m)r(x)K_n(x)K_m(x) \quad \textbf{(11.33)}$$

Integrating Eq. 11.33 with respect to x over the domain of interest yields

$$- \int_a^b \frac{d}{dx}\left[p(x)\frac{dK_n(x)}{dx} \right]K_m(x)\, dx + \int_a^b \frac{d}{dx}\left[p(x)\frac{dK_m(x)}{dx} \right]K_n(x)\, dx$$

$$= (\xi_n - \xi_m)\int_a^b r(x)K_n(x)K_m(x)\, dx \quad \textbf{(11.34)}$$

Carrying out integration by parts on the two integrals in the LHS of Eq. 11.34 gives

$$- \left[p(x)K_m(x)\frac{dK_n(x)}{dx} \right]_a^b + \left[p(x)K_n(x)\frac{dK_m(x)}{dx} \right]_a^b$$

$$= (\xi_n - \xi_m)\int_a^b r(x)K_n(x)K_m(x)\, dx \quad \textbf{(11.35)}$$

Finally, substituting the homogeneous boundary conditions (Eq. 11.32) into Eq. 11.35, terms cancel and what remains is the orthogonality condition

$$\int_a^b r(x)K_n(x)K_m(x)\, dx = 0 \qquad \text{for} \qquad n \neq m \quad \textbf{(11.36)}$$

which is the inner product for the general case (cf. Eq. 11.21).

We say that $K_n(x)$ is orthogonal to $K_m(x)$ with respect to the weighting function $r(x)$. This orthogonality condition provides us a way to define the inner product (or the integral transform) for the Hilbert space. This integral tranform on any function $g(x, t)$ is defined as

$$\langle g(x,t), K_n \rangle = \int_a^b r(x)g(x,t)K_n(x)\, dx \quad \textbf{(11.37)}$$

where $g(x, t)$ is any continuous function. With this definition of integral transform (inner product), the coordinate K_n of the Hilbert space is then orthogonal to all other coordinates, and the so-called unit vector of the K_n coordinate, as mentioned previously, is simply

$$\phi_n(x) = \frac{K_n(x)}{[\langle K_n, K_n \rangle]^{1/2}} \qquad (11.38)$$

where it is easy to see that $\langle \phi_n, \phi_n \rangle = 1$, which implies normalization. With this definition of the unit vector, an arbitrary function y can be represented by the following series expansion

$$y = \sum_{n=1}^{\infty} \frac{\langle y, K_n \rangle}{[\langle K_n, K_n \rangle]^{1/2}} \phi_n(x) \qquad \text{for} \qquad a < x < b \qquad (11.39)$$

where $\langle y, K_n \rangle / [\langle K_n, K_n \rangle]^{1/2}$ is viewed as the scaled projection of the function y onto the coordinate K_n. It is also called the Fourier constant in many textbooks. Equation 11.39 can be proved by assuming that y can be expressed as a linear combination of K_n; then after applying the orthogonality condition (Eq. 11.36), we obtain the expression previously given.

Next, we will prove that the eigenvalues are positive. Let us start from Eq. 11.35, and if we take $\xi_n = \alpha_n + i\beta_n$ and let $\bar{\xi}_m$ be the conjugate of ξ_n, so that we have

$$\bar{\xi}_m = \alpha_n - i\beta_n$$

Eq. 11.35 then becomes

$$-\left[p(x)\overline{K_n}(x) \frac{dK_n(x)}{dx} \right]_a^b + \left[p(x)K_n(x) \frac{d\overline{K_n}(x)}{dx} \right]_a^b$$

$$= 2i\beta_n \int_a^b r(x) K_n(x) \overline{K_n}(x)\, dx \qquad (11.40)$$

where $\overline{K_n}$ is the complex conjugate of $K_n(x)$.

Using the homogeneous boundary conditions (Eqs. 11.32), Eq. 11.40 becomes

$$2i\beta_n \int_a^b r(x) K_n^2(x)\, dx = 0 \qquad (11.41)$$

Now since $r(x)$ is assumed positive over the domain (a, b), the integral of Eq. 11.41 is therefore positive definite. Hence, we must have

$$\beta_n = 0 \qquad (11.42)$$

This proves that the eigenvalues of the Sturm–Liouville system are real, not complex numbers.

We have just established that the eigenvalues are real. Next, we wish to show that they are positive. Multiplying Eq. 11.29 by K_n and integrating the result with respect to x from a to b, we obtain

$$\xi_n = \frac{\int_a^b p(x)\left[\dfrac{dK_n(x)}{dx}\right]^2 dx + \int_a^b q(x)[K_n(x)]^2 dx}{\int_a^b r(x)[K_n(x)]^2 dx} \tag{11.43}$$

Since $p(x)$ and $r(x)$ are positive functions and $q(x)$ is nonnegative (i. e., $q(x)$ can be zero), it is clear from Eq. 11.43 that all integrals are positive definite, hence ξ_n must be positive.

The above analysis (from Eq. 11.31 to Eq. 11.43) has established the following theorem.

Theorem 11.1

For a Sturm–Liouville system defined as:

$$\frac{d}{dx}\left[p(x)\frac{dK(x)}{dx}\right] - q(x)K(x) + \xi r(x)K(x) = 0$$

subject to

$$A_1 K(a) + A_2\frac{dK(a)}{dx} = B_1 K(b) + B_2\frac{dK(b)}{dx} = 0$$

and $p(x)$, $q(x)$, $r(x)$, and $dp(x)/dx$ are continuous; $p(x)$ and $r(x)$ are positive; and $q(x)$ is nonnegative over the domain (a, b), the eigenfunction K_n will have the following orthogonality conditions:

$$\int_a^b r(x) K_n(x) K_m(x)\, dx = 0 \qquad \text{for} \qquad n \neq m$$

and the eigenvalues are real and positive.

The eigenfunctions $K_n(x)$ together with the definition of integral transform (inner product) defined in Eq. 11.37 will define a Hilbert space, and any arbitrary function y can be represented as a series as

$$y \doteq \sum_{n=1}^{\infty} \frac{\langle y, K_n\rangle}{\langle K_n, K_n\rangle} K_n(x) \tag{11.44}$$

which defines the *inverse transform*.

If the function y satisfies the same homogeneous boundary conditions as those for the eigenfunctions, then the series representation (Eq. 11.44) will converge uniformly to y for all x in the domain $[a, b]$. Proof of this can be found in Churchill (1958) using the Green's function approach.

To generalize the integral transform approach, let us consider the following operator.

$$L = \frac{1}{r(x)} \cdot \frac{d}{dx}\left[p(x)\frac{d}{dx} \right] - \frac{q(x)}{r(x)} \qquad (11.45)$$

This implies the eigenfunctions satisfy $LK_n = -\xi_n K_n$, by comparison with Eq. 11.29. Here, $r(x), p(x), q(x)$ are continuous functions and $p(x), r(x) > 0$, and $q(x)$ is nonnegative, and the following two boundary operators are defined:

$$M(\cdot) = \left[A_1(\cdot) + A_2\frac{d(\cdot)}{dx} \right]_a \qquad (11.46a)$$

$$N(\cdot) = \left[B_1(\cdot) + B_2\frac{d(\cdot)}{dx} \right]_b \qquad (11.46b)$$

where $M(\cdot)$ is a boundary operator at the point $x = a$, and $N(\cdot)$ is the boundary operator at the point $x = b$. Henceforth, it will be understood that M operates at point a and N at point b.

Suppose the problem we wish to solve is the ODE

$$Ly(x) = f(x) \qquad (11.47a)$$

where $f(x)$ is some continuous forcing function. Further, suppose Eq. 11.47a is subject to the following boundary conditions

$$My = 0 \quad \text{and} \quad Ny = 0 \qquad (11.47b)$$

that is,

$$A_1 y(a, t) + A_2\frac{dy(a, t)}{dx} = 0 \quad \text{and} \quad B_1 y(b, t) + B_2\frac{dy(b, t)}{dx} = 0$$

If we now take the approach of Example 11.1 as follows:

1. Multiplying the LHS and RHS of Eq. 11.47a by $K_n(x)$.
2. Integrating the result twice with respect to x from a to b.
3. Making use of the boundary conditions (11.47b) and then removing the unwelcome terms.

We then obtain the following associated eigenproblem for Eq. 11.47, which is the Sturm–Liouville system

$$LK_n = -\xi_n K_n$$
$$MK_n = 0 \quad \text{and} \quad NK_n = 0 \qquad (11.48)$$

In the process of deriving the eigenproblem, the integral transform (inner product) evolves naturally and is given as in Eq. 11.37, and the transform of

Eq. 11.47a is simply,

$$\langle Ly, K_n \rangle = \langle f(x), K_n \rangle \tag{11.49}$$

It is not difficult to prove (see Problems), using integration by parts, the following formula

$$\langle Ly, K_n \rangle = \langle y, LK_n \rangle \tag{11.50}$$

If an operator L satisfies the above relation, it is called a *self-adjoint operator*. This self-adjoint property is important in the derivation of the Sturm–Liouville theorem. In fact, it was the critical component in the derivation of the orthogonality condition (11.36).

Performing the self-adjoint operation (11.50) on the LHS of Eq. 11.49, we see directly

$$\langle y, LK_n \rangle = \langle f(x), K_n \rangle \tag{11.51}$$

But from the definition of the Sturm–Liouville eigenvalue problem (Eq.11.48), the LHS of Eq. 11.51 gives

$$\langle y, -\xi_n K_n \rangle = -\xi_n \langle y, K_n \rangle$$

hence, we now have

$$-\xi_n \langle y, K_n \rangle = \langle f(x), K_n \rangle$$

which can be solved directly for $\langle y, K_n \rangle$

$$\langle y, K_n \rangle = -\frac{1}{\xi_n} \langle f(x), K_n \rangle$$

Thus, the self-adjoint property allows direct solution for the transform $\langle y, K_n \rangle$.

The inversion will then be given as shown in Eq. 11.39, that is,

$$y = -\sum_{n=1}^{\infty} \frac{\langle f(x), K_n \rangle}{\xi_n \langle K_n, K_n \rangle} K_n(x) \tag{11.52}$$

The solution is complete when the elementary integrals $\langle K_n, K_n \rangle$ and $\langle f, K_n \rangle$ are inserted.

One can see that the integral transform indeed facilitates the resolution of ODE boundary value problems and also partial differential equations comprised of Sturm–Liouville operators (e.g., Eq. 11.45). The simplicity of such operational methods lead to algebraic solutions and also give a clearer view on how the solution is represented in Hilbert space. Moreover, students may find that the Sturm–Liouville integral transform is a faster and fail-safe way of getting the solution. Thus, Eq. 11.52 represents the solution to an almost infinite variety of ordinary differential equations, as we see in more detail in the homework section.

The application of the Sturm–Liouville integral transform using the general linear differential operator (11.45) has now been demonstrated. One of the important new components of this analysis is the *self-adjoint* property defined in Eq. 11.50. The linear differential operator is then called a *self-adjoint differential operator*.

Before we apply the Sturm–Liouville integral transform to practical problems, we should inspect the self-adjoint property more carefully. Even when the linear differential operator (Eq. 11.45) possesses self-adjointness, the self-adjoint property is not complete since it actually depends on the type of boundary conditions applied. The homogeneous boundary condition operators, defined in Eq. 11.46, are fairly general and they lead naturally to the self-adjoint property. This self-adjoint property is only correct when the boundary conditions are unmixed as defined in Eq. 11.46, that is, conditions at one end do not involve the conditions at the other end. If the boundary conditions are mixed, then the self-adjoint property may not be applicable.

EXAMPLE 11.2

Now we consider application of the transform technique to a transient problem of heat conduction or Fickian diffusion in a cylinder (Fig. 11.4). The slab problem (Example 11.1) was dealt with in Section 11.2.2.

A transient heat or mass balance equation in a cylinder has the following form

$$\frac{\partial y}{\partial t} = \frac{1}{x}\frac{\partial}{\partial x}\left(x\frac{\partial y}{\partial x}\right) \equiv Ly \tag{11.53}$$

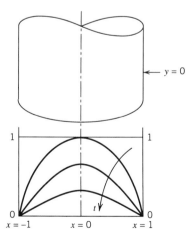

Figure 11.4 Temperature profiles in a cylinder.

subject to the following initial and boundary conditions

$$t = 0; \quad y = 1 \tag{11.54}$$

$$x = 0; \quad \frac{\partial y}{\partial x} = 0 \tag{11.55}$$

$$x = 1; \quad y = 0 \tag{11.56}$$

To apply the Sturm–Liouville integral transform, we follow the same procedure as described in Example 11.1; that is,

1. Multiply Eq. 11.53 by $xK_n(x)$ and integrate the result with respect to x from 0 to 1.
2. Apply the boundary conditions (11.55) and (11.56) and remove the unwelcome terms.

We then obtain the following associated eigenproblem, which defines the kernel K_n of the transform

$$\frac{1}{x}\frac{d}{dx}\left(x\frac{dK_n}{dx}\right) + \xi_n^2 K_n(x) = 0 \tag{11.57a}$$

and the requirements similar to Eqs. 11.8 and 11.9 are

$$x = 0; \quad \frac{dK_n}{dx} = 0 \tag{11.57b}$$

$$x = 1; \quad K_n = 0 \tag{11.57c}$$

In symbolic form, we could write Eq. 11.57a as $LK_n = -\xi_n^2 K_n(x)$.

From the process outlined, not only do we obtain the equations for the kernel (Eq. 11.57), but also the definition of the integral transform, which evolved naturally to become

$$\langle y, K_n \rangle = \int_0^1 xy(x)K_n \, dx \tag{11.58}$$

where the natural weighting function for cylinders is x (for spheres it would be x^2, and for slabs, unity).

You may note that the eigenvalue in Eq. 11.57a is specified as ξ_n^2. It is also clear that the solution of Eq. 11.57 is

$$K_n = J_0(\xi_n x) \tag{11.59}$$

and the eigenvalues arise naturally from the boundary condition $x = 1$; $K_n = 0$, so that the transcendental relation allows computation of ξ_n:

$$J_0(\xi_n) = 0 \tag{11.60}$$

Again, the multiplicative constant for K_n is taken as unity. The properties of Bessel function are discussed in Chapter 3.

Applying the integral transform defined in Eq. 11.58 to Eq. 11.53, we have first

$$\frac{d\langle y, K_n \rangle}{dt} = \langle Ly, K_n \rangle \qquad (11.61)$$

Now, applying the self-adjoint property as

$$\langle Ly, K_n \rangle = \langle y, LK_n \rangle$$

we get

$$\frac{d\langle y, K_n \rangle}{dt} = \langle y, LK_n \rangle = \langle y, -\xi_n^2 K_n \rangle$$

where we have replaced $LK_n = -\xi_n^2 K_n$. Now making use of the definition of the integral transform (Eq. 11.58), the above equation becomes

$$\frac{d\langle y, K_n \rangle}{dt} = \int_0^1 xy\left(-\xi_n^2 K_n\right) dx = -\xi_n^2 \int_0^1 xyK_n \, dx = -\xi_n^2 \langle y, K_n \rangle \quad (11.62)$$

Thus, the self-adjoint property accomplishes the steps from 11.10 to 11.16 in a single step! The initial condition for Eq. 11.62 is obtained by taking the transform of Eq. 11.54; that is,

$$t = 0; \quad \langle y, K_n \rangle = \langle 1, K_n \rangle = \int_0^1 1 x K_n(x) \, dx = \frac{J_1(\xi_n)}{\xi_n} \qquad (11.63)$$

The solution of Eq. 11.62 subject to the initial condition (11.63) is

$$\langle y, K_n \rangle = \langle 1, K_n \rangle e^{-\xi_n^2 t} \qquad (11.64)$$

Finally, the inverse of $\langle y, K_n \rangle$ is simply,

$$y = \sum_{n=1}^{\infty} \frac{\langle y, K_n \rangle}{\langle K_n, K_n \rangle} K_n(x) = \sum_{n=1}^{\infty} \frac{\langle 1, K_n \rangle}{\langle K_n, K_n \rangle} K_n(x) e^{-\xi_n^2 t} \qquad (11.65)$$

where $K_n(x) = J_0(\xi_n x)$, and the inner product for $n = m$ is

$$< K_n, K_n > = \int_0^1 x J_0^2(\xi_n x) \, dx = \frac{J_1^2(\xi_n)}{2}$$

The solution methodology for the Sturm–Liouville integral transform is now quite straightforward and of course yields the same results as the separation of variables method, as the reader can verify.

The solution form obtained in Eq. 11.65 for cylindrical coordinates is identical in structure for other coordinates, that is, rectangular and spherical. The only difference among them is the definition of the eigenproblem. We tabulate below the value for K_n, $\langle 1, K_n \rangle$, and $\langle K_n, K_n \rangle$ for the three geometries: slab, cylinder, and sphere, respectively.

Slab

$$K_n = \cos(\xi_n x), \qquad \xi_n = \left(n - \frac{1}{2}\right)\pi, \qquad \langle 1, K_n \rangle = \frac{\sin(\xi_n)}{\xi_n},$$

$$(11.66a)$$

$$\langle K_n, K_n \rangle = \frac{1}{2}$$

Cylinder

$$K_n = J_0(\xi_n x), \qquad J_0(\xi_n) = 0, \qquad \langle 1, K_n \rangle = \frac{J_1(\xi_n)}{\xi_n}, \qquad \langle K_n, K_n \rangle = \frac{J_1^2(\xi_n)}{2}$$

$$(11.66b)$$

Sphere

$$K_n = \frac{\sin(\xi_n x)}{x}, \qquad \xi_n = n\pi, \qquad \langle 1, K_n \rangle = -\frac{\cos(\xi_n)}{\xi_n}, \qquad \langle K_n, K_n \rangle = \frac{1}{2}$$

$$(11.66c)$$

11.2.3   Inhomogeneous Boundary Conditions

In the previous section, we developed the finite integral transform for a general Sturm–Liouville system. Homogeneous boundary conditions were used in the analysis up to this point. Here, we would like to discuss cases where the boundary conditions are not homogeneous, and determine if complications arise which impede the inversion process.

If the boundary conditions are not homogeneous, they can be rendered so by rearranging the dependent variable as discussed in Chapter 1. To show this, we use an example which follows the nomenclature of the previous section

$$\frac{\partial y}{\partial t} = Ly \tag{11.67}$$

$$t = 0; \qquad y = y_0(x) \tag{11.68}$$

$$My = \alpha \tag{11.69a}$$

$$Ny = \beta \tag{11.69b}$$

where L, M, and N are operators of the type defined in Eqs. 11.45 and 11.46.

We can decompose $y(x, t)$ into two parts (such as deviation variables used in Chapters 9 and 10)

$$y(x, t) = Y(x, t) + u(x) \tag{11.70}$$

Substitution of Eq. 11.70 into Eqs. 11.67 to 11.69 yields

$$\frac{\partial Y}{\partial t} = LY + Lu \tag{11.71}$$

$$t = 0; \quad Y + u(x) = y_0(x) \tag{11.72}$$

$$M(Y + u) = \alpha \tag{11.73a}$$

$$N(Y + u) = \beta \tag{11.73b}$$

To make the boundary conditions for Y homogeneous, we could define the following auxiliary equation for u, which is simply the steady-state solution

$$Lu = 0; \quad M(u) = \alpha; \quad N(u) = \beta \tag{11.74}$$

Having defined u as in Eq. 11.74, the governing equations for the new dependent variable Y become

$$\frac{\partial Y}{\partial t} = LY \tag{11.75}$$

$$t = 0; \quad Y = y_0(x) - u(x) \tag{11.76}$$

$$M(Y) = N(Y) = 0 \tag{11.77}$$

This new set of equations for Y now can be readily solved by either the method of separation of variables or the Sturm–Liouville integral transform method. We must also find $u(x)$, but this is simply described by an elementary ODE ($Lu = 0$), so the inhomogeneous boundary conditions (11.74) are not a serious impediment.

EXAMPLE 11.3

We wish to investigate the solution of the following problem

$$\frac{\partial y}{\partial \tau} = \nabla^2 y \tag{11.78a}$$

subject to the initial and boundary conditions

$$\tau = 0; \quad y = 0 \tag{11.78b}$$

$$x = 0; \quad \frac{\partial y}{\partial x} = 0 \tag{11.78c}$$

$$x = 1; \quad \frac{\partial y}{\partial x} = Bi(1 - y) \tag{11.78d}$$

where the operator ∇^2 is the Laplacian operator defined as

$$\nabla^2 = \frac{1}{x^s} \frac{\partial}{\partial x} \left(x^s \frac{\partial}{\partial x} \right) \tag{11.79}$$

with s being the shape factor of the domain. It takes a value of $0, 1$, or 2 for slab, cylindrical, or spherical coordinates.

The model Eqs. 11.78 describe the concentration distribution in a particle where adsorption is taking place with a linear adsorption isotherm. The external film mass transfer is reflected in the dimensionless Biot (Bi) number. The model equations can also describe heat conduction in a solid object and the heat transfer coefficient is reflected in the dimensionless Bi parameter.

We note that the boundary condition (11.78d) is inhomogeneous. To render the boundary conditions homogeneous, we need to solve the steady-state problem

$$\nabla^2 u = 0 \qquad (11.80a)$$

subject to

$$x = 0; \quad \frac{du}{dx} = 0 \quad \text{and} \quad x = 1; \quad \frac{du}{dx} = \text{Bi}(1 - u) \qquad (11.80b)$$

Solution of Eqs. 11.80 is simply

$$u = 1 \qquad (11.81)$$

Thus the new dependent variable Y is defined as in Eq. 11.70; that is,

$$Y = y - 1 \qquad (11.82)$$

Substitution of Eq. 11.82 into Eq. 11.78 would yield the following set of equations for Y having homogeneous boundary conditions

$$\frac{\partial Y}{\partial \tau} = \nabla^2 Y \qquad (11.83a)$$

$$\tau = 0; \qquad Y = -1 \qquad (11.83b)$$

$$x = 0; \qquad \frac{\partial Y}{\partial x} = 0 \qquad (11.83c)$$

$$x = 1; \qquad \frac{\partial Y}{\partial x} + \text{Bi} \cdot Y = 0 \qquad (11.83d)$$

This new set of equations now can be solved readily by the finite integral transform method. Using the procedure outlined in the last two examples, the following integral transform is derived as

$$\langle Y, K_n \rangle = \int_0^1 x^s Y(x, \tau) K_n(x) \, dx \qquad (11.84)$$

where the kernel of the transform is defined from the following associated eigenproblem

$$\nabla^2 K_n(x) + \xi_n^2 K_n(x) = 0 \tag{11.85a}$$

$$x = 0; \qquad \frac{dK_n}{dx} = 0 \tag{11.85b}$$

$$x = 1; \qquad \frac{dK_n}{dx} + \mathrm{Bi}\, K_n = 0 \tag{11.85c}$$

Using the procedure described in the earlier sections, the solution for Y is readily seen to be

$$Y = -\sum_{n=1}^{\infty} \frac{\langle 1, K_n \rangle e^{\xi_n^2 \tau}}{\langle K_n, K_n \rangle} K_n(x) \tag{11.86}$$

and hence

$$y = 1 - \sum_{n=1}^{\infty} \frac{\langle 1, K_n \rangle e^{-\xi_n^2 \tau}}{\langle K_n, K_n \rangle} K_n(x) \tag{11.87}$$

For three different shapes, the expressions for $K_n(x)$, ξ_n, $\langle 1, K_n \rangle$ and $\langle K_n, K_n \rangle$ are

Slab

$$K_n(x) = \cos(\xi_n x) \tag{11.88a}$$

$$\xi_n \sin(\xi_n) = \mathrm{Bi}\cos(\xi_n) \tag{11.88b}$$

$$\langle 1, K_n \rangle = \frac{\sin(\xi_n)}{\xi_n} \tag{11.88c}$$

$$\langle K_n, K_n \rangle = \frac{1}{2}\left[1 + \frac{\sin^2(\xi_n)}{\mathrm{Bi}}\right] \tag{11.88d}$$

Cylinder

$$K_n = J_0(\xi_n x) \tag{11.89a}$$

$$\xi_n J_1(\xi_n) = \mathrm{Bi}\cdot J_0(\xi_n) \tag{11.89b}$$

$$\langle 1, K_n \rangle = \frac{J_1(\xi_n)}{\xi_n} \tag{11.89c}$$

$$\langle K_n, K_n \rangle = \frac{J_1^2(\xi_n)}{2}\left[1 + \left(\frac{\xi_n}{\mathrm{Bi}}\right)^2\right] \tag{11.89d}$$

Sphere

$$K_n = \frac{\sin(\xi_n x)}{x} \tag{11.90a}$$

$$\xi_n \cos(\xi_n) = (1 - \text{Bi}) \sin(\xi_n) \tag{11.90b}$$

$$\langle 1, K_n \rangle = \frac{[\sin(\xi_n) - \xi_n \cos(\xi_n)]}{\xi_n^2} \tag{11.90c}$$

$$\langle K_n, K_n \rangle = \frac{1}{2}\left[1 + \frac{\cos^2(\xi_n)}{(\text{Bi} - 1)} \right] \tag{11.90d}$$

The concentration distribution is defined in Eq. 11.87, and of practical interest is the volumetric average. The volumetric average is defined as

$$y_{\text{avg}} = (s + 1) \int_0^1 x^s y(x, \tau) \, dx \tag{11.91}$$

Substitution of Eq. 11.87 into Eq. 11.91 yields the following general solution for the volumetric average

$$y_{\text{avg}} = 1 - (s + 1) \sum_{n=1}^{\infty} \frac{\langle 1, K_n \rangle^2}{\langle K_n, K_n \rangle} \cdot e^{-\xi_n^2 \tau} \tag{11.92}$$

Of further interest is the half-time of the process, that is, the time at which the average concentration is half of the equilibrium concentration. This is found by simply setting y_{avg} equal to $1/2$, and solving the resulting equation for the half-time $\tau_{0.5}$. For example, when $\text{Bi} \to \infty$, the following solution for the half-time is obtained

$$\tau_{0.5} = \begin{cases} 0.19674 & \text{for } s = 0 \\ 0.06310 & \text{for } s = 1 \\ 0.03055 & \text{for } s = 2 \end{cases} \tag{11.93}$$

The average concentration (or temperature) and the half-time solutions are particularly useful in adsorption and heat transfer studies, as a means to extract parameters from experimental measurements.

EXAMPLE 11.4

As an alternative to the previous example, we can also solve the problems with inhomogeneous boundary conditions by direct application of the finite integral transform, without the necessity of homogenizing the boundary conditions. To demonstrate this, we consider the following transient diffusion and reaction problem for a catalyst particle of either slab, cylindrical, or spherical shape. The dimensionless mass balance equations in a catalyst particle with a first order chemical reaction are

$$\frac{\partial y}{\partial \tau} = \nabla^2 y - \phi^2 y \tag{11.94a}$$

$$\tau = 0; \qquad y = 0 \tag{11.94b}$$

$$x = 0; \qquad \frac{\partial y}{\partial x} = 0 \tag{11.94c}$$

$$x = 1; \qquad y = 1 \tag{11.94d}$$

where the Laplacian operator ∇^2 is defined as in Eq. 11.79, and ϕ is the Thiele modulus.

If we applied the previous procedure, we must solve the steady-state equation

$$\nabla^2 u - \phi^2 u = 0 \tag{11.95a}$$

$$x = 0; \quad \frac{du}{dx} = 0 \quad \text{and} \quad x = 1; \quad u = 1 \tag{11.95b}$$

and then the finite integral transform is applied on the following equation for $Y(y = Y + u)$

$$\frac{\partial Y}{\partial \tau} = \nabla^2 Y - \phi^2 Y \tag{11.96a}$$

$$\tau = 0; \qquad Y = -u(x) \tag{11.96b}$$

$$x = 0; \qquad \frac{\partial Y}{\partial x} = 0 \tag{11.96c}$$

$$x = 1; \qquad Y = 0 \tag{11.96d}$$

The solutions for the steady-state concentration u are

$$u = \frac{\cosh(\phi x)}{\cosh(\phi)} \tag{11.97a}$$

$$u = \frac{I_0(\phi x)}{I_0(\phi)} \tag{11.97b}$$

$$u = \frac{\sinh(\phi x)}{x \sinh(\phi)} \tag{11.97c}$$

for slab, cylinder, and sphere, respectively.

To find Y, we apply the finite integral transform to Eqs. 11.96a, where we again solve the following associated eigenproblem

$$\nabla^2 K_n - \phi^2 K_n + \xi_n^2 K_n = 0 \tag{11.98a}$$

$$x = 0; \qquad \frac{dK_n}{dx} = 0 \tag{11.98b}$$

$$x = 1; \qquad K_n = 0 \tag{11.98c}$$

and then follow the procedure of the previous example to find the solution. However, it is possible to attack the problem directly, as we show next.

To avoid the process of homogenization, we simply apply the integral transform directly to the original Eq. 11.94. For a given differential operator ∇^2 and the boundary conditions defined in Eqs. 11.94c and d, the kernel of the transform is defined from the following eigenproblem

$$\nabla^2 K_n + \xi_n^2 K_n = 0 \qquad (11.99a)$$

$$x = 0; \qquad \frac{dK_n}{dx} = 0 \qquad (11.99b)$$

$$x = 1; \qquad K_n = 0 \qquad (11.99c)$$

This particular eigenproblem was obtained by simply applying the operator ∇^2 on K_n and setting it equal to $-\xi_n^2 K_n$. The boundary condition of this eigenfunction is the homogeneous version of Eqs. 11.94c and 11.94d.

The several solutions for this eigenproblem itself have already been given in Eqs. 11.88, 11.89, and 11.90 for slab, cylinder, and sphere, respectively.

The integral transform for this inhomogeneous problem is defined as before

$$\langle y, K_n \rangle = \int_0^1 x^s y(x, \tau) K_n(x)\, dx \qquad (11.100)$$

where the general weighting function is x^s.

Now applying the integral transform directly to the original equation Eq.11.94, we obtain

$$\frac{d\langle y, K_n \rangle}{d\tau} = \xi_n^2 [\langle 1, K_n \rangle - \langle y, K_n \rangle] - \phi^2 \langle y, K_n \rangle \qquad (11.101)$$

where the first two terms in the RHS of Eq. 11.101 arise from the integration by parts, as illustrated earlier in Eq. 11.8.

The initial condition of Eq. 11.101 is obtained by taking the transform of Eq. 11.94b; that is,

$$\tau = 0; \qquad \langle y, K_n \rangle = 0 \qquad (11.102)$$

The solution of Eq. 11.101 subject to the initial condition (11.102) is

$$\langle y, K_n \rangle = \langle 1, K_n \rangle \frac{\xi_n^2}{\left(\xi_n^2 + \phi^2\right)} - \frac{\langle 1, K_n \rangle \xi_n^2}{\left(\xi_n^2 + \phi^2\right)} e^{-\xi_n^2 \tau} \qquad (11.103)$$

The first term on the RHS of Eq. 11.103 can be rearranged to give

$$\langle y, K_n \rangle = \left[\langle 1, K_n \rangle - \langle 1, K_n \rangle \frac{\phi^2}{\left(\xi_n^2 + \phi^2\right)} \right] - \frac{\langle 1, K_n \rangle \xi_n^2}{\left(\xi_n^2 + \phi^2\right)} e^{-\xi_n^2 \tau} \qquad (11.104)$$

The inverse of Eq. 11.104 can now be readily obtained as

$$y = \sum_{n=1}^{\infty} \frac{\langle y, K_n \rangle}{\langle K_n, K_n \rangle} K_n(x) \qquad (11.105a)$$

or

$$y = \sum_{n=1}^{\infty} \frac{\langle 1, K_n \rangle}{\langle K_n, K_n \rangle} K_n(x) - \phi^2 \sum_{n=1}^{\infty} \frac{\langle 1, K_n \rangle K_n(x)}{(\xi_n^2 + \phi^2)\langle K_n, K_n \rangle}$$

$$- \sum_{n=1}^{\infty} \frac{\langle 1, K_n \rangle \xi_n^2 e^{-\xi_n^2 \tau}}{(\xi_n^2 + \phi^2)\langle K_n, K_n \rangle} \qquad (11.105b)$$

The first series on the RHS of Eq. 11.105b is in fact the series representation of the function "1," that is, the solution can now be rewritten as

$$y = 1 - \phi^2 \sum_{n=1}^{\infty} \frac{\langle 1, K_n \rangle K_n(x)}{(\xi_n^2 + \phi^2)\langle K_n, K_n \rangle} - \sum_{n=1}^{\infty} \frac{\langle 1, K_n \rangle \xi_n^2 e^{-\xi_n^2 \tau}}{(\xi_n^2 + \phi^2)\langle K_n, K_n \rangle} \qquad (11.106)$$

where $K_n(x), \xi_n, \langle 1, K_n \rangle, \langle K_n, K_n \rangle$ are defined in Eqs. 11.88, 11.89, and 11.90 for slab, cylinder, and sphere, respectively.

The solution obtained by the second method is quite straightforward and faster than the traditional way, even though the conventional way could yield the steady-state solution in an analytical form (Equation 11.97), yet the second method yields the steady-state solution in a series form (the first two terms of Equation 11.106). The series in the second set of terms on the RHS of Eq. 11.106 has a slow convergence property. For the summation done by a computer, this is not really a hurdle, but the series can be further rearranged to have faster convergence rate. Interested readers should refer to Do and Bailey (1981) and Johnston and Do (1987) for this convergence enhancement.

11.2.4 Inhomogeneous Equations

The last section deals with the case of nonhomogeneous boundary conditions. Now we will study the problem when the equation itself is inhomogeneous.

Now if the partial differential equation is inhomogeneous of the form

$$\frac{\partial y}{\partial \tau} = Ly - f(x) \qquad (11.107)$$

$$\tau = 0; \qquad y = y_0(x) \qquad (11.108)$$

$$M(y) = N(y) = 0 \qquad (11.109)$$

A new dependent variable Y can be introduced as in Eq. 11.70. Substituting Eq. 11.70 into Eqs. 11.107 to 11.109 gives

$$\frac{\partial Y}{\partial \tau} = LY + Lu - f(x) \tag{11.110}$$

$$\tau = 0; \quad Y + u(x) = y_0(x) \tag{11.111}$$

$$M(Y + u) = N(Y + u) = 0 \tag{11.112}$$

Thus, if we define u as the steady-state solution

$$Lu(x) - f(x) = 0 \quad \text{and} \quad M(u) = N(u) = 0 \tag{11.113}$$

the new set of equations for Y are homogeneous and are the same as Eqs. 11.75 to 11.77, which then can be readily solved by either the separation of variables method or the finite integral transform method. However, we can also attack this problem directly without introducing deviation from steady state, as we show next.

As in the previous section, we can obtain the solution for y by directly applying the integral transform to Eq. 11.107. The kernel of the transform is obtained from an eigenproblem, which is defined by taking the operator L on K_n and setting the result equal to $-\xi_n^2 K_n$; that is,

$$LK_n(x) + \xi_n^2 K_n(x) = 0 \tag{11.114a}$$

$$M(K_n) = N(K_n) = 0 \tag{11.114b}$$

The application of the transform to Eq. 11.107 would yield

$$\frac{d\langle y, K_n \rangle}{d\tau} = \langle Ly, K_n \rangle - \langle f(x), K_n \rangle \tag{11.115}$$

Using the self-adjoint property (obtained by integration by parts of the first term on the RHS of Eq. 11.115), we have

$$\frac{d\langle y, K_n \rangle}{d\tau} = \langle y, LK_n \rangle - \langle f(x), K_n \rangle \tag{11.116}$$

Next, using the definition of the kernel, Eq. 11.116 becomes

$$\frac{d\langle y, K_n \rangle}{d\tau} = -\xi_n^2 \langle y, K_n \rangle - \langle f(x), K_n \rangle \tag{11.117}$$

The initial condition of Eq. 11.117 is obtained by taking the transform of Eq. 11.108

$$\tau = 0; \quad \langle y, K_n \rangle = \langle y_0(x), K_n \rangle \tag{11.118}$$

The solution of the I-factor ODE in Eq. 11.117, subject to the initial condition (11.118), is

$$\langle y, K_n \rangle = -\frac{1}{\xi_n^2} \langle f(x), K_n \rangle$$

$$+ \left[\langle y_0(x), K_n \rangle + \frac{1}{\xi_n^2} \langle f(x), K_n \rangle \right] e^{-\xi_n^2 \tau} \qquad (11.119)$$

from which the inverse can be obtained as

$$y = -\sum_{n=1}^{\infty} \frac{\langle f(x), K_n \rangle K_n(x)}{\xi_n^2 \langle K_n, K_n \rangle}$$

$$+ \sum_{n=1}^{\infty} \frac{\left[\langle y_0(x), K_n \rangle + \frac{1}{\xi_n^2} \langle f(x), K_n \rangle \right] e^{-\xi_n^2 \tau}}{\langle K_n, K_n \rangle} K_n(x) \qquad (11.120)$$

The first term on the RHS of Eq. 11.120 is the series representation of the steady-state solution u, defined in Eq. 11.113.

11.2.5 Time-Dependent Boundary Conditions

Even when the boundary conditions involve time-dependent functions, the method described in the previous section can be used to good effect. Let us demonstrate this by solving a mass transfer problem in a particle when the bulk concentration varies with time.

EXAMPLE 11.5

The mass balance equations in dimensionless form are

$$\frac{\partial y}{\partial \tau} = \nabla^2 y \qquad (11.121)$$

$$\tau = 0; \qquad y = 0 \qquad (11.122)$$

$$x = 0; \qquad \frac{\partial y}{\partial x} = 0 \qquad (11.123a)$$

$$x = 1; \qquad y = e^{-\tau} \qquad (11.123b)$$

where ∇^2 is defined in Eq. 11.79. Equation 11.123b is the time-dependent boundary condition.

The boundary condition (11.123b) represents the situation where the particle is exposed to a bulk concentration, which decays exponentially with time.

The general integral transform for any coordinate system is

$$\langle y, K_n \rangle = \int_0^1 x^s y(x, \tau) K_n(x) \, dx \qquad (11.124)$$

where the kernel $K_n(x)$ is defined as in the eigenproblem equation 11.99.
Taking the integral transform of Eq. 11.121 would give

$$\frac{d\langle y, K_n \rangle}{d\tau} = \langle \nabla^2 y, K_n \rangle \qquad (11.125)$$

Carrying out the integration by parts of the first term on the RHS of Eq. 11.125 gives

$$\frac{d\langle y, K_n \rangle}{d\tau} = \left[x^s K_n \frac{\partial y}{\partial x} - x^s y \frac{dK_n}{dx} \right]_0^1 + \int_0^1 x^s (\nabla^2 K_n) y \, dx \qquad (11.126)$$

Using the boundary conditions for y (Eqs. 11.123) and the boundary conditions for K_n (Eqs. 11.99b, c) along with Eq. 11.99 ($\nabla^2 K_n = -\xi_n^2 K_n$), then Eq. 11.126 becomes

$$\frac{d\langle y, K_n \rangle}{d\tau} = -\frac{dK_n(1)}{dx} e^{-\tau} - \xi_n^2 \langle y, K_n \rangle \qquad (11.127)$$

Multiplying Eq. 11.99a by x^s and integrating the result from 0 to 1 gives

$$-\frac{dK_n(1)}{dx} = \xi_n^2 \langle 1, K_n \rangle \qquad (11.128)$$

in which the condition (11.99b) has been used.
Using Eq. 11.128, Eq. 11.127 can be rewritten as

$$\frac{d\langle y, K_n \rangle}{d\tau} = \xi_n^2 \langle 1, K_n \rangle e^{-t} - \xi_n^2 \langle y, K_n \rangle \qquad (11.129)$$

Rearrange Eq. 11.129 as

$$\frac{d\langle y, K_n \rangle}{d\tau} + \xi_n^2 \langle y, K_n \rangle = \xi_n^2 \langle 1, K_n \rangle e^{-\tau} \qquad (11.130)$$

The initial condition for Eq. 11.130 can be obtained by taking the transform of Eq. 11.122

$$\tau = 0; \qquad \langle y, K_n \rangle = 0 \qquad (11.131)$$

Equation 11.130 is a first order ordinary differential equation with exponential forcing function (*I*-factor form). The methods from Chapter 2 show

$$\langle y, K_n \rangle = \xi_n^2 \langle 1, K_n \rangle \frac{e^{-\tau} - e^{-\xi_n^2 \tau}}{(\xi_n^2 - 1)} \tag{11.132}$$

The inverse of the previous equation is

$$y = \sum_{n=1}^{\infty} \frac{\xi_n^2 \langle 1, K_n \rangle \left[e^{-\tau} - e^{-\xi_n^2 \tau} \right] K_n(x)}{(\xi_n^2 - 1) \langle K_n, K_n \rangle} \tag{11.133}$$

The mean concentration in the particle (of practical interest) is then given by

$$\bar{y}(\tau) = (s + 1) \sum_{n=1}^{\infty} \frac{\xi_n^2 \langle 1, K_n \rangle^2 \left[e^{-\tau} - e^{-\xi_n^2 \tau} \right]}{(\xi_n^2 - 1) \langle K_n, K_n \rangle} \tag{11.134}$$

where all eigenproperties are given in Eqs. 11.88, 11.89, and 11.90 for slab, cylinder, and sphere, respectively. For example, for a slab particle, the explicit solution is

$$\bar{y}(\tau) = 2 \sum_{n=1}^{\infty} \frac{\left[e^{-\tau} - e^{-\xi_n^2 \tau} \right]}{(\xi_n^2 - 1)}$$

Figure 11.5 shows a plot of this mean concentration versus time. It rises from zero (initial state) to some maximum value and then decays to zero because of the depletion of the external source.

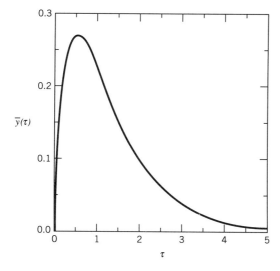

Figure 11.5 Plot of the mean concentration (Eq. 11.134) versus τ.

11.2.6   Elliptic Partial Differential Equations

We have demonstrated the application of the finite integral transform to a number of parabolic partial differential equations. These are important because they represent the broadest class of time-dependent PDEs dealt with by chemical engineers. Now we wish to illustrate its versatility by application to elliptic differential equations, which are typical of steady-state diffusional processes (heat, mass, momentum), in this case for two spatial variables. We have emphasized the parabolic PDEs, relative to the elliptic ones, because many texts and much mathematical research has focussed too long on elliptic PDEs.

EXAMPLE 11.6

Consider the problem introduced in Chapter 1, connected with modeling the cooling of a bath owing to protruding cylindrical rods. The model equations in dimensionless form were given as

$$\frac{1}{\xi}\frac{\partial}{\partial \xi}\left(\xi\frac{\partial u}{\partial \xi}\right) + \Delta^2\frac{\partial^2 u}{\partial \zeta^2} = 0 \tag{11.135}$$

$$\xi = 0; \qquad \frac{\partial u}{\partial \xi} = 0 \tag{11.136a}$$

$$\xi = 1; \qquad \frac{\partial u}{\partial \xi} = -\operatorname{Bi} u \tag{11.136b}$$

$$\zeta = 0; \qquad u = 1 \tag{11.137a}$$

$$\zeta = 1; \qquad \frac{\partial u}{\partial \zeta} = 0 \tag{11.137b}$$

We note early that the boundary conditions are homogeneous in the ξ-domain (Eq. 11.136). So if we take the integral transform with respect to this variable, the transform is defined as

$$\langle u, K_n \rangle = \int_0^1 \xi u(\xi, \zeta) K_n(\xi)\, d\xi \tag{11.138}$$

where $K_n(\xi)$ is the kernel of the transform and is defined in terms of the associated eigenproblem

$$\frac{1}{\xi}\frac{d}{d\xi}\left[\xi\frac{dK_n}{d\xi}\right] + \beta_n^2 K_n(\xi) = 0 \tag{11.139a}$$

$$\xi = 0; \qquad \frac{dK_n}{d\xi} = 0 \tag{11.139b}$$

$$\xi = 1; \qquad \frac{dK_n}{d\xi} = -\operatorname{Bi} K_n \tag{11.139c}$$

where, by comparison with the Sturm–Liouville equation, the weighting function must be ξ.

The solution of this associated eigenproblem is:

$$K_n(x) = J_0(\beta_n \xi) \tag{11.140}$$

where the eigenvalues β_n are determined by inserting Eq. 11.140 into Eq. 11.139c to get the transcendental relation

$$\beta_n J_1(\beta_n) = \text{Bi} J_0(\beta_n) \tag{11.141}$$

Taking the integral transform of Eq. 11.135, we have

$$\left\langle \frac{1}{\xi} \frac{d}{d\xi}\left(\xi \frac{du}{d\xi}\right), K_n \right\rangle + \Delta^2 \left\langle \frac{\partial^2 u}{\partial \zeta^2}, K_n \right\rangle = 0$$

Since the integral transform is with respect to ξ, the second term in the LHS then becomes

$$\left\langle \frac{1}{\xi} \frac{d}{d\xi}\left(\xi \frac{du}{d\xi}\right), K_n \right\rangle + \Delta^2 \frac{d^2}{d\zeta^2}\langle u, K_n \rangle = 0 \tag{11.142a}$$

Now let us consider the first term in the LHS of Eq. 11.142a and utilize the definition of the transform as given in Eq. 11.138. We then have

$$\left\langle \frac{1}{\xi} \frac{d}{d\xi}\left(\xi \frac{du}{d\xi}\right), K_n \right\rangle = \int_0^1 \xi\left[\frac{1}{\xi} \frac{d}{d\xi}\left(\xi \frac{du}{d\xi}\right)\right] K_n\, d\xi \tag{11.142b}$$

Carrying out the integration by parts twice on the integral in the RHS of Eq. 11.142b gives

$$\left\langle \frac{1}{\xi} \frac{d}{d\xi}\left(\xi \frac{du}{d\xi}\right), K_n \right\rangle = \int_0^1 \xi u\left[\frac{1}{\xi} \frac{d}{d\xi}\left(\xi \frac{dK_n}{d\xi}\right)\right] d\xi$$

$$\left\langle \frac{1}{\xi} \frac{d}{d\xi}\left(\xi \frac{du}{d\xi}\right), K_n \right\rangle = \left\langle u, \frac{1}{\xi} \frac{d}{d\xi}\left(\xi \frac{dK_n}{d\xi}\right) \right\rangle \tag{11.142c}$$

which is in fact the self-adjoint property.

Using the definition of the eigenproblem (Eq 11.139a), Eq. 11.142c becomes

$$\left\langle \frac{1}{\xi} \frac{d}{d\xi}\left(\xi \frac{du}{d\xi}\right), K_n \right\rangle = \int_0^1 \xi u\left[-\beta_n^2 K_n\right] d\xi = -\beta_n^2 \int_0^1 \xi u K_n\, d\xi$$

that is,

$$\left\langle \frac{1}{\xi} \frac{d}{d\xi}\left(\xi \frac{du}{d\xi}\right), K_n \right\rangle = -\beta_n^2 \langle u, K_n \rangle \tag{11.142d}$$

Substituting Eq. 11.142d into Eq. 11.142a, we get

$$-\beta_n^2 \langle u, K_n \rangle + \Delta^2 \frac{d^2 \langle u, K_n \rangle}{d\zeta^2} = 0 \tag{11.143}$$

Taking the integral transform of Eq. 11.137, we have

$$\zeta = 0; \quad \langle u, K_n \rangle = \langle 1, K_n \rangle \tag{11.144a}$$

$$\zeta = 1; \quad \frac{d\langle u, K_n \rangle}{d\zeta} = 0 \tag{11.144b}$$

Equation 11.143 is a second order ordinary differential equation, and its general solution is

$$\langle u, K_n \rangle = a \exp\left(\frac{\beta_n \zeta}{\Delta}\right) + b \exp\left(-\frac{\beta_n \zeta}{\Delta}\right) \tag{11.145}$$

Applying the transformed boundary conditions (11.144), the integration constants a and b can be found. The final solution written in terms of hyperbolic functions is

$$\langle u, K_n \rangle = \langle 1, K_n \rangle \frac{\cosh\left[\frac{\beta_n(1-\zeta)}{\Delta}\right]}{\cosh\left(\frac{\beta_n}{\Delta}\right)} \tag{11.146}$$

Knowing the projection $\langle u, K_n \rangle$ on the basis function $K_n(\xi)$, the vector $\mathbf{u}$ can be constructed as a linear combination as was done previously to yield the inverse transform

$$u = \sum_{n=1}^{\infty} \frac{\langle u, K_n \rangle}{\langle K_n, K_n \rangle} K_n(x) \tag{11.147a}$$

or in terms of specific functions

$$u(\xi, \zeta) = \sum_{n=1}^{\infty} \frac{\langle 1, K_n \rangle}{\langle K_n, K_n \rangle} \frac{\cosh\left[\frac{\beta_n(1-\zeta)}{\Delta}\right]}{\cosh\left(\frac{\beta_n}{\Delta}\right)} K_n(\xi) \tag{11.147b}$$

Equation 11.147b yields the solution of u in terms of ξ and ζ. Here,

$$\langle 1, K_n \rangle = \int_0^1 \zeta K_n(\xi) \, d\xi = \int_0^1 \xi J_0(\beta_n \xi) \, d\xi = \frac{J_1(\beta_n)}{\beta_n} \tag{11.148a}$$

and

$$\langle K_n, K_n \rangle = \int_0^1 \xi K_n^2(\xi) \, d\xi = \frac{J_1^2(\beta_n)}{2}\left[1 + \left(\frac{\beta_n}{\text{Bi}}\right)^2\right] \tag{11.148b}$$

The quantity of interest, flux through the base of the rod, which was needed in Chapter 1, is

$$I = \int_0^1 \xi\left[-\frac{\partial u(\xi, 0)}{\partial \zeta}\right] d\xi \tag{11.149}$$

Differentiate Eq. 11.147b with respect to ζ, evaluate this at $\zeta = 0$, and substitute the final result into Eq. 11.149 to get

$$I = \sum_{n=1}^{\infty} \frac{\langle 1, K_n \rangle^2}{\langle K_n, K_n \rangle}\left(\frac{\beta_n}{\Delta}\right)\tanh\left(\frac{\beta_n}{\Delta}\right) \tag{11.150}$$

Substituting Eqs. 11.148 into Eq. 11.150 then yields

$$I = \frac{2}{\Delta} \sum_{n=1}^{\infty} \frac{\tanh\left(\dfrac{\beta_n}{\Delta}\right)}{\beta_n \left[1 + \left(\dfrac{\beta_n}{\text{Bi}}\right)^2\right]} \tag{11.151}$$

Since the original partial differential equation is elliptic, we could have taken the finite integral transform with respect to ζ instead of ξ. If we now define

$$v = 1 - u \tag{11.152}$$

so that the boundary conditions in ζ are homogeneous, we will have the following partial differential equation in terms of v.

$$\frac{1}{\xi}\frac{\partial}{\partial \xi}\left(\xi \frac{\partial v}{\partial \xi}\right) + \Delta^2 \frac{\partial^2 v}{\partial \zeta^2} = 0 \tag{11.153}$$

$$\xi = 0; \qquad \frac{\partial v}{\partial \xi} = 0 \tag{11.154a}$$

$$\xi = 1; \qquad \frac{\partial v}{\partial \xi} = \text{Bi}(1 - v) \tag{11.154b}$$

$$\zeta = 0; \qquad v = 0 \tag{11.155a}$$

$$\zeta = 1; \qquad \frac{\partial v}{\partial \zeta} = 0 \tag{11.155b}$$

Now, let us define the finite integral transform with respect to ζ as

$$\langle v, T_n \rangle = \int_0^1 v(\xi, \zeta) T_n(\zeta)\, d\zeta \tag{11.156}$$

and then we proceed with the usual integral transform procedure to obtain the solution for v, and hence, the heat flux. This will be left as a homework exercise.

We now return to the problem of heat removal from a bath of hot solvent by way of protruding steel rods (Section 1.6). Various levels of modeling were carried out with the simplest level (level 1) being the case where the heat loss is controlled by the heat transfer through the gas film surrounding the steel rod. The rate of heat loss of this level 1 is (Eq. 1.55c)

$$Q_1 = 2\pi R h_G L_2 (T_1 - T_0) \tag{11.157}$$

The highest level we dealt with in Chapter 1 is level 4, in which we considered the axial as well as radial heat conduction in the segment of the steel rod exposed to the atmosphere. The rate of heat loss is given in Eq. 1.89, and is written here in terms of integral flux I (Eq. 11.149)

$$Q_4 = \frac{2\pi R^2 k}{L_2}(T_1 - T_0) I \tag{11.158}$$

To investigate the significance of the radial and axial conduction in the steel rod, we investigate the following ratio of heat fluxes obtained for levels 1 and 4

$$\text{Ratio} = \frac{Q_4}{Q_1} = \frac{\Delta^2}{Bi} I \qquad (11.159)$$

where $\Delta = R/L_2$ and $Bi = h_G R/k$.

Figure 11.6 shows this ratio versus Δ with the Biot as the parameter in the curves. The radial and axial heat conduction in the steel rod is insignificant when the ratio is close to one. This is possible when (i) the parameter Biot number is very small or (ii) the geometric factor Δ is much greater than about 1. Let us investigate these two possibilities. The first simply means that the heat transfer through the gas film surrounding the steel rod is much smaller than the heat conduction in the steel rod. Hence, the heat transfer is controlled by the gas film as we would expect on physical grounds. Thus, the level 4 solution is reduced to a level 1 solution. The second possibility is that the geometric factor Δ is much greater than about one.

To summarize the method of finite integral transform, we list (as follows) the key steps in the application of this technique to solve PDE.

Step 1 Arrange the equation in the format such that the linear differential operator has the form as given in Eq. 11.45.

Step 2 Define the associated eigenproblem (given in Eq. 11.48) with the associated homogeneous boundary conditions. This eigenproblem defines the kernel of the transform.

Step 3 Define the integral transform (as given in Eq. 11.37) and apply this transform to the governing equation of Step 1.

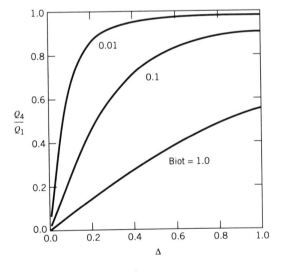

Figure 11.6 Plot of Q_4/Q_1 versus Δ with Biot as parameter.

Step 4   Integrate the result of Step 3 and use the boundary conditions of the equation as well as the eigenproblem; an equation for the image of y, that is, $\langle y, K_n \rangle$ should be obtained.

Step 5   Solve the equation for the image $\langle y, K_n \rangle$

Step 6   Obtain the inverse y from the image by using either Eq. 11.24 or 11.39.

We have presented the method of Sturm–Liouville integral transforms, and applied it to a number of problems in chemical engineering. Additional reading on this material can be found in Sneddon (1972) or Tranter (1958).

11.3   GENERALIZED STURM–LIOUVILLE INTEGRAL TRANSFORM

In the previous sections, we showed how the finite integral transform technique can be applied to single linear partial differential equations having finite spatial domain. The Sturm–Liouville integral transform was presented with necessary concepts of eigenproblem and orthogonality to facilitate the solution of single partial or ordinary differential equations. In this section, we shall deal with coupled linear partial differential equations also having a finite spatial domain, and we set out to show that the same procedure can be applied if we introduce the concept of function space, with elements having state variables as components. This will require representation using vector analysis.

11.3.1   Introduction

We illustrated in the previous sections that a single equation can be solved by either the method of separation of variables or the method of finite integral transform. The integral transform is more evolutionary in format and requires less intuition and practice than the method of separation of variables. In this section, however, we will show how to apply a generalized form of the Sturm–Liouville integral transform to coupled PDEs. With this generalized integral transform, large systems of linear partial differential equations can be solved, and the methodology follows naturally from the Sturm–Liouville integral transform presented earlier. Although the procedure is developed in an evolutionary format, its application is not as straightforward as the Laplace transform (Chapters 9 and 10) if the problem has a timelike variable.

   The approach we shall take is to teach by example, as we did in the finite Sturm–Liouville integral transform. We start by using the batch adsorber problem to illustrate the procedure. A comprehensive account of the method can be found in the book by Ramkrishna and Amundson (1987).

11.3.2   The Batch Adsorber Problem

To apply Sturm–Liouville transforms to coupled PDEs, we shall need to introduce a new concept: function space. We introduce this methodology by way of a practical example: batch adsorption in a vessel (Fig. 11.7).

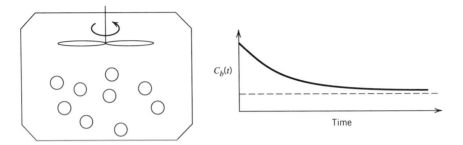

Figure 11.7 Batch adsorber.

EXAMPLE 11.7

Let us start with the derivation of the mass balance equations for a batch adsorber containing spherical adsorbent particles. The solute molecules diffuse from the bulk reservoir into the particles through a network of pores within the particle. The diffusion process is generally described by a Fickian type equation

$$J = -D_e \frac{\partial C}{\partial r} \tag{11.160}$$

where D_e is the effective diffusivity of the solute in the particle, and the flux J has units of moles transported per unit total cross-sectional area per unit time.

During the diffusion of solute molecules through the network of pores, some of the solute molecules are adsorbed onto the interior surface of the particle. This process of adsorption is normally very fast relative to the diffusion process, and so we can model it as local equilibrium between the solute in the pore fluid and the solute bound to the interior surface of the particle. This partitioning is often referred to as the *adsorption isotherm*, which implies constant temperature conditions. When the solute concentration in the pore is low enough, the relationship becomes linear (Henry's law); hence, mathematically, we can write

$$C_\mu = KC \tag{11.161}$$

where C is the solute concentration in the pore, C_μ is the concentration on the interior surface, and K is the slope of the linear isotherm, which represents the strength of the adsorption. We take the units of C_μ as moles per volume of particle (excluding the pore volume), so typical units would be moles/cm^3 solid.

We are now in a position to derive the mass balance equation inside the particle. Carrying out the mass balance equation over a small element of thickness Δr at the position r (Fig. 11.8), we obtain for an element of spherical particle

$$4\pi r^2 J|_r - 4\pi r^2 J|_{r+\Delta r} = \frac{\partial}{\partial t}\left[4\pi r^2 \Delta r\, \varepsilon C + 4\pi r^2 \Delta r(1-\varepsilon)C_\mu\right] \tag{11.162}$$

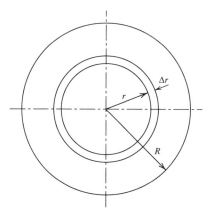

Figure 11.8 Shell element of adsorbent particle.

The porosity ε enters into the accumulation terms, thereby accounting separately for hold-up in fluid and solid phases.

Dividing Eq. 11.162 by $4\pi\Delta r$, and taking the limit when the element shrinks to zero gives

$$\varepsilon\frac{\partial C}{\partial t} + (1 - \varepsilon)\frac{\partial C_\mu}{\partial t} = -\frac{1}{r^2}\cdot\frac{\partial}{\partial r}(r^2 J) \qquad (11.163)$$

Now substituting Eqs. 11.160 and 11.161 into Eq. 11.163 gives

$$[\varepsilon + (1 - \varepsilon)K]\frac{\partial C}{\partial t} = D_e\frac{1}{r^2}\frac{\partial}{\partial r}\left(r^2\frac{\partial C}{\partial r}\right) \qquad (11.164)$$

If diffusion only occurred in the pore fluid, we would replace D_e with εD_{pore}. As it now stands, diffusion is allowed in both pore and solid phases (i.e., surface diffusion of adsorbed molecules may occur).

This equation describes the change of the solute concentration in the void space of the particle. Basically it states that the rate of change of mass hold-up in the particle (LHS of Eq. 11.164) is balanced by the rate of intraparticle diffusion (RHS). Since the mass balance equation (Eq. 11.164) is second order with respect to the spatial variable, we shall need two boundary conditions to complete the mass balance on the particle. One condition will be at the center of the particle where symmetry must be maintained. The other is at the exterior surface of the particle. If the stirred pot containing the particles is well mixed, then the fluid at the surface of particles is the same composition as the well-mixed fluid. These physical requirements are written mathematically as

$$r = 0; \qquad \frac{\partial C}{\partial r} = 0 \qquad (11.165a)$$

$$r = R; \qquad C = C_b \qquad (11.165b)$$

where C_b is the solute concentration in the reservoir. If the pot is not so well stirred, we would need to account for film resistance at the particle surface.

Note that the solute concentration changes with time because the reservoir is finite (i.e., the solute capacity is finite). To learn how this solute concentration changes with time, we must also carry out a mass balance for the fluid contained in the reservoir (i.e., excluding the particles which carry their own pore fluid).

Let the (dry) mass of the particle be m_p and the particle density be ρ_p, hence, the number of particles in the reservoir is

$$N_p = \frac{\left(\dfrac{m_p}{\rho_p}\right)}{\dfrac{4\pi R^3}{3}}$$

(11.166)

The influx into a single particle is equal to the product of the flux at the exterior surface and the area of the exterior surface; that is,

$$M = (4\pi R^2)\left[D_e \frac{\partial C(R,t)}{\partial r}\right]$$

(11.167)

So, the total mass loss from the reservoir is the product of the above two equations; that is,

$$\frac{3}{R}\left(\frac{m_p}{\rho_p}\right)D_e \frac{\partial C(R,t)}{\partial r}$$

(11.168)

Knowing this mass loss, the total mass balance equation for the solute in the reservoir is

$$V\frac{dC_b}{dt} = -\frac{3}{R}\left(\frac{m_p}{\rho_p}\right)D_e \frac{\partial C(R,t)}{\partial r}$$

(11.169)

where V is the fluid reservoir volume excluding the particle volume. This equation describes the concentration changes in the reservoir. The initial condition for this equation is

$$t = 0; \quad C_b = C_{b0}$$

(11.170)

If we assume that the particle is initially filled with inert solvent (i.e., free of solute), the initial condition for C then is

$$t = 0; \quad C = 0$$

(11.171)

Before we demonstrate the method of generalized integral transforms, we convert the dimensional mass balance equations into dimensionless form. This is achieved by defining the following dimensionless variables and parameters

$$A = \frac{C}{C_{b0}}; \quad A_b = \frac{C_b}{C_{b0}}; \quad x = \frac{r}{R}; \quad \tau = \frac{D_e t}{R^2[\varepsilon + (1 - \varepsilon)K]} \quad (11.172)$$

$$B = \frac{\left(\dfrac{m_p}{\rho_p}\right)[\varepsilon + (1 - \varepsilon)K]}{V} \quad (11.173)$$

These new variables allow the transport equations to be represented in much more simplified form

$$\frac{\partial A}{\partial \tau} = \frac{1}{x^2}\frac{\partial}{\partial x}\left(x^2\frac{\partial A}{\partial x}\right) \quad (11.174a)$$

$$\frac{dA_b}{d\tau} = -\frac{3}{B} \cdot \frac{\partial A(1,\tau)}{\partial x} \quad (11.174b)$$

subject to the following initial and boundary conditions

$$\tau = 0; \quad A = 0; \quad A_b = 1 \quad (11.175)$$

$$x = 0; \quad \frac{\partial A}{\partial x} = 0 \quad (11.176a)$$

$$x = 1; \quad A = A_b \quad (11.176b)$$

The previous set of equations describes the concentration changes in spherically shaped particles. For particles of slab and cylindrical shape, Eqs. 11.174 are simply replaced by

$$\frac{\partial A}{\partial \tau} = \frac{1}{x^s}\frac{\partial}{\partial x}\left(x^s\frac{\partial A}{\partial x}\right) \quad (11.177a)$$

$$\frac{dA_b}{d\tau} = -\frac{(s + 1)}{B}\frac{\partial A(1,\tau)}{\partial x} \quad (11.177b)$$

subject to the initial condition (11.175) and the boundary conditions (11.176). Here, the shape factor s can take a value of 0, 1, or 2 to account for slab, cylindrical, or spherical geometries, respectively. This structure allows very general solutions to evolve.

The parameter B is dimensionless, and it is the ratio of the mass of adsorbate in the reservoir at equilibrium to the total mass of adsorbate in the particle (also at equilibrium). Generally, this parameter is of order of unity for a well-designed experiment. If this parameter is much larger than unity (i.e., very large reservoir), then there will be practically no change in the reservoir concentration. When it is much less than unity, most of the solute in the reservoir will be adsorbed by the particle, hence, the bulk concentration will be close to zero at equilibrium.

Equations 11.174 to 11.176 completely define the physical system for a batch adsorber. By suitable change of symbols, those equations can also describe the heating or cooling of a solid in a finite liquid bath, with A being the dimensionless temperature of the solid object and A_b being the dimensionless temperature in the reservoir and B being the ratio of heat capacitances for the two phases. It can also be used to describe leaching from particles, as for liquid–solid extraction, when the boundary conditions are suitably modified.

The batch adsorber problem will yield a finite steady-state concentration for both the individual particle and also within the well-mixed reservoir. This steady-state concentration can be readily obtained from Eqs. 11.174 as follows.

Multiplying Eq. 11.174a by $3x^2\,dx$ and integrating the result from 0 to 1, we obtain the following equation for average concentration

$$\frac{d\overline{A}}{d\tau} = -3\frac{\partial A(1,\tau)}{\partial x} \tag{11.178}$$

where the mean concentration in the particle is defined as

$$\overline{A} = 3\int_0^1 x^2 A\,dx \tag{11.179}$$

Combining Eq. 11.174b with Eq. 11.178 yields

$$\frac{d\overline{A}}{d\tau} + B\frac{dA_b}{d\tau} = 0 \tag{11.180}$$

This is the differential form of the overall mass balance equation combining both phases. Integrating this equation with respect to time and using the initial condition (11.175) yields the following integral overall mass balance equation

$$\overline{A} + BA_b = B \tag{11.181}$$

As time approaches infinity, the fluid concentration in the adsorbent is equal to the concentration in the reservoir. Let this concentration be A_∞. Then we have

$$A_\infty = \frac{B}{(1+B)} \tag{11.182}$$

We now define the following new variables, which are the deviation of the concentration variables from their corresponding steady-state values

$$y = A - A_\infty \tag{11.183a}$$

$$y_b = A_b - A_\infty \tag{11.183b}$$

the new mass balance equations will become

$$\frac{\partial y}{\partial \tau} = \frac{1}{x^2} \cdot \frac{\partial}{\partial x}\left(x^2\frac{\partial y}{\partial x}\right) \tag{11.184a}$$

$$\frac{dy_b}{d\tau} = -\frac{3}{B}\frac{\partial y(1,\tau)}{\partial x} \tag{11.184b}$$

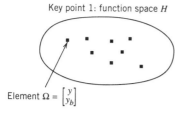

Key point 1: function space H

Element $\Omega = \begin{bmatrix} y \\ y_b \end{bmatrix}$

Figure 11.9 Rule 1 of the generalized integral transform.

subject to the following initial and boundary conditions in dimensionless form

$$\tau = 0; \quad y = -\frac{B}{(1+B)}, \quad y_b = \frac{1}{(1+B)} \tag{11.185}$$

$$x = 0; \quad \frac{\partial y}{\partial x} = 0 \tag{11.186a}$$

$$x = 1; \quad y = y_b \tag{11.186b}$$

This set of equations (Eqs. 11.184) can be solved readily by the Laplace transform method, as was done in Problem 10.17. Here we are going to apply the method of generalized integral transforms to solve this set of equations.

To apply this technique, it is necessary to define a function space. This is needed because we are dealing with two state variables (y and y_b), and they are coupled through Eqs. 11.184. Let us denote this function space as **H** (Fig. 11.9).

Inspecting the two differential equations (Eqs. 11.184), we see a logical choice for an element of the function space as simply the vector

$$\mathbf{\Omega} = [y, y_b]^{\mathrm{T}} \tag{11.187}$$

where superscript T denotes transpose of a vector. The operation on state variables using vector algebra is detailed in Appendix B.

Having defined our space, we need to designate a suitable operator, which manipulates variables in this function space. This operator is similar to integral or derivative operators in the case of scalar functions. By inspecting the RHS of Eqs. 11.184, we deduce the following operator for the function space of elements $\mathbf{\Omega}$

$$\mathbf{L} = \begin{bmatrix} \dfrac{1}{x^2} \dfrac{\partial}{\partial x}\left(x^2 \dfrac{\partial}{\partial x}\right) & 0 \\[2ex] -\dfrac{3}{B} \lim_{x \to 1} \dfrac{\partial}{\partial x} & 0 \end{bmatrix} \tag{11.188}$$

Defining this matrix operator is the second key step in the methodology of the generalized integral transform method. The operator **L** applied onto an element $\mathbf{\Omega}$ will give another element also lying in the same space (Fig. 11.10).

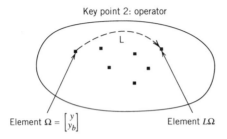

Figure 11.10 Rule 2 of the generalized integral transform.

With these definitions of the element and the operator, the original equations Eqs. 11.184 can be put into the very compact form

$$\frac{d\mathbf{\Omega}}{d\tau} = \mathbf{L}\mathbf{\Omega} \tag{11.189}$$

which of course is just another way to represent Eqs. 11.184.

Thus, so far we have created a function space **H**, which has the elements **Ω** along with an operator **L** operating in that space. Now we need to stipulate properties for this space because not all combinations of y and y_b can belong to the space. To belong to this space, the components y and y_b of the element **Ω** must satisfy the boundary conditions (Eqs. 11.186). This is the third key point of the methodology; that is, properties must be imposed on all elements of the space (Fig. 11.11).

The initial condition from Eq. 11.185 can also be written in the compact form

$$\tau = 0; \quad \mathbf{\Omega} = \mathbf{\Omega}_0 = \left[-\frac{B}{(1+B)}, \frac{1}{(1+B)} \right]^T \tag{11.190}$$

Thus, we have cast the variables in the original equations into a format of a functional space. To apply the generalized integral transform, we need to endow our function space with an inner product; that is, an operation between two

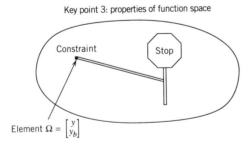

Figure 11.11 Rule 3 of the generalized integral transform.

elements in the function space **H**. This is our fourth key point. The inner product for the function space is similar to the conventional dot product of the Euclidean vector space.

Obviously, we can define an inner product to have any form at this point, but there will be only one definition of the inner product that will make the operator **L** self-adjoint, a property discussed earlier in the application of integral transforms. The inner product (i.e., the operation between two elements) is not known at this stage, but it will arise naturally from the analysis of the associated eigenproblem.

EIGENPROBLEM

The form of the associated eigenproblem for Eq. 11.189 is derived in exactly the same way as in the last section. One simply uses the operator **L**, operates on the so-called eigenfunction **K**, and sets the result to equal to $-\xi\mathbf{K}$. Thus, the eigenproblem is

$$\mathbf{LK} = -\xi \cdot \mathbf{K} \tag{11.191}$$

where the components of this eigenfunction are W and D

$$\mathbf{K} = [W, D]^T \tag{11.192}$$

We can view the vector **K** as special elements of the function space **H**. Since the eigenfunction **K** belongs to the function space **H**, its components must satisfy the properties set out for this space (recall our third key point), that is, they must satisfy the following conditions

$$x = 0; \qquad \frac{dW}{dx} = 0 \tag{11.193a}$$

$$x = 1; \qquad W = D \tag{11.193b}$$

As we have observed with the Sturm–Liouville integral transform and we will observe later for this generalized integral transform, there will arise an infinite set of eigenvalues and an infinite set of corresponding eigenfunctions. We then rewrite Eqs. 11.191 to 11.193 as follows to represent the nth values

$$\mathbf{LK}_n = -\xi_n\mathbf{K}_n \tag{11.194a}$$

$$x = 0; \qquad \frac{dW_n}{dx} = 0 \tag{11.194b}$$

$$x = 1; \qquad W_n = D_n \tag{11.194c}$$

To obtain the analytical form for the eigenfunction and its eigenvalue, we need to solve Eqs. 11.194 by inserting $\mathbf{K}_n = [W_n, D_n]^T$. Written in component form, Eq. 11.194a becomes

$$\frac{1}{x^2}\frac{d}{dx}\left(x^2\frac{dW_n}{dx}\right) + \xi_n W_n = 0 \tag{11.195a}$$

$$\frac{3}{B}\frac{dW_n(1)}{dx} = \xi_n D_n \tag{11.195b}$$

Equation 11.195a is a linear second order ordinary differential equation. The methods taught in Chapters 2 and 3 can be used to solve this elementary equation. The solution for the boundary conditions stated (taking the multiplicative constant as unity) is

$$W_n = \frac{\sin\left(\sqrt{\xi_n}\, x\right)}{x} \tag{11.196}$$

Equation 11.196 is the first component of the eigenfunction $\mathbf{K}_n$. To obtain the second component D_n, we can use either Eq. 11.194c or Eq. 11.195b. We use Eq. 11.194c to yield the solution for D_n

$$D_n = \sin\left(\sqrt{\xi_n}\right) \tag{11.197}$$

Thus, the eigenfunction $\mathbf{K}_n = [W_n, D_n]^T$ is determined with its components W_n and D_n defined in Eqs. 11.196 and 11.197. To completely determine this eigenfunction $\mathbf{K}_n$, we need to find the corresponding eigenvalue. This can be accomplished by using Eq. 11.195b, and we obtain the following transcendental equation for the many eigenvalues

$$\sqrt{\xi_n}\cot\sqrt{\xi_n} - 1 = \left(\frac{B}{3}\right)\xi_n \tag{11.198}$$

Figure 11.12 shows plots of the LHS and RHS of Eq. 11.198 versus $\sqrt{\xi}$. The intersection between these two curves yields the spectrum of countably infinite eigenvalues.

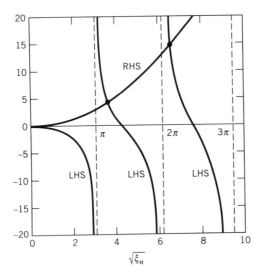

Figure 11.12 Plots of LHS and RHS of the transcendental equation 11.198 versus $\sqrt{\xi_n}$.

INNER PRODUCT

As we mentioned earlier, the fourth point of the development of this method is the inner product, which is the necessary ingredient to allow inversion. To obtain the inner product, just as in the case of Sturm–Liouville transform, we will start from the definition of the eigenproblem (Eqs. 11.195).

Multiplying Eq. 11.195a by $x^2 W_m$ (where $m \neq n$) yields the equation

$$W_m(x) \frac{d}{dx}\left[x^2 \frac{dW_n}{dx} \right] + \xi_n x^2 W_n W_m = 0 \qquad \textbf{(11.199)}$$

Integrating this equation over the whole domain of interest, we have

$$\int_0^1 W_m(x) \frac{d}{dx}\left[x^2 \frac{dW_n}{dx} \right] dx + \xi_n \int_0^1 x^2 W_n W_m \, dx = 0 \qquad \textbf{(11.200)}$$

Carrying out the integration by parts of the first term on the RHS of Eq. 11.200, we obtain

$$W_m(1) \frac{dW_n(1)}{dx} - \int_0^1 x^2 \frac{dW_n}{dx} \frac{dW_m}{dx} \, dx + \xi_n \int_0^1 x^2 W_n W_m \, dx = 0 \qquad \textbf{(11.201)}$$

Because n and m are arbitrary integers, we can interchange them in Eq. 11.200 to yield the equation

$$W_n(1) \frac{dW_m(1)}{dx} - \int_0^1 x^2 \frac{dW_m}{dx} \frac{dW_n}{dx} \, dx + \xi_m \int_0^1 x^2 W_m W_n \, dx = 0 \qquad \textbf{(11.202)}$$

Subtracting Eq. 11.202 from Eq. 11.201, we have

$$(\xi_n - \xi_m) \int_0^1 x^2 W_n W_m \, dx + W_m(1) \frac{dW_n(1)}{dx} - W_n(1) \frac{dW_m(1)}{dx} = 0 \qquad \textbf{(11.203)}$$

Next, we multiply Eq. 11.195b by D_m and obtain the result

$$-D_m \frac{dW_n(1)}{dx} + \frac{B}{3} \xi_n D_n D_m = 0 \qquad \textbf{(11.204)}$$

Replacing D_m in the first term with (obtained from Eq. 11.194c)

$$D_m = W_m(1) \qquad \textbf{(11.205)}$$

to obtain

$$-W_m(1) \frac{dW_n(1)}{dx} + \frac{B}{3} \xi_n D_n D_m = 0 \qquad \textbf{(11.206)}$$

Now we interchange n and m in Eq. 11.206 to obtain a new equation similar in form to Eq. 11.206. Then we subtract this equation from Eq. 11.206 to get

$$\frac{B}{3}(\xi_n - \xi_m) D_n D_m + W_n(1) \frac{dW_m(1)}{dx} - W_m(1) \frac{dW_n(1)}{dx} = 0 \quad \textbf{(11.207)}$$

Finally, adding Eqs. 11.203 and 11.207 to eliminate the cross term $W_n(1) \cdot dW_m(1)/dx$, we obtain the equation

$$(\xi_n - \xi_m)\left[\int_0^1 x^2 W_n W_m \, dx + \frac{B}{3} D_n D_m\right] = 0 \quad \textbf{(11.208)}$$

Since the eigenvalues corresponding to n and m $(n \neq m)$ are different, Eq. 11.208 gives us information on exactly how we should define the inner product. This inner product leads to a natural notion of distance between elements of the function space **H**. If we define two vectors

$$\mathbf{u} = \begin{bmatrix} u_1 \\ u_2 \end{bmatrix} \quad \text{and} \quad \mathbf{v} = \begin{bmatrix} v_1 \\ v_2 \end{bmatrix} \quad \textbf{(11.209)}$$

as elements of the function space **H**, the inner product should be defined for the present problem as

$$\langle \mathbf{u}, \mathbf{v} \rangle = \int_0^1 x^2 u_1 v_1 \, dx + \frac{B}{3} u_2 v_2 \quad \textbf{(11.210)}$$

Thus, we see from Eq. 11.210 that the inner product of two vector elements **u** and **v** is a scalar quantity, which may be regarded as a measure of length between these two vector elements.

Of course, we can choose an inner product of any form we wish, but Eq. 11.210 is the only form that leads to the orthogonality conditions between two eigenfunctions, and for the present problem

$$\langle \mathbf{K}_n, \mathbf{K}_m \rangle = 0 \quad \text{for } n \neq m \quad \textbf{(11.211)}$$

Thus, we have endowed the function space for the current problem with an inner product, defined as in Eq. 11.210. Since the eigenfunctions are orthogonal to each other with respect to the inner product, they will form a complete basis for other elements of the space. This is a general approach, and the obvious choices allow an orthogonality condition to evolve.

SELF-ADJOINT PROPERTY

The next important property we need to prove is the self-adjoint property for the differential operator defined in Eq. 11.188. Let **u** and **v** be elements of the function space, i.e., they must satisfy the boundary values

$$\frac{\partial u_1(0)}{\partial x} = 0; \quad u_2 = u_1(1) \quad \textbf{(11.212)}$$

$$\frac{\partial v_1(0)}{\partial x} = 0; \quad v_2 = v_1(1) \quad \textbf{(11.213)}$$

These represent symmetry at centerline and equality of boundary concentrations.

We need to validate the following self-adjoint property

$$\langle \mathbf{Lu}, \mathbf{v} \rangle = \langle \mathbf{u}, \mathbf{Lv} \rangle \tag{11.214}$$

Let us consider first the LHS of Eq. 11.214. By definition of the inner product (Eq. 11.210), we have

$$\langle \mathbf{Lu}, \mathbf{v} \rangle = \int_0^1 x^2 \left[\frac{1}{x^2} \frac{\partial}{\partial x} \left(x^2 \frac{\partial u_1}{\partial x} \right) \right] v_1 \, dx - \frac{B}{3} \frac{3}{B} \frac{\partial u_1(1)}{\partial x} v_2 \tag{11.215}$$

or

$$\langle \mathbf{Lu}, \mathbf{v} \rangle = \int_0^1 \frac{\partial}{\partial x} \left(x^2 \frac{\partial u_1}{\partial x} \right) v_1 \, dx - \frac{\partial u_1(1)}{\partial x} v_2 \tag{11.216}$$

Carrying out the integration by parts of the first term of the RHS of Eq. 11.216 gives

$$\langle \mathbf{Lu}, \mathbf{v} \rangle = v_1(1) \frac{\partial u_1(1)}{dx} - \frac{\partial v_1(1)}{\partial x} u_1(1)$$
$$+ \int_0^1 x^2 \left[\frac{1}{x^2} \frac{\partial}{\partial x} \left(x^2 \frac{\partial v_1}{\partial x} \right) \right] u_1 \, dx - \frac{\partial u_1(1)}{\partial x} v_2 \tag{11.217}$$

Now we combine the three nonintegral terms in the RHS of Eq. 11.217 and make use of Eqs. 11.212 and 11.213. We then obtain the result

$$\langle \mathbf{Lu}, \mathbf{v} \rangle = \int_0^1 x^2 \left[\frac{1}{x^2} \frac{\partial}{\partial x} \left(x^2 \frac{\partial v_1}{\partial x} \right) \right] u_1 \, dx - \frac{\partial u_1(1)}{\partial x} v_2 \tag{11.218}$$

The RHS is the definition of $\langle \mathbf{u}, \mathbf{Lv} \rangle$. Thus, we have proved the self-adjoint property for the differential operator $\mathbf{L}$.

TRANSFORM EQUATIONS

Having defined the eigenfunctions (i.e., the orthogonal basis), the inner product and the self-adjoint property, we are ready to apply the integral transform to the physical system (Eq. 11.189).

Taking the inner product of Eq. 11.189 and Eq. 11.190 with one of the eigenfunction $\mathbf{K}_n$ (in the manner of the Sturm–Liouville integral transform, this action is regarded as the projection of the vector $\mathbf{\Omega}$ onto the basis $\mathbf{K}_n$), we have

$$\frac{d}{d\tau} \langle \mathbf{\Omega}, \mathbf{K}_n \rangle = \langle \mathbf{L\Omega}, \mathbf{K}_n \rangle \tag{11.219}$$

$$\tau = 0; \quad \langle \mathbf{\Omega}, \mathbf{K}_n \rangle = \langle \mathbf{\Omega}_0, \mathbf{K}_n \rangle \tag{11.220}$$

The inner product $\langle \mathbf{\Omega}, \mathbf{K}_n \rangle$ can be treated as the projection or image of $\mathbf{\Omega}$ onto the basis $\mathbf{K}_n$. Now, we make use of the self-adjoint property (Eq. 11.214) on the RHS of Eq. 11.219 to get

$$\frac{d}{d\tau} \langle \mathbf{\Omega}, \mathbf{K}_n \rangle = \langle \mathbf{\Omega}, \mathbf{LK}_n \rangle \tag{11.221}$$

Using the definition of the eigenproblem (Eq. 11.194a), Eq. 11.221 becomes:

$$\frac{d}{d\tau} \langle \mathbf{\Omega}, \mathbf{K}_n \rangle = -\xi_n \langle \mathbf{\Omega}, \mathbf{K}_n \rangle \tag{11.222}$$

from which the solution is

$$\langle \mathbf{\Omega}, \mathbf{K}_n \rangle = \langle \mathbf{\Omega}_0, \mathbf{K}_n \rangle e^{-\xi_n \tau} \tag{11.223}$$

Thus, to solve the problem up to this point we have used the inner product, the eigenproblem and the self-adjoint property of the linear operator. It is recalled that the actions taken so far are identical to the Sturm–Liouville integral transform treated in the last section. The only difference is the element. In the present case, we are dealing with multiple elements, so the vector (rather than scalar) methodology is necessary.

INVERSE

Knowing the projection or image $\langle \mathbf{\Omega}, \mathbf{K}_n \rangle$ (Eq. 11.223), we can reconstruct the original vector by using the following series expansion in terms of coordinate vectors $\mathbf{K}_n$

$$\mathbf{\Omega} = \sum_{n=1}^{\infty} \delta_n \mathbf{K}_n \tag{11.224}$$

where δ_n is a linear coefficient. These coordinate vectors (eigenfunctions) form a complete set in the function space, so the determination of the linear coefficients δ_n is unique.

Applying the orthogonality condition (i.e., by taking the inner product of Eq. 11.224 with the eigenfunction basis where the inner product is defined in Eq. 11.210), we obtain the following expression for the linear coefficient

$$\delta_n = \frac{\langle \mathbf{\Omega}, \mathbf{K}_n \rangle}{\langle \mathbf{K}_n, \mathbf{K}_n \rangle} \tag{11.225}$$

Hence, the final solution is

$$\mathbf{\Omega} = \sum_{n=1}^{\infty} \langle \mathbf{\Omega}, \mathbf{K}_n \rangle \left[\frac{\mathbf{K}_n}{\langle \mathbf{K}_n, \mathbf{K}_n \rangle} \right] \tag{11.226}$$

If we now treat

$$\frac{\mathbf{K}_n}{[\langle \mathbf{K}_n, \mathbf{K}_n \rangle]^{1/2}} \tag{11.227}$$

as the nth normalized basis function, the solution vector $\boldsymbol{\Omega}$ is the summation of all its projections $\langle \boldsymbol{\Omega}, \mathbf{K}_n \rangle / [\langle \mathbf{K}_n, \mathbf{K}_n \rangle]^{1/2}$ multiplied by the nth normalized basis function (Eq. 11.227).

Next, we substitute the known solution of the projection or image $\langle \boldsymbol{\Omega}, \mathbf{K}_n \rangle$ given in Eq. 11.223 into Eq. 11.226 to get

$$\boldsymbol{\Omega} = \sum_{n=1}^{\infty} \frac{\langle \boldsymbol{\Omega}_0, \mathbf{K}_n \rangle}{\langle \mathbf{K}_n, \mathbf{K}_n \rangle} e^{-\xi_n \tau} \mathbf{K}_n \tag{11.228}$$

Our final task is to evaluate the various inner products appearing in the solution, $\langle \boldsymbol{\Omega}_0, \mathbf{K}_n \rangle$ and $\langle \mathbf{K}_n, \mathbf{K}_n \rangle$. Using the definition of the inner product (Eq. 11.210), we have

$$\langle \boldsymbol{\Omega}_0, \mathbf{K}_n \rangle = \int_0^1 x^2 \left[-\frac{B}{(1+B)} \right] K_n \, dx + \frac{B}{3} \left[\frac{1}{(1+B)} \right] D_n \tag{11.229}$$

where $\boldsymbol{\Omega}_0$ denotes initial state vector, and

$$\langle \mathbf{K}_n, \mathbf{K}_n \rangle = \int_0^1 x^2 K_n^2 \, dx + \frac{B}{3} D_n^2 \tag{11.230}$$

where W_n and D_n are given in Eqs. 11.196 and 11.197. Straightforward evaluation of these integrals gives

$$\langle \boldsymbol{\Omega}_0, \mathbf{K}_n \rangle = -\frac{\left[\sin\left(\sqrt{\xi_n} \right) - \sqrt{\xi_n} \cos\left(\sqrt{\xi_n} \right) \right]}{\xi_n} \tag{11.231}$$

and

$$\langle \mathbf{K}_n, \mathbf{K}_n \rangle = \frac{1}{2} - \frac{1}{4\sqrt{\xi_n}} \sin\left(2\sqrt{\xi_n} \right) + \frac{B}{3} \sin^2\left(\sqrt{\xi_n} \right) \tag{11.232}$$

Knowing the vector given in Eq. 11.228, the concentration in the particle is the first component and the concentration in the reservoir is the second component of the vector.

The solution for the bulk concentration A_b can be obtained from the second component of the vector defined in Eq. 11.228, that is,

$$A_b = \frac{B}{(1+B)} + \sum_{n=1}^{\infty} \frac{\langle \boldsymbol{\Omega}_0, \mathbf{K}_n \rangle e^{-\xi_n \tau}}{\langle \mathbf{K}_n, \mathbf{K}_n \rangle} D_n \tag{11.233}$$

Substituting Eqs. 11.231 and 11.232 into Eq. 11.233 gives the following explicit solution for the bulk fluid concentration (compare with Problems 10.16 and 10.17).

$$A_b = \frac{B}{(1 + B)} + \frac{B}{3} \sum_{n=1}^{\infty} \frac{e^{-\xi_n \tau}}{\left[\frac{1}{2} + \frac{B}{2} + \frac{B^2}{18} \xi_n \right]} \tag{11.234}$$

To evaluate the series on the RHS of Eq. 11.234, we need to first determine the eigenspectrum (defined by the transcendental equation 11.198). This is a nonlinear equation and can be easily handled by the methods presented in Appendix A. Figure 11.12 shows that the nth eigenvalue $\sqrt{\xi_n}$ lies between $n\pi$ and $(n + 1)\pi$. We, therefore, can start with a bisection method to get close to the solution and then switch to the Newton–Raphson method to get the quadratic convergence. Note that when eigenvalues are large, the difference between eigenvalues is very close to π. This additional information is also useful in determining large eigenvalues.

Figure 11.13 illustrates the behavior of $A_b = C_b/C_0$ versus τ. This family of plots is very useful for those who would like to use a batch adsorber to determine the effective diffusivity of solute in an adsorbent particle. With a few modifications of symbols, it could also be used to determine the thermal conductivity of solid. In an experiment, the bulk fluid concentration is monitored as function of time, t. Since the partition coefficient K is available from equilibrium experiments, that is, the slope of the linear isotherm is known, the parameter B can be calculated. Knowing B, the curve in Fig. 11.13 corresponding to this value of B will be the characteristic curve for the given system. From

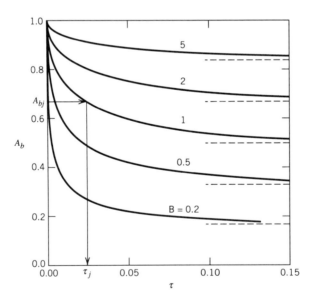

Figure 11.13 Plots of A_b versus nondimensional τ with B as the varying parameter.

this curve we can obtain the dimensionless time τ_j (see the sample in Fig. 11.13) from every measured A_{bj} at the time t_j.

Now, since the dimensionless time τ is related to the real time t by Eq. 11.172, we simply plot τ_j versus t_j and this should give a straight line through the origin with a slope of

$$\frac{D_e}{R^2[\varepsilon + (1 - \varepsilon)K]} \qquad (11.235)$$

Thus, knowing the particle characteristics, R and ε, and the slope of the linear isotherm, we can readily determine the effective diffusivity D_e. This method can be conveniently carried out with a simple linear plot, without recourse to any numerical optimization procedures (provided, of course, that the capacity ratio B is known beforehand).

This example summarizes all the necessary steps for the generalized integral transform. To recap the technique, we list below the specific steps to apply this technique.

Step 1   Define a function space with stipulated elements.
Step 2   Define the operator to manipulate elements of the space.
Step 3   Define the properties for the space (decided by *BC*s).
Step 4   Define an inner product, and derive the orthogonality condition and the self-adjoint property.
Step 5   Use superposition of solutions to define the inversion back to the Ω domain.

In the final step, we use the self-adjoint property and the definition of the eigenproblem to solve for the image (or projection of our variables on the nth eigenvalue coordinate). Knowing the image, the inverse can be found in a straightforward manner using the orthogonality condition.

We have applied the methodology for the generalized integral transform to the batch adsorber. Although the approach shows striking similarity with the conventional Sturm–Liouville transform, no special representation of the operator **L** has been proposed. For some partial differential equations, the operator is decomposed before the generalized integral transform can be applied. The book by Ramkrishna and Amundson (1986) gives additional details for these special circumstances.

11.4   REFERENCES

1. Abramowitz, M., and I. A. Stegun, "Handbook of Mathematical Functions," U. S. Government Printing Office, Washington, D.C. (1964).
2. Churchill, R. V. *Operational Mathematics*. McGraw Hill, New York (1958).
3. Do, D. D., and J. E. Bailey, "Useful Representations of Series Solutions for Transport Problems," *Chem. Eng. Sci.*, 36, 1811–1818 (1981).
4. Johnston, P. R., and D. D. Do, "A Concept of Basic Functions in the Series Representations of Transport Problems," *Chem. Eng. Commun.*, 60, 343–359 (1987).
5. Ramkrishna, D., and N. R. Amundson. *Linear Operator Methods in Chemical Engineering*, Prentice Hall, New Jersey (1987).

6. Sneddon, I. N. *The Use of Integral Transform*. McGraw Hill, New York (1972).
7. Tranter, C. J. *Integral Transforms in Mathematical Physics*. Chapman and Hall Ltd., London (1966).

11.5   PROBLEMS

11.1$_2$. A general form of the second order differential operator defined as

$$L = \frac{1}{r_0(x)}\left[p_0(x)\frac{d^2}{dx^2} + p_1(x)\frac{d}{dx} + p_2(x)\right]$$

may not be self-adjoint. The function p_0 is a positive function over the whole domain of interest. To make use of the nice property of self-adjointness, the above operator must be converted to a self-adjoint form.

(a) Multiply the denominator and the numerator of the above operator by $\gamma(x)$ to show that

$$L = \frac{1}{r_0(x)\gamma(x)}\left[p_0(x)\gamma(x)\frac{d^2}{dx^2} + p_1(x)\gamma(x)\frac{d}{dx} + p_2(x)\gamma(x)\right]$$

Rearrange the above equation as

$$L = \frac{1}{r_0(x)\gamma(x)}\left\{p_0(x)\gamma(x)\left[\frac{d^2}{dx^2} + \frac{p_1(x)}{p_0(x)}\frac{d}{dx}\right] + p_2(x)\gamma(x)\right\}$$

and compare with the following differentiation formula

$$\frac{d}{dx}\left(I\frac{d}{dx}\right) = I\left[\frac{d^2}{dx^2} + \frac{1}{I}\frac{dI}{dx}\frac{d}{dx}\right]$$

to show that

$$I = p_0(x)\gamma(x)$$

$$\frac{1}{I}\frac{dI}{dx} = \frac{p_1(x)}{p_0(x)}$$

(b) Show that the solution for $\gamma(x)$ will take the form

$$\gamma(x) = \frac{1}{p_0(x)}\exp\left[\int_a^x \frac{p_1(s)}{p_0(s)}ds\right]$$

(c) With this definition of the positive function $\gamma(x)$ (since $p_0(x)$ is positive), show that the linear differential operator is

$$L = \frac{1}{r_0(x)\gamma(x)}\left\{\frac{d}{dx}\left[p_0(x)\gamma(x)\frac{d}{dx}\right] + p_2(x)\gamma(x)\right\}$$

(d) Define

$$r_0(x)\gamma(x) = r(x) > 0$$

$$p_0(x)\gamma(x) = p(x) > 0$$

and

$$p_2(x)\gamma(x) = -q(x) \le 0$$

and show that the above differential operator is exactly the same as the self-adjoint Sturm–Liouville differential operator, defined in Eq. 11.45.

(e) If the boundary conditions are

$$A_1 y(a) + A_2 \frac{dy(a)}{dx} = 0$$

$$B_1 y(b) + B_2 \frac{dy(b)}{dx} = 0$$

show that the new operator obtained in part (d) is self-adjoint.

11.2₂. Consider the following second order differential operator

$$L = \frac{d^2}{dx^2} + \frac{1}{x}\frac{d}{dx}$$

Comparing it with form of the operator of Problem 11.1, we have

$$r_0(x) = 1 > 0; \quad p_0(x) = 1 > 0; \quad p_1(x) = \frac{1}{x}$$

(a) Use the method of Problem 11.1 to show that the function $\gamma(x)$ is

$$\gamma(x) = \exp\left[\int_a^x \frac{1}{s}\,ds\right] = \frac{x}{a}$$

(b) Then show that the new operator is

$$L = \frac{1}{x}\frac{d}{dx}\left(x\frac{d}{dx}\right)$$

which is the usual Laplacian operator in cylindrical coordinates. Show that it is self-adjoint with the two boundary conditions

$$x = 0; \quad \frac{dy}{dx} = 0$$

$$x = a; \quad y = 0$$

11.3₃. Repeat Problem 10.1 by using the method of finite integral transform. Compare the solution obtained by this method with that obtained by the method of separation of variables.

11.4*. A very long, cylindrical stainless steel rod having a diameter of 1.5 meters is heated to 200°C until the temperature is uniform throughout. The rod then is quickly cooled under a strong flow of air at 25°C. Under these conditions the heat transfer coefficient of the film surrounding the rod is calculated as 150 W m^{-2} K^{-1}. The thermal properties of the stainless steel are $k = 15$ W m^{-1} K^{-1} and $k/\rho C_p = 4 \times 10^{-6}$ m^2/sec.

(a) Obtain the analytical solution by the method of finite integral transform and determine the time it takes for the center to reach 50°C.

(b) What will be the surface temperature when the center reaches 50°C?

(c) If the air flow is extremely high (i.e., ideal), what will be the surface temperature and how long will it take for the center to reach 50°C? Compare this time with that in part (a).

11.5*. A hot slab of aluminum is quickly quenched in a water bath of temperature 50°C. The heat transfer coefficient of water film surrounding the slab is 10,000 W m^{-2} K^{-1}. This is a one-dimensional problem, since the slab is thin. Ignore losses from the edges.

(a) Obtain the transient solution by the method of finite integral transforms.

(b) If the initial temperature of the slab is 500°C, calculate the center temperature after 30 sec and 1 minute exposure in the water bath. The slab half thickness is 0.05 m. Properties of the aluminum are:

$$k = 200 \text{ W m}^{-1} \text{ K}^{-1}; \quad C_p = 1020 \text{ J kg}^{-1} \text{ K}^{-1}; \quad \rho = 2700 \text{ kg m}^{-3}$$

(c) After 30 sec exposure in water bath, the slab is quickly removed and allowed to cool to a flow of air of 25°C with a heat transfer coefficient of 150 W m^{-2} K^{-1}. What will be the center and surface temperatures after 5 minutes of exposure in air.

11.6*. The partial differential equation (Eq. 11.153), rewritten as follows, has homogeneous boundary conditions in ζ.

$$\frac{1}{\xi} \frac{\partial}{\partial \xi} \left(\xi \frac{\partial v}{\partial \xi} \right) + \Delta^2 \frac{\partial^2 v}{\partial \zeta^2} = 0$$

$$\xi = 0; \qquad \frac{\partial v}{\partial \xi} = 0$$

$$\xi = 1; \qquad \frac{\partial v}{\partial \xi} = \text{Bi} \, (1 - v)$$

$$\zeta = 0; \qquad v = 0$$

$$\zeta = 1; \qquad \frac{\partial v}{\partial \zeta} = 0$$

(a) Use the usual procedure of finite integral transforms to show that the transform with respect to ζ is defined as

$$\langle v, T_n \rangle = \int_0^1 v(\xi, \zeta) T_n(\zeta) \, d\zeta$$

where T_n is defined from the associated eigenproblem

$$\frac{d^2 T_n}{d\zeta^2} + \gamma_n^2 T_n = 0$$

$$\zeta = 0; \qquad T_n = 0$$

$$\zeta = 1; \qquad \frac{dT_n}{d\zeta} = 0$$

(b) Show that the solution of the eigenproblem of part (a) is

$$T_n(\zeta) = \cos[\gamma_n(1 - \zeta)]$$

where the eigenvalues are given as

$$\gamma_n = \left(n - \frac{1}{2}\right)\pi$$

with $n = 1, 2, 3, \ldots$

(c) Take the integral transform of the partial differential equation and two boundary conditions in ξ and show that the differential equation for $\langle v, T_n \rangle$ and its boundary conditions are

$$\frac{1}{\xi}\frac{d}{d\xi}\left[\xi\frac{d\langle v, T_n \rangle}{d\xi}\right] - \Delta^2 \gamma_n^2 \langle v, T_n \rangle = 0$$

$$\xi = 0; \qquad \frac{d\langle v, T_n \rangle}{d\xi} = 0$$

$$\xi = 1; \frac{d\langle v, T_n \rangle}{d\xi} = \text{Bi}[\langle 1, T_n \rangle - \langle v, T_n \rangle]$$

(d) Use the Bessel function discussed in Chapter 3 to show that the solution for $\langle v, T_n \rangle$ of part (c) is:

$$\langle v, T_n \rangle = a I_0(\Delta \gamma_n \xi) + b K_0(\Delta \gamma_n \xi)$$

where a and b are constants of integration.

(e) Use the transformed boundary conditions to show that

$$a = \frac{\langle 1, T_n \rangle}{\left[I_0(\Delta \gamma_n) + \dfrac{1}{\text{Bi}}\Delta \gamma_n I_1(\Delta \gamma_n)\right]}; \qquad b = 0$$

hence, the complete solution for $\langle v, T_n \rangle$ is

$$\langle v, T_n \rangle = \frac{\langle 1, T_n \rangle I_0(\Delta \gamma_n \xi)}{\left[I_0(\Delta \gamma_n) + \dfrac{1}{\text{Bi}}\Delta \gamma_n I_1(\Delta \gamma_n)\right]}$$

(f) Start from the eigenproblem to show that the orthogonality condition is

$$\int_0^1 T_n(\gamma_n \zeta) T_m(\gamma_m \zeta)\, d\zeta = 0 \qquad n \neq m$$

(g) Use the orthogonality condition to show that the inverse v is given by

$$v = \sum_{n=1}^{\infty} \frac{\langle 1, K_n \rangle I_0(\Delta\gamma_n \xi) \cos[\gamma_n(1 - \zeta)]}{\langle K_n, K_n \rangle \left[I_0(\Delta\gamma_n) + \dfrac{1}{\mathrm{Bi}} \Delta\gamma_n I_1(\Delta\gamma_n) \right]}$$

11.7*. (a) To show that the eigenvalue of Example 11.7 is real valued, assume the eigenvalue ξ is a complex number as

$$\xi_n = \alpha_n + i\beta_n$$

and let the mth eigenvalue be the conjugate of the nth eigenvalue; that is,

$$\xi_m = \overline{\xi_n} = \alpha_n - i\beta_n$$

Substitute these two relations into Eq. 11.208 to show that

$$2i\beta_n \left[\int_0^1 x^2 W_n^2\, dx + \frac{B}{3} D_n^2 \right] = 0$$

hence, show

$$\beta_n = 0$$

(b) Part (a) shows that the eigenvalue is real valued. Now show that the eigenvalue is positive.
Hint: Multiply Eq. 11.195a by $x^2 W_n\, dx$ and carry out the integration by multiplying Eq. 11.195b with D_n and then adding them in the following way:

$$\xi_n \left[\int_0^1 x^2 W_n^2\, dx + \frac{B}{3} D_n^2 \right] = -\int_0^1 W_n \frac{d}{dx}\left(x^2 \frac{dW_n}{dx} \right) dx + D_n \frac{dW_n(1)}{dx}$$

Carry out the integration by parts of the first term on the RHS and use Eq. 11.194c to see $W_n(1) = D_n$, then solve for ξ_n to get

$$\xi_n = \frac{\displaystyle\int_0^1 x^2 \left(\frac{dW_n}{dx} \right)^2 dx}{\displaystyle\int_0^1 x^2 W_n^2\, dx + \frac{B}{3} D_n^2}$$

11.8*. Apply the generalized integral transform method to solve the following problem

$$\frac{\partial y_1}{\partial \tau} = \beta^2 \frac{\partial^2 y_1}{\partial x^2} \qquad \text{for} \qquad 0 < x < \delta$$

$$\frac{\partial y_2}{\partial \tau} = \frac{\partial^2 y_2}{\partial x^2} \qquad \text{for} \qquad \delta < x < 1$$

subject to the following boundary conditions

$$x = \delta; \quad y_1 = y_2; \quad \beta \frac{\partial y_1}{\partial x} = \frac{\partial y_2}{\partial x}$$

$$x = 0; \qquad \frac{\partial y_1}{\partial x} = 0$$

$$x = 1; \qquad y_2 = 0$$

The initial condition is assumed to take the form

$$\tau = 0; \qquad y_1 = y_2 = 1$$

(a) Show that the function space **H** for this problem will contain elements having the following form

$$\mathbf{\Omega} = \begin{bmatrix} y_1 \\ y_2 \end{bmatrix}$$

and the operator operating on this space will be

$$\mathbf{L} = \begin{bmatrix} \beta^2 \dfrac{\partial^2}{\partial x^2} & 0 \\ 0 & \dfrac{\partial^2}{\partial x^2} \end{bmatrix}$$

(b) Then show that the equation can be cast into the form

$$\frac{\partial \mathbf{\Omega}}{\partial \tau} = \mathbf{L}\mathbf{\Omega}$$

with the initial condition

$$\mathbf{\Omega}_0 = \begin{bmatrix} y_1(0) \\ y_2(0) \end{bmatrix} = \begin{bmatrix} 1 \\ 1 \end{bmatrix}$$

(c) Show that the eigenproblem for the above problem is

$$\mathbf{L}\mathbf{K}_n = -\xi_n^2 \mathbf{K}_n$$

where the eigenfunction has two components as

$$\mathbf{K}_n = \begin{bmatrix} W_n \\ D_n \end{bmatrix}$$

(d) Show that the homogeneous boundary conditions for W_n and D_n are

$$x = \delta; \quad W_n = D_n; \quad \beta^2 \frac{dW_n}{dx} = \frac{dD_n}{dx}$$

$$x = 0; \qquad \frac{dW_n}{dx} = 0$$

$$x = 1; \qquad D_n = 0$$

(e) Solve the eigenproblem and show that W_n and D_n will take the form

$$W_n = \cos\left(\frac{\xi_n}{\beta} x\right)$$

$$D_n = A \sin(\xi_n x) + B \cos(\xi_n x)$$

where A and B are given by

$$A = \cos\left(\frac{\xi_n}{\beta}\delta\right) \sin(\xi_n \delta) - \beta \sin\left(\frac{\xi_n}{\beta}\delta\right) \cos(\xi_n \delta)$$

$$B = \cos\left(\frac{\xi_n}{\beta}\delta\right) \cos(\xi_n \delta) + \beta \sin\left(\frac{\xi_n}{\beta}\delta\right) \sin(\xi_n \delta)$$

and show that the eigenvalue is determined from the following transcendental equation

$$\left[\cos\left(\frac{\xi_n}{\beta}\delta\right) \sin(\xi_n \delta) - \beta \sin\left(\frac{\xi_n}{\beta}\delta\right) \cos(\xi_n \delta)\right] \sin(\xi_n)$$

$$+ \left[\cos\left(\frac{\xi_n}{\beta}\delta\right) \cos(\xi_n \delta) + \beta \sin\left(\frac{\xi_n}{\beta}\delta\right) \sin(\xi_n \delta)\right] \cos(\xi_n) = 0$$

(f) Start from the eigenproblem to show that the inner product, which defines the integral transform, is

$$\langle \mathbf{u}, \mathbf{v} \rangle = \int_0^\delta u_1 v_1 \, dx + \int_\delta^1 u_2 v_2 \, dx$$

for

$$\mathbf{u} = \begin{bmatrix} u_1 \\ u_2 \end{bmatrix}; \qquad \mathbf{v} = \begin{bmatrix} v_1 \\ v_2 \end{bmatrix}$$

(g) With this definition of inner product, show that the eigenvalues are orthogonal to each other; that is,

$$\langle \mathbf{K}_n, \mathbf{K}_m \rangle = 0 \qquad \text{for} \qquad n \neq m$$

(h) Prove the self-adjoint property

$$\langle \mathbf{Lu}, \mathbf{v} \rangle = \langle \mathbf{u}, \mathbf{Lv} \rangle$$

(i) Now solve the problem and show that the solution is

$$\mathbf{\Omega} = \sum_{n=1}^{\infty} \frac{\langle \mathbf{\Omega}, \mathbf{K}_n \rangle}{\langle \mathbf{K}_n, \mathbf{K}_n \rangle} \mathbf{K}_n$$

where

$$\langle \mathbf{\Omega}, \mathbf{K}_n \rangle = \langle \mathbf{\Omega}_0, \mathbf{K}_n \rangle \exp\left(-\xi_n^2 \tau \right)$$

with

$$\langle \mathbf{\Omega}_0, \mathbf{K}_n \rangle = \int_0^\delta (1) W_n \, dx + \int_\delta^1 (1) D_n \, dx$$

Chapter 12

Approximate and Numerical Solution Methods for PDEs

In the previous three chapters, we described various analytical techniques to produce practical solutions for linear partial differential equations. Analytical solutions are most attractive because they show explicit parameter dependences. In design and simulation, the system behavior as parameters change is quite critical. When the partial differential equations become nonlinear, numerical solution is the necessary last resort. Approximate methods are often applied, even when an analytical solution is at hand, owing to the complexity of the exact solution. For example, when an eigenvalue expression requires trial–error solutions in terms of a parameter (which also may vary), then the numerical work required to successfully use the analytical solution may become more intractable than a full numerical solution would have been. If this is the case, solving the problem directly by numerical techniques is attractive since it may be less prone to human error than the analytical counterpart.

In this chapter, we will present several alternatives, including: polynomial approximations, singular perturbation methods, finite difference solutions and orthogonal collocation techniques. To successfully apply the polynomial approximation, it is useful to know something about the behavior of the exact solution. Next, we illustrate how perturbation methods, similar in scope to Chapter 6, can be applied to partial differential equations. Finally, finite difference and orthogonal collocation techniques are discussed since these are becoming standardized for many classic chemical engineering problems.

12.1 POLYNOMIAL APPROXIMATION

Polynomial approximation involves two key steps. One is the selection of the form of the solution, normally presented as a polynomial expression in the spatial variable with time dependent coefficients. The second step is to convert the differential equation into an integral form. This means that if the governing equation is a differential heat balance equation, then one necessary step is to convert this into an integral heat balance equation, on which the polynomial approximation will be applied. Since the process is in essence a form of "averaging," the final approximate solution is not unique. The advantage of the polynomial approximation technique is the simplicity of the final result and the ease with which it can be used for study of complex phenomenon (Rice 1982). The method evolves quite naturally, as we show next in a classical example.

EXAMPLE 12.1

Find an approximate solution, valid for long times, to the linear parabolic partial differential equation, describing mass or heat transport from/to a sphere with constant physical properties, such as thermal conductivity and diffusion coefficient

$$\frac{\partial y}{\partial \tau} = \frac{1}{x^2} \frac{\partial}{\partial x} \left(x^2 \frac{\partial y}{\partial x} \right) \tag{12.1a}$$

subject to the following initial and boundary conditions

$$\tau = 0; \quad y = 0 \tag{12.1b}$$

$$x = 0; \quad \frac{\partial y}{\partial x} = 0 \tag{12.1c}$$

$$x = 1; \quad y = 1 \tag{12.1d}$$

For mass transfer, the dimensionless variables are

$$x = \frac{r}{R}; \quad \tau = \frac{Dt}{R^2}; \quad y = \frac{C_0 - C}{C_0 - C_B}$$

where C_0 is the initial composition, C_B is bulk fluid composition at the boundary $r = R$, and D represents diffusivity.

This linear partial differential equation was solved analytically in Chapters 10 and 11 using Laplace transform, separation of variables, and finite integral transform techniques, respectively. A solution was given in the form of infinite series

$$y(x, \tau) = 1 - \frac{2}{x} \sum_{n=1}^{\infty} \frac{(-\cos \xi_n) \sin(\xi_n x)}{\xi_n} \exp\left(-\xi_n^2 \tau \right); \quad \xi_n = n\pi \tag{12.2}$$

The solution for the average concentration, which is needed later for comparison with the approximate solutions, is

$$\bar{y} = 3 \int_0^1 x^2 y(x, \tau) \, dx = \frac{6}{\pi^2} \sum_{n=1}^{\infty} \frac{\exp\left(-n^2 \pi^2 \tau \right)}{n^2} \tag{12.3}$$

Except for short times, the series on the RHS of Eq. 12.3 converges quite fast, hence, only a few terms are needed. For short times, the following solution, obtained from the Laplace transform analysis, is particularly useful

$$\bar{y} \approx \left(\frac{6}{\sqrt{\pi}} \right) \sqrt{\tau} = 3.38514\sqrt{\tau} \tag{12.4}$$

We wish to generate a compact, polynomial approximations to this elementary problem. As a first approximation, we assume that the solution will take the following parabolic profile

$$y_a(x, \tau) = a_1(\tau) + a_2(\tau)x^2 \tag{12.5}$$

where $a_1(\tau)$ and $a_2(\tau)$ are unknown coefficients, which are functions of time only. The linear term with respect to x was not included in light of the symmetry condition at the center of the particle. This means that the center condition is automatically satisfied by Eq. 12.5.

Next, we integrate the differential mass (heat) balance equation by multiplying the LHS and RHS of Eq. 12.1a with $x^2 dx$ and integrate over the whole domain of interest (i.e., [0, 1]), to find

$$\frac{d\bar{y}}{d\tau} = 3\left[\frac{\partial y}{\partial x} \right]_{x=1} \tag{12.6}$$

It is useful to note that the volumetric mean concentration is defined as

$$\bar{y} = \frac{\int_0^1 x^2 y \, dx}{\int_0^1 x^2 \, dx} = 3\int_0^1 x^2 y \, dx \tag{12.7}$$

Knowing the form of the assumed solution (Eq. 12.5), we can obtain the two relations

$$\bar{y}_a = 3\int_0^1 x^2 y_a(x, \tau) \, dx = a_1(\tau) + \frac{3}{5}a_2(\tau) \tag{12.8}$$

$$\left[\frac{\partial y_a}{\partial x} \right]_{x=1} = 2a_2(\tau) \tag{12.9}$$

Next, substitute the assumed solution into the boundary condition at the particle surface, to see

$$y_a(1, \tau) = a_1(\tau) + a_2(\tau) = 1 \tag{12.10}$$

Subtracting Eq. 12.10 from Eq. 12.8 yields

$$1 - \bar{y}_a = \frac{2}{5}a_2(\tau) \tag{12.11}$$

Finally eliminating $a_2(\tau)$ from Eqs. 12.9 and 12.11, we obtain

$$\left[\frac{\partial y_a}{\partial x}\right]_{x=1} = 5(1 - \bar{y}_a) \qquad (\mathbf{12.12})$$

If we now assume that the overall mass balance equation (Eq. 12.6) is satisfied by the approximate solution Eq. 12.5, then substituting Eq. 12.12 into that overall mass balance equation we have

$$\frac{d\bar{y}_a}{d\tau} = 15(1 - \bar{y}_a) \qquad (\mathbf{12.13})$$

The initial condition for this equation is

$$\tau = 0; \qquad \bar{y}_a = 0 \qquad (\mathbf{12.14})$$

The solution to Eq. 12.13 is straightforward

$$\bar{y}_a = 1 - \exp(-15\tau) \qquad (\mathbf{12.15})$$

Knowing the mean concentration, the coefficients $a_1(\tau)$ and $a_2(\tau)$ are determined from (using Eqs. 12.8 and 12.10)

$$a_2(\tau) = \frac{5}{2}(1 - \bar{y}_a) = \frac{5}{2}\exp(-15\tau) \qquad (\mathbf{12.16}a)$$

$$a_1(\tau) = 1 - a_2(\tau) = 1 - \frac{5}{2}\exp(-15\tau) \qquad (\mathbf{12.16}b)$$

Now, the approximate solution $y_a(x,\tau)$ can be written rather compactly as

$$y_a(x,\tau) = \left[1 - \frac{5}{2}\exp(-15\tau)\right] + \frac{5}{2}\exp(-15\tau)x^2 \qquad (\mathbf{12.17})$$

Thus, we see that the procedure for obtaining the approximate solution is as follows:

1. Assume an expression for the solution as a polynomial in terms of the spatial variable with the coefficients as a function of time.
2. Integrate the differential equation over the whole domain of interest to obtain an "average" integral equation, which is satisfied by the approximate solution.
3. Solve the equation for the mean concentration.
4. Knowing the mean concentration, determine the time-dependent coefficients of the assumed polynomial solution, and finally obtain the complete polynomial representation.

Observing Steps 3 and 4, we see that the procedure yields the mean (average) concentration first and then the polynomial approximate solution. This is in reverse order to the usual analytical procedure (such as the separation of variables or integral transforms) where we obtain the solution $y(x,\tau)$ first before the mean concentration can be determined.

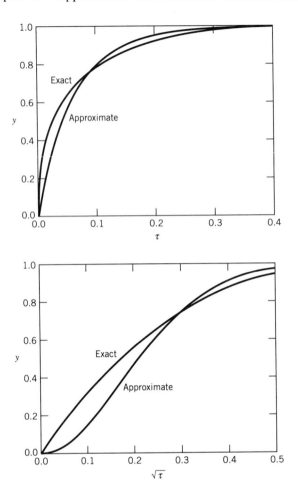

Figure 12.1 Plots of exact and approximate average concentrations.

Figure 12.1 shows plots of the average concentrations obtained from the analytical (Eq. 12.3) and polynomial approximate procedure (Eq. 12.15) versus τ.

Inspection of the curves shows that the two solutions cross each other. This is expected since the polynomial approximate solution satisfies the mass balance equation only in some average sense. Now, the exact solution behaves roughly as $\sqrt{\tau}$ for small times (see Eq. 12.4), so Fig. 12.1 also shows plots of the average concentrations versus $\sqrt{\tau}$. The polynomial solution fails to exhibit the linear $\sqrt{\tau}$ dependence at short times, but rather it takes a sigmoidal shape.

We may have expected poor performance at short times, since we may have recognized from our earlier work that many terms are necessary for the short time period. To see this more clearly, we plot the exact solution $y(x, \tau)$ (Eq. 12.2) and the polynomial approximate solution $y_a(x, \tau)$ (Eq. 12.17) at two

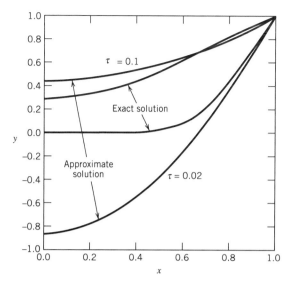

Figure 12.2 Exact and approximate concentrations profiles.

values of times, 0.02 and 0.1. It is seen in Fig. 12.2 that the concentration profile obtained from the approximate solution has negative values over a portion of the spatial domain for $\tau = 0.02$.

We inspect the exact concentration profiles in Fig. 12.3 for a few values of time and note that the profiles are quite sharp and they move into the interior of the particle like a penetration front; that is, there is a time dependent position in the particle domain where the concentration tends to zero. This physical reasoning on the solution behavior may help to develop another scheme for approximate solution, as we show next, which is applicable to short times.

Before we develop this new scheme, it is convenient to transform the spherical coordinate problem by applying the following change of variables

$$y = \frac{v}{x} \qquad (12.18)$$

With this transformation, we have

$$\frac{dy}{dx} = \frac{1}{x}\frac{dv}{dx} - \frac{v}{x^2} \qquad (12.19a)$$

$$\frac{d}{dx}\left(x^2\frac{dy}{dx}\right) = \frac{d}{dx}\left[x^2\left(\frac{1}{x}\frac{dv}{dx} - \frac{v}{x^2}\right)\right] = x\frac{d^2v}{dx^2} \qquad (12.19b)$$

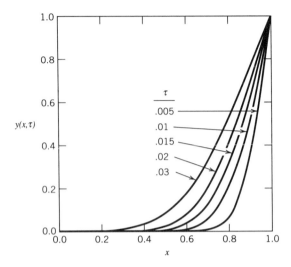

Figure 12.3 Exact concentration profiles at $\tau =$ 0.005, 0.01, 0.015, 0.02, and 0.03.

and the mass balance equation Eq. 12.1 becomes

$$\frac{\partial v}{\partial \tau} = \frac{\partial^2 v}{\partial x^2} \tag{12.20}$$

The initial and boundary conditions on v are

$$\tau = 0; \quad v = 0 \tag{12.21}$$

$$x = 0; \quad y \text{ is finite.} \therefore v \sim x \tag{12.22a}$$

$$x = 1; \quad v = 1 \tag{12.22b}$$

For the new set of equations in terms of v, we propose the new polynomial approximate solution for v to take the form

$$v_a = a_0(\tau) + a_1(\tau)x + a_2(\tau)x^2 \tag{12.23}$$

where x lies between the penetration front $X(\tau)$ and the particle surface ($x = 1$). Since the approximate solution is valid only in this new subdomain, we need to introduce the following condition for v_a at the position $X(\tau)$

$$x = X(\tau); \quad v_a = 0 \tag{12.24}$$

that is, the concentration at the penetration front is zero. Further, since the penetration front is still an unknown at this stage, we need to define one more equation for the complete formulation of the approximate problem. This is introduced by assuming that at the penetration front, $X(\tau)$, the slope of the profile for v_a is zero, that is,

$$x = X(\tau); \qquad \frac{\partial v_a}{\partial x} = 0 \qquad (12.25)$$

Using the conditions (Eqs. 12.22b, 12.24, and 12.25), and following the usual procedure, the approximate solution becomes

$$v_a = \left[\frac{x - X(\tau)}{1 - X(\tau)}\right]^2 \qquad (12.26)$$

where the penetration front $X(\tau)$ is still an unknown function of time.

Next, we integrate Eq. 12.20 with respect to x from $X(\tau)$ to 1 and obtain

$$\frac{d}{d\tau}\int_{X(\tau)}^{1} v\,dx = \left[\frac{\partial v}{\partial x}\right]^{1}_{x = X(\tau)} \qquad (12.27)$$

Now we assume that the approximate solution, v_a, satisfies the integral mass balance relation Eq. 12.27, and we substitute Eq. 12.26 into Eq. 12.27 to obtain the following differential equation for the penetration front

$$\frac{d}{d\tau}[1 - X(\tau)] = \frac{6}{1 - X(\tau)} \qquad (12.28a)$$

The initial condition for the penetration front will be the position at the particle surface

$$\tau = 0; \qquad X(0) = 1 \qquad (12.28b)$$

The solution of Eq. 12.28 is clearly seen to be

$$X(\tau) = 1 - 2\sqrt{3\tau} \qquad (12.29)$$

Knowing the solution for v_a from Eq. 12.26, the solution for y_a is

$$y_a(x,\tau) = \frac{1}{x}\left[\frac{x - X(\tau)}{1 - X(\tau)}\right]^2 \qquad (12.30)$$

We can use the polynomial solution to obtain the mean concentration

$$\bar{y}_a = \frac{\int_{X(\tau)}^{1} x^2 y_a\,dx}{\int_{0}^{1} x^2\,dx} = 3\int_{X(\tau)}^{1} x^2 y_a\,dx = \frac{3}{4}[1 - X(\tau)]^2 + X(\tau)[1 - X(\tau)]$$

$$(12.31)$$

Substituting the solution for the penetration front (Eq. 12.29) into Eq. 12.31, we

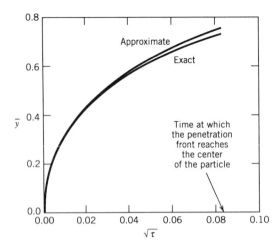

Figure 12.4 Plots of the mean concentration versus τ.

obtain a solution *valid for short times*

$$\bar{y}_a = 2\sqrt{3\tau} - 3\tau \tag{12.32}$$

Plots of this approximate solution for the mean concentration as well as the exact solution are shown in Fig. 12.4.

It is seen that the new approximate solution agrees rather well with the exact solution for short times. It even has the correct $\sqrt{\tau}$ behavior at small time. At very small times, the approximate solution behaves like

$$\bar{y}_a \approx 2\sqrt{3\tau} = 3.4641\sqrt{\tau} \tag{12.33}$$

which compares well with the exact solution at small times

$$\bar{y} \approx \left(\frac{6}{\sqrt{\pi}}\right)\sqrt{\tau} = 3.38514\sqrt{\tau} \tag{12.34}$$

All seems well, but the polynomial approximate solution starts to fail when the penetration front reaches the center of the particle. Hence, this polynomial solution is valid only for τ less than $1/12$ (obtained by setting Eq. 12.29 to zero), at which time the net uptake is about 75%.

As we have mentioned in the introduction, the polynomial solution is certainly not unique and depends on choices made early by the analyst. We have observed that the first approximate solution (parabolic profile approximation) agrees with the exact solution in the average sense over the whole time course (reflected by the cross over between the two solutions). The second approximation, using the notion of penetration front, agrees rather well with the exact solution but fails when uptake is more than 75%. We now present below another method of finding a polynomial approximation, which may help to bridge the gap between the previous two solutions.

In this new formulation, we propose an arbitrary power law expression for the approximate solution (Do and Mayfield 1987)

$$y_a(x,\tau) = a_0(\tau) + a_n(\tau)x^{n(\tau)} \tag{12.35}$$

in light of the profiles of the exact solution (Fig. 12.3). Here, the approximate profile follows a polynomial of degree n and this degree n is a *function of time*. From Eq.12.35, we obtain

$$\bar{y}_a(\tau) = a_0(\tau) + \left[\frac{3}{3 + n(\tau)}\right]a_n(\tau) \tag{12.36a}$$

$$\left[\frac{\partial y_a}{\partial x}\right]_{x=1} = n(\tau)a_n(\tau) \tag{12.36b}$$

If we substitute Eq. 12.36 into the integral mass balance equation (Eq. 12.6), we obtain

$$\frac{d\bar{y}_a}{d\tau} = 3[3 + n(\tau)](1 - \bar{y}_a) \tag{12.37}$$

Note, so far we have not stipulated a value for the exponent $n(\tau)$. This time-dependent function can be estimated by substituting the exact solution (Eq. 12.3) into Eq. 12.37, yielding the result for $n(\tau)$

$$n(\tau) = \frac{\dfrac{1}{3}\left[\dfrac{d\bar{y}(\tau)}{d\tau}\right]}{1 - \bar{y}(\tau)} - 3 \tag{12.38}$$

Figure 12.5a shows a behavior of n versus τ. It is seen that this exponent is a very large number when time is small and then it decreases rapidly and eventually reaches an asymptote of 0.2899. The large number for n at small times is indeed reflected by the sharp profiles, and when times are larger, the profiles are more shallow, and this naturally leads to smaller values for the exponent n.

The asymptote for the exponent n, $n_\infty = 0.2899$, can be easily derived by comparison with the exact solution (Eq. 12.3). When times are large, the infinite series in the RHS of Eq. 12.3 can be truncated to only one term; that is,

$$\bar{y}(\tau) \approx 1 - \frac{6}{\pi^2}\exp(-\pi^2\tau) \tag{12.39}$$

Substituting the asymptotic solution into Eq. 12.38, we obtain the limiting value for n

$$n_\infty = \frac{\pi^2}{3} - 3 = 0.2899 \tag{12.40}$$

confirming the numerical observation shown in Fig. 12.5.

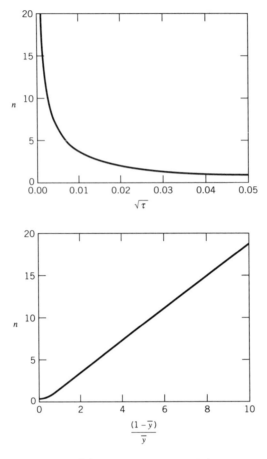

Figure 12.5 (*a*) Plot of n versus τ. (*b*) Plot of n versus $(1 - \bar{y})/\bar{y}$.

If we now plot the exponent n versus

$$\frac{(1 - \bar{y})}{\bar{y}} \tag{12.41}$$

as shown in Fig. 12.5*b*, we see that the curve exhibits a linear pattern over most of the domain, except when times are very large. The linear expression that best represents the linear part of this plot is

$$n = \alpha + \beta\left(\frac{1 - \bar{y}}{\bar{y}}\right) \quad \text{where} \quad \alpha = -0.591 \quad \text{and} \quad \beta = 1.9091 \tag{12.42}$$

When this is inserted into Eq. 12.37, we find

$$\frac{d\bar{y}_a}{d\tau} = 3\left[3 + \alpha + \beta\left(\frac{1 - \bar{y}_a}{\bar{y}_a}\right)\right](1 - \bar{y}_a) \tag{12.43a}$$

The initial condition for Eq. 12.43a is

$$\tau = 0; \qquad \bar{y}_a = 0 \qquad \qquad (\mathbf{12.43}b)$$

Before we integrate this equation, let us investigate this equation at small times. At small times, Eq. 12.43a is approximated by

$$\frac{d\bar{y}_a}{d\tau} \approx \frac{3\beta}{\bar{y}_a} \qquad \qquad (\mathbf{12.44})$$

of which the solution is

$$\bar{y}_a(\tau) \approx \sqrt{6\beta}\,\sqrt{\tau} = 3.3845\sqrt{\tau} \qquad \qquad (\mathbf{12.45})$$

This solution has the correct $\sqrt{\tau}$ dependence, and it compares very well with the exact solution (Eq. 12.4).

Now, we return to the solution of Eq. 12.43, which is finally

$$\ln\left(\frac{1}{1 - \bar{y}_a}\right) + \frac{\beta}{(3 + \alpha - \beta)} \ln\left(\frac{\beta}{\beta + (3 + \alpha - \beta)\bar{y}_a}\right) = 3(3 + \alpha)\tau \quad (\mathbf{12.46})$$

Curves for the mean concentrations of the exact and approximate solutions are shown in Fig. 12.6. Good agreement between these two solutions is observed over the whole course of practical time scales. While the approximate solution is now quite compact, the form it takes is not explicit in y but only in τ, a serious shortcoming.

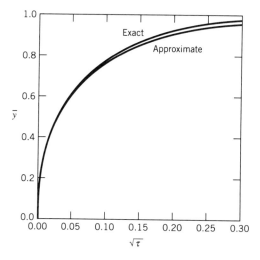

Figure 12.6 Plots of the mean concentration versus τ.

EXAMPLE 12.2 *A PROBLEM WITH SEMI-INFINITE DOMAIN*

We have tested polynomial approximation methods for a finite spatial domain, using several variations on the polynomial form. In this example, we will show how the polynomial approximation can also be carried out on problems having semi-infinite domain.

We consider again the classical problem of mass or heat transfer in a semi-infinite domain. Let us consider a mass transfer problem where a gas is dissolved at the interface and then diffuses into liquid of unlimited extent. Ignoring the gas film mass transfer resistance, the mass balance equations in nondimensional form are

$$\frac{\partial y}{\partial \tau} = \frac{\partial^2 y}{\partial x^2} \tag{12.47}$$

subject to the following initial and boundary conditions

$$\tau = 0; \quad y = 0 \tag{12.48}$$

$$x = 0; \quad y = 1 \tag{12.49a}$$

$$x \to \infty; \quad y = 0; \quad \frac{\partial y}{\partial x} = 0 \tag{12.49b}$$

The time scale (τ) is written as the product Dt, where D is dissolved gas diffusivity, and x is the actual distance from the interface. Here, y denotes C/C^* where C^* is the gas solubility.

The exact solution can be obtained by the combination of variables (Chapter 10), shown as follows.

$$y = 1 - \text{erf}\left[\frac{x}{2\sqrt{\tau}}\right] = \text{erfc}\left[\frac{x}{2\sqrt{\tau}}\right] \tag{12.50}$$

It is useful for comparison purposes to calculate the mass flux at the gas–liquid interface for exposure τ

$$\text{flux} = -D\frac{\partial C}{\partial x}\bigg|_{x=0} = \frac{2}{\sqrt{\pi}}\frac{DC^*}{2\sqrt{Dt}} = C^*\sqrt{\frac{D}{\pi t}}$$

In terms of original variables, this is written

$$-\frac{\partial y}{\partial x}\bigg|_{x=0} = \frac{1}{\sqrt{\pi \tau}} \tag{12.51}$$

Similarly, the time average (net) mass transfer into the liquid medium is

$$\text{Average Mass Transfer} = \frac{1}{t_e}\int_0^{t_e}\left[-D\frac{\partial C}{\partial x}\bigg|_{x=0}\right]dt = 2C^*\sqrt{\frac{D}{\pi t_e}}$$

Or, written in terms of $\tau_e = Dt_e$, this is

$$\int_0^{\tau_e} -\frac{\partial y}{\partial x}\bigg|_{x=0} d\tau = 2\sqrt{\frac{\tau_e}{\pi}} \tag{12.52}$$

which is the net penetration following exposure time t_e, since $\tau_e = Dt_e$ has units length squared.

In our formulation for the polynomial approximation solution, we will assume that there is a penetration front moving into the bulk liquid medium; that is, there is an effective mass transfer domain where most of the mass diffusion is taking place. Let this mass penetration front be denoted $X(\tau)$, then beyond this front, there will be no dissolved gas; that is,

$$x \geq X(\tau); \qquad y_a(x,\tau) = 0 \tag{12.53a}$$

Following the procedure in the last example, we will also introduce the condition that the slope of the concentration profile tends to zero at the position $X(\tau)$

$$x = X(\tau); \qquad \frac{\partial y_a}{\partial x} = 0 \tag{12.53b}$$

Our next step is to convert the differential mass balance equation in the liquid domain into an integral form. This is accomplished by integrating Eq. 12.47 with respect to x from the gas-liquid interface ($x = 0$) to the penetration front ($x = X(\tau)$). In so doing, we find the interesting result

$$\frac{d\bar{y}}{d\tau} = -\left[\frac{\partial y}{\partial x}\right]_{x=X(\tau)} \tag{12.54}$$

where the LHS is the rate of change of the mean concentration over the effective domain of mass transfer. This mean concentration is defined for cartesian coordinates as

$$\bar{y}(\tau) = \int_0^{X(\tau)} y(x,\tau)\, dx \tag{12.55}$$

Now we are at the crucial step in the polynomial approximation technique. We assume that the concentration over the effective domain of mass transfer is to take the form

$$y_a(x,\tau) = a_0(\tau) + a_1(\tau)x + a_2(\tau)x^2 \tag{12.56}$$

where again the subscript a denotes the approximate solution.

Using the conditions at the gas-liquid interface (Eq. 12.49a) and at the penetration front (Eq. 12.53), the polynomial approximation expression can be written as

$$y_a(x, \tau) = \left[1 - \frac{x}{X(\tau)}\right]^2 \tag{12.57}$$

which is a quadratic expression.

Having the form for the polynomial, we now force it to satisfy the overall mass balance equation (Eq. 12.54); hence, we have the equation for the penetration front

$$\frac{dX(\tau)}{d\tau} = \frac{6}{X(\tau)} \tag{12.58a}$$

The initial condition for the penetration front is the position at the gas-liquid interface, that is,

$$\tau = 0; \qquad X(0) = 0 \tag{12.58b}$$

Hence, the solution for the penetration front is

$$X(\tau) = 2\sqrt{3\tau} \tag{12.59}$$

Knowing the solution for the penetration front, the solution for the concentration profile is obtained from (Eq. 12.57)

$$y_a(x, \tau) = \left[1 - \frac{x}{2\sqrt{3\tau}}\right]^2 \tag{12.60}$$

This concentration profile allows us to obtain the mass flux and the average mass transfer rate up to time τ

$$-\frac{\partial y}{\partial x}\bigg|_{x=0} = \frac{1}{\sqrt{3\tau}} \tag{12.61a}$$

$$\int_0^{\tau_e} \left[-\frac{\partial y}{\partial x}\bigg|_{x=0}\right] d\tau = 2\sqrt{\frac{\tau_e}{3}} \tag{12.61b}$$

Curves showing y versus τ are presented in Fig. 12.7, where the exact solutions are also shown for comparison. Good agreement between the two solutions is observed.

Now, we extend the number of terms in the polynomial approximate solution to see if improved accuracy can be obtained. By keeping four terms, instead of three, the polynomial approximation solution can be written as

$$y_a(x, \tau) = a_0(\tau) + a_1(\tau)x + a_2(\tau)x^2 + a_3(\tau)x^3 \tag{12.62}$$

which is a cubic profile in the effective domain of mass transfer ($0 < x < X(\tau)$).

Since there are four terms in the polynomial approximation solution, we shall need one more condition, in addition to the ones imposed in the last trial

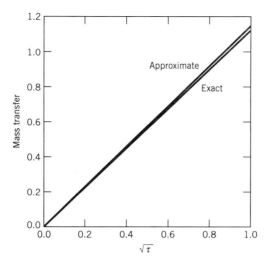

Figure 12.7 Plots of mass transfer versus $\sqrt{\tau}$.

(Eq. 12.49a and 12.53). It is reasonable to focus on the penetration front again, and we shall take the added condition

$$x = X(\tau); \qquad \frac{\partial^2 y_a}{\partial x^2} = 0 \qquad (12.63)$$

This forces curvature of the profiles to move quickly toward zero.

Using the conditions at the gas-liquid interface and at the penetration front, we finally obtain the following expression for the concentration profile:

$$y_a(x, \tau) = \left[1 - \frac{x}{X(\tau)}\right]^3 \qquad (12.64)$$

Substituting this expression into the integral mass balance equation (Eq. 12.54), and integrating the result we have the solution for the penetration front

$$X(\tau) = 2\sqrt{6\tau} \qquad (12.65)$$

Therefore, the gradient and penetration are

$$-\left.\frac{\partial y_a}{\partial x}\right|_{x=0} = \frac{3}{2\sqrt{6\tau}} \qquad (12.66a)$$

$$\int_0^{\tau_e} \left[-\left.\frac{\partial y_a}{\partial x}\right|_{x=0}\right] d\tau = \left(\frac{3}{\sqrt{6}}\right)\sqrt{\tau_e} \qquad (12.66b)$$

It is interesting to see that the quadratic solutions (Eqs. 12.61) are more accurate than the cubic solutions (Eqs. 12.66). For example, the percentage relative error between the cubic approximation solution and the exact solution is 8.6% compared to 2.3% for the quadratic solutions. This example provides a warning to the analyst; additional terms may not be worth the effort.

12.2   SINGULAR PERTURBATION

The method of singular perturbation taught in Chapter 6 for application to ODEs can also be applied to partial differential equations. The methodology for partial differential equations is not as straightforward as for ODEs, and ingenuity is often needed for a successful outcome. We illustrate singular perturbation in the following example for adsorption in a porous slab particle with a rectangular (irreversible) isotherm.

EXAMPLE 12.3

Consider a slab-shaped porous particle into which solute molecules diffuse through the tortuous network of pores. Along their path of diffusion, they adsorb onto the internal solid surface with an irreversible rate, modeled as

$$R_{\text{ads}} = k_a C(C_{\mu s} - C_\mu) \tag{12.67}$$

where k_a is the rate constant for adsorption, C is the solute intraparticle concentration (mole/cc of fluid volume) in the internal pore at the position r, C_μ is the solute concentration on the surface (mole/cc of particle volume excluding the fluid volume) and $C_{\mu s}$ is its maximum concentration. The units of R_{ads} are moles adsorbed per unit particle volume, per unit time. The rate expression shows that uptake ceases as $C \to 0$, and also when the surface is saturated as $C_\mu \to C_{\mu s}$.

The rate of diffusion of solute molecules through the pore network is assumed to take the following "Fickian" type diffusion formula

$$J = -D_e \frac{\partial C}{\partial r} \tag{12.68}$$

where the flux J has the units of mole transported per unit cross-sectional area per unit time. The diffusion through the pore network is assumed to be the only mode of transport of molecules into the interior of the particle. In general, the adsorbed molecules also diffuse along the surface (surface diffusion), but in the present case we assume that the adsorbed molecules are bound so strongly on the surface that their mobility can be neglected compared to the mobility of the free pore molecules.

Knowing the rate of adsorption (i.e., the rate of mass removal from the internal fluid phase) and the rate of diffusion into the particle, we are now ready to derive a mass balance equation for the free species in a thin shell having a thickness of Δr at the position r. The mass balance equation on this thin element is

$$SJ|_r - SJ|_{r+\Delta r} - S\Delta r R_{\text{ads}} = \frac{\partial}{\partial t}(S\Delta r \varepsilon C) \qquad (12.69)$$

where the term in the RHS is the accumulation term of the free species in the element, and the first two terms in the LHS are the diffusion terms in and out of the element, whereas the last term is the removal term from the fluid phase. Here, S is the cross-sectional area, which is constant for slab geometry.

Dividing Eq. 12.69 by $S\Delta r$ and taking the limit when the element shrinks to zero (i.e., $\Delta r \to 0$), we have the following differential mass balance equation

$$\varepsilon \frac{\partial C}{\partial t} = -\frac{\partial J}{\partial r} - R_{\text{ads}} \qquad (12.70)$$

Substituting the flux equation (Eq. 12.68) and the equation for the adsorption rate (Eq. 12.67) into the above mass balance equation, we obtain

$$\varepsilon \frac{\partial C}{\partial t} = D_e \frac{\partial^2 C}{\partial r^2} - k_{\text{ads}} C(C_{\mu s} - C_\mu) \qquad (12.71)$$

At this point, we recognize the system to be nonlinear, owing to the adsorption terms.

Equation 12.71 includes the adsorption term, which involves the concentration of the adsorbed species, C_μ. So to complete the formulation of the mass balance inside the particle, we need to write the mass balance relation for the adsorbed species. Since we assume that there is no mobility of the adsorbed species, the mass balance is basically the balance between the mass uptake and the accumulation, that is,

$$\frac{\partial}{\partial t}\left[S\Delta r(1 - \varepsilon)C_\mu\right] = \left[S\Delta r R_{\text{ads}}\right] \qquad (12.72)$$

Dividing this equation by $S\Delta r$ and using Eq. 12.67, we rewrite the above equation as

$$(1 - \varepsilon)\frac{\partial C_\mu}{\partial t} = k_{\text{ads}} C(C_{\mu s} - C_\mu) \qquad (12.73)$$

Eqs. 12.71 and 12.73 now completely define the differential mass balance inside the particle. To know how the concentrations of the free species as well as the adsorbed species evolve, we need to specify initial and boundary conditions. We assume initially the particle is free from solute of any form, and the fluid surrounding the particle is thoroughly stirred so that there is no film mass transfer resistance. Mathematically, these conditions are

$$t = 0; \quad C = C_\mu = 0 \qquad (12.74)$$

$$r = 0; \quad \frac{\partial C}{\partial r} = 0 \quad \text{and} \quad r = R; \quad C = C_0 \qquad (12.75)$$

where C_0 is the constant external bulk concentration and R represents the distance from the centerline of the slab-shaped particle whose thickness is $2R$.

To eliminate excess baggage, we nondimensionalize the mass balance equations by defining the following nondimensional variables and parameters

$$y = \frac{C}{C_0}; \quad y_\mu = \frac{C_\mu}{C_{\mu s}}; \quad \tau = \frac{D_e t}{R^2 \left[(1 - \varepsilon) \dfrac{C_{\mu s}}{C_0} \right]}; \quad x = \frac{r}{R} \quad (\textbf{12.76})$$

$$\sigma = \frac{\varepsilon C_0}{(1 - \varepsilon) C_{\mu s}}; \quad \mu = \frac{D_e}{k_{ads} C_{\mu s} R^2} \quad (\textbf{12.77})$$

With these definitions of variables and parameters, the nondimensional mass balance equations take the uncluttered form containing only two (dimensionless) parameters

$$\sigma \mu \frac{\partial y}{\partial \tau} = \mu \frac{\partial^2 y}{\partial x^2} - y(1 - y_\mu) \quad (\textbf{12.78}a)$$

$$\mu \frac{\partial y_\mu}{\partial \tau} = y(1 - y_\mu) \quad (\textbf{12.78}b)$$

subject to the following nondimensional initial and boundary conditions

$$\tau = 0; \quad y = y_\mu = 0 \quad (\textbf{12.79})$$

$$x = 0; \quad \frac{\partial y}{\partial x} = 0 \quad \text{and} \quad x = 1; \quad y = 1 \quad (\textbf{12.80})$$

Once again, remember that these mass balance equations are nonlinear (because of the nonlinear adsorption expression); hence, they can not be solved by direct analytical means. Generally, numerical methods must be used to solve this problem. However, when the adsorption rate is much faster than the diffusion rate (i.e., $\mu \ll 1$), conventional numerical methods will run into some difficulty owing to instability. This arises because of the steepness of the profiles generated. To achieve convergence with conventional numerical schemes, a large number of discretization points must be used for the spatial domain. The fine mesh is necessary to observe such a very sharp change in the profile.

Perhaps we can exploit in a positive way the smallness of the parameter $\mu(\mu \ll 1)$ and use singular perturbation to advantage. Before we proceed with the formality of the singular perturbation structure, let us have a look at the remaining parameter, σ. We rewrite this parameter as a capacity ratio

$$\sigma = \frac{\varepsilon C_0}{(1 - \varepsilon) C_{\mu s}} = \frac{V \varepsilon C_0}{V(1 - \varepsilon) C_{\mu s}} \quad (\textbf{12.81})$$

where the numerator is the mass holdup in the fluid phase within the particle and the denominator is the holdup on the adsorbed phase. In most practical adsorbents this ratio is very small. For example, the following parameters are typical of a practical gas phase separation process

$$\varepsilon = 0.4; \quad C_0 = 0.000001 \text{ mole/cc}; \quad C_{\mu s} = 0.001 \text{ mole/cc}$$

Using these values, the value of σ is 0.000667, which is a very small number indeed. This means that when the adsorption rate is much faster than the diffusion rate into the particle, we have two small parameters at our disposal, μ and σ.

Now we can proceed with the delineation of the formal structure for a singular perturbation approach. For given x and τ, we write y and y_μ as asymptotic expansions, and the first few terms are

$$y(x, \tau; \mu, \sigma) = y_0(x, \tau) + \mu y_1(x, \tau) + o(\mu) \tag{12.82a}$$

$$y_\mu(x, \tau; \mu, \sigma) = y_{\mu 0}(x, \tau) + \mu y_{\mu 1}(x, \tau) + o(\mu) \tag{12.82b}$$

We next substitute these expansions into Eqs. 12.78 and then observe the order of magnitude of all terms. This leads to a sequence of infinite subproblems. We write below the first two subproblems

$$O(1): \quad y_0(1 - y_{\mu 0}) = 0 \tag{12.83}$$

$$O(\mu): \quad \frac{\partial^2 y_0}{\partial x^2} - y_1(1 - y_{\mu 0}) + y_0 y_{\mu 1} = 0 \tag{12.84a}$$

$$\frac{\partial y_{\mu 0}}{\partial \tau} = y_1(1 - y_{\mu 0}) - y_0 y_{\mu 1} \tag{12.84b}$$

Equation 12.83 is the leading order (or zero-order) subproblem, and Eqs. 12.84 are first order subproblems. We start with the leading order problem. Inspecting Eq. 12.83, we note that there are two possibilities (i.e., two solutions). One solution is

$$y_0^{I} = 0 \quad \text{and} \quad y_{\mu 0}^{I} \neq 1 \tag{12.85}$$

and the other is

$$y_0^{II} \neq 0 \quad \text{and} \quad y_{\mu 0}^{II} = 1 \tag{12.86}$$

The existence of the two solutions simply means that the spatial domain $(0, 1)$ must be divided into two subdomains, and each solution is valid only in one such subdomain. If we stipulate a demarcation point dividing these two subdomains as $X(\tau)$, the two subdomains are $(0, X(\tau))$ and $(X(\tau), 1)$. Since the supply of solute comes from the exterior surface of the particle, it is obvious that the solution set (set I) having zero intraparticle fluid concentration must be applicable in the subdomain $(0, X(\tau))$, which is the inner core. The other solution set having the saturation adsorbed phase concentration $C_{\mu s}$ ($y_{\mu 0} = 1$) is valid in the subdomain $(X, 1)$, which is the outer shell. Figure 12.8 illustrates graphically these subdomains.

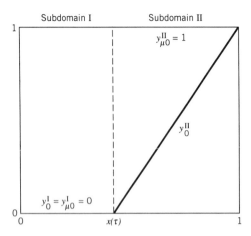

Figure 12.8 Schematic diagram of the two subdomains.

Let us consider the subdomain $(0, X)$ first. We replace the set applicable to this domain (Eq. 12.85) into the first order subproblem Eqs. 12.84, and obtain

$$\frac{\partial y_{\mu 0}^{I}}{\partial \tau} = 0 \qquad (12.87)$$

This, with the zero initial condition Eq. 12.79, simply implies that

$$y_{\mu 0}^{I} = 0 \qquad (12.88)$$

Thus, the leading order solutions in the subdomain $(0, X)$ (Eqs. 12.85 and 12.88) suggest that in the inner core of the particle, there is no solute of any form, either in the free or adsorbed state.

Now, we turn to the other subdomain $(X, 1)$, the outer shell. Substitute Eq. 12.86 into the first order subproblem Eqs. 12.84, and obtain

$$\frac{\partial^2 y_0^{II}}{\partial x^2} = 0 \qquad \text{for} \qquad x \in (X(\tau), 1) \qquad (12.89)$$

This equation, when written in dimensional form, is

$$D_e \frac{\partial^2 C}{\partial r^2} = 0 \qquad \text{for} \qquad x \in (X(\tau), 1) \qquad (12.90)$$

This equation simply states that in the outer subdomain the movement of the free species is controlled *purely* by diffusion. This is so because the outer subdomain is saturated with adsorbed species ($y_{\mu 0} = 1$) and there is no further adsorption taking place in this subdomain. Hence, free molecules diffuse through this outer shell without any form of mass retardation.

To solve Eq. 12.89, we must impose boundary conditions. One condition is that at the exterior particle surface where $y(1, \tau) = 1$ (Eq. 12.80) and the other must be at the demarcation position, $X(\tau)$. Since we know that the concentration of the free species is zero in the inner core, the appropriate boundary condition at this demarcation position is modelled as

$$x = X(\tau); \qquad y_0^{\text{II}} = 0 \qquad\qquad (\mathbf{12.91})$$

Solution of Eq. 12.89 subject to boundary conditions Eqs. 12.80 and 12.91 is obtained by successive integration to be

$$y_0^{\text{II}} = a_0(\tau)x + a_1(\tau)$$

hence, when the boundary conditions are applied, we finally obtain

$$y_0^{\text{II}} = \frac{x - X(\tau)}{1 - X(\tau)} \qquad \text{for} \qquad x \in (X(\tau), 1) \qquad (\mathbf{12.92})$$

Thus far, we know that the domain is broken up into two subdomains and the solutions for these two subdomains are given in Eqs. 12.85, 12.86, 12.88, and 12.92 as a function of the demarcation position, $X(\tau)$. At this stage, we have no knowledge regarding the form $X(\tau)$ must take. Even if we solve more higher order subproblems, we will not obtain any information about the demarcation function $X(\tau)$. The movement of this front is dictated by adsorption, which occurs only in the neighborhood of the demarcation position (so-called adsorption front), and the solutions obtained so far are not valid when we are close to the adsorption front. These solutions (Eqs. 12.85, 12.86, 12.88, and 12.92) are therefore called the outer solutions, because they are strictly valid only in regions far removed from the point $X(\tau)$.

To reveal the behavior near the adsorption front, we must enlarge the spatial scale around that front. This can be achieved by stretching (or magnifying) the spatial scale around this point as

$$\xi = \frac{[x - X(\tau)]}{\mu^{\gamma}} \qquad\qquad (\mathbf{12.93})$$

where ξ is called the *inner variable* and μ^{γ} represents the relative thickness of the adsorption front.

Thus, to obtain the relevant equations applicable in the neighborhood of the adsorption front, we need to make the variables transformation

$$(x, \tau) \rightarrow (\xi, \tau) \qquad\qquad (\mathbf{12.94})$$

In the manner of Chapter 10, we write the total derivative

$$dy = \left(\frac{\partial y}{\partial \xi}\right)_{\tau} d\xi + \left(\frac{\partial y}{\partial \tau}\right)_{\xi} d\tau = \left(\frac{\partial y}{\partial x}\right)_{\tau} dx + \left(\frac{\partial y}{\partial \tau}\right)_{x} d\tau \qquad (\mathbf{12.95})$$

and by equating like coefficients where

$$d\xi = \frac{1}{\mu^\gamma} dx - \frac{dX}{d\tau} \frac{d\tau}{\mu^\gamma}$$

and taking the front velocity as

$$Z = \frac{dX}{d\tau}$$

we obtain

$$\left(\frac{\partial y}{\partial \tau}\right)_x = \left(\frac{\partial y}{\partial \tau}\right)_\xi - \frac{1}{\mu^\gamma} Z(\tau) \left(\frac{\partial y}{\partial \xi}\right)_\tau \qquad (12.96a)$$

$$\left(\frac{\partial y_\mu}{\partial \tau}\right)_x = \left(\frac{\partial y_\mu}{\partial \tau}\right)_\xi - \frac{1}{\mu^\gamma} Z(\tau) \left(\frac{\partial y_\mu}{\partial \xi}\right)_\tau \qquad (12.96b)$$

and for the second derivative

$$\left(\frac{\partial^2 y}{\partial x^2}\right)_\tau = \frac{1}{\mu^{2\gamma}} \left(\frac{\partial^2 y}{\partial \xi^2}\right)_\tau \qquad (12.96c)$$

The velocity of the wave front

$$Z(\tau) = \frac{dX(\tau)}{d\tau} \qquad (12.96d)$$

is not necessarily constant.

Substituting Eqs. 12.96 into Eqs. 12.78 to effect the transformation, we obtain the following equations which are valid in the neighborhood of the adsorption front

$$-\sigma\mu^{1-\gamma} Z(\tau) \frac{\partial y}{\partial \xi} + \sigma\mu \frac{\partial y}{\partial \tau} = \mu^{1-2\gamma} \frac{\partial^2 y}{\partial \xi^2} - y(1 - y_\mu) \qquad (12.97a)$$

$$-\mu^{1-\gamma} Z(\tau) \frac{\partial y_\mu}{\partial \xi} + \mu \frac{\partial y_\mu}{\partial \tau} = y(1 - y_\mu) \qquad (12.97b)$$

where we have dropped the subscripts ξ and τ from the partial derivatives for the sake of simplicity of notation.

In the neighborhood of the adsorption front, the concentration of the free species is very low and the concentration of the adsorbed species varies from unity in the outer shell to zero in the inner core (see Fig. 12.9 for the behavior of the inner solutions). Thus, we will assume that the inner solutions have the asymptotic expansions

$$y^{(i)} = \mu^\lambda \left[y_0^{(i)} + o(\mu) \right] \qquad (12.98a)$$

$$y_\mu^{(i)} = y_{\mu 0}^{(i)} + o(\mu) \qquad (12.98b)$$

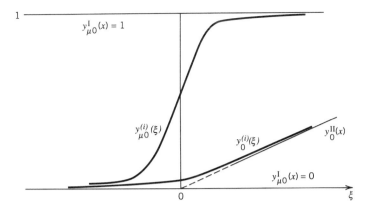

Figure 12.9 Behavior of the inner solutions.

If we substitute these expansions into Eq. 12.97, we then obtain the equations

$$-\sigma\mu^{1-\gamma+\lambda}Z(\tau)\frac{\partial y_0^{(i)}}{\partial\xi} + \sigma\mu^{1+\lambda}\frac{\partial y_0^{(i)}}{\partial\tau} = \mu^{1-2\gamma+\lambda}\frac{\partial^2 y_0^{(i)}}{\partial\xi^2} - \mu^{\lambda}y_0^{(i)}\left(1 - y_{\mu 0}^{(i)}\right)$$

$$(\textbf{12.99}a)$$

$$-\mu^{1-\gamma}Z(\tau)\frac{\partial y_{\mu 0}^{(i)}}{\partial\xi} + \mu\frac{\partial y_{\mu 0}^{(i)}}{\partial\tau} = \mu^{\lambda}y_0^{(i)}\left(1 - y_{\mu 0}^{(i)}\right) \qquad (\textbf{12.99}b)$$

If we now match the diffusion term (the first term on the RHS of Eq. 12.99a) with the adsorption term (the second term), and match the accumulation of the adsorbed species with the adsorption term in Eq. 12.99b, we obtain the two equations for γ and λ

$$1 - 2\gamma + \lambda = \lambda \qquad (\textbf{12.100}a)$$

$$1 - \gamma = \lambda \qquad (\textbf{12.100}b)$$

This results in

$$\lambda = \gamma = \frac{1}{2} \qquad (\textbf{12.101})$$

This result implies that the thickness of the adsorption front is of order of $\mu^{0.5}$. With these values of λ and γ, Eqs. 12.99 then become

$$-\sigma\mu Z(\tau)\frac{\partial y_0^{(i)}}{\partial\xi} + \sigma\mu^{3/2}\frac{\partial y_0^{(i)}}{\partial\tau} = \mu^{1/2}\frac{\partial^2 y_0^{(i)}}{\partial\xi^2} - \mu^{1/2}y_0^{(i)}\left(1 - y_{\mu 0}^{(i)}\right) \quad (\textbf{12.102}a)$$

$$-\mu^{1/2}Z(\tau)\frac{\partial y_{\mu 0}^{(i)}}{\partial\xi} + \mu\frac{\partial y_{\mu 0}^{(i)}}{\partial\tau} = \mu^{1/2}y_0^{(i)}\left(1 - y_{\mu 0}^{(i)}\right) \qquad (\textbf{12.102}b)$$

from which we can obtain the leading order subproblem for the inner solutions

$$\frac{\partial^2 y_0^{(i)}}{\partial \xi^2} - y_0^{(i)}\left(1 - y_{\mu 0}^{(i)}\right) = 0 \qquad (\mathbf{12.103}a)$$

$$-Z(\tau)\frac{\partial y_{\mu 0}^{(i)}}{\partial \xi} = y_0^{(i)}\left(1 - y_{\mu 0}^{(i)}\right) \qquad (\mathbf{12.103}b)$$

Eliminating the nonlinear adsorption terms between Eqs. 12.103a and 12.103b, we obtain

$$-Z(\tau)\frac{\partial y_{\mu 0}^{(i)}}{\partial \xi} = \frac{\partial^2 y_0^{(i)}}{\partial \xi^2} \qquad (\mathbf{12.104})$$

To find the conditions for the inner solutions, we have to match with the outer solutions in the limits of ξ approaching to $+\infty$ (matching with the outer solutions in the outer shell) and to $-\infty$ (matching with the outer solutions in the inner core). Proceeding with the matching, we obtain the following conditions for the inner solutions (see Fig. 12.9)

$$\xi \to \infty; \quad y_{\mu 0}^{(i)} = 1; \quad \frac{\partial y_0^{(i)}}{\partial \xi} = \frac{\partial y_0^{II}}{\partial x} = \frac{1}{1 - X(\tau)} \qquad (\mathbf{12.105}a)$$

$$\xi \to -\infty; \quad y_{\mu 0}^{(i)} = 0; \quad \frac{\partial y_0^{(i)}}{\partial \xi} = 0 \qquad (\mathbf{12.105}b)$$

Integrating Eq. 12.104 with respect to ξ from $-\infty$ to $+\infty$ and using the boundary conditions (12.105), we obtain the following ordinary differential equation for the adsorption front position, $X(\tau)$

$$\frac{dX(\tau)}{d\tau} = -\frac{1}{1 - X(\tau)} \qquad (\mathbf{12.106})$$

The initial condition of this adsorption front will be at the particle exterior surface

$$\tau = 0; \quad X(0) = 1 \qquad (\mathbf{12.107})$$

Therefore, the adsorption position as a function of time can be obtained as integration of Eq. 12.106

$$X(\tau) = 1 - \sqrt{2\tau} \qquad (\mathbf{12.108})$$

Knowing this front as a function of time, the fractional amount adsorbed up to time τ is simply

$$F = \int_{X(\tau)}^1 y_\mu dx = \sqrt{2\tau} \quad \text{for} \quad \tau \in \left[0, \frac{1}{2}\right] \qquad (\mathbf{12.109})$$

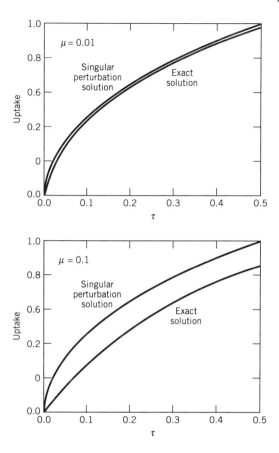

Figure 12.10 Plots of uptake versus time.

Equation 12.109 is plotted in Fig. 12.10. Also shown in the figure are the numerical calculations using the orthogonal collocation method (to be discussed in the next section). It is remarkable to see that the singular perturbation result agrees extremely well with the numerically exact solution even with $\mu = 0.01$. Note that the thickness of the neighborhood of the adsorption front is $\mu^{1/2}$. In Fig. 12.10, we also show the comparison between the singular perturbation solution and the numerically exact solution for $\mu = 0.1$. The agreement is not as satisfactory as the previous case. Additional computations show good agreement when $\mu = 0.001$. Thus, the singular perturbation result is quite reasonable when μ is less than 0.01.

We have shown an example of the application of the singular perturbation technique to PDE. Obviously, even for this simple example, the application still requires some knowledge of the solution behavior. Its application is not as straightforward as that for ODEs, but we present it here for the serious readers who may have practical problems in setting up the procedure. More often than not, when an analytical means is not available to solve the problem, we suggest that numerical methods should be tried first to get an overall picture of the system behavior. Often, this leads to enough insight to try an analytical approach such as singular perturbation. The rewards for analytical methods are simplicity, and compactness, which are highly valued in the scientific community.

Numerical procedures, as we show presently, pose little difficulty when nondimensional parameters are of order of unity. However, some difficulty in convergence arises when parameters become small or large. In this case, the singular perturbation method can be tried to isolate the sharpness of the solution behavior. It is difficult to imagine, however, that numerical methods could ever lead to the simple result $X(\tau) = 1 - \sqrt{2\tau}$.

We end this chapter by noting that the application of the singular perturbation method to partial differential equations requires more ingenuity than its application to ODEs. However, as in the case of ODEs, the singular perturbation solutions can be used as a tool to explore parametric dependencies and, as well, as a valuable check on numerical solutions. The book by Cole (1968) provides a complete and formal treatment of partial differential equations by the singular perturbation method.

12.3   FINITE DIFFERENCE

The finite difference method is one of the oldest methods to handle differential equations. Like the orthogonal collocation method discussed in Chapter 8, this technique involves the replacement of continuous variables, such as y and x, with discrete variables, that is, instead of obtaining a solution which is continuous over the whole domain of interest, the finite difference method will yield values at discrete points, chosen by the analyst.

In the finite difference method, a derivative at a discrete point, say point x_j, is evaluated using the information about discrete variables close to that point x_j (local information). This is in contrast to the orthogonal collocation method for which a derivative is evaluated by using information from all discrete variables. This is the reason why the collocation method is more stable than the finite difference procedure. However, the finite difference technique, owing to the fact that it utilizes only the local information, is readily applied to handle many problems of awkward geometry. As long as the grid size is properly chosen, the finite difference approach leads to stable solutions. This section will describe the procedure for the finite difference representation in solving differential equations.

Let us consider the Taylor series of a function $y(x + \Delta x)$ evaluated around the point x, as

$$y(x + \Delta x) = y(x) + \frac{dy(x)}{dx}\Delta x + \frac{d^2y(x)}{dx^2}(\Delta x)^2 + \cdots \quad \textbf{(12.110)}$$

Now, if the second and higher order terms are ignored, we have

$$\frac{dy(x)}{dx} \approx \frac{y(x + \Delta x) - y(x)}{\Delta x} \quad \textbf{(12.111)}$$

which simply states that the slope of the function y at point x is approximated by the slope of the line segment joining between two points $(x, y(x))$ and $(x + \Delta x, y(x + \Delta x))$. Thus, the first derivative is calculated using the local discrete points, and this formula is correct to the first order since the first term omitted in Eq. 12.110 is of order of $(\Delta x)^2$.

Similarly, if we expand the function $y(x - \Delta x)$ around the point x using the Taylor series method, we would obtain the formula

$$y(x - \Delta x) = y(x) - \frac{dy(x)}{dx}\Delta x + \frac{d^2y(x)}{dx^2}(\Delta x)^2 + \cdots \quad \textbf{(12.112)}$$

Again, when the second and higher order terms are ignored, the first derivative can be calculated from another formula, using discrete points $(x, y(x))$ and $(x - \Delta x, y(x - \Delta x))$, as

$$\frac{dy(x)}{dx} \approx \frac{y(x) - y(x - \Delta x)}{\Delta x} \quad \textbf{(12.113)}$$

Like Eq. 12.111, this approximation of the first derivative at the point x requires only values close to that point.

Another formula for approximating the first derivative can be obtained by subtracting Eq. 12.110 from 12.112; we obtain

$$\frac{dy(x)}{dx} \approx \frac{y(x + \Delta x) - y(x - \Delta x)}{2\Delta x} \quad \textbf{(12.114)}$$

This formula is more accurate than the last two (Eqs. 12.111 and 12.113) because the first term truncated in deriving this equation contains $(\Delta x)^2$, compared to (Δx) in the last two formulas.

To obtain the discrete formula for the second derivative, we add Eq. 12.110 and 12.112 and ignore the third order and higher order terms to obtain

$$\frac{d^2y(x)}{dx^2} \approx \frac{y(x + \Delta x) - 2y(x) + y(x - \Delta x)}{(\Delta x)^2} \quad \textbf{(12.115)}$$

The first term truncated in deriving this formula contains $(\Delta x)^2$; hence, the error for this formula is comparable to Eq. 12.114 obtained for the first derivative.

We can proceed in this way to obtain higher order derivatives, but only first and second derivatives are needed to handle most engineering problems. For cylindrical and spherical coordinates, the shifted position procedure in Problem 12.8 shows how to deal with the Laplacian operators.

12.3.1 Notations

To simplify the analysis, it is convenient to define a few short-hand notations just as we did in the other techniques. We have shown in Chapter 8 that, without loss of generality, the spatial domain can be taken to be $[0, 1]$ as any domain $[a, b]$ can be readily transformed to $[0, 1]$.

Let the domain be divided into N equal intervals with the length of each interval, called the *grid size*, given as

$$\Delta x = \frac{1}{N}$$

We define the point i as the point having the coordinate $i(\Delta x)$, and denote that point x_i; that is,

$$x_i = i(\Delta x) \qquad \text{for} \qquad i = 0, 1, 2, \ldots, N$$

This means $x_0 = 0$, $x_N = 1$, $x_{i+1} = (i + 1)\Delta x$, $x_{i-1} = (i - 1)\Delta x$, etc.

The variable y corresponding to the point x_i is denoted as y_i; that is, $y(x_i) = y_i$. In terms of these shorthand notations, the three previous formulas for approximating first derivatives are, respectively,

$$\left(\frac{dy}{dx}\right)_i \approx \frac{y_{i+1} - y_i}{\Delta x} \tag{12.116a}$$

$$\left(\frac{dy}{dx}\right)_i \approx \frac{y_i - y_{i-1}}{\Delta x} \tag{12.116b}$$

$$\left(\frac{dy}{dx}\right)_i \approx \frac{y_{i+1} - y_{i-1}}{2\Delta x} \tag{12.116c}$$

and the approximating formula for the second derivative is

$$\left(\frac{d^2y}{dx^2}\right)_i \approx \frac{y_{i+1} - 2y_i + y_{i-1}}{(\Delta x)^2} \tag{12.117}$$

Equations 12.116a, b are first order correct, whereas the last two are second order correct.

12.3.2   Essence of the Method

It is best to illustrate the essence of the method by way of a simple example. An iterative solution is sought which, in the limit of small grid size, should converge very closely to the exact solution.

EXAMPLE 12.4

Let us start with the elementary second order differential equation

$$\frac{d^2y}{dx^2} - 2\frac{dy}{dx} - 10y = 0 \qquad \text{for} \qquad 0 < x < 1 \tag{12.118}$$

This relation can be used to describe a chemical reaction occurring in a fixed bed. The equation, of course, can be handled by techniques developed in Chapters 2 and 3, but we will solve it numerically for the purpose of demonstration.

We first divide the domain into N intervals, with the grid size being $\Delta x = 1/N$. The ith point is the point having the coordinate $i(\Delta x)$. Eq. 12.118 is valid at any point within the domain $[0, 1]$; thus, if we evaluate it at the point x_i, we have

$$\left.\frac{d^2y}{dx^2}\right|_{x_i} - 2\left.\frac{dy}{dx}\right|_{x_i} - 10y|_{x_i} = 0 \qquad (12.119)$$

Using the discrete formula for the first and second derivatives, (Eqs. 12.116c and 12.117), which are both second order correct, we obtain

$$\frac{y_{i+1} - 2y_i + y_{i-1}}{(\Delta x)^2} - 2\frac{y_{i+1} - y_{i-1}}{2\Delta x} - 10y_i = 0 \qquad (12.120a)$$

Simplifying this equation gives the finite-difference equation

$$\alpha y_{i-1} - \beta y_i - \gamma y_{i+1} = 0 \qquad (12.120b)$$

where

$$\alpha = \frac{1}{(\Delta x)^2} + \frac{1}{\Delta x}; \quad \beta = \frac{2}{(\Delta x)^2} + 10; \quad \gamma = \frac{1}{\Delta x} - \frac{1}{(\Delta x)^2} \qquad (12.120c)$$

This discrete equation Eq. 12.120b can be solved using the calculus of finite difference (Chapter 5), to give a general solution in terms of arbitrary constants.[1] Boundary conditions are necessary to complete the problem, if we wish to develop an iterative solution. The remainder of the procedure depends on the form of the specified boundary conditions. To show this, we choose the following two boundary conditions

$$x = 0; \qquad y = 1 \qquad (12.121)$$

$$x = 1; \qquad \frac{dy}{dx} = 0 \qquad (12.122)$$

where the boundary condition at $x = 0$ involves the specification of the value of y while the other involves the first derivative.

Evaluating Eq. 12.120b at the point x_1 (i.e., the first point next to the left boundary) would give

$$\alpha y_0 - \beta y_1 - \gamma y_2 = 0$$

---

[1]This finite difference equation has the general analytical solution

$$y_i = A(r_1)^i + B(r_2)^i$$

where

$$r_{1,2} = \frac{-\beta \pm \sqrt{\beta^2 + 4\alpha\gamma}}{2\gamma}$$

But since we specify the value of y at $x = 0$ as $y_0 = 1$, the above equation will become

$$-\beta y_1 - \gamma y_2 = -\alpha \qquad (12.123)$$

To deal with the other boundary condition at $x = 1$, we apply the second-order correct approximation for the first derivative at $x = 1$; that is, Eq. 12.122 becomes

$$\frac{y_{N+1} - y_{N-1}}{2\Delta x} \approx 0 \quad \therefore \quad y_{N+1} \approx y_{N-1}$$

Note that the point x_{N+1} is a *fictitious* point, since it is outside the domain of interest.

Evaluating Eq. 12.120b at $i = N$ (i.e., at the boundary) will yield

$$\alpha y_{N-1} - \beta y_N - \gamma y_{N+1} = 0$$

but since $y_{N+1} = y_{N-1}$, the previous equation will become

$$(\alpha - \gamma) y_{N-1} - \beta y_N = 0 \qquad (12.124)$$

Thus, our finite difference equations are Eq. 12.123, Eq. 12.120b for $i = 2, 3, \ldots, N - 1$, and Eq. 12.124, totalling N equations, and we have exactly N unknown discrete variables to be found ($y_1, y_2, \ldots, y_N$). Since the starting equation we choose is linear, the resulting set of N equations are linear algebraic equations, which can be handled by the usual matrix methods given in Appendix B. The matrix formed by this set of linear equations has a special format, called the tridiagonal matrix, which will be considered in the next section.

To summarize the finite difference method, all we have to do is to replace all derivatives in the equation to be solved by their appropriate approximations to yield a finite difference equation. Next, we deal with boundary conditions. If the boundary condition involves the specification of the variable y, we simply use its value in the finite difference equation. However, if the boundary condition involves a derivative, we need to use the fictitious point which is outside the domain to effect the approximation of the derivative as we did in the above example at $x = 1$. The final equations obtained will form a set of algebraic equations which are amenable to analysis by methods such as those in Appendix A. If the starting equation is linear, the finite difference equation will be in the form of tridiagonal matrix and can be solved by the Thomas algorithm presented in the next section.

12.3.3   Tridiagonal Matrix and the Thomas Algorithm

The tridiagonal matrix is a square matrix in which all elements not on the diagonal line and the two lines parallel to the diagonal line are zero. Elements on these three lines may or may not be zero. For example, a 7 x 7 tridiagonal matrix will have the following format

$$
\begin{bmatrix}
+ & + & 0 & 0 & 0 & 0 & 0 \\
+ & + & + & 0 & 0 & 0 & 0 \\
0 & + & + & + & 0 & 0 & 0 \\
0 & 0 & + & + & + & 0 & 0 \\
0 & 0 & 0 & + & + & + & 0 \\
0 & 0 & 0 & 0 & + & + & + \\
0 & 0 & 0 & 0 & 0 & + & +
\end{bmatrix}
\tag{12.125}
$$

with the symbol $+$ denoting value which may or may not be zero.

With this special form of the tridiagonal matrix, the Thomas algorithm (Von Rosenberg 1969) can be used to effect a solution. The algorithm for a square $N \times N$ tridiagonal matrix is simple and is given as

The equations are

$$
a_i y_{i-1} + b_i y_i + c_i y_{i+1} = d_i \qquad \text{for } 1 < i < N
$$

$$
\text{with } a_1 = c_N = 0
$$

Step 1.   Compute:

$$
\beta_i = b_i - a_i c_{i-1}/\beta_{i-1} \qquad \text{with } \beta_1 = b_1
$$

$$
\gamma_i = (d_i - a_i \gamma_{i-1})/\beta_i \qquad \text{with } \gamma_1 = d_1/b_1
$$

Step 2.   The values of the dependent variables are calculated from

$$
y_N = \gamma_N \quad \text{and} \quad y_i = \gamma_i - c_i y_{i+1}/\beta_i
$$

Now, coming back to the problem at hand in the last section, let us choose five intervals; that is, $N = 5$ and $\Delta x = 1/5 = 0.2$. The five linear equations are (written in matrix form)

$$
\begin{bmatrix}
-\beta & -\gamma & 0 & 0 & 0 \\
\alpha & -\beta & -\gamma & 0 & 0 \\
0 & \alpha & -\beta & -\gamma & 0 \\
0 & 0 & \alpha & -\beta & -\gamma \\
0 & 0 & 0 & (\alpha - \gamma) & -\beta
\end{bmatrix}
\cdot
\begin{bmatrix}
y_1 \\ y_2 \\ y_3 \\ y_4 \\ y_5
\end{bmatrix}
=
\begin{bmatrix}
-\alpha \\ 0 \\ 0 \\ 0 \\ 0
\end{bmatrix}
\tag{12.126}
$$

The first equation comes from the boundary condition at $x = 0$ (Eq. 12.123), the next four come from Eq. 12.120b, and the last comes from the boundary condition at $x = 1$ (Eq. 12.124). Solving this set of equations using the Thomas algorithm just presented will yield solutions for discrete values y_1 to y_5. Table 12.1 shows these approximate values and the exact solution, which is

Table 12.1 Comparison Between the Approximate and Exact Solutions

x	y_{approx}	y_{exact}	Relative Error%
0.2	0.635135	0.630417	0.75
0.4	0.405405	0.399566	1.46
0.6	0.263514	0.258312	2.01
0.8	0.182432	0.178912	1.97
1.0	0.152027	0.151418	0.40

obtained using complementary solution methods taught in Chapter 2:

$$y = \frac{r_2 e^{r_2} e^{r_1 x} - r_1 e^{r_1} e^{r_2 x}}{r_2 e^{r_2} - r_1 e^{r_1}}; \qquad r_{1,2} = 1 \pm \sqrt{11} \qquad (12.127)$$

An increase in the number of intervals (smaller grid size) will yield approximate solutions closer to the exact solution, as one may expect.

12.3.4   Linear Parabolic Partial Differential Equations

The preliminary steps for solving partial differential equations by finite difference methods is similar to the last section. We shall now illustrate this for parabolic partial differential equations.

EXAMPLE 12.5

Let us start with a linear equation written in nondimensional form as

$$\frac{\partial y}{\partial t} = \frac{\partial^2 y}{\partial x^2} \qquad (12.128\,a)$$

$$x = 0; \quad y = 1 \qquad (12.128\,b)$$

$$x = 1; \quad y = 0 \qquad (12.128\,c)$$

This equation describes many transient heat and mass transfer processes, such as the diffusion of a solute through a slab membrane with constant physical properties. The exact solution, obtained by either the Laplace transform, separation of variables (Chapter 10) or the finite integral transform (Chapter 11), is given as

$$y = 1 - x - \frac{2}{\pi} \sum_{n=1}^{\infty} \frac{\sin(n\pi x)}{n} e^{-n^2 \pi^2 t} \qquad (12.129)$$

We will use this exact solution later for evaluating the efficiency of the finite difference solutions.

First, we divide the membrane spatial domain into N equal intervals with the grid size being $\Delta x = 1/N$. Next, evaluating Eq. 12.128a at $N - 1$ interior points (i.e., points within the domain), we have

$$\frac{\partial y_i}{\partial t} = \frac{\partial^2 y}{\partial x^2}\bigg|_{x_i} \approx \frac{y_{i+1} - 2y_i + y_{i-1}}{(\Delta x)^2} \qquad \text{for} \qquad i = 1, 2, \ldots, N - 1 \quad (12.130)$$

in which we have used Eq. 12.117 for the approximation of the second partial derivative. Note that $x_i = i(\Delta x)$ and $y_i = y(x_i)$.

Since the values of y at the boundary points are known in this problem (i.e., $y_0 = 1$ and $y_N = 0$), the above equation written for $i = 1$ and $i = N - 1$ will become

$$\frac{\partial y_1}{\partial t} \approx \frac{y_2 - 2y_1 + 1}{(\Delta x)^2} \qquad\qquad\qquad (12.131)$$

$$\frac{\partial y_{N-1}}{\partial t} \approx \frac{-2y_{N-1} + y_{N-2}}{(\Delta x)^2} \qquad\qquad\qquad (12.132)$$

Thus, Eq. 12.131, Eq. 12.130 for $i = 2, 3, \ldots, N - 2$, and Eq. 12.132 will form a set of $N - 1$ equations with $N - 1$ unknowns ($y_1, y_2, \ldots, y_{N-1}$). This set of coupled ordinary differential equations can be solved by any of the integration solvers described in Chapter 7. Alternately, we can apply the same finite difference procedure to the time domain, for a completely iterative solution. This is essentially the Euler, backward Euler and the Trapezoidal rule discussed in Chapter 7, as we shall see.

We use the grid size in the time domain as Δt and the index j to count the time, such that

$$t_j = j(\Delta t) \qquad\qquad\qquad (12.133)$$

The variable at the grid point x_i and at the time t_j is denoted as $y_{i,j}$. The $N - 1$ ordinary differential equations (Eqs. 12.130, 12.131, and 12.132) are valid at times greater than 0; thus, evaluating those equations at an arbitrary time t_j would give the approximations for time derivative as

$$\frac{\partial y_{1,j}}{\partial t} \approx \frac{y_{2,j} - 2y_{1,j} + 1}{(\Delta x)^2} \qquad\qquad\qquad (12.134a)$$

$$\frac{\partial y_{i,j}}{\partial t} \approx \frac{y_{i+1,j} - 2y_{i,j} + y_{i-1,j}}{(\Delta x)^2} \qquad \text{for} \qquad i = 2, 3, \ldots, N - 2 \quad (12.134b)$$

$$\frac{\partial y_{N-1,j}}{\partial t} \approx \frac{-2y_{N-1,j} + y_{N-2,j}}{(\Delta x)^2} \qquad\qquad\qquad (12.134c)$$

If we use the *forward difference* in time (an analogue to Eq. 7.72) as

$$\frac{\partial y_{i,j}}{\partial t} \approx \frac{y_{i,j+1} - y_{i,j}}{\Delta t} \tag{12.135}$$

then Eq. 12.134 will become

$$\frac{y_{1j,+1} - y_{1,j}}{\Delta t} \approx \frac{y_{2,j} - 2y_{1,j} + 1}{(\Delta x)^2} \tag{12.136a}$$

$$\frac{y_{i,j+1} - y_{i,j}}{\Delta t} \approx \frac{y_{i+1,j} - 2y_{i,j} + y_{i-1,j}}{(\Delta x)^2} \qquad \text{for} \qquad i = 2, 3, \dots, N-2$$

$$\tag{12.136b}$$

$$\frac{y_{N-1,j+1} - y_{N-1,j}}{\Delta t} \approx \frac{-2y_{N-1,j} + y_{N-2,j}}{(\Delta x)^2} \tag{12.136c}$$

All the values of y at the jth time are known, and hence, the previous equation can be written *explicitly* in terms of the unknown $y_{i,j+1}$ at the time t_{j+1} as

$$y_{1,j+1} \approx \frac{\Delta t}{(\Delta x)^2}(y_{2,j} + 1) + \left[1 - \frac{2\Delta t}{(\Delta x)^2}\right]y_{1,j} \tag{12.137a}$$

$$y_{i,j+1} \approx \frac{\Delta t}{(\Delta x)^2}(y_{i+1,j} + y_{i-1,j}) + \left[1 - \frac{2\Delta t}{(\Delta x)^2}\right]y_{i,j} \tag{12.137b}$$

$$\text{for} \qquad i = 2, 3, \dots, N-2$$

$$y_{N-1,j+1} \approx \frac{\Delta t}{(\Delta x)^2}y_{N-2,j} + \left[1 - \frac{2\Delta t}{(\Delta x)^2}\right]y_{N-1,j} \tag{12.137c}$$

To solve this set of equations, we need initial conditions. Let us assume that the membrane is initially solute free. Hence, the initial conditions for this discrete set of variables are

$$y_{i,0} = 0 \qquad \text{for} \qquad i = 1, 2, \dots, N-1 \tag{12.138}$$

All the terms in the right-hand side of Eq. 12.137 are known, and hence, this makes the forward difference in time rather attractive. However, as in the Euler method (Chapter 7), this forward difference scheme in time suffers the same handicap, that is, it is unstable if the grid size is not properly chosen. Using the stability analysis (von Rosenberg 1969), the criterion for stability (see Problem 12.12) is

$$\frac{\Delta t}{(\Delta x)^2} \le \frac{1}{2} \tag{12.139}$$

which plays a similar role as the condition $h\lambda < 2$ for the explicit Euler method of Chapter 7. This restriction is quite critical, since to reduce the error in the x domain (i.e., the profile at a given time) the grid size Δx must be small; therefore, the time step size to satisfy the above criterion must also be small for stability.

To resolve the problem of stability, we approach the problem in the same way we did in the backward Euler method. We evaluate Eq. 12.130 at the *unknown time level* t_{j+1} and use the following *backward difference* formula for the time derivative term (which is first order correct)

$$\frac{\partial y_{i,j+1}}{\partial t} \approx \frac{y_{i,j+1} - y_{i,j}}{\Delta t} \qquad (12.140)$$

we obtain

$$\frac{y_{i,j+1} - y_{i,j}}{\Delta t} \approx \frac{y_{i+1,j+1} - 2y_{i,j+1} + y_{i-1,j+1}}{(\Delta x)^2} \qquad \text{for} \qquad i = 1, 2, \ldots, N-1 \qquad (12.141)$$

Rearranging this equation in the form of the tridiagonal matrix, we have

$$y_{i-1,j+1} - \left[2 + \frac{(\Delta x)^2}{\Delta t}\right] y_{i,j+1} + y_{i+1,j+1} = -\frac{(\Delta x)^2}{\Delta t} y_{i,j} \qquad (12.142)$$

for $i = 1, 2, 3, \ldots, N-1$.

When $i = 1$, the above equation can be written as

$$-\left[2 + \frac{(\Delta x)^2}{\Delta t}\right] y_{1,j+1} + y_{2,j+1} = -\frac{(\Delta x)^2}{\Delta t} y_{1,j} - 1 \qquad (12.143)$$

in which we have made use of the boundary condition (12.128b); that is,

$$y_{0,j} = 1 \qquad \text{for all } j \qquad (12.144)$$

Similarly, for $i = N - 1$, we have

$$y_{N-2,j+1} - \left[2 + \frac{(\Delta x)^2}{\Delta t}\right] y_{N-1,j+1} = -\frac{(\Delta x)^2}{\Delta t} y_{N-1,j} \qquad (12.145)$$

in which we have made use of the other boundary condition at $x = 1$; that is,

$$y_{N,j} = 0 \qquad \text{for all } j \qquad (12.146)$$

If we arrange the derived finite difference equations in the following order: Eq. 12.143, Eq. 12.142 for $i = 2, 3, \ldots, N-2$, and Eq. 12.145, they will form a set of equations in tridiagonal form, which can be solved by the Thomas algorithm presented in Section 12.3.3. Comparing these equations with the equation format for the Thomas algorithm, we must have

$$a_i = 1; \qquad b_i = -\left[2 + \frac{(\Delta x)^2}{\Delta t}\right]; \qquad c_i = 1 \qquad (12.147a)$$

$$d_i = -\frac{(\Delta x)^2}{\Delta t} y_{i,j} - \delta_{i1}; \qquad a_1 = 0; \qquad c_{N-1} = 0 \qquad (12.147b)$$

where δ_{i1} is the Kronecker delta function. With this new scheme of backward in time (similar to the backward Euler method), there will be no restriction on the size of Δt for stability (see Problem 12.13). Since it is stable, one can choose the step size of any magnitude to minimize the truncation error. As we have discussed in Chapter 7, even though the backward Euler method is very stable, the formula used in the approximation of the time derivative is only first order correct (Eq. 12.140). So we shall need a better scheme, which will utilize the second order correct formula for the time derivative.

The second order correct formula for the time derivative is derived by using the Taylor series of $y(t + \Delta t)$ around the point $t + \Delta t/2$ shown as

$$y(t + \Delta t) \approx y(t + \Delta t/2) + \left(\frac{\partial y}{\partial t}\right)_{t+\Delta t/2} \frac{\Delta t}{2} + \frac{1}{2!}\left(\frac{\partial^2 y}{\partial t^2}\right)_{t+\Delta t/2}\left(\frac{\Delta t}{2}\right)^2 + \cdots$$

$$(12.148a)$$

Similarly, we can write for any function $y(t)$ as a Taylor series expansion around the time $(t + \Delta t/2)$

$$y(t) \approx y(t + \Delta t/2) - \left(\frac{\partial y}{\partial t}\right)_{t+\Delta t/2} \frac{\Delta t}{2} + \frac{1}{2!}\left(\frac{\partial^2 y}{\partial t^2}\right)_{t+\Delta t/2}\left(\frac{\Delta t}{2}\right)^2 + \cdots$$

$$(12.148b)$$

Subtracting these two equations will yield the following equation for the approximation of the time derivative at the time $t_{j+1/2}$ ($t_{j+1/2} = t + \Delta t/2$)

$$\left(\frac{\partial y}{\partial t}\right)_{t_{j+1/2}} \approx \frac{y_{j+1} - y_j}{\Delta t} \qquad (12.149)$$

The first term omitted in the previous formula contains $(\Delta t)^2$; thus, the formula is of second order correct. This approximation is the famous *Crank and Nicolson equation* (Crank and Nicolson 1947).

Now let us evaluate the mass balance equation (Eq. 12.128a) at the time $t_{j+1/2}$ and use the above formula for the approximation of the time derivative, we have

$$\frac{y_{i,j+1} - y_{i,j}}{\Delta t} \approx \left(\frac{\partial^2 y}{\partial x^2}\right)_{i,j+1/2} \qquad (12.150)$$

The second derivative in the RHS, if approximated by the procedure derived in the last two methods, will be given by

$$\left(\frac{\partial^2 y}{\partial x^2}\right)_{i,j+1/2} \approx \frac{y_{i-1,j+1/2} - 2y_{i,j+1/2} + y_{i+1,j+1/2}}{(\Delta x)^2} \qquad \textbf{(12.151)}$$

The values of y at $t_{j+1/2}$ are not what we are after. What we want are y at t_{j+1} in terms of what we already know at t_j. To get around this, we assume that the step size is sufficiently small such that the following arithmetic mean for the second order spatial derivative is applicable

$$\left(\frac{\partial^2 y}{\partial x^2}\right)_{i,j+1/2} \approx \frac{1}{2}\left[\left(\frac{\partial^2 y}{\partial x^2}\right)_{i,j} + \left(\frac{\partial^2 y}{\partial x^2}\right)_{i,j+1}\right] \qquad \textbf{(12.152)}$$

This is very similar to the trapezoidal rule taught in Chapter 7.

Using Eq. 12.117 for the approximation of the second derivative, Eq. 12.152 becomes

$$\left(\frac{\partial^2 y}{\partial x^2}\right)_{i,j+1/2} \approx \frac{1}{2}\left[\frac{y_{i+1,j} - 2y_{i,j} + y_{i-1,j}}{(\Delta x)^2} + \frac{y_{i+1,j+1} - 2y_{i,j+1} + y_{i-1,j+1}}{(\Delta x)^2}\right]$$

$$\textbf{(12.153)}$$

Next, substitute this equation into Eq. 12.150, and we have

$$\frac{y_{i,j+1} - y_{i,j}}{\Delta t} \approx \frac{1}{2}\left[\frac{y_{i+1,j} - 2y_{i,j} + y_{i-1,j}}{(\Delta x)^2} + \frac{y_{i+1,j+1} - 2y_{i,j+1} + y_{i-1,j+1}}{(\Delta x)^2}\right]$$

$$\textbf{(12.154)}$$

This equation is now second order correct in both x and t. Remember that all values of y at the jth time step are known and so we can obtain the values at t_{j+1}. Rearranging the previous equation in the form of tridiagonal matrix, we have

$$y_{i-1,j+1} - \left[2 + \frac{2(\Delta x)^2}{\Delta t}\right]y_{i,j+1} + y_{i+1,j+1}$$

$$\textbf{(12.155)}$$

$$\approx -y_{i-1,j} + \left[2 - \frac{2(\Delta x)^2}{\Delta t}\right]y_{i,j} - y_{i+1,j}$$

for $i = 1, 2, \ldots, N - 1$, and $j = 0, 1, 2, 3, \ldots$

For $i = 1$ (i.e., the first discrete point adjacent to the left boundary) and $i = N - 1$ (the last point adjacent to the right boundary), the previous equation involves the boundary values $y_{0,j}$, $y_{0,j+1}$, $y_{N,j}$, and $y_{N,j+1}$. The boundary condition (Eqs. 12.128b, c) provides these values; that is,

$$y_{0,j} = y_{0,j+1} = 1 \tag{12.156a}$$

$$y_{N,j} = y_{N,j+1} = 0 \tag{12.156b}$$

After these boundary conditions are substituted into Eq. 12.155 for $i = 1$ and $i = N - 1$, we get

$$-\left[2 + \frac{2(\Delta x)^2}{\Delta t}\right] y_{1,j+1} + y_{2,j+1} \approx -2 + \left[2 - \frac{2(\Delta x)^2}{\Delta t}\right] y_{1,j} - y_{2,j}$$

$$\tag{12.157}$$

$$y_{N-2,j+1} - \left[2 + \frac{2(\Delta x)^2}{\Delta t}\right] y_{N-1,j+1} \approx -y_{N-2,j} + \left[2 - \frac{2(\Delta x)^2}{\Delta t}\right] y_{N-1,j}$$

$$\tag{12.158}$$

When the equations are put in the order: Eq. 12.157, Eq. 12.155 (for $i = 2, 3, \ldots, N - 2$) and Eq. 12.158, we obtain a $(N - 1, N - 1)$ matrix of tridiagonal form with $N - 1$ unknown discrete variables y_1 to y_{N-1}. This can be solved by the Thomas algorithm of Section 12.3.3.

Note that the backward difference in time method and the Crank–Nicolson method both yield finite difference equations in the form of tridiagonal matrix, but the latter involves computations of three values of $y(y_{i-1}, y_i,$ and $y_{i+1})$ of the previous time t_j, whereas the former involves only y_i.

Since the time derivative used in the Crank–Nicolson method is second order correct, its step size can be larger and hence more efficient (see Fig. 12.11c). Moreover, like the backward difference method, the Crank–Nicolson is stable in both space and time.

Another useful point regarding the Crank–Nicolson method is that the forward and backward differences in time are applied successively. To show this, we evaluate the finite difference equation at the $(j + 2)$th time by using the backward formula; that is,

$$y_{i-1,j+2} - \left[2 + \frac{(\Delta x)^2}{\Delta t}\right] y_{i,j+2} + y_{i+1,j+2} = -\frac{(\Delta x)^2}{\Delta t} y_{i,j+1} \tag{12.159}$$

If we evaluate the values of y at the $(j + 1)$th time in the RHS of Eq. 12.159 using the forward difference formula in time, we would obtain

$$y_{i-1,j+2} - \left[2 + \frac{(\Delta x)^2}{\Delta t}\right] y_{i,j+2} + y_{i+1,j+2}$$

$$\tag{12.160}$$

$$= -y_{i-1,j} + \left[2 - \frac{2(\Delta x)^2}{(\Delta t)}\right] y_{i,j} - y_{i+1,j}$$

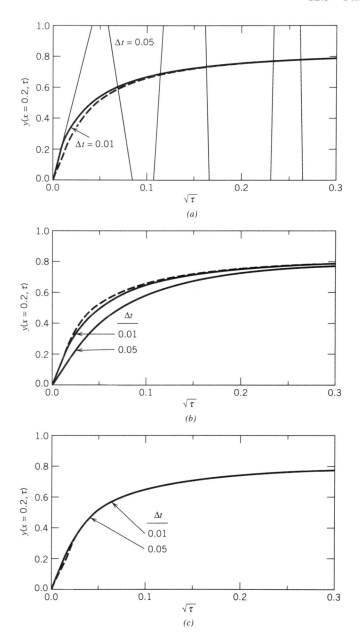

Figure 12.11 Plots of y_1 versus time for: (a) forward difference; (b) backward difference; (c) Crank–Nicolson.

Thus, we see that this formula is basically the Crank–Nicolson formula written for the time step size of $2(\Delta t)$ (compare this with Eq. 12.155 where the step size is Δt).

Figures 12.11 shows plots of $y_1 = y(x = 0.2)$ as a function of time. Computations from the forward difference scheme are shown in Fig. 12.11a, while those of the backward difference and the Crank–Nicolson schemes are shown in Figs. 12.11b and c, respectively. Time step sizes of 0.01 and 0.05 are used as parameters in these three figures. The exact solution (Eq. 12.129) is also shown in these figures as dashed lines. It is seen that the backward difference and the Crank–Nicolson methods are stable no matter what step size is used, whereas the forward difference scheme becomes unstable when the stability criterion of Eq. 12.139 is violated. With the grid size of $\Delta x = 0.2$, the maximum time step size for stability of the forward difference method is $\Delta t = (\Delta x)^2/2 = 0.02$.

12.3.5 Nonlinear Parabolic Partial Differential Equations

We have demonstrated the method of finite differences for solving linear parabolic partial differential equations. But the utility of the numerical method is best appreciated when we deal with nonlinear equations. In this section, we will consider a nonlinear parabolic partial differential equation, and show how to deal with the nonlinear terms.

EXAMPLE 12.6

Let us again consider the same problem of mass transfer through a membrane of Example 12.5 but this time we account for the fact that diffusivity is a function of the intramembrane concentration. The nondimensional equation can be written as

$$\frac{\partial y}{\partial t} = \frac{\partial}{\partial x}\left[f(y)\frac{\partial y}{\partial x}\right] \tag{12.161}$$

Expanding the RHS will give

$$\frac{\partial y}{\partial t} = f(y)\frac{\partial^2 y}{\partial x^2} + f'(y)\left(\frac{\partial y}{\partial x}\right)^2 \tag{12.162}$$

As in Example 12.5, if we divide the domain $[0, 1]$ into N intervals, evaluate the above equation at the $(j + 1/2)$th time, and use the second order correct formula for both time and space derivatives, we have the following finite difference formulas for the derivatives

$$\left(\frac{\partial y}{\partial t}\right)_{i,j+1/2} \approx \frac{y_{i,j+1} - y_{i,j}}{\Delta t} \tag{12.163}$$

$$\left(\frac{\partial^2 y}{\partial x^2}\right)_{i,j+1/2} \approx \frac{1}{2}\left[\frac{y_{i-1,j+1} - 2y_{i,j+1} + y_{i+1,j+1}}{(\Delta x)^2} + \frac{y_{i-1,j} - 2y_{i,j} + y_{i+1,j}}{(\Delta x)^2}\right] \tag{12.164}$$

$$\left(\frac{\partial y}{\partial x}\right)_{i,j+1/2} \approx \frac{1}{2}\left[\frac{y_{i+1,j+1} - y_{i-1,j+1}}{2(\Delta x)} + \frac{y_{i+1,j} - y_{i-1,j}}{2(\Delta x)}\right] \tag{12.165}$$

The first equation comes from Eq. 12.149, the second from Eq. 12.153, and the last is the Crank–Nicolson analog for the first spatial derivative, similar in form to the second spatial derivative Eq. 12.153 derived earlier.

We have dealt with the derivative terms in Eq. 12.162, let us now turn to dealing with $f(y)$ and $f'(y)$. Evaluating these at the discrete point i and time $(j + 1/2)$, we have

$$f(y)_{i,j+1/2} = f(y_{i,j+1/2}) \qquad (12.166)$$

and

$$f'(y)_{i,j+1/2} = f'(y_{i,j+1/2}) \qquad (12.167)$$

The complication we have here is the unknown $y_{i,j+1/2}$ appearing in the above two equations. One simple way of resolving this problem is to use the known values of y at the jth time; that is,

$$f(y)_{i,j+1/2} \approx f(y_{i,j}) \qquad (12.168a)$$

$$f'(y)_{i,j+1/2} \approx f'(y_{i,j}) \qquad (12.168b)$$

then substitute these equations together with the equations for the derivatives (Eqs. 12.163 to 12.165) into Eq. 12.162 and use the information on the boundary values (Eqs. 12.128b, c) to obtain a set of equations of tridiagonal matrix form. Solving this using the Thomas algorithm will yield values of y at the $(j + 1)$th time. Note, however, that to obtain this solution, $f(y)$ and $f'(y)$ were approximated using values of y at the jth time. To improve the solution, we use the better approximation for $f(y)$ and $f'(y)$ by applying the averaging formulas

$$f(y)_{i,j+1/2} \approx f\left(\frac{y_{i,j} + y_{i,j+1}^*}{2}\right) \qquad (12.169a)$$

$$f'(y)_{i,j+1/2} \approx f'\left(\frac{y_{i,j} + y_{i,j+1}^*}{2}\right) \qquad (12.169b)$$

where $y_{i,j+1}^*$ is obtained from the last approximation. We can, of course, repeat this process with the iteration scheme to solve for $y^{(k+1)}$ at the $(k + 1)$th iteration

$$f(y)_{i,j+1/2} \approx f\left(\frac{y_{i,j} + y_{i,j+1}^{(k)}}{2}\right) \qquad (12.170)$$

with

$$y_{i,j+1}^{(0)} = y_{i,j} \qquad (12.171)$$

The iteration process can be stopped when some convergence criterion is satisfied. The following one is commonly used to stop the iteration sequence

$$\left| \frac{y_{i,j+1}^{(k+1)} - y_{i,j+1}^{(k)}}{y_{i,j+1}^{(k+1)}} \right| < \varepsilon \qquad \text{for all } i$$

where ε is some predetermined small value.

There are a number of variations of such methods in the literature. They basically differ in the way the nonlinear coefficients are evaluated. For example, for the function

$$f(y)_{i,j+1/2} = f(y_{i,j+1/2}) \qquad (12.172)$$

the value of $y_{i,j+1/2}$ in the arguement can be found by applying the finite difference formula to the equation with a time step of $\Delta t/2$, and using $f(y_{i,j})$ and $f'(y_{i,j})$ to approximate $f(y)$ and $f'(y)$, respectively. Other variations can be found in Douglas (1961) and Von Rosenberg (1969). There exists considerable scope for student-generated averaging techniques, of the type shown previously.

12.3.6   Elliptic Equations

Treatment of elliptic equations is essentially the same as that for parabolic equations. To show this, we demonstrate the technique with the following example wherein two finite spatial variables exist.

EXAMPLE 12.7

Consider a rectangle catalyst of two dimensions in which a first order chemical reaction is taking place. The nondimensional equation is

$$\frac{\partial^2 y}{\partial x^2} + \frac{\partial^2 y}{\partial z^2} - \phi^2 y = 0 \qquad (12.173a)$$

where ϕ^2 is the typical Thiele modulus, and $y = C/C_0$ with C_0 being the bulk concentration.

The boundary conditions to this problem are assumed to take the form

$$x = 0; \qquad \frac{\partial y}{\partial x} = 0 \qquad (12.173b)$$

$$x = 1; \qquad y = 1 \qquad (12.173c)$$

$$z = 0; \qquad \frac{\partial y}{\partial z} = 0 \qquad (12.173d)$$

$$z = 1; \qquad y = 1 \qquad (12.173e)$$

Due to the symmetry of the catalyst, we consider only the positive quadrant (i.e., x and z are positive). Let us divide the domains x and z into N intervals. The coordinate of the x domain is $i(\Delta x)$ and that of the z domain is $j(\Delta z)$, where

$$\Delta x = \frac{1}{N} = h; \qquad \Delta z = \frac{1}{N} = h$$

Next, we evaluate the differential equation at the point (i, j)

$$\left(\frac{\partial^2 y}{\partial x^2}\right)_{i,j} + \left(\frac{\partial^2 y}{\partial z^2}\right)_{i,j} - \phi^2 y_{i,j} = 0 \qquad (12.174)$$

for $i, j = 1, 2, \ldots, N - 1$.

Using the approximation formula for the second derivative (Eq. 12.117), Eq. 12.174 will become

$$\left(y_{i-1,j} - 2y_{i,j} + y_{i+1,j}\right) + \left(y_{i,j-1} - 2y_{i,j} + y_{i,j+1}\right) - h^2\phi^2 y_{i,j} = 0 \quad (12.175)$$

One method for solving this set of equations is the Alternate-Direction-Implicit method. The basic idea of this approach utilizes the fact that the elliptic PDE (Eq. 12.173a) is the steady-state solution of the associated transient partial differential equation

$$\frac{\partial y}{\partial t} = \alpha \left[\frac{\partial^2 y}{\partial x^2} + \frac{\partial^2 y}{\partial z^2} - \phi^2 y\right] \qquad (12.176)$$

where α is some arbitrary constant.

Now, let us use the finite difference to approximate the previous parabolic PDE. Suppose that we know values of y at the time t_k and wish to use the finite difference approach to find their values at the next time t_{k+1}. The time derivative $\partial y / \partial t$ evaluated at the time t_{k+1} is approximated by

$$\left(\frac{\partial y}{\partial t}\right)_{i,j}^{(k+1)} = \frac{y_{i,j}^{(k+1)} - y_{i,j}^{(k)}}{\Delta t} \qquad (12.177a)$$

where the subscripts i and j denote the positions in the x and z domains, respectively, whereas the upperscript k represents the time t_k.

If we approximate the second order spatial derivatives $\partial^2 y / \partial x^2$ at the time $(k + 1)$ by the values of y at the time t_{k+1} and the other derivative $\partial^2 y / \partial z^2$ by the values of y at the time t_k, we have

$$\left(\frac{\partial^2 y}{\partial x^2}\right)_{i,j}^{(k+1)} = \frac{y_{i-1,j}^{(k+1)} - 2y_{i,j}^{(k+1)} + y_{i+1,j}^{(k+1)}}{(\Delta x)^2} \qquad (12.177b)$$

$$\left(\frac{\partial^2 y}{\partial z^2}\right)_{i,j}^{(k)} = \frac{y_{i,j-1}^{(k)} - 2y_{i,j}^{(k)} + y_{i,j+1}^{(k)}}{(\Delta z)^2} \qquad (12.177c)$$

The purpose of the last step is to evaluate the value of y in the x direction while assuming that the spatial derivative in z is approximated by the values of discrete y at the old time t_k.

The reaction term $\phi^2 y$ can be approximated by the arithmetic mean between the values at time t_k and t_{k+1}; that is,

$$\phi^2 y = \frac{\phi^2}{2}\left(y_{i,j}^{(k)} + y_{i,j}^{(k+1)}\right) \qquad (12.177d)$$

Substituting Eqs. 12.177 into Eq. 12.176, we have

$$\rho\left[y_{i,j}^{(k+1)} - y_{i,j}^{(k)}\right] = \left(y_{i-1,j}^{(k+1)} - 2y_{i,j}^{(k+1)} + y_{i+1,j}^{(k+1)}\right)$$

$$+ \left(y_{i,j-1}^{(k)} - 2y_{i,j}^{(k)} + y_{i,j+1}^{(k)}\right) \qquad (12.178a)$$

$$- \frac{\phi^2 h^2}{2}\left(y_{i,j}^{(k+1)} + y_{i,j}^{(k)}\right)$$

where

$$\rho = \frac{h^2}{\alpha \Delta t}; \qquad h = \Delta x = \Delta z \qquad (12.178b)$$

Rearranging Eq. 12.178a by grouping all terms at time t_{k+1} to one side and terms at time t_k to the other side, we then have

$$y_{i-1,j}^{(k+1)} - \left(2 + \rho + \frac{1}{2}h^2\phi^2\right)y_{i,j}^{(k+1)} + y_{i+1,j}^{(k+1)}$$

$$\qquad (12.179a)$$

$$= -y_{i,j-1}^{(k)} + \left(2 - \rho + \frac{1}{2}h^2\phi^2\right)y_{i,j}^{(k)} - y_{i,j+1}^{(k)}$$

Now, if we apply Eq. 12.176 at the next time t_{k+2} and approximate the derivatives as

$$\left(\frac{\partial y}{\partial t}\right)_{i,j}^{(k+2)} = \frac{y_{i,j}^{(k+2)} - y_{i,j}^{(k+1)}}{\Delta t}$$

$$\left(\frac{\partial^2 y}{\partial x^2}\right)_{i,j}^{(k+1)} = \frac{y_{i-1,j}^{(k+1)} - 2y_{i,j}^{(k+1)} + y_{i+1,j}^{(k+1)}}{(\Delta x)^2}$$

$$\left(\frac{\partial^2 y}{\partial z^2}\right)_{i,j}^{(k+2)} = \frac{y_{i,j-1}^{(k+2)} - 2y_{i,j}^{(k+2)} + y_{i,j+1}^{(k+2)}}{(\Delta z)^2}$$

In opposite to what we did in the last time step, here we approximate the spatial derivative in z at the time t_{k+2}, whereas the spatial derivative in x is approximated using old values at the time t_{k+1}.

The reaction term is approximated by the arithmetic mean between values at t_{k+1} and t_{k+2}; that is,

$$\phi^2 y = \frac{\phi^2}{2}\left(y_{i,j}^{(k+1)} + y_{i,j}^{(k+2)}\right)$$

Substituting these approximation formulas into Eq. 12.176, we obtain the equation (after some algebraic manipulation)

$$y_{i,j-1}^{(k+2)} - \left(2 + \rho + \frac{1}{2}h^2\phi^2\right)y_{i,j}^{(k+2)} + y_{i,j+1}^{(k+2)}$$
$$= -y_{i-1,j}^{(k+1)} + \left(2 - \rho + \frac{1}{2}h^2\phi^2\right)y_{i,j}^{(k+1)} - y_{i+1,j}^{(k+1)} \qquad (12.179b)$$

Equations 12.179a and b are iteration equations known as the *Peaceman–Rachford Alternating-Direction-Implicit* scheme (1955). Basically, this iteration scheme assumes an initial guess of all points and the process is started by keeping j constant and applying Eq. 12.179a to find the next estimate along the ith coordinate. This is achieved with the Thomas algorithm since the matrix is tridiagonal. Knowing this, we apply Eq. 12.179b to find the next estimate along the jth coordinate and the process continues until the method provides convergent solutions. Because of the alternation between directions, the method was given the name Alternating-Direction-Implicit method.

Note the definition of the parameter ρ in Eq. 12.178b. Since our main concern is to solve elliptic equation, that is, the steady-state solution to Eq. 12.176, the solution can be obtained effectively by choosing a sequence of parameters ρ_k (Peaceman and Rachford 1955). The optimum sequence of this parameter is only found for simple problems. For other problems, it is suggested that this optimum sequence is obtained by numerical experiments. However, if the scalar parameter ρ is kept constant, the method has been proved to converge for all values of ρ. Interested readers should refer to Wachspress and Habetler (1960) for further exposition of this method.

Dealing with nonlinear elliptic equations is more difficult than linear equations, as one would expect. All schemes proposed to handle nonlinear elliptic equations are of the iterative type. Douglas (1959) has proposed a useful way to deal with nonlinearities.

Table 12.2 Collocation Formulas

Variables	N + 1 interpolation points N interior points & 1 boundary point		N + 2 interpolation points N interior points & 2 boundary points	Equation number
	Boundary point at x = 0	Boundary point at x = 1		
Interpolation points	$x_0 = 0;\ x_1, x_2, \ldots, x_N$ are root of $J_N^{(\alpha,\beta)}(x) = 0$	$x_1, x_2, \ldots, x_N$ are root of $J_N^{(\alpha,\beta)}(x) = 0$; $x_{N+1} = 1$	$x_0 = 0, x_1, x_2, \ldots, x_N$ are root of $J_N^{(\alpha,\beta)}(x) = 0$; $x_1,$ and $x_{N+1} = 1$	12.180a
Number of points	$N + 1$		$N + 2$	12.180b
Lagrangian interpolation polynomial	$y_N(x) = \sum_{j=1}^{N+1} l_j(x) y_j$		$y_{N+1}(x) = \sum_{j=1}^{N+2} l_j(x) y_j$	12.180c
Lagrangian building blocks	$l_j(x) = \prod_{\substack{i=1 \\ i\neq j}}^{N+1} \frac{(x-x_i)}{(x_j - x_i)}$		$l_j(x) = \prod_{\substack{i=1 \\ i\neq j}}^{N+2} \frac{(x-x_i)}{(x_j - x_i)}$	12.180d
Derivatives	$\dfrac{dy_N(x_i)}{dx} = \sum_{j=1}^{N+1} A_{ij} y_j;\quad \dfrac{d^2 y_N(x_i)}{dx^2} = \sum_{j=1}^{N+1} B_{ij} y_j$		$\dfrac{dy_{N+1}(x_i)}{dx} = \sum_{j=1}^{N+2} A_{ij} y_j;\quad \dfrac{d^2 y_{N+1}(x_i)}{dx^2} = \sum_{j=1}^{N+2} B_{ij} y_j$	12.180e
Matrices A & B	$A_{ij} = \dfrac{dl_j(x_i)}{dx};\quad B_{ij} = \dfrac{d^2 l_j(x_i)}{dx^2};\quad i,j = 1,2,\ldots, N+1$		$A_{ij} = \dfrac{dl_j(x_i)}{dx};\quad B_{ij} = \dfrac{d^2 l_j(x_i)}{dx^2};\quad i,j = 1,2,\ldots, N+2$	12.180f
Desired integral	$\int_0^1 [x^{\beta'}(1-x)^{\alpha'}] y_N(x)\, dx \approx \sum_{j=1}^{N+1} w_j y_j$		$\int_0^1 [x^{\beta'}(1-x)^{\alpha'}] y_{N+1}(x)\, dx \approx \sum_{j=1}^{N+2} w_j y_j$	12.180g
Quadrature weights	$w_j = \int_0^1 [x^{\beta'}(1-x)^{\alpha'}] l_j(x)\, dx$		$w_j = \int_0^1 [x^{\beta'}(1-x)^{\alpha'}] l_j(x)\, dx$	12.180h
α and β of the Jacobian polynomial	$\alpha = \alpha';\quad \beta = \beta' + 1$	$\alpha = \alpha' + 1;\quad \beta = \beta'$	$\alpha = \alpha' + 1;\quad \beta = \beta' + 1$	12.180i

12.4 ORTHOGONAL COLLOCATION FOR SOLVING PDEs

In Chapters 7 and 8, we presented numerical methods for solving ODEs of initial and boundary value type. The method of orthogonal collocation discussed in Chapter 8 can be also used to solve PDEs. For elliptic PDEs with two spatial domains, the orthogonal collocation is applied on both domains to yield a set of algebraic equations, and for parabolic PDEs the collocation method is applied on the spatial domain (domains if there are more than one) resulting in a set of coupled ODEs of initial value type. This set can then be handled by the methods provided in Chapter 7.

We will illustrate the application of orthogonal collocation to a number of examples. Elliptic PDEs will be dealt with first and typical parabolic equations occurring in chemical engineering will be considered next.

Before we start with orthogonal collocation, it is worthwhile to list in Table 12.2 a number of key formulas developed in Chapter 8, since they will be needed in this section. Table 12.2 shows the various formula for the few basic properties of the orthogonal collocation method.

We can use Table 12.2 as follows. Suppose the number of collocation points is chosen as $N + 2$, where N is the number of the interior interpolation points and 2 represents two boundary points. The Lagrangian interpolation polynomial is thus defined as in Eq. 12.180c with the building blocks $l_j(x)$ given in Eq. 12.180d. The first and second derivatives at interpolation points are given in Eq. 12.180e, and the integral with the weighting function $x^{\beta'}(1 - x)^{\alpha'}$ is given in Eq. 12.180g. The optimal parameters α and β for the Jacobi polynomial are given in Eq. 12.180i. Thus, on a practical level, this table will prove quite useful.

12.4.1 Elliptic PDE

We now show the application of the orthogonal collocation method to solve an elliptic partial differential equation.

EXAMPLE 12.8

The level 4 of modeling the cooling of a solvent bath using cylindrical rods presented in Chapter 1 gave rise to an elliptic partial differential equation. These equations were derived in Chapter 1, and are given here for completeness.

$$\frac{1}{\xi} \frac{\partial}{\partial \xi} \left(\xi \frac{\partial u}{\partial \xi} \right) + \Delta^2 \frac{\partial^2 u}{\partial \zeta^2} = 0 \tag{12.181}$$

$$\xi = 0; \qquad \frac{\partial u}{\partial \xi} = 0 \tag{12.182a}$$

$$\xi = 1; \qquad \frac{\partial u}{\partial \xi} = -\text{Bi} \cdot u \tag{12.182b}$$

$$\zeta = 0; \qquad u = 1 \tag{12.182c}$$

$$\zeta = 1; \qquad \frac{\partial u}{\partial \zeta} = 0 \tag{12.182d}$$

This set of equations has been solved analytically in Chapter 11 using the finite integral transform method. Now, we wish to apply the orthogonal collocation method to investigate a numerical solution. First, we note that the problem is symmetrical in ξ at $\xi = 0$ and as well as in ζ at $\zeta = 1$. Therefore, to make full use of the symmetry properties, we make the transformations

$$y = (1 - \zeta)^2 \qquad \text{and} \qquad z = \xi^2 \tag{12.183}$$

which is a transformation we have consistently employed when dealing with problems having symmetry. With this transformation, Eq. 12.181 then becomes

$$4z \frac{\partial^2 u}{\partial z^2} + 4 \frac{\partial u}{\partial z} + \Delta^2 \left[4y \frac{\partial^2 u}{\partial y^2} + 2 \frac{\partial u}{\partial y} \right] = 0 \tag{12.184}$$

The boundary conditions at $\xi = 0$ and $\zeta = 1$ are not needed, owing to the transformation. Hence, the remaining boundary conditions for Equation 12.184 written in terms of y and z variables are

$$z = 1; \qquad \frac{2}{\text{Bi}} \frac{\partial u}{\partial z} + u = 0 \tag{12.185a}$$

and

$$y = 1; \qquad u = 1 \tag{12.185b}$$

Now we have two new spatial domains, z and y. The first step in the orthogonal collocation scheme is to choose interpolation points. We shall choose N interior collocation points in the z domain and M points in the y domain. Thus, the total number of interpolation points in the z domain is $N + 1$, including the point at $z = 1$, and that in the y domain is $M + 1$, including the point at $y = 1$.

If we use the index i to describe the ith point in the z domain and k to denote the kth point in the y domain, the heat balance equation (Eq. 12.184) must be satisfied at the point (i, k) for $i = 1, 2, \ldots, N$ and $k = 1, 2, \ldots, M$ (i.e., the interior points). Evaluating Eq. 12.184 at the (i, k) point, we have

$$4z_i \frac{\partial^2 u}{\partial z^2}\bigg|_{i,k} + 4 \frac{\partial u}{\partial z}\bigg|_{i,k} + \Delta^2 \left[4y_k \frac{\partial^2 u}{\partial y^2}\bigg|_{i,k} + 2 \frac{\partial u}{\partial y}\bigg|_{i,k} \right] = 0 \tag{12.186}$$

for $i = 1, 2, \ldots, N$ and $k = 1, 2, \ldots, M$.

Using the approximation formula for the first and second order derivative equations (Eq. 12.180e), we have

$$\left.\frac{\partial u}{\partial z}\right|_{i,k} = \sum_{j=1}^{N+1} Az(i,j)u(j,k) \qquad (12.187a)$$

$$\left.\frac{\partial^2 u}{\partial z^2}\right|_{i,k} = \sum_{j=1}^{N+1} Bz(i,j)u(j,k) \qquad (12.187b)$$

$$\left.\frac{\partial u}{\partial y}\right|_{i,k} = \sum_{l=1}^{M+1} Ay(k,l)u(i,l) \qquad (12.187c)$$

$$\left.\frac{\partial^2 u}{\partial y^2}\right|_{i,k} = \sum_{l=1}^{M+1} By(k,l)u(i,l) \qquad (12.187d)$$

where **Az** and **Bz** are first and second order derivative matrices in the z domain, and **Ay** and **By** are derivative matrices in the y domain. Note again that once the interpolation points are chosen in the z and y domains, these derivative matrices are known.

Substituting Eqs. 12.187 into Eq. 12.186, we have

$$\sum_{j=1}^{N+1} Cz(i,j)u(j,k) + \Delta^2 \sum_{l=1}^{M+1} Cy(k,l)u(i,l) = 0 \qquad (12.188a)$$

where

$$Cz(i,j) = 4z_i Bz(i,j) + 4Az(i,j) \qquad \text{and}$$

$$Cy(k,l) = 4y_k By(k,l) + 4Ay(k,l) \qquad (12.188b)$$

The last terms of the two series contain the boundary values. By taking them out of the series, we have

$$\sum_{j=1}^{N} Cz(i,j)u(j,k) + Cz(i,N+1)u(N+1,k)$$

$$+ \Delta^2 \left[\sum_{l=1}^{M} Cy(k,l)u(i,l) + Cy(k,M+1)u(i,M+1) \right] = 0 \qquad (12.189)$$

Note that the boundary condition (Eq. 12.185b) gives

$$u(i, M+1) = 1 \qquad (12.190)$$

and from the boundary condition (Eq. 12.185a), we have

$$\left.\frac{2}{Bi}\frac{\partial u}{\partial z}\right|_{N+1,k} + u(N+1,k) = 0 \qquad (12.191)$$

Using the first derivative formula (Eq. 12.180e) at the boundary point ($i = N + 1$), the previous equation can be written in terms of the discrete values u given as

$$\frac{2}{\text{Bi}} \sum_{j=1}^{N+1} \text{Az}(N + 1, j)u(j, k) + u(N + 1, k) = 0 \qquad (\mathbf{12.192})$$

from which $u(N + 1, k)$ can be solved directly as

$$u(N + 1, k) = \frac{-\dfrac{2}{\text{Bi}} \displaystyle\sum_{j=1}^{N} \text{Az}(N + 1, j)u(j, k)}{\left[1 + \dfrac{2}{\text{Bi}}\text{Az}(N + 1, N + 1)\right]} \qquad (\mathbf{12.193})$$

Substitute Eqs. 12.190 and 12.193 into Eq. 12.189 to yield

$$\sum_{j=1}^{N} D(i, j)u(j, k) + \Delta^2\left[\sum_{l=1}^{M} \text{Cy}(k, l)u(i, l) + \text{Cy}(k, M + 1)\right] = 0 \quad (\mathbf{12.194})$$

for $i = 1, 2, \ldots, N$ and $k = 1, 2, \ldots, M$, where the matrix D is given by

$$D(i, j) = \text{Cz}(i, j) - \frac{\left[\dfrac{2}{\text{Bi}}\text{Cz}(i, N + 1)\text{Az}(N + 1, j)\right]}{\left[1 + \dfrac{2}{\text{Bi}}\text{Az}(N + 1, N + 1)\right]} \qquad (\mathbf{12.195})$$

for $i = 1, 2, \ldots, N$ and $k = 1, 2, \ldots, M$.

Equation 12.194 represents $M \times N$ coupled algebraic equations with the same number of unknown $u(i,k)$, $i = 1, 2, \ldots, N$ and $k = 1, 2, \ldots, M$. These equations can be solved using any of the algebraic solution methods described in Appendix A. Before doing this, we will introduce the global indexing scheme such that the variable u with two indices (because of the two coordinates) is mapped into a single vector with one counting index. This is done for the purpose of programming and subsequent computation. We define a new variable **Y** as

$$Y[(k - 1)N + i] = u(i, k) \qquad (\mathbf{12.196})$$

Equation 12.194 then becomes

$$\sum_{j=1}^{N} D(i, j)Y[(k - 1)N + j]$$

$$+ \Delta^2\left[\sum_{l=1}^{M} \text{Cy}(k, l)Y[(l - 1)N + i] + \text{Cy}(k, M + 1)\right] = 0 \qquad (\mathbf{12.197})$$

for $i = 1, 2, \ldots, N$ and $k = 1, 2, \ldots, M$.

For example, if we choose $N = M = 2$, the unknown vector **Y** of Eq. 12.196 is

$$\begin{bmatrix} Y(1) \\ Y(2) \\ Y(3) \\ Y(4) \end{bmatrix} = \begin{bmatrix} u(1,1) \\ u(2,1) \\ u(1,2) \\ u(2,2) \end{bmatrix} \tag{12.198}$$

and Eq. 12.197 written in component form is

$$D(1,1)Y(1) + D(1,2)Y(2) + \Delta^2[Cy(1,1)Y(1)$$
$$+ Cy(1,2)Y(3) + Cy(1,3)] = 0$$
$$D(2,1)Y(1) + D(2,2)Y(2) + \Delta^2[Cy(1,1)Y(2)$$
$$+ Cy(1,2)Y(4) + Cy(1,3)] = 0$$
$$D(1,1)Y(3) + D(1,2)Y(4) + \Delta^2[Cy(2,1)Y(1)$$
$$+ Cy(2,2)Y(3) + Cy(2,3)] = 0$$
$$D(2,1)Y(3) + D(2,2)Y(4) + \Delta^2[Cy(2,1)$$
$$+ Cy(2,2)Y(4) + Cy(2,3)] = 0$$

$$\tag{12.199}$$

Since our original elliptic equation is linear, the resulting discretized equation Eq. 12.199 is also linear. Therefore, if we define

$$\mathbf{E} = \begin{bmatrix} D(1,1) + \Delta^2 Cy(1,1) & D(1,2) & \Delta^2 Cy(1,2) & 0 \\ D(2,1) & D(2,2) + \Delta^2 Cy(1,1) & 0 & \Delta^2 Cy(1,2) \\ \Delta^2 Cy(2,1) & 0 & D(1,1) + \Delta^2 Cy(2,2) & D(1,2) \\ 0 & \Delta^2 Cy(2,1) & D(2,1) & D(2,2) + \Delta^2 Cy(2,2) \end{bmatrix} \tag{12.200a}$$

and

$$\mathbf{b} = \begin{bmatrix} -\Delta^2 Cy(1,3) \\ -\Delta^2 Cy(1,3) \\ -\Delta^2 Cy(2,3) \\ -\Delta^2 Cy(2,3) \end{bmatrix} \tag{12.200b}$$

Equation 12.199 can be written in compact vector format

$$\mathbf{EY} = \mathbf{b} \tag{12.201}$$

from which the unknown vector **Y** can be readily obtained by using matrix algebra. The solution for **Y** is

$$\mathbf{Y} = \mathbf{E}^{-1}\mathbf{b}$$

12.4.2 Parabolic PDE: Example 1

We saw in the last example for the elliptic PDE that the orthogonal collocation was applied on two spatial domains (sometime called double collocation). Here, we wish to apply it to a parabolic PDE. The heat or mass balance equation used in Example 11.3 (Eq. 11.55) is used to demonstrate the technique. The difference between the treatment of parabolic and elliptic equations is significant. The collocation analysis of parabolic equations leads to coupled ODEs, in contrast to the algebraic result for the elliptic equations.

EXAMPLE 12.9

The nondimensional heat or mass balance equations are

$$\frac{\partial y}{\partial \tau} = \frac{1}{x}\frac{\partial}{\partial x}\left(x\frac{\partial y}{\partial x}\right) \qquad (12.202a)$$

$$\tau = 0; \qquad y = 0 \qquad (12.202b)$$

$$x = 0; \qquad \frac{\partial y}{\partial x} = 0 \qquad (12.202c)$$

$$x = 1; \qquad y = 1 \qquad (12.202d)$$

The quantity of interest is the mean concentration or temperature, which is calculated from the integral

$$I = 2\int_0^1 xy\,dx \qquad (12.203)$$

We note that this problem is symmetrical at $x = 0$. Therefore, the application of the symmetry transformation

$$u = x^2 \qquad (12.204)$$

is appropriate. In terms of this new independent variable $u = x^2$, the mass (heat) balance equations (Eqs. 12.202) become

$$\frac{\partial y}{\partial \tau} = 4u\frac{\partial^2 y}{\partial u^2} + 4\frac{\partial y}{\partial u} \qquad (12.205a)$$

$$u = 1; \qquad y = 1 \qquad (12.205b)$$

The mean concentration or temperature I in terms of the new independent u is

$$I = \int_0^1 y\,du \qquad (12.206)$$

The weighting function of the above integral is

$$W(u) = u^{\beta'}(1 - u)^{\alpha'}; \qquad \alpha' = \beta' = 0 \tag{12.207}$$

Therefore, to use the Radau quadrature with the exterior point ($u = 1$) included, the N interior collocation points are chosen as roots of the Jacobi polynomial $J_N^{(\alpha, \beta)}$ with $\alpha = 1$ and $\beta = 0$. Once $N + 1$ interpolation points are chosen, the first and second order derivative matrices are known.

Evaluating Eq. 12.205a at the interior collocation point i gives

$$\frac{\partial y_i}{\partial \tau} = 4u_i \frac{\partial^2 y}{\partial u^2}\bigg|_i + 4\frac{\partial y}{\partial u}\bigg|_i \tag{12.208}$$

for $i = 1, 2, \ldots, N$.

The first and second order derivatives at the collocation point i are given by the formula

$$\frac{\partial y}{\partial u}\bigg|_i = \sum_{j=1}^{N+1} A_{ij} y_j; \qquad \frac{\partial^2 y}{\partial u^2}\bigg|_i = \sum_{j=1}^{N+1} B_{ij} y_j \tag{12.209}$$

where **A** and **B** are given in Eqs. 12.180f.

When we substitute Eq. 12.209 into Eq. 12.208, this equation is obtained

$$\frac{dy_i}{d\tau} = \sum_{j=1}^{N+1} C_{ij} y_j \tag{12.210}$$

where the matrix **C** is defined as

$$C_{ij} = 4u_i B_{ij} + 4A_{ij} \tag{12.211}$$

Next, we take the last term of the series out of the summation and make use of the boundary condition, $y(1) = y_{N+1} = 1$, and obtain the equation

$$\frac{dy_i}{d\tau} = \sum_{j=1}^{N} C_{ij} y_j + C_{i, N+1} \tag{12.212}$$

for $i = 1, 2, 3, \ldots, N$.

The above equation represents N coupled ODEs and they are readily solved by using the methods discussed in Chapter 7, such as the Runge–Kutta–Gill method. Once the concentration y is obtained from the integration of Eq. 12.212, the mean concentration or temperature I is calculated from the quadrature formula

$$I = \sum_{j=1}^{N+1} w_j y_j \tag{12.213}$$

where the quadrature weights are given in Eq. 12.180h.

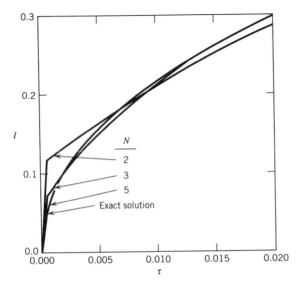

Figure 12.12 Plots of I versus τ for $N = 2, 3, 5$, and 10.

Figure 12.12 shows plots of the mean concentration or temperature versus τ for $N = 2, 3, 5$, and 10. The plots for $N = 5$ and 10 practically superimpose the exact solution given in Eq. 11.87. This example shows the simplicity of the collocation method, and the number of collocation points needed for many problems is usually less than ten. A larger number of points may be needed to handle sharp gradients in the mass transfer zone.

12.4.3   Coupled Parabolic PDE: Example 2

EXAMPLE 12.10

To further demonstrate the simplicity and the straightforward nature of the orthogonal collocation method, we consider the adsorption problem dealt with in Section 12.2 where the singular perturbation approach was used. The nondimensional mass balance equations are

$$\sigma\mu\frac{\partial y}{\partial \tau} = \mu\frac{\partial^2 y}{\partial x^2} - y\left(1 - y_\mu\right) \qquad (12.214a)$$

$$\mu\frac{\partial y_\mu}{\partial \tau} = y\left(1 - y_\mu\right) \qquad (12.214b)$$

subject to the nondimensional initial and boundary conditions

$$\tau = 0; \qquad y = y_\mu = 0 \qquad (12.214c)$$

$$x = 0; \quad \frac{\partial y}{\partial x} = 0 \quad \text{and} \quad x = 1; \quad y = 1 \qquad (12.214d)$$

Again, we note the symmetry condition at $x = 0$ and make the symmetry transformation $u = x^2$. In terms of this new variable, the mass balance equations are

$$\sigma\mu\frac{\partial y}{\partial \tau} = \mu\left[4u\frac{\partial^2 y}{\partial u^2} + 2\frac{\partial y}{\partial u}\right] - y(1 - y_\mu) \qquad (\mathbf{12.215}a)$$

$$\mu\frac{\partial y_\mu}{\partial \tau} = y(1 - y_\mu) \qquad (\mathbf{12.215}b)$$

The initial and boundary conditions are

$$\tau = 0; \qquad y = y_\mu = 0 \qquad (\mathbf{12.216})$$

$$u = 1; \qquad y = 1 \qquad (\mathbf{12.217})$$

Our objective here is to obtain the mean adsorbed concentration versus time. In nondimensional form, this mean concentration is given by

$$\bar{y}_\mu = \int_0^1 y_\mu \, dx \qquad (\mathbf{12.218})$$

Written in terms of the u variable, the above equation becomes

$$\bar{y}_\mu = \frac{1}{2}\int_0^1 u^{-1/2} y_\mu \, du \qquad (\mathbf{12.219})$$

The weighting function in the above integral is $W(u) = u^{-1/2}(1 - u)^0$, that is, $\alpha' = 0$ and $\beta' = -1/2$. Therefore, the N interior collocation point must be chosen as the root of the Nth degree Jacobi polynomial $J_N^{(1,-1/2)}$ (see Eq. 12.180i) and the $(N + 1)$th point is the point $u = 1$.

The mass balance equation (Eqs. 12.215a) is valid in the interior of the domain; thus, we evaluate it at the ith interior collocation point ($i = 1, 2, 3, \ldots, N$) and obtain

$$\sigma\mu\frac{\partial y(i)}{\partial \tau} = \mu\left[4u(i)\frac{\partial^2 y}{\partial u^2}\bigg|_i + 2\frac{\partial y}{\partial u}\bigg|_i\right] - y(i)\left[1 - y_\mu(i)\right] \qquad (\mathbf{12.220})$$

for $i = 1, 2, 3, \ldots, N$.

Equation 12.215b, on the other hand, is valid at all points including the point at the boundary; that is, $i = N + 1$. Evaluating this equation at the interpolation point i, we have

$$\mu\frac{\partial y_\mu(i)}{\partial \tau} = y(i)\left[1 - y_\mu(i)\right] \qquad (\mathbf{12.221})$$

for $i = 1, 2, 3, \ldots, N, N + 1$.

Using the following formula for the first and second derivatives

$$\frac{\partial y}{\partial u}\bigg|_i = \sum_{j=1}^{N+1} A(i,j)y(j) \tag{12.222a}$$

$$\frac{\partial^2 y}{\partial u^2}\bigg|_i = \sum_{j=1}^{N+1} B(i,j)y(j) \tag{12.222b}$$

and substituting them into Eqs. 12.220, we obtain

$$\sigma\mu\frac{\partial y(i)}{\partial\tau} = \mu\sum_{j=1}^{N+1} C(i,j)y(j) - y(i)\big[1 - y_\mu(i)\big] \tag{12.223}$$

for $i = 1, 2, \ldots, N$, where

$$C(i,j) = 4u(i)B(i,j) + 2A(i,j) \tag{12.224}$$

Taking the last term of the series in Eq. 12.223 out of the summation sign, we obtain

$$\sigma\mu\frac{\partial y(i)}{\partial\tau} = \mu\left[\sum_{j=1}^{N} C(i,j)y(j) + C(i,N+1)y(N+1)\right] - y(i)\big[1 - y_\mu(i)\big] \tag{12.225}$$

From the boundary condition (Eq. 12.217), we have

$$y(N+1) = 1 \tag{12.226}$$

Using this information, Eq. 12.225 becomes

$$\sigma\mu\frac{\partial y(i)}{\partial\tau} = \mu\left[\sum_{j=1}^{N} C(i,j)y(j) + C(i,N+1)\right] - y(i)\big[1 - y_\mu(i)\big] \tag{12.227}$$

The equation for the adsorbed species (Eq. 12.221) can be written for interior points ($i = 1, 2, \ldots, N$)

$$\mu\frac{\partial y_\mu(i)}{\partial\tau} = y(i)\big[1 - y_\mu(i)\big] \tag{12.228}$$

and for the boundary point ($i = N + 1$)

$$\mu\frac{\partial y_\mu(N+1)}{\partial\tau} = y(N+1)\big[1 - y_\mu(N+1)\big] \tag{12.229}$$

But since $y(N + 1) = 1$, the above equation then becomes:

$$\mu \frac{\partial y_\mu(N + 1)}{\partial \tau} = \left[1 - y_\mu(N + 1)\right] \qquad (12.230)$$

Equations 12.227 together with 12.228 and 12.230 represent $2N + 1$ equations and we have exactly the same number of unknowns, $y(1), y(2), \ldots, y(N)$, $y_\mu(1)$, $y_\mu(2), \ldots, y_\mu(N), y_\mu(N + 1)$.

If we now introduce a new vector $\mathbf{Y}$, a new global indexing scheme evolves for the purpose of programming as

$$Y(i) = y(i) \qquad \text{for} \qquad i = 1, 2, \ldots, N \qquad (12.231a)$$

$$Y(N + i) = y_\mu(i) \qquad \text{for} \qquad i = 1, 2, \ldots, N, N + 1 \qquad (12.231b)$$

Equations 12.227, 12.228, and 12.230 then become

$$\sigma\mu \frac{\partial Y(i)}{\partial \tau} = \mu \left[\sum_{j=1}^{N} C(i, j)Y(j) + C(i, N + 1)\right] - Y(i)\left[1 - Y(N + i)\right] \qquad (12.232a)$$

$$\mu \frac{\partial Y(N + i)}{\partial \tau} = Y(i)\left[1 - Y(N + i)\right] \qquad (12.232b)$$

for $i = 1, 2, 3, \ldots, N$, and

$$\mu \frac{\partial Y(2N + 1)}{\partial \tau} = \left[1 - Y(2N + 1)\right] \qquad (12.232c)$$

Equations 12.232 can be solved by any of the numerical schemes presented in Chapter 7. After the vector $\mathbf{Y}$ is known, the mean concentration defined in Equation 12.219 can be written as the quadrature

$$\bar{y}_\mu = \frac{1}{2} \sum_{j=1}^{N+1} w(j)y_\mu(j) \qquad (12.233)$$

where $w(j)$ are known quadrature weights.

Plots of the mean concentration versus time for $\sigma = 0.001$ and $\mu = 0.01$ and 0.1 are shown in Fig. 12.10 along with a comparison of the singular perturbation solution.

12.5   ORTHOGONAL COLLOCATION ON FINITE ELEMENTS

The previous section showed how straightforward the orthogonal collocation can be when solving partial differential equations, particularly parabolic and elliptic equations. We now present a variation of the orthogonal collocation method, which is useful in solving problems with a sharp variation in the profiles.

The method taught in Chapter 8 (as well as in Section 12.4) can be applied over the whole domain of interest $[0, 1]$ (any domain $[a, b]$ can be easily transformed into $[0, 1)$), and it is called the global orthogonal collocation method. A variation of this is the situation where the domain is split into many subdomains and the orthogonal collocation is then applied on each subdomain. This is particularly useful when dealing with sharp profiles and, as well, it leads to reduction in storage for efficient computer programming.

In this section, we will deal with the orthogonal collocation on finite elements, that is, the domain is broken down into a number of subdomains (called elements) and the orthogonal collocation is then applied on each element. At the junctions of these elements, we impose the obvious physical conditions on the continuity of concentration (or temperature) and the continuity of mass flux (or heat flux).

We will demonstrate the orthogonal collocation on finite element by using the example of diffusion and reaction in a catalyst, a problem discussed in Chapter 8. When the reaction rate is very high compared to diffusion, the concentration profile in the particle is very sharp, and if the traditional orthogonal collocation method is applied as we did in Chapter 8, a large number of collocation points is required to achieve a reasonable accuracy. Using this approach in problems having sharp gradients can be a very expensive exercise because of the excessive number of collocation points needed.

To alleviate this problem of sharp gradients, we present in this section the orthogonal collocation on finite elements, where the domain is broken down into many subdomains and orthogonal collocation is then applied on each element. This flexibility will allow us to *concentrate collocation points in the region where the sharp gradient* is expected. In regions where the gradients are shallow we need only a few points. This new method is called the orthogonal collocation on finite elements.

EXAMPLE 12.11

It is useful to illustrate collocation on finite elements by treating the diffusion and reaction problem described in Chapter 8. The governing equations written in nondimensional form are

$$\frac{d^2 y}{dx^2} - \phi^2 y = 0 \qquad (12.234a)$$

subject to

$$x = 0; \qquad \frac{dy}{dx} = 0 \qquad (12.234b)$$

$$x = 1; \qquad y = 1 \qquad (12.234c)$$

The first step is to split the domain $[0, 1]$ into many subdomains. For the sake of demonstration, we use only two subdomains. Let w be the point that splits the domain $[0, 1]$ into two subdomains $[0, w]$ and $[w, 1]$. Next, we denote y_1 to

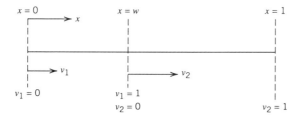

Figure 12.13 Two elements in the domain $[0, 1]$.

describe y in the first domain, and y_2 is the solution in the second subdomain. Of course, y_1 and y_2 must satisfy the mass balance equation (Eq. 12.234a), which is valid at all points within the domain $[0, 1]$. Before we apply the orthogonal collocation in each subdomain, we must normalize the domains to have a range of $[0, 1]$. This can be easily achieved by denoting a new coordinate for the subdomain 1 as v_1 defined as

$$v_1 = \frac{x}{w} \tag{12.235}$$

Similarly, we denote a new coordinate for the subdomain 2 as v_2 and it is defined as

$$v_2 = \frac{x - w}{1 - w} \tag{12.236}$$

Thus, we see immediately that v_1 and v_2 range between 0 and 1 in their respective domains (see Fig. 12.13).

In terms of v_1, the mass balance equations in the subdomain 1 are (noting that $dv_1/dx = 1/w$)

$$\frac{d^2 y_1}{dv_1^2} - w^2 \phi^2 y_1 = 0 \tag{12.237}$$

subject to the condition at the center of the particle Eq. 12.234b

$$v_1 = 0; \qquad \frac{dy_1}{dv_1} = 0 \tag{12.238}$$

Similarly, the mass balance equations in the subdomain 2 written in terms of v_2 are [also noting that $dv_2/dx = 1/(1 - w)$]

$$\frac{d^2 y_2}{dv_2^2} - (1 - w)^2 \phi^2 y_2 = 0 \tag{12.239}$$

subject to the condition at the exterior surface of the particle Eq. 12.234c

$$v_2 = 1; \qquad y_2 = 1 \tag{12.240}$$

Equations 12.237 and 12.239 are two second order differential equations, and therefore for the complete formulation we must have four conditions. Equations 12.238 and 12.240 provide two, and hence, we require two more conditions. These are obtained by invoking the continuity of concentration and mass flux at the junction of the two subdomains, that is,

$$y_1(x = w^-) = y_2(x = w^+) \tag{12.241}$$

and

$$\frac{dy_1(x = w^-)}{dx} = \frac{dy_2(x = w^+)}{dx} \tag{12.242}$$

Written in terms of their respective variables v_1 and v_2, we have

$$y_1(v_1 = 1) = y_2(v_2 = 0) \tag{12.243}$$

and

$$\frac{1}{w}\left[\frac{dy_1}{dv_1}\right]_{v_1=1} = \frac{1}{(1-w)}\left[\frac{dy_2}{dv_2}\right]_{v_2=0} \tag{12.244}$$

Thus, the complete formulation of equations is given in Eqs. 12.237 to 12.240, 12.243, and 12.244. The quantity of interest is the nondimensional reaction rate, defined as

$$I = \int_0^1 y\, dx \tag{12.245}$$

When written in terms of variables appropriate for the two subdomains, we have

$$I = w\int_0^1 y_1\, dv_1 + (1-w)\int_0^1 y_2\, dv_2 \tag{12.246}$$

We are now ready to apply the orthogonal collocation to each subdomain. Let us start with the first subdomain. Let N be the number of *interior* collocation points; hence, the total number of collocation points will be $N + 2$, including the points $v_1 = 0$ (center of particle) and $v_1 = 1$ (junction point between the two subdomains). The mass balance equation (Eq. 12.237) will be valid only for the interior collocation points, that is,

$$\left[\frac{d^2y_1}{dv_1^2}\right]_i - w^2\phi^2 y_1(i) = 0 \qquad \text{for} \qquad i = 2, 3, \ldots, N + 1 \tag{12.247}$$

where $y_1(i)$ is the value of y_1 at the collocation point i. The first term on the LHS is the second order derivative at the collocation point i. From Eq. 12.180e, this second order derivative can be written as

$$\left[\frac{d^2 y_1}{dv_1^2}\right]_i = \sum_{j=1}^{N+2} B1(i,j) y_1(j) \qquad (12.248)$$

where **B1** is the second order derivative matrix for the subdomain 1 having a dimension of $(N + 2, N + 2)$. Remember that this matrix is fixed once the collocation points are known. Eq. 12.247 then becomes

$$\sum_{j=1}^{N+2} B1(i,j) y_1(j) - w^2 \phi^2 y_1(i) = 0 \qquad \text{for} \qquad i = 2, 3, \ldots, N + 1 \quad (12.249)$$

The boundary condition at $v_1 = 0$ (Eq. 12.238) is rewritten as

$$\left[\frac{dy_1}{dv_1}\right]_{v_1=0} = 0 \qquad (12.250)$$

Using the formula for the first derivative (Eq. 12.180e), this equation becomes

$$\sum_{j=1}^{N+2} A1(1,j) y_1(j) = 0 \qquad (12.251)$$

where the first order derivative matrix **A1** has a dimension of $(N + 2, N + 2)$. Again, just like the second order derivative matrix **B1**, this matrix is also known once the collocation points are chosen.

Before we consider the collocation analysis of the junction point, $x = w$, ($v_1 = 1$ or $v_2 = 0$), we consider the collocation analysis of the equations in the subdomain 2. If we choose M (M can be different from N used for the subdomain 1) as the number of interior collocation points of the subdomain 2, the total number of interpolation points will be $M + 2$ including the two end points ($v_2 = 0$ the junction point, and $v_2 = 1$ the exterior surface of the particle). Evaluating the mass balance equation (Eq. 12.239) at the interior collocation point i ($i = 2, 3, \ldots, M + 1$), we obtain

$$\left[\frac{d^2 y_2}{dv_2^2}\right]_i - (1 - w)^2 \phi^2 y_2(i) = 0 \qquad (12.252)$$

Using the second order derivative formula (Eq. 12.180e), the first term can be written in terms of the second order derivative matrix **B2** $(M + 2, M + 2)$. Equation 12.252 then becomes

$$\sum_{j=1}^{M+2} B2(i,j) y_2(j) - (1 - w)^2 \phi^2 y_2(i) = 0 \qquad \text{for} \qquad i = 2, 3, \ldots, M + 1$$

$$(12.253)$$

The boundary condition at the exterior particle surface Eq. 12.240 is

$$y_2(M + 2) = 1 \tag{12.254}$$

Now we turn to the conditions at the junction of the two subdomains. The first condition of continuity of concentration (Eq. 12.243) is written in terms of the collocation variables as

$$y_1(N + 2) = y_2(1) \tag{12.255}$$

And the second condition of continuity of flux becomes

$$\frac{1}{w}\left[\frac{dy_1}{dv_1}\right]_{N+2} = \frac{1}{(1-w)}\left[\frac{dy_2}{dv_2}\right]_1 \tag{12.256}$$

that is, the LHS is evaluated at the last collocation point of the first subdomain, whereas the second term is evaluated at the first point of the second subdomain. Using the first order derivative formula (Eq. 12.180e), the previous equation becomes

$$\frac{1}{w}\sum_{j=1}^{N+2} A1(N + 2, j)y_1(j) = \frac{1}{(1-w)}\sum_{j=1}^{M+2} A2(1, j)y_2(j) \tag{12.257}$$

where $A1(N + 2, N + 2)$ and $A2(M + 2, M + 2)$ are the first order derivative matrices of the first and second subdomains, respectively.

Thus, we have now completed the orthogonal collocation treatment of finite elements. We have an equal number of equations and unknown variables. Let us now rewrite our collocation equations (Eqs. 12.249, 12.251, 12.253, 12.254, and 12.256) in the following order:

$$\sum_{j=1}^{N+2} A1(1, j)y_1(j) = 0 \tag{12.258a}$$

$$\sum_{j=1}^{N+2} B1(i, j)y_1(j) - w^2\phi^2 y_1(i) = 0 \qquad \text{for} \qquad i = 2, 3, \ldots, N + 1$$

$$\tag{12.258b}$$

$$\frac{1}{w}\sum_{j=1}^{N+2} A1(N + 2, j)y_1(j) = \frac{1}{(1-w)}\sum_{j=1}^{M+2} A2(1, j)y_2(j) \tag{12.258c}$$

$$\sum_{j=1}^{M+2} B2(i, j)y_2(j) - (1 - w)^2\phi^2 y_2(i) = 0 \qquad \text{for} \qquad i = 2, 3, \ldots, M + 1$$

$$\tag{12.258d}$$

$$y_2(M + 2) = 1 \tag{12.258e}$$

Note, the continuity condition of concentration at the junction, $y_1(N + 2) = y_2(1)$, is not listed as part of Eqs. 12.158, and we shall see that this is taken care of in the global indexing scheme.

Since we have two sets of variables, $y_1 = [y_1(1), y_1(2), \ldots, y_1(N + 2)]^T$ and $y_2 = [y_2(1), y_2(2), \ldots, y_2(M + 2)]^T$, it is convenient to define a global indexing scheme, which maps these two sets into one vector set and also is convenient for programming, as follows:

$$Y(i) = y_1(i) \qquad \text{for} \qquad i = 1, 2, \ldots, N + 2 \qquad (12.259a)$$

$$Y(N + 1 + i) = y_2(i) \qquad \text{for} \qquad i = 1, 2, \ldots, M + 2 \quad (12.259b)$$

With this definition of the new vector **Y**, the concentrations for $y_1(N + 2)$ and $y_2(1)$ at the junction are mapped into the same variable $Y(N + 2)$. This means that the continuity of concentration condition at the junction (Eq. 12.255) is *automatically satisfied* by this mapping. The vector **Y** has $N + M + 3$ components, that is, $N + M + 3$ unknowns.

With the new definition of vector **Y**, Eqs. 12.258 can be written as

$$\sum_{j=1}^{N+2} A1(1, j)Y(j) = 0 \qquad (12.260a)$$

$$\sum_{j=1}^{N+2} B1(i, j)Y(j) - w^2\phi^2 Y(i) = 0 \quad \text{for} \quad i = 2, 3, \ldots, N + 1 \quad (12.260b)$$

$$\frac{1}{w} \sum_{j=1}^{N+2} A1(N + 2, j)Y(j) = \frac{1}{(1 - w)} \sum_{j=1}^{M+2} A2(1, j)Y(N + 1 + j) \quad (12.260c)$$

$$\sum_{j=1}^{M+2} B2(i, j)Y(N + 1 + j) \quad - (1 - w)^2 \phi^2 Y(N + 1 + i) = 0$$
$$(12.260d)$$

$$\text{for} \quad i = 2, 3, \ldots, M + 1$$

$$Y(N + M + 3) = 1 \qquad (12.260e)$$

Equations 12.260 represent $N + M + 3$ equations, and we have exactly the same number of unknown variables.

In general, the set of Eqs. 12.260 is a set of nonlinear algebraic equations, which can be solved by the Newton's method taught in Appendix A. For the present problem, since the governing equations are linear (Eq. 12.234), the resulting discretized equations (Eqs. 12.260) are also linear and we will show, as follows, how these linear equations can be cast into the matrix–vector format. We must first arrange Eqs. 12.260 into the familiar form of summation from 1 to $N + M + 3$ because our unknown vector **Y** has a dimension of $N + M + 3$.

We first shift the index j in the second series in Eq. 12.260c and the first series of Eq. 12.260d by $N + 1$ as

$$\sum_{j=1}^{M+2} A2(1, j)Y(N + 1 + j) = \sum_{j=N+2}^{N+M+3} A2(1, j - N - 1)Y(j) \quad \textbf{(12.261)}$$

$$\sum_{j=1}^{M+2} B2(i, j)Y(N + 1 + j) = \sum_{j=N+2}^{N+M+3} B2(i, j - N - 1)Y(j) \quad \textbf{(12.262)}$$

for $i = 2, 3, \ldots, M + 1$.

The idea of this shift is to bring the counting index for **Y** into the format of $Y(j)$ instead of $Y(j + N + 1)$. With this shift in j, Eqs. 12.260c and 12.260d then become

$$\frac{1}{w} \sum_{j=1}^{N+2} A1(N + 2, j)Y(j) - \frac{1}{(1 - w)} \sum_{j=N+2}^{N+M+3} A2(1, j - N - 1)Y(j) = 0$$

$$\textbf{(12.263)}$$

and

$$\sum_{j=N+2}^{N+M+3} B2(i, j - N - 1)Y(j) - (1 - w)^2 \phi^2 Y(N + 1 + i) = 0 \quad \textbf{(12.264)}$$

for $i = 2, 3, \ldots, M + 1$.

Now we note that Eq. 12.260a is written for $i = 1$ (the first point of the first subdomain), Eq. 12.260b is for $i = 2, 3, \ldots, N + 1$ (the N interior points in the first subdomain), Eq. 12.263 is for $i = N + 2$ (the junction point). To convert Eq. 12.264 into the new index $i = N + 3, N + 4, \ldots, N + M + 2$ (i.e., the M interior points for the second subdomain) for the purpose of continuous counting of the index i, we need to shift the index i of Eq. 12.264 by $N + 1$, and as a result we have the equation

$$\sum_{j=N+2}^{N+M+3} B2(i - N - 1, j - N - 1)Y(j) - (1 - w)^2 \phi^2 Y(i) = 0 \quad \textbf{(12.265)}$$

for $i = N + 3, N + 4, \ldots, N + M + 2$.

Thus Eq. 12.265 is valid for $i = N + 3, N + 4, \ldots, N + M + 2$ (for the M interior collocation points in the second subdomain). Finally, Eq. 12.260e is valid for $i = N + M + 3$, the last point of the second subdomain. Figure 12.14 shows the setup of all collocation equations.

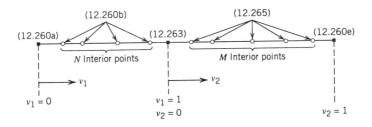

Figure 12.14 Two subdomains and the distribution of equations

Now the final format of equations that we wish to solve is

For $i = 1$,

$$\sum_{j=1}^{N+2} A1(1, j)Y(j) = 0 \qquad (12.266a)$$

For $i = 2, 3, \ldots, N + 1$,

$$\sum_{j=1}^{N+2} B1(i, j)Y(j) - w^2\phi^2 Y(i) = 0 \qquad (12.266b)$$

For $i = N + 2$,

$$\frac{1}{w} \sum_{j=1}^{N+2} A1(N + 2, j)Y(j) - \frac{1}{(1 - w)} \sum_{j=N+2}^{N+M+3} A2(1, j - N - 1)Y(j) = 0$$

$$(12.266c)$$

For $i = N + 3, N + 4, \ldots, N + M + 2$,

$$\sum_{j=N+2}^{N+M+3} B2(i - N - 1, j - N - 1)Y(j) - (1 - w)^2\phi^2 Y(i) = 0 \quad (12.266d)$$

For $i = N + M + 3$,

$$Y(N + M + 3) = 1 \qquad (12.266e)$$

Note the continuous counting of the index in Eq. 12.266a to 12.266e.
If we now define a new function $H(i)$ as

$$H(i) = \begin{cases} 0 & \text{if } i \le 0 \\ 1 & \text{if } i > 0 \end{cases} \qquad (12.267)$$

Equation 12.266a can be now written with a summation having the index j running from 1 to $N + M + 3$ instead of from 1 to $N + 2$.

For $i = 1$,

$$\sum_{j=1}^{N+M+3} A1(1, j) H(N + 3 - j) Y(j) = 0 \qquad (\mathbf{12.268\,}a)$$

The idea for this is to convert Eqs. 12.266 into the full vector–matrix format, that is, the index of **Y** is j and the counting of this index runs from 1 to $N + M + 3$, the full dimension of the unknown matrix **Y**. Likewise, we can do this for Eqs. 12.266b to e, and obtain

For $i = 2, 3, \ldots, N + 1$,

$$\sum_{j=1}^{N+M+3} \left[B1(i, j) - w^2 \phi^2 \delta(i, j) \right] H(N + 3 - j) Y(j) = 0 \quad (\mathbf{12.268\,}b)$$

For $i = N + 2$,

$$\sum_{j=1}^{N+M+3} \left[\frac{1}{w} A1(N + 2, j) H(N + 3 - j) \right.$$

$$\left. - \frac{1}{(1 - w)} A2(1, j - N - 1) H(j - N - 1) \right] Y(j) = 0 \quad (\mathbf{12.268\,}c)$$

For $i = N + 3, N + 4, \ldots, N + M + 2$,

$$\sum_{j=1}^{N+M+3} \left[B2(i - N - 1, j - N - 1) - (1 - w)^2 \phi^2 \delta(i, j) \right]$$

$$H(j - N - 1) Y(j) = 0 \qquad (\mathbf{12.268\,}d)$$

For $i = N + M + 3$,

$$Y(N + M + 3) = 1 \qquad (\mathbf{12.268\,}e)$$

where $\delta(i, j)$ is the *Kronecker delta function* and is defined as

$$\delta(i, j) = \begin{cases} 0 & \text{if} & i \neq j \\ 1 & \text{if} & i = j \end{cases} \qquad (\mathbf{12.269})$$

With the summation now carried from 1 to $N + M + 3$, we can see clearly Eqs. 12.268 can be readily converted into the simple vector–matrix format

$$\mathbf{DY} = \mathbf{F} \qquad (\mathbf{12.270})$$

where

$$\mathbf{Y} = \left[Y(1), Y(2), Y(3), \ldots, Y(N + M + 3) \right]^T \qquad (\mathbf{12.271\,}a)$$

$$\mathbf{F} = [0, 0, 0, \ldots, 1]^T \qquad (\mathbf{12.271\,}b)$$

and the matrix **D** of dimension $(N + M + 3, N + M + 3)$ is given by

For $i = 1$,

$$D(1, j) = A1(1, j) H(N + 3 - j) \tag{12.272a}$$

For $i = 2, 3, \ldots, N + 1$,

$$D(i, j) = \left[B1(i, j) - w^2 \phi^2 \delta(i, j) \right] H(N + 3 - j) \tag{12.272b}$$

For $i = N + 2$,

$$D(N + 2, j) = \frac{1}{w} A1(N + 2, j) H(N + 3 - j)$$

$$- \frac{1}{(1 - w)} A2(1, j - N - 1) H(j - N - 1) \tag{12.272c}$$

For $i = N + 3, N + 4, \ldots, N + M + 2$,

$$D(i, j) = \left[B2(i - N - 1, j - N - 1) - (1 - w)^2 \phi^2 \delta(i, j) \right] H(j - N - 1) \tag{12.272d}$$

For $i = N + M + 3$,

$$D(N + M + 3, j) = H(j - N - M - 2) \tag{12.272e}$$

The linear equation (Eq. 12.272) can be readily solved by matrix inversion to find the solution for the vector **Y**. Knowing this vector **Y**, the quantity of interest I is calculated in Eq. 12.246. It is written in terms of the quadrature as

$$I = w \sum_{j=1}^{N+2} w_1(j) y_1(j) + (1 - w) \sum_{j=1}^{M+2} w_2(j) y_2(j) \tag{12.273}$$

Written in terms of **Y**, we have

$$I = w \sum_{i=1}^{N+2} w_1(j) Y(j) + (1 - w) \sum_{j=1}^{M+2} w_2(j) Y(N + 1 + j) \tag{12.274}$$

This completes the analysis on finite elements.

If we wished to write the matrix D, we choose the number of interior collocation point in the subdomain 1 as $N = 1$ and that in the second subdomain as $M = 2$. The matrix $\mathbf{D}$ is then given by

$$\mathbf{D} = \begin{bmatrix} A1(1,1) & A1(1,2) & A1(1,3) & 0 & 0 & 0 \\ B1(2,1) & B1(2,2) - w^2\phi^2 & B1(2,3) & 0 & 0 & 0 \\ A1(3,1)/w & A1(3,2)/w & A1(3,3)/w - A2(1,1)/(1-w) & -A2(1,2)/(1-w) & -A2(1,3)/(1-w) & -A2(1,4)/(1-w) \\ 0 & 0 & B2(2,1) & B2(2,2) - (1-w)^2\phi^2 & B2(2,3) & B2(2,4) \\ 0 & 0 & B2(3,1) & B2(3,2) & B2(3,3) - (1-w)^2\phi^2 & B2(3,4) \\ 0 & 0 & 0 & 0 & 0 & 1 \end{bmatrix}$$

$$(\mathbf{12.275})$$

with $\mathbf{Y}$ and $\mathbf{F}$ given by

$$\mathbf{Y} = \begin{bmatrix} Y(1) \\ Y(2) \\ Y(3) \\ Y(4) \\ Y(5) \\ Y(6) \end{bmatrix} \quad \text{and} \quad \mathbf{F} = \begin{bmatrix} 0 \\ 0 \\ 0 \\ 0 \\ 0 \\ 1 \end{bmatrix} \qquad (\mathbf{12.276})$$

Figure 12.15 shows plots of the concentration profile (y vs x) using the orthogonal collocation on finite elements method for $\phi = 100$ (very fast reaction). It is seen that due to the very fast reaction relative to diffusion, the profile is very sharp and is concentrated near the surface of the catalyst. Also, on the same figure are plots of the nondimensional chemical reaction rate, I (Eq. 12.245), versus the number of the interior collocation point of the second subdomain (M). The number of the interior collocation point in the first domain is chosen as 1 since the profile in that subdomain is very flat and close to zero. The location w between the two subdomains is the varying parameter in

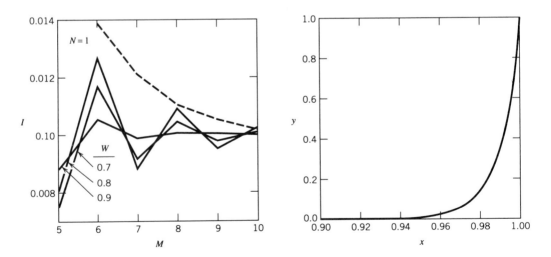

Figure 12.15 Plots if I versus the number of interior collocation number M, and the concentration profile versus x.

Fig. 12.15. Also shown in the figure are plots of the numerical solution using the global orthogonal collocation method (Example 8.4), shown as a dashed line. The exact solution for the nondimensional reaction rate is $\tanh(\phi)/\phi = 0.01$. It is seen in the figure that the results using the orthogonal collocation on finite elements with $w = 0.9$ agree well with the exact solution for $M = 6$. On the other hand, the global orthogonal collocation solution is of comparable accuracy only when the number of interior collocation points used is more than 10. It is noted that, however, that the global collocation method uses the transformation $u = x^2$ in the analysis. What this means is that the global collocation method uses only even polynomials in x, whereas the orthogonal collocation on finite elements use both odd and even polynomials, thus doubling the number of equations without increasing the accuracy. However, it is clear in this simple example where sharp gradients exist that the orthogonal collocation on finite elements is clearly more advantageous.

The application of the orthogonal collocation on finite elements is straightforward as we have illustrated in the above example of linear ordinary differential equation. The resulting set of discretized equations is a set of linear equations (Eqs. 12.260), which is amenable to matrix approach by using the global indexing procedure described from Eqs. 12.261 to 12.272. For nonlinear ordinary differential equations, the resulting set of discretized equations will be a set of nonlinear algebraic equations, which can be handled by the Newton's approach discussed in Appendix A.

The orthogonal collocation on finite elements can also be applied to partial differential equations as straightforward as we did in the last example for ODE. If the partial differential equations are linear, the resulting set of equations will be a set of coupled linear ordinary differential equations. On the other hand, if the equations are nonlinear, the discretized equations are coupled nonlinear ordinary differential equations. In either case, these sets of coupled ordinary differential equations can be soived effectively with integration solvers taught in Chapter 7. More of this can be found in Finlayson (1980).

12.6   REFERENCES

1. Cole, J. D., *Perturbation Methods in Applied Mathematics*, Blaisdell, Waltham, Massachusetts (1968).
2. Crank, J., and P. Nicolson, "A Practical Method for Numerical Evaluation of Solutions of Partial Differential Equations of the Heat Conduction Type," *Proc. Camb. Phil. Soc.*, 43, 50–67 (1947).
3. Do, D. D., and P. L. J. Mayfield. "A New Simplified Model for Adsorption in a Single Particle," *AIChE Journal*, 33, 1397–1400 (1987).
4. Douglas, J., Jr., "A Survey of Numerical Methods for Parabolic Differential Equations," in *Advances in Computers*, Vol. 6, edited by Franz L. Alt, Academic Press, New York (1961).
5. Douglas, J., Jr., D. W. Peaceman, and H. H. Rachford, Jr., "A Method for Calculating Multi-dimensional Immiscible Displacement," *J. Petrol. Trans.*, 216, 297 (1959).

6. Finlayson, B. *Nonlinear Analysis in Chemical Engineering*. McGraw Hill, New York (1980).

7. Michelsen, M., "A Fast Solution Technique for a Class of Linear Partial Differential Equations," *Chem. Eng. Journal*, 18, 59–65 (1979).

8. Peaceman, D. W. and Rachford, H. H., "The Numerical Solution of Parabolic and Elliptic Differential Equations," *J. Soc. Industrial Appl. Math.*, 3, 28–41 (1955).

9. Rice, R. G., "Approximate Solutions for Batch, Packed Tube and Radial Flow Adsorbers-Comparison with Experiment," *Chem. Eng. Sci.*, 37, 83 (1982).

10. Von Rosenberg, D. U., *Methods for the Numerical Solution of Partial Differential Equations*. Elsevier, New York (1969).

11. Wachspress, E. L., and Habetler, G. J., "An Alternating-Direction-Implicit Iteration Technique," *J. Soc. Industrial Appl. Math.*, 8, 403–424 (1960).

12.7 PROBLEMS

12.1₃. Apply the polynomial approximation technique, as in Example 12.1, for the cases:

(a) *Rectangular coordinates*

$$\frac{\partial y}{\partial \tau} = \frac{\partial^2 y}{\partial x^2}$$

(b) *Cylindrical coordinates*

$$\frac{\partial y}{\partial \tau} = \frac{1}{x} \frac{\partial}{\partial x}\left(x \frac{\partial y}{\partial x}\right)$$

with the same initial and boundary conditions as given in Eq. 12.1.

12.2₃. Replace the fixed boundary conditions in Example 12.1 with a boundary condition (12.1*d*) of the resistance type

$$x = 1; \qquad \frac{\partial y}{\partial x} = \text{Bi}(1 - y)$$

where $\text{Bi} = k_c R/D$ and as before $y = (C_0 - C)/(C_0 - C_B)$, and find the polynomial approximate solution.

12.3₃. The following equations describe a constant heat flux problem into a semi-infinite slab material

$$\frac{\partial y}{\partial \tau} = \frac{\partial^2 y}{\partial x^2}$$

$$\tau = 0; \qquad y = 0$$

$$x = 0; \qquad -\frac{\partial y}{\partial x} = 1$$

$$x \to \infty; \qquad \frac{\partial y}{\partial x} \to 0$$

The exact solution to this problem (obtained by Laplace transform) is

$$y(x, \tau) = 2\sqrt{\frac{\tau}{\pi}} \exp\left(-\frac{x^2}{4\tau}\right) - x \, \text{erfc}\left(\frac{x}{2\sqrt{\tau}}\right)$$

and the surface temperature is

$$y(0, \tau) = \sqrt{\frac{4\tau}{\pi}}$$

(a) Assuming that the penetration depth is $X(\tau)$ and the temperature and heat flux at the penetration front are zero (see Eqs. 12.24 and 12.25), show that the integral heat balance equation is

$$\int_0^{X(\tau)} y(x, \tau) \, dx = \tau$$

(b) Next, assume that the temperature profile is parabolic, then prove that the expression for the temperature distribution must be

$$y(x, \tau) = \frac{X(\tau)}{2}\left[1 - \frac{x}{X(\tau)}\right]^2$$

(c) Show that the approximate solution for the penetration depth is

$$X(\tau) = \sqrt{6\tau}$$

(d) Derive the solution for the temperature at $x = 0$ and compare your results with the exact solution.

12.4₃. Modeling the heating of a semi-infinite slab with a film resistance at the surface gives rise to the nondimensional equations

$$\frac{\partial y}{\partial \tau} = \frac{\partial^2 y}{\partial x^2}$$

$$\tau = 0; \qquad y = 0$$

$$x = 0; \qquad \frac{\partial y}{\partial x} = \text{Bi}_h(y - 1)$$

The exact solution to the problem is

$$y = \text{erfc}\left(\frac{x}{2\sqrt{\tau}}\right) - \exp\left(\text{Bi}_h x + \text{Bi}_h^2 \tau\right)\text{erfc}\left[\frac{x}{2\sqrt{\tau}} + \text{Bi}_h\sqrt{\tau}\right]$$

(a) Use the quadratic approximation for the temperature distribution, using penetration front $X(\tau)$, to find the approximate solution

$$y(x,\tau) \approx \frac{Bi_h X(\tau)}{2 + Bi_h X(\tau)} \left[1 - \frac{x}{X(\tau)} \right]$$

where the penetration X is determined from the implicit equation

$$(Bi_h X)^2 + 4Bi_h X - 8 \ln \left(1 + \frac{Bi_h X}{2} \right) = 12Bi_h^2 \tau$$

(b) Derive the approximate surface temperature and compare it with the exact solution.

12.5$_3$. Example 12.4 considers a problem of first order chemical reaction in a fixed bed. The boundary condition at the bed entrance does not account for the axial diffusion of mass. To account for this, the boundary conditions at the two ends should be (Danckwerts boundary conditions)

$$x = 0; \qquad \frac{dy}{dx} = 2(y - 1)$$

$$x = 1; \qquad \frac{dy}{dx} = 0$$

with the mass balance equation the same as Eq. 12.118.

(a) Apply suitable finite difference representation (second order correct for first and second derivatives, Eqs. 12.116c and 12.117) to show that the mass balance equation takes the form

$$\alpha y_{i-1} - \beta y_i - \gamma y_{i+1} = 0$$

where α, β, and γ are given in Eq. 12.120c.

(b) Use fictitious points one step in front of the bed and one step after the bed to show that the two boundary conditions written in finite difference form are

$$\frac{y_1 - y_{-1}}{2 \Delta x} \approx 2(y_0 - 1)$$

$$\frac{y_{N+1} - y_{N-1}}{2 \Delta x} \approx 0$$

where y_{-1} and y_{N+1} are discrete values of y of the points before and after the bed, respectively.

(c) Make use of the finite difference approximation for the boundary conditions to show that the finite difference equations describing the system behavior are

$$-(4\alpha \Delta x + \beta) y_0 + (\alpha - \gamma) y_1 = -4\alpha \Delta x$$

$$\alpha y_{i-1} - \beta y_i - \gamma y_{i+1} = 0 \qquad \text{for} \qquad i = 1, 2, \ldots, N - 1$$

and

$$(\alpha - \gamma)y_{N-1} - \beta y_N = 0$$

(d) Use the Thomas algorithm to solve the finite difference equations in part c.

12.6₃. Modeling of a second order chemical reaction in catalyst of slab geometry with constant physical and chemical properties will give rise to the equation

$$\frac{d^2y}{dx^2} - \phi^2 y^2 = 0$$

where ϕ^2 is the square of the Thiele modulus, which describes the relative strength between the reaction rate to the diffusion rate. Assume constant temperature conditions exist.

(a) Show that suitable variables for nondimensional analysis are

$$y = \frac{C}{C_0}; \quad x = \frac{r}{R}; \quad \phi^2 = \frac{kC_0R^2}{D_e}$$

where C_0 is the bulk fluid concentration, R is the half length of the catalyst, D_e is the effective diffusivity, and k is the chemical reaction rate constant per unit catalyst volume.

(b) If there is symmetry at the center and a stagnant film surrounds the catalyst, then show that the boundary conditions should take the form

$$x = 0; \quad \frac{dy}{dx} = 0$$

$$x = 1; \quad \frac{dy}{dx} = \text{Bi}(1 - y)$$

where $\text{Bi} = k_c R/D_e$.

(c) Apply finite differencing to the mass balance equation and write the approximation to the mass balance equation as

$$y_{i-1} - 2y_i + y_{i+1} - \phi^2(\Delta x)^2 y_i^2 \approx 0$$

(d) Similarly, show that the finite difference approximation of the two boundary conditions are

$$\frac{y_1 - y_{-1}}{2\Delta x} \approx 0$$

$$\frac{y_{N+1} - y_{N-1}}{2\Delta x} \approx \text{Bi}(1 - y_N)$$

where y_{-1} and y_{N+1} are discrete values at two fictitious points outside the domain.

(e) Eliminate the values of y at fictitious points to show that the final finite difference equations decribing the system are

$$-2y_0 - \phi^2(\Delta x)^2 y_0^2 + 2y_1 = 0$$

$$y_{i-1} - 2y_i - \phi^2(\Delta x)^2 y_i^2 + y_{i+1} = 0 \quad \text{for} \quad i = 1, 2, \ldots, N-1$$

$$2y_{N-1} - (2 + 2Bi\,\Delta x)y_N - \phi^2(\Delta x)^2 y_N^2 = -2Bi\,\Delta x$$

(f) For $\phi^2 = 10$, $Bi = 100$, and $N = 3$, the above set of four equations is a coupled set of nonlinear algebraic equations. Apply the Newton technique of Appendix A to solve for the discrete values of concentration.

12.7₃. It was demonstrated in the text that the number of equations to be solved using the method of finite difference depends on the type of boundary conditions. For example, if the functional values are specified at two end points of a boundary value problem, the number of equations to be solved is $N - 1$. On the other hand, if the functional value is specified at only one point and the other end involves a first derivative, then the number of equations is N. This homework problem will show that the variation in the number of equations can be avoided if the locations of the discrete point are shifted as

$$x_i = \left(i - \frac{1}{2}\right)\Delta x$$

instead of the conventional way of defining $x_i = i\Delta x$. With this new way of defining the point, the first point will be $\Delta x/2$ from the left boundary, and the Nth point will also be $\Delta x/2$ from the right boundary. Figure 12.16 shows the locations of all discrete points.

Consider the following set of equations

$$\frac{d^2 y}{dx^2} - 2\frac{dy}{dx} - 10y = 0$$

$$x = 0; \quad \frac{dy}{dx} = 2(y - 1)$$

$$x = 1; \quad \frac{dy}{dx} = 0$$

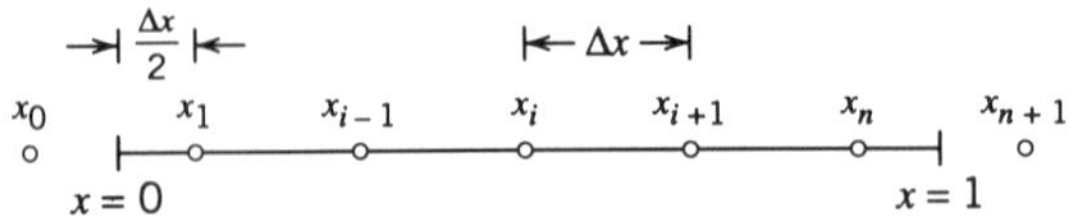

Figure 12.16

(a) Apply finite difference approximations to the governing equation to show that it will take the form

$$\alpha y_{i-1} - \beta y_i - \gamma y_{i+1} = 0$$

where

$$\alpha = \frac{1}{\Delta x} + \frac{1}{(\Delta x)^2}; \quad \beta = \frac{2}{(\Delta x)^2} + 10; \quad \gamma = \frac{1}{\Delta x} - \frac{1}{(\Delta x)^2}$$

(b) Next apply the finite difference to the first boundary condition (i.e., the condition at $x = 0$) to show that

$$\frac{y_1 - y_0}{\Delta x} \approx 2(y|_{x=0} - 1)$$

where y_0 is the value of y at the fictitious point x_0, which is $\Delta x/2$ outside the left boundary as shown in the above figure, and $y|_{x=0}$ is the value of y at the point $x = 0$. But note that there is no grid point at $x = 0$. To resolve this, we need to approximate this value based on values at two sides of the position $x = 0$. To approximate this value, we take the average of the values of y at the two sides of $x = 0$, that is,

$$y|_{x=0} \approx \frac{y_1 + y_0}{2}$$

Use this to substitute into the finite difference relation for the boundary condition at $x = 0$ to solve for y_0. Show that this excercise takes the form

$$y_0 \approx \left(\frac{1 - \Delta x}{1 + \Delta x} \right) y_1 + \frac{2 \Delta x}{1 + \Delta x}$$

(c) Apply the same procedure at in part b to the right boundary condition to show that

$$y_{N+1} \approx y_N$$

where y_{N+1} is the value of y at the fictitious point $\Delta x/2$ outside the right boundary.

(d) Insert the values of y_0 and y_{N+1} obtained in parts b and c into the finite difference analog of the governing equation to yield N equations with N unknowns $(y_1, y_2, \ldots, y_N)$. Solve the resulting N linear equations using the Thomas algorithm. Use $N = 3, 5$, and 10 and compare the approximate solutions with the exact solution.

This problem shows how useful the shifted position strategy is in solving differential equations. It does not matter what form the boundary conditions take, yet the number of equations will always be N. The only minor disadvantage of this procedure is that the values of y at the boundaries are not given directly from the solution of the finite difference equations. They must be determined from the average of the last interior value and the exterior value, as described in part b.

12.8$_2$. The *shifted position procedure* discussed in Problem 12.7 is also useful when dealing with problems with radial coordinates, such as problems having cylindrical and spherical geometries. To show this, solve the following problem of first order chemical reaction in a spherical catalyst having the dimensionless form

$$\frac{d^2y}{dx^2} + \frac{2}{x}\frac{dy}{dx} - 9y = 0$$

$$x = 0; \qquad \frac{dy}{dx} = 0$$

$$x = 1; \qquad y = 1$$

(a) Apply the second order correct analog to the derivatives of the governing equation to show that the finite difference approximation is

$$\frac{y_{i-1} - 2y_i + y_{i+1}}{(\Delta x)^2} + \frac{2}{x_i}\frac{y_{i+1} - y_{i-1}}{2\Delta x} - 9y_i \approx 0$$

where $x_i = (i - 1/2)\Delta x$.

(b) Next apply the finite difference approximation to the boundary conditions at two end points to show that

$$y_0 \approx y_1; \qquad \frac{y_{N+1} + y_N}{2} \approx 1$$

where y_0 and y_{N+1} are values of y at two fictitious points $\Delta x/2$ outside the boundaries.

(c) Solve the N resulting linear difference equations and compare it with the exact solution

$$y = \frac{1}{x}\frac{\cosh(3x)}{\cosh(3)}$$

12.9$_3$. Apply the conventional layout of the discrete points, that is, $x_i = i\Delta x$, to solve the linear parabolic partial differential equation

$$\frac{\partial y}{\partial t} = \frac{\partial^2 y}{\partial x^2}$$

subject to the boundary conditions

$$x = 0; \qquad y = 1$$

$$x = 1; \qquad \frac{dy}{dx} + 10y = 0$$

(a) Apply the second order correct analog to the second order spatial derivative, and the first order correct forward formula for the time derivative to show that the value of y at the position x_i and the time t_{j+1} is given explicitly as

$$y_{i,j+1} = y_{i,j} + \frac{\Delta t}{(\Delta x)^2}\left(y_{i-1,j} - 2y_{i,j} + y_{i+1,j}\right)$$

(b) The boundary condition at the left boundary is trivial. To handle the condition at the right boundary, apply the finite difference to show that

$$\frac{y_{N+1,j} - y_{N-1,j}}{2(\Delta x)} + 10y_{N,j} = 0$$

where y_{N+1} is the value of y at the fictitious point ouside the right boundary. Using this finite difference analog of the boundary condition at $x = 1$ to eliminate y_{N+1} from the finite difference equation, and hence finally obtain N equations for $y_{1,j+1}, y_{2,j+1}, \ldots, y_{N,j+1}$, which can be solved explicitly.

(c) Choose $N = 5$ (i.e., $\Delta x = 0.2$), and compute the solutions in part b with $\Delta t = 0.01, 0.02$, and 0.05. Discuss the results obtained.

12.10₃. Use the backward difference in the approximation of time derivative to solve Problem 12.9.

(a) Show that the finite difference analog to the governing equation becomes

$$y_{i-1,j+1} - \left[2 + \frac{(\Delta x)^2}{\Delta t}\right]y_{i,j+1} + y_{i+1,j+1} = -\frac{(\Delta x)^2}{\Delta t}y_{i,j}$$

that is, the values of y at the time t_{j+1} are written in terms of the values of y at t_j.

(b) Treat the boundary conditions the same way as in Problem 12.9, and prove that the final N equations will take the form of tridiagonal matrix.

(c) Use the Thomas algorithm to solve the finite difference equations for $\Delta x = 0.2$ and $\Delta t = 0.01, 0.02$ and 0.05. From the computed results, discuss the stability of the problem.

12.11₃. Solve Problem 12.9 using the Crank–Nicolson method, and show that the final N finite difference equations have the tridiagonal matrix form. Compute the results and discuss the stability of the simulations.

12.12*. This homework problem will illustrate the stability of the three different methods of approximating time derivatives in solving the problem

$$\frac{\partial y}{\partial t} = \frac{\partial^2 y}{\partial x^2}$$

subject to

$$x = 0; \quad y = 1 \quad \text{and} \quad x = 1; \quad y = 0$$

(a) Apply the Taylor series expansions to $y(x + \Delta x, t)$ and $y(x - \Delta x, t)$ around the point (x, t) and use the results of these two expansion to show that

$$\left(\frac{\partial^2 y}{\partial x^2}\right)_{x,t} = \frac{y(x + \Delta x, t) - 2y(x, t) + y(x - \Delta x, t)}{(\Delta x)^2}$$

$$- \frac{2}{4!}\left(\frac{\partial^4 y}{\partial x^4}\right)_{x,t}(\Delta x)^2 + \cdots$$

(b) Similarly, use the Taylor series expansion of $y(x, t + \Delta t)$ around the point (x, t) to show that

$$\left(\frac{\partial y}{\partial t}\right)_{x,t} = \frac{y(x, t + \Delta t) - y(x, t)}{\Delta t} - \frac{1}{2!}\left(\frac{\partial^2 y}{\partial t^2}\right)_{x,t}\Delta t + \cdots$$

(c) Substitute these equations in parts a and b into the governing equation, and use the notations

$$y(x, t) = y_{i,j}; \quad y(x + \Delta x, t) = y_{i+1,j}; \quad y(x, t + \Delta t) = y_{i,j+1}$$

to obtain the equation

$$\frac{y_{i,j+1} - y_{i,j}}{\Delta t} - \frac{1}{2!}\left(\frac{\partial^2 y}{\partial t^2}\right)_{i,j}\Delta t + \cdots$$

$$= \frac{y_{i-1,j} - 2y_{i,j} + y_{i+1,j}}{(\Delta x)^2} - \frac{2}{4!}\left(\frac{\partial^4 y}{\partial x^4}\right)_{i,j}(\Delta x)^2 + \cdots$$

(d) If z is the finite difference solution in approximating y; that is,

$$\frac{z_{i,j+1} - z_{i,j}}{\Delta t} = \frac{z_{i-1,j} - 2z_{i,j} + z_{i+1,j}}{(\Delta x)^2}$$

then show that the error of the finite difference solution, defined as $e = y - z$, is given by the equation

$$\frac{e_{i,j+1} - e_{i,j}}{\Delta t} - \frac{1}{2!}\left(\frac{\partial^2 y}{\partial t^2}\right)_{i,j}\Delta t + \cdots$$

$$= \frac{e_{i-1,j} - 2e_{i,j} + e_{i+1,j}}{(\Delta x)^2} - \frac{2}{4!}\left(\frac{\partial^4 y}{\partial x^4}\right)_{i,j}(\Delta x)^2 + \cdots$$

(e) Neglecting the small order terms in part d to show that the finite difference solution for the error e is

$$\frac{e_{i,j+1} - e_{i,j}}{\Delta t} = \frac{e_{i-1,j} - 2e_{i,j} + e_{i+1,j}}{(\Delta x)^2}$$

and also show that $e_{0,j} = e_{N,j} = 0$.

(f) By virtue of the fact that the finite difference equation and the governing equation have homogeneous boundary conditions, assume that the error will take the form

$$e_{i,j} = K_i T_j$$

This is identical to the separation of variables method described in Chapter 10. Substitute this into the finite difference equation for the error to show that

$$\frac{T_{j+1}}{T_j} = \frac{\Delta t}{(\Delta x)^2}\left[\frac{K_{i-1} - 2K_i + K_{i+1}}{K_i}\right] + 1$$

The LHS is a function of j (i.e., time), whereas the RHS is a function of i (i.e., x); therefore, the only possibility that these two can be equated is that they must equal to the same number. Let this number be λ. Hence, show that the equation for K will take the form

$$K_{i-1} - 2K_i + K_{i+1} + \frac{(\Delta x)^2}{\Delta t}(1 - \lambda)K_i = 0$$

with $K_0 = K_N = 0$.

(g) One possible solution that satisfies the homogeneous boundary conditions is

$$K_i = \sin(\pi p x_i)$$

Substitute this into the equation for K and finally show λ will be

$$\lambda = 1 - \frac{4\,\Delta t}{(\Delta x)^2}\sin^2\left(\frac{\pi p\,\Delta x}{2}\right)$$

(h) Now, since $T_{j+1}/T_j = \lambda$, hence, to prevent the error from exploding, the constraint on λ must be $\lambda < 1$; that is,

$$\left|1 - \frac{4\,\Delta t}{(\Delta x)^2}\sin^2\left(\frac{\pi p\,\Delta x}{2}\right)\right| < 1$$

from which show that the final constraint on the step sizes in time (so that the finite difference solution is bounded) will be

$$\frac{\Delta t}{(\Delta x)^2} < \frac{1}{2}$$

This is the stability criterion for the forward difference scheme in time.

12.13*. For the backward difference in time, follow the same procedure as in Problem 12.12 to show that the finite difference equation for the error in solving the same problem is

$$\frac{e_{i,j+1} - e_{i,j}}{\Delta t} = \frac{e_{i-1,j+1} - 2e_{i,j+1} + e_{i+1,j+1}}{(\Delta x)^2}$$

subject to $e_{0,j} = e_{N,j} = 0$.

(a) Assume $e_{i,j}$ takes the form

$$e_{i,j} = K_i T_j$$

and substitute this into the finite difference equation for the error to show that

$$\frac{T_{j+1}}{T_j} = \lambda$$

and

$$K_{i-1} - 2K_i + K_{i+1} + \frac{(1 - \lambda)(\Delta x)^2}{\lambda \, \Delta t} K_i = 0$$

(b) Let $K_i = \sin(\pi p x_i)$ and substitute it into the equation for K to show that λ will take the form

$$\lambda = \frac{1}{1 + \dfrac{4 \, \Delta t}{(\Delta x)^2} \sin^2\left(\dfrac{\pi p \, \Delta x}{2}\right)}$$

which is always less than 1. This means that the backward formula is always stable for any step size in time used.

12.14$_3$. Repeat Problem 12.13 for the Crank–Nicolson method to show that λ will take the form

$$\lambda = \frac{1 - 2\dfrac{\Delta t}{(\Delta x)^2} \sin^2\left(\dfrac{\pi p \, \Delta x}{2}\right)}{1 + 2\dfrac{\Delta t}{(\Delta x)^2} \sin^2\left(\dfrac{\pi p \, \Delta x}{2}\right)}$$

which is also always less than 1. Thus, the Crank–Nicolson method is always stable.

12.15*. The modeling of cooling a fluid in a laminar pipe flow was considered in Example 10.3. Neglecting the axial heat conduction relative to the convection term, the following heat balance equation describes the temperature change inside the pipe

$$2v_0\left[1 - \left(\frac{r}{R}\right)^2\right]\frac{\partial T}{\partial z} = \alpha\frac{1}{r}\frac{\partial}{\partial r}\left(r\frac{\partial T}{\partial r}\right)$$

where $\alpha = k/\rho C_p$ defines thermal diffusivity, v_0 is the mean velocity, R is the pipe radius, and r and z are radial and axial coordinates, respectively.

The initial and boundary conditions are given by

$$z = 0; \quad T = T_0$$

$$r = 0; \quad \frac{\partial T}{\partial r} = 0$$

$$r = R; \quad T = T_w$$

(a) Define the nondimensional variables as in Eq. 10.216 and then show that the governing equation will become

$$(1 - \xi^2)\frac{\partial T}{\partial \zeta} = \frac{1}{\xi}\frac{\partial}{\partial \xi}\left(\xi\frac{\partial T}{\partial \xi}\right)$$

(b) Show that the bulk mean temperature at any point along the pipe is given by

$$\overline{T}(\zeta) = 4\int_0^1 \xi(1 - \xi^2)T(\xi, \zeta)\, d\xi$$

(c) Note the symmetry condition at the center of the pipe, then use the usual transformation of $u = \xi^2$ to show that the governing equation will take the form

$$(1 - u)\frac{\partial T}{\partial \zeta} = 4u\frac{\partial^2 T}{\partial u^2} + 4\frac{\partial T}{\partial u}$$

and that the conditions will be

$$\zeta = 0; \quad T = T_0$$

$$u = 1; \quad T = T_w$$

(d) Apply the orthogonal collocation by assuming the function $y(\xi, \zeta)$ takes the form of Lagrangian interpolation polynomial

$$y(u, \zeta) = l_{N+1}(u)y(u_{N+1}, \zeta) + \sum_{j=1}^{N} l_j(u)y(u_j, \zeta)$$

with $N + 1$ interpolation points, where the first N points are zeros of the Jacobi polynomial $J_N^{(0,0)}(u) = 0$ and the $(N + 1)$th point is $u_{N+1} = 1$. The functions $l_j(u)$ are Nth degree Lagrangian polynomial, defined in Eq. 8.90.

Substitute this approximation polynomial into the governing equation and force the residual to be zero at the collocation points (u_1 to u_N) and show that the N discretized equations are

$$\frac{\partial T_i}{\partial \zeta} = C_{i,\,N+1}T_w + \sum_{j=1}^{N} C_{i,\,j}T_j \qquad \text{for} \qquad i = 1, 2, \ldots, N$$

where $C_{i,\,j}$ is defined as

$$C_{i,\,j} = \frac{4u_i B_{i,\,j} + 4A_{i,\,j}}{1 - u_i}$$

with A_{ij} and B_{ij} are first and second derivative matrices, respectively, defined in Eqs. 8.102 and 8.103.

(e) The N coupled ordinary differential equations have initial condition

$$\zeta = 0; \quad T_i = T_0 \quad \text{for} \quad i = 1, 2, \ldots, N$$

This set of equations is linear and is susceptible to linear analysis. Use the vector–matrix approach of Appendix B to show that the solution is

$$\mathbf{T} = T_0\mathbf{U} \exp{(\mathbf{K}\zeta)}\mathbf{U}^{-1}\mathbf{I}$$

where $\mathbf{I}$ is the identity matrix, $\mathbf{K}$ is the diagonal matrix with eigenvalues on the diagonal line, $\mathbf{U}$ is the matrix of eigenvectors of $\mathbf{C}$.

(f) Once T is found in part e, show that the mean bulk temperature is obtained from the Gauss–Jacobi quadrature

$$\overline{T}(\zeta) = \sum_{j=1}^{N} w_j\big[2(1 - u_j)T(u_j, \zeta)\big]$$

where w_j are Gauss–Jacobi quadrature weights.

This cooling of fluid in a pipe with wall heat transfer resistance was solved by Michelsen (1979) using the method of orthogonal collocation. This problem without the wall resistance is a special case of the situation dealt with by Michelsen, and is often referred to as the Graetz problem (see Example 10.3 and Problem 3.4).

12.16*. Example 10.2 considers the modeling of a CVD process in a parallel flat plate system. The entrance length problem was analytically dealt with by the method of combination of variables. Here, we assume that the chemical reaction at the plate follows a nonlinear reaction, and apply the orthogonal collocation to solve this problem numerically.

(a) Show that the governing equation and conditions take the form

$$v_{max}\left[1 - \left(\frac{x}{B}\right)^2\right]\frac{\partial C}{\partial z} = D_A\frac{\partial^2 C}{\partial x^2}$$

$$x = 0; \quad \frac{\partial C}{\partial x} = 0$$

$$x = B; \quad -D_A\frac{\partial C}{\partial x} = f(C)$$

where $f(C)$ is the nonlinear chemical rate per unit surface area.

(b) Reduce the set of equations in part a to following nondimensional form such that the x-directed coordinate is normalized to $(0, 1)$

$$(1 - \xi^2)\frac{\partial C}{\partial \zeta} = \frac{\partial^2 C}{\partial \xi^2}$$

(c) Note the symmetry condition in the ξ-directed coordinate, apply the usual transformation $u = \xi^2$, where $\xi = x/B$, to reduce the governing equation to a new set of equations

$$(1 - u)\frac{\partial C}{\partial \zeta} = 4u\frac{\partial^2 C}{\partial u^2} + 2\frac{\partial C}{\partial u}$$

$$u = 1; \quad \frac{\partial C}{\partial u} = G(C)$$

What is the form of G written in terms of $f(C)$ and other variables?

(d) Apply the orthogonal collocation and show that the discretized equations take the form

$$\frac{dC_i}{d\zeta} = C_{i,N+1}C_{N+1} + \sum_{j=1}^{N} C_{i,j}C_j \quad \text{for} \quad i = 1, 2, \ldots, N$$

and

$$\sum_{j=1}^{N} A_{N+1,j}C_j + A_{N+1,N+1}C_{N+1} = G(C_{N+1})$$

where

$$C_{i,j} = \frac{4u_i B_{i,j} + 2A_{i,j}}{1 - u_i}$$

Here, **A** and **B** are first and second order derivative matrices.

(e) The set of discretized equations in part d is nonlinear because of the nonlinear reaction. Develop a scheme by applying methods of Chapter 7 to show how these equations are integrated.

Appendix A

Review of Methods for Nonlinear Algebraic Equations

This appendix presents a number of solution methods to solve systems of algebraic equations. We will start with the basic techniques, such as bisection and successive substitution, and then discuss one of the most widely used Newton–Raphson methods. These methods are useful in solving roots of polynomials such as $f(x) = 0$.

A.1 THE BISECTION ALGORITHM

The bisection algorithm is the most simple method to locate a root of a nonlinear algebraic equation if we know the domain $[a, b]$ which bounds the root. This implies that the functional values at $x = a$ and $x = b$ must be different in sign (see Fig. A.1) and if the function is continuous, which is often the case in chemical engineering problems, then there will be at least one root lying inside that domain $[a, b]$. This domain is often known, especially when we solve transcendental equations for determining eigenvalues (Chapters 10 and 11), and therefore this algorithm is quite suitable as a starter. Following this initiation step, a much faster convergence (Newton–Raphson) method is used at some point to speed up convergence. We shall discuss Newton–Raphson in Section A.3.

Given a continuous function $f(x)$, defined in the interval $[a, b]$ with $f(a)$ and $f(b)$ being of opposite sign, then there exists a value p ($a < p < b$), for which $f(p) = 0$. There may be more than one value of p.

The method calls for a repeated halving of the subinterval $[a, b]$ and at each stop, locating the "half" region containing p.

To begin the process of repeated halving, we make a first iteration to find p (call this p_1) and let p_1 be the midpoint of $[a, b]$; that is,

$$p_1 = \frac{(a_1 + b_1)}{2}$$

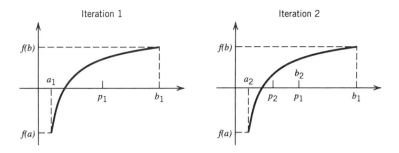

Figure A.1 Graphical representation of the bisection technique.

where

$$a_1 = a \quad \& \quad b_1 = b$$

If $f(p_1) = 0$ then p_1 is the solution. If not then $f(p)$ has the same sign as either $f(a_1)$ or $f(b_1)$. If $f(p_1)$ and $f(a_1)$ have the same sign, then $p \in (p_1, b_1)$ (i.e., p lies between p_1 and b_1) and we set $a_2 = p_1$ and $b_2 = b_1$. If $f(p_1)$ and $f(b_1)$ are of the same sign, then $p \in (a_1, p_1)$, and we set $a_2 = a_1$ and $b_2 = p_1$. We then apply the same procedure to the subinterval $[a_2, b_2]$ until a convergence test is satisfied. Figure A.1 shows the graphical representation of this iteration.

Some convergence tests could be used to stop the iteration. Given a tolerance $\varepsilon > 0$, we generate $p_1, p_2, \ldots, p_n$ until one of the following conditions is met

$$|p_n - p_{n-1}| < \varepsilon \tag{A.1a}$$

$$\frac{|p_n - p_{n-1}|}{|p_n|} < \varepsilon \quad |p_n| \neq 0 \tag{A.1b}$$

$$|f(p_n)| < \varepsilon \tag{A.1c}$$

Difficulties may arise using any of these stopping criteria, especially the criterion (A.1a). If there is no knowledge beforehand regarding the function f or the approximate location of the exact root p, Eq. A.1b is the recommended stopping criterion.

Drawbacks of the bisection method are:

1. An interval $[a, b]$ must be found with the product $f(a) \cdot f(b)$ being negative.
2. The convergence rate is rather slow (relative to the Newton's method).

Despite the drawbacks of the bisection listed above, it has two advantages:

1. The method is very simple.
2. The method always converges.

The key advantage of the technique is, if an interval $[a, b]$ is found, the method always converges. For this reason, it is often used as a "starter" for the more efficient root-seeking methods described later.

A.2   THE SUCCESSIVE SUBSTITUTION METHOD

The bisection method presented in the previous section can only be used when an interval $[a, b]$ is known. However, if this is not possible, the present and the following sections (Newton–Raphson) will prove to be useful.

Let us start with the single nonlinear algebraic equation

$$F(x) = 0 \qquad \text{(A.2)}$$

which we will generalize later for a set of nonlinear equations.

The underlying principle of the successive substitution method is to arrange the given equation to the form

$$x = f(x) \qquad \text{(A.3)}$$

One way of doing this is to simply add x to the LHS and RHS of Eq. A.2 as

$$x = x + F(x) \overset{\text{def}}{=} f(x) \qquad \text{(A.4)}$$

The iteration scheme for the successive substitution method to search for a root is defined as

$$x^{(k+1)} = x^{(k)} + F(x^{(k)}) = f(x^{(k)}) \qquad \text{(A.5)}$$

where the superscript k denotes the iteration number. It is expressed in bracketed form to distinguish it from powers.

Application of the successive method is quite simple. We choose (by guessing) a value for $x^{(0)}$ and calculate $x^{(1)}$ from the iteration equation (A.5). Repeating the procedure, we obtain $x^{(2)}$ and $x^{(3)}$, and so on. The method may or may not converge, but the method is direct and very easy to apply, despite the uncertainty of convergence. We will demonstrate the method and its uncertainty in the following example.

Consider the elementary equation

$$F(x) = x^2 - 4 = 0 \qquad \text{(A.6)}$$

for which the exact solution is $x = \pm 2$.

The iteration equation for this example can be initiated by adding x to both sides of Eq. A.6; that is,

$$x = x^2 - 4 + x = f(x) \qquad \text{(A.7)}$$

The iteration scheme for this equation is

$$x^{(k+1)} = \left(x^{(k)}\right)^2 - 4 + \left(x^{(k)}\right) \qquad \text{(A.8)}$$

Starting with an initial guess of $x^{(0)} = 3$, we get the following diverging behavior: $x^{(1)} = 8$, $x^{(2)} = 68$, $x^{(3)} = 4688$. The method does not converge. Even if we try $x^{(0)} = 2.01$, which is very close to the exact solution, again the method does not converge. This method exhibits a divergence behavior, even when the initial guess is close to the exact solution. Before we discuss conditions to cause this method to converge, we rearrange Eq. A.6 by dividing it by $(-2x)$ and obtain

$$G(x) = \frac{2}{x} - \frac{x}{2} = 0 \qquad \text{(A.9)}$$

The successive substitution is applied to this equation, and this equation is obtained

$$x^{(k+1)} = x^{(k)} + \frac{2}{x^{(k)}} - \frac{x^{(k)}}{2} = \frac{x^{(k)}}{2} + \frac{2}{x^{(k)}} \qquad \text{(A.10)}$$

Starting with $x^{(0)} = 4$, we obtain $x^{(1)} = 2.5$, $x^{(2)} = 2.05$, $x^{(3)} = 2.00061$, and $x^{(4)} = 2.000000093$. The iteration process for the *new arrangement* converges rapidly in a few iterations!

Obviously, the new arrangement for the iteration equation is better than the original iteration equation (Eq. A.8). Before we discuss the conditions for convergence, let us practice this method on a set of coupled nonlinear algebraic equations.

Consider a set of nonlinear algebraic equations.

$$F_1(x_1, x_2, \ldots, x_n) = 0$$
$$F_2(x_1, x_2, \ldots, x_n) = 0$$
$$\vdots \qquad\qquad\qquad\qquad \text{(A.11)}$$
$$F_n(x_1, x_2, \ldots, x_n) = 0$$

The number of algebraic equations is the same as the number of unknowns x, where

$$\mathbf{x} = [x_1, x_2, \ldots, x_n]^T \qquad \text{(A.12)}$$

The above set of algebraic equations can be written in a more compact form as

$$F_i(\mathbf{x}) = 0 \qquad \text{(A.13)}$$

for $i = 1, 2, \ldots, n$, or in vector form, write as

$$\mathbf{F}(\mathbf{x}) = \mathbf{0} \qquad \text{(A.14a)}$$

where

$$\mathbf{F} = [F_1, F_2, \ldots, F_n]^T \quad ; \quad \mathbf{0} = [0, 0, \ldots, 0]^T \qquad \text{(A.14b)}$$

Our objective for a solution is to find $\mathbf{x}$ such that $F_i(\mathbf{x}) = 0$ for $i = 1, 2, \ldots, n$.

By adding x_i to the ith equation (Eq. A.13), we have

$$x_i + F_i(\mathbf{x}) = x_i \qquad \text{(A.15)}$$

for $i = 1, 2, \ldots, n$.

The iteration scheme for successive substitution is defined as follows:

$$x_i^{(k+1)} = x_i^{(k)} + F_i(\mathbf{x}^{(k)}) \tag{A.16}$$

for $i = 1, 2, \ldots, n$.

Application of the successive substitution method is simply choosing (by guessing) a value for $\mathbf{x}^{(0)}$ and calculate $\mathbf{x}^{(1)}$ according to Eq. A.16 for $i = 1$, $2, \ldots, n$. Repeating the procedure, we obtain $\mathbf{x}^{(2)}$, $\mathbf{x}^{(3)}$, and determine if the method will converge. Similar to the case of single equation, the convergence of the iteration scheme is uncertain. We need to establish conditions which can tell us when the method will converge.

The following convergence theorem (sometimes called the contraction mapping theorem) will provide this information.

Theorem A.1

Let $\boldsymbol{\alpha}$ be the solution of $\alpha_i = f_i(\boldsymbol{\alpha})$, for $i = 1, 2, \ldots, n$. Assume that given an $h > 0$ there exists a number $0 < \mu < 1$ such that

$$\sum_{j=1}^{n} \left| \frac{\partial f_i}{\partial x_j} \right| \leq \mu \quad \text{for} \quad |x_i - \alpha_i| < h \quad i = 1, 2, \ldots, n$$

and if the iteration equation is

$$x_i^{(k)} = f_i(\mathbf{x}^{(k-1)})$$

then $x_i(k)$ converges to α_i as k increases.

An elementary proof for this theorem may be found in Finlayson (1980).

The condition for convergence is *conservative*; that is, if the condition is satisfied, the iteration process will converge. However, nothing is said about when the condition is not met. In such a case, the iteration process *may converge or diverge*.

For a one-dimensional problem $\alpha = f(\alpha)$, the previous contraction mapping theorem is reduced as follows. For a given $h > 0$, there exists a number $\mu(0 < \mu < 1)$ such that $|df/dx| < \mu$ for $|x - \alpha| < h$, then the following iteration

$$x^{(k)} = f(x^{(k-1)}) \tag{A.17}$$

will converge to α.

This can be shown graphically in Fig. A.2 with plots of functions x and $f(x)$. The solution α is simply the intersection of these two functions. It is recognized that if the slope of the function $f(x)$ is less than that of the function x (i.e., $|df/dx| < 1$) in the neighborhood of α, we see in Fig. A.2 that any point in that neighborhood will map itself into the same domain (see the direction of arrows in Fig. A.2). This explains why the theory was given the name of contraction mapping.

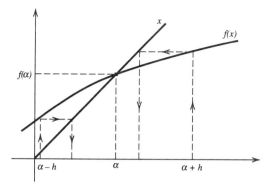

Figure A.2 Graphical representation of one-dimensional contraction mapping.

Therefore, with any starting point in that neighborhood, the iteration process will converge to the solution α.

The advantages of the successive substitution are

1. There is no need to find the interval $[a, b]$ as in the bisection method.
2. The method is very simple to apply.

And the disadvantages are

1. There is no guarantee of convergence (the contracting mapping theorem must be applied, but it is conservative).
2. A rather slow convergence rate (linear convergence) persists.

A.3   THE NEWTON–RAPHSON METHOD

The Newton–Raphson method is one of the most effective methods to solve nonlinear algebraic equations. We first illustrate the method with a single equation, and then generalize it to coupled algebraic equations.

Let us assume that the function $F(x)$ is at least twice differentiable over a domain $[a, b]$. Let $x^* \in [a, b]$ be an approximation to α, where α is the exact solution, such that $F'(x^*) \neq 0$ and $|x^* - \alpha|$ is small.

Consider the Taylor series of $F(x)$ expanded around the value x^*, and keep only up to the first order derivative term

$$F(x) \approx F(x^*) + \frac{dF(x^*)}{dx}(x - x^*) + \cdots \tag{A.18}$$

Since α is the exact solution, that is, $F(\alpha) = 0$, we substitute $x = \alpha$ into Eq. A.18 and obtain the result

$$0 \cong F(x^*) + \frac{dF(x^*)}{dx}(\alpha - x^*) \tag{A.19}$$

in which we have neglected all the terms higher than $(\alpha - x^*)^2$.

Solving for α from Eq. A.19, we obtain

$$\alpha \approx x^* - \frac{F(x^*)}{\dfrac{dF(x^*)}{dx}} \tag{A.20}$$

Eq. A.20 is the basis for the algorithm in the Newton–Raphson method. The iteration process is defined as

$$x^{(k+1)} = x^{(k)} - \frac{F(x^{(k)})}{\dfrac{dF(x^{(k)})}{dx}} \tag{A.21}$$

provided, of course, that

$$\frac{dF(x^{(k)})}{dx} \neq 0 \tag{A.22}$$

during the iteration process. Figure A.3 shows the Newton–Raphson method graphically.

Coming back to the example we used earlier

$$F(x) = x^2 - 4 \tag{A.23}$$

we now apply the Newton–Raphson scheme and obtain the following iteration scheme using Eq. A.21:

$$x^{(k+1)} = x^{(k)} - \frac{\left(x^{(k)}\right)^2 - 4}{2x^{(k)}} = \frac{x^{(k)}}{2} + \frac{2}{x^{(k)}} \tag{A.24}$$

A comparison with Eq. A.10 shows we were using the Newton–Raphson scheme by the artificial device of division by x.

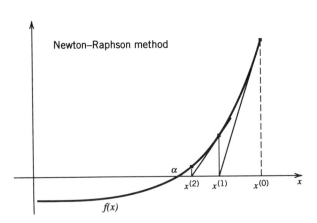

Figure A.3 Graphical representation of the Newton–Raphson method.

The Newton–Raphson method is a very powerful method, but it has the disadvantage that the derivative is needed. This can be a problem when one deals with complex coupled nonlinear algebraic equations.

To circumvent this problem, one can estimate the derivative $F(x^{(k)})$ by the following finite-difference approximation

$$\frac{dF(x^{(k)})}{dx} \approx \frac{F(x^{(k)}) - F(x^{(k-1)})}{x^{(k)} - x^{(k-1)}} \tag{A.25}$$

Using this approximation, the Newton–Raphson iteration equation becomes

$$x^{(k+1)} = x^{(k)} - \frac{F(x^{(k)})(x^{(k)} - x^{(k-1)})}{F(x^{(k)}) - F(x^{(k-1)})} \tag{A.26}$$

This formula is often called the Secant method, and to initiate it one needs two initial guesses. Normally, these two guesses are very close to each other. Figure A.4 shows the secant iteration method graphically.

Now we turn to presenting Newton's method for coupled nonlinear algebraic equations, such as portrayed in Eq. A.14. Newton's method for this set of equations is defined in terms of the Jacobian, which is the matrix of derivatives

$$\mathbf{x}^{(k+1)} = \mathbf{x}^{(k)} - \mathbf{J}^{-1}(\mathbf{x}^{(k)}) \cdot \mathbf{F}(\mathbf{x}^{(k)}) \tag{A.27}$$

for $k = 0, 1, \ldots$, and $\mathbf{J}$ is the matrix defined as

$$\mathbf{J}(\mathbf{x}) = \begin{bmatrix} \dfrac{\partial F_1}{\partial x_1} & \dfrac{\partial F_1}{\partial x_2} & \cdots & \dfrac{\partial F_1}{\partial x_n} \\ \dfrac{\partial F_2}{\partial x_1} & \dfrac{\partial F_2}{\partial x_2} & \cdots & \dfrac{\partial F_2}{\partial x_n} \\ \cdot & \cdot & & \cdot \\ \dfrac{\partial F_n}{\partial x_1} & \dfrac{\partial F_n}{\partial x_2} & \cdots & \dfrac{\partial F_n}{\partial x_n} \end{bmatrix}. \tag{A.28}$$

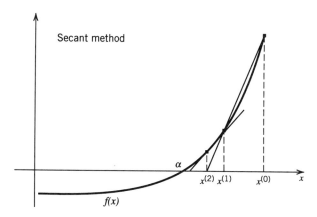

Figure A.4 Graphical representation of the Secant method.

The matrix $\mathbf{J}$ is called the Jacobian matrix. With an initial guess close to the exact solution, Newton's method is expected to give a quadratic convergence, provided of course that the Jacobian $\mathbf{J}$ exists. To illustrate the method for coupled equations, we inspect the example,

$$F_1 = 3x_1^2 - x_2^2$$

$$F_2 = 3x_1 x_2^2 - x_1^3 - 1$$

The Jacobian matrix $\mathbf{J}$ is given as

$$\mathbf{J} = \begin{bmatrix} \dfrac{\partial F_1}{\partial x_1} & \dfrac{\partial F_1}{\partial x_2} \\[2mm] \dfrac{\partial F_2}{\partial x_1} & \dfrac{\partial F_2}{\partial x_2} \end{bmatrix} = \begin{bmatrix} 6x_1 & -2x_2 \\ 3x_2^2 - 3x_1^2 & 6x_1 x_2 \end{bmatrix}$$

The iteration scheme for Newton's method is given in Eq. A.27, and is repeated here as

$$\mathbf{x}^{(k+1)} = \mathbf{x}^{(k)} - \mathbf{y}$$

where $\mathbf{y}$ is the solution of the set of linear equations

$$\mathbf{J}(\mathbf{x}^{(k)})\mathbf{y} = \mathbf{F}(\mathbf{x}^{(k)})$$

which can be solved using any standard linear equation package, such as MathCad.

We start with the initial guess $x_1^{(0)} = x_2^{(0)} = 1$, from which we can calculate the Jacobian and F as

$$\mathbf{J} = \begin{bmatrix} 6 & -2 \\ 0 & 6 \end{bmatrix} \quad \text{and} \quad \mathbf{F} = \begin{bmatrix} 2 \\ 1 \end{bmatrix}$$

Knowing the Jacobian and $\mathbf{F}$ we can use the iteration equation to calculate $\mathbf{x}^{(1)}$; that is,

$$\mathbf{x}^{(1)} = \begin{bmatrix} 1 \\ 1 \end{bmatrix} - \begin{bmatrix} 6 & -2 \\ 0 & 6 \end{bmatrix}^{-1} \cdot \begin{bmatrix} 2 \\ 1 \end{bmatrix} = \begin{bmatrix} \dfrac{11}{18} \\[2mm] \dfrac{5}{6} \end{bmatrix}$$

Knowing $\mathbf{x}^{(1)}$ of the first iteration, we can calculate the Jacobian and the function $\mathbf{F}$ as

$$\mathbf{J} = \begin{bmatrix} 3.6666 & -1.6667 \\ 0.9630 & 3.0556 \end{bmatrix} \quad \text{and} \quad \mathbf{F} = \begin{bmatrix} 0.4259 \\ 0.04493 \end{bmatrix}$$

and then using the iteration equation to calculate the second iteration

$$\mathbf{x}^{(2)} = \begin{bmatrix} \dfrac{11}{18} \\ \dfrac{5}{6} \end{bmatrix} - \begin{bmatrix} 3.6667 & -1.6667 \\ 0.9630 & 3.0556 \end{bmatrix}^{-1} \cdot \begin{bmatrix} 0.4259 \\ 0.04493 \end{bmatrix} = \begin{bmatrix} 0.5037 \\ 0.8525 \end{bmatrix}$$

Repeating the procedure, the method will converge to the exact solution $x_1 = 0.5$ and $x_2 = 0.5\sqrt{3}$.

A weakness of Newton's method is the analytical evaluation of the Jacobian matrix. In many practical situations this is somewhat inconvenient and often tedious. This can be overcome by using finite difference to approximate the partial derivative; that is,

$$\frac{\partial F_i}{\partial x_j} \approx \frac{F_i(\mathbf{x} + \mathbf{e}_j h) - F_i(\mathbf{x})}{h} \tag{A.29}$$

where $\mathbf{e}_j$ is the vector where the only nonzero element is the jth component, which is equal to unity. In the RHS of Eq. A.29, only the jth component of the vector $\mathbf{x}$ is increased by a small number h.

A.4   RATE OF CONVERGENCE

We have presented several techniques to handle nonlinear algebraic equations, but nothing has been said about the rate of convergence; that is, how fast the iteration process yields a convergent solution. We discuss this point in the next section.

Definition of Speed of Convergence

Let $\{\alpha_n\}$ be a sequence that converges to α and thereby define the error as $e_n = \alpha_n - \alpha$ for each $n \geq 0$. If positive constants λ and β exist with

$$\lim_{n \to \infty} \frac{|\alpha_{n+1} - \alpha|}{|\alpha_n - \alpha|^\beta} = \lim_{n \to \infty} \frac{|e_{n+1}|}{|e_n|^\beta} = \lambda \tag{A.30}$$

then $\{\alpha_n\}$ is said to converge to α of order β, with asymptotic error represented by the constant λ.

A sequence with a large order of convergence will converge more rapidly than that with low order. The asymptotic constant λ will affect the speed of convergence, but it is not as important as the order. However, for first order methods, λ becomes quite important.

If $\beta = 1$, convergence of the iteration scheme is called linear and it is called quadratic when $\beta = 2$.

We now illustrate that the successive iteration of the function

$$x = f(x)$$

has a linear convergence. Let us assume that $f(x)$ maps the interval $[a, b]$ into itself (i.e., for any values of x lying between a and b, the operation f will yield a value $f(x)$, which also lies between a and b) and a positive $\mu(0 < \mu < 1)$ exists such that $|f'(x)| < \mu$ for all $x \in [a, b]$. The convergence theorem proved earlier indicates that $f(x)$ has a unique fixed point $\alpha \in [a, b]$ and if $\alpha_0 \in [a, b]$ the sequence $\{\alpha_n\}$ will converge to α. The convergence rate will be shown to be linear provided that $f'(\alpha) \neq 0$; thus, we have

$$e_{n+1} = \alpha_{n+1} - \alpha = f(\alpha_n) - f(\alpha)$$
$$= f'(\xi_n)(\alpha_n - \alpha) = f'(\xi_n)e_n \tag{A.31}$$

where ξ_n lies between α_n and α. Since $\{\alpha_n\}$ will converge to α, the sequence $\{\xi_n\}$ will also converge to α.

Assuming that $f'(x)$ is continuous on $[a, b]$, we then have

$$\lim_{n \to \infty} f'(\xi_n) = f'(\alpha)$$

Thus, from Eq. A.31, we have

$$\lim_{n \to \infty} \frac{|e_{n+1}|}{|e_n|} = \lim_{n \to \infty} |f'(\xi_n)| = |f'(\alpha)|$$

Hence, the successive iteration exhibits a linear convergence and the asymptotic constant is $|f'(\alpha)|$, provided that $f'(\alpha) \neq 0$. Higher order convergence can occur when $f'(\alpha) = 0$. An example of this is Eq. A.10, where $f(x) = x/2 + 2/x$, so

$$f'(x) = \frac{1}{2} - \frac{2}{x^2}$$

and $\alpha = 2$. It is clear that $f'(\alpha) = f'(2) = 0$, and therefore, the iteration process of Eq. A.10 has higher order convergence than 1. We have shown earlier that this iteration scheme is in fact a Newton–Raphson scheme. It clearly sustains a quadratic convergence. The following theorem can be proved (Burden and Faires 1989).

Theorem A.2

Let α be a solution of $x = f(x)$. Suppose that $f'(\alpha) = 0$ and f'' is continuous in an open interval containing α. Then, there exists a δ such that for

$$\alpha_0 \in [\alpha - \delta, \alpha + \delta]$$

the sequence $\{\alpha_n\}$ defined by

$$\alpha_{n+1} = f(\alpha_n)$$

is quadratically convergent.

A.5   MULTIPLICITY

So far, we have shown various solution techniques for obtaining simple roots of nonlinear algebraic equations. Next, we will address the issue of multiple roots and a method of handling roots having multiplicity higher than one.

Multiplicity

A solution α of $F(x)$ is said to be a zero of multiplicity m of F if $F(x)$ can be written as:

$$F(x) = (x - \alpha)^m q(x) \tag{A.32}$$

where

$$\lim_{x \to \alpha} q(x) = c \neq 0 \tag{A.33}$$

The function $q(x)$ is basically the portion of the function $F(x)$ that does not contribute to the zero α of the function $F(x)$. Here, we imply a "zero" as simply one of the values of x, which produces $F(x) = 0$.

To handle the problem of multiple roots, we define a function

$$\mu(x) = \frac{F(x)}{F'(x)} \tag{A.34}$$

If α is a root of multiplicity $m \geq 1$ and $F(x) = (x - \alpha)^m q(x)$, Eq. A.34 then becomes

$$\mu(x) = \frac{(x - \alpha)^m q(x)}{m(x - \alpha)^{m-1} q(x) + (x - \alpha)^m q'(x)} \tag{A.35}$$

or

$$\mu(x) = \frac{(x - \alpha) q(x)}{m q(x) + (x - \alpha) q'(x)} \tag{A.36}$$

also has a root at α, but of multiplicity one, because $q(\alpha) \neq 0$.

The Newton–Raphson method can then be applied to the function $\mu(x)$ of Eq. A.36 to give

$$x = f(x) = x - \frac{\mu(x)}{\mu'(x)} \tag{A.37}$$

or

$$x = f(x) = x - \frac{F(x) F'(x)}{[F'(x)]^2 - [F(x) F''(x)]} \tag{A.38}$$

If $f(x)$ defined in Eq. A.38 has the required continuity condition, the iteration applied to f will be quadratically convergent *regardless of the multiplicity* of the root. In practice, the drawback is the requirement of a second derivative $F''(x)$.

A.6   ACCELERATING CONVERGENCE

Any sequence that is linear convergent can be accelerated by a method called the Aitken $\triangle^2$ method.

Let $\{\alpha_n\}$ be a linearly convergent sequence with a limit α. By definition, we have

$$\lim_{n \to \infty} \frac{|e_{n+1}|}{|e_n|} = \lambda, \qquad 0 < \lambda < 1 \tag{A.39}$$

where $e_n = \alpha_n - \alpha$.

Given a sequence $\{\alpha_n\}$ we now wish to construct a new sequence $\{\delta_n\}$, which will converge at a rate faster than the original sequence $\{\alpha_n\}$.

For n sufficiently large, and if we assume that all components in the sequence e_n have the *same* sign, then

$$e_{n+1} \approx \lambda e_n \tag{A.40}$$

Hence,

$$\alpha_{n+1} \approx \alpha + \lambda(\alpha_n - \alpha) \tag{A.41}$$

By increasing the index by 1, we have the equation

$$\alpha_{n+2} \approx \alpha + \lambda(\alpha_{n+1} - \alpha) \tag{A.42}$$

Eliminating λ between Eqs. A.41 and A.42, we obtain the equation for α

$$\alpha \approx \alpha_n - \frac{(\alpha_{n+1} - \alpha_n)^2}{(\alpha_{n+2} - 2\alpha_{n+1} + \alpha_n)} \tag{A.43}$$

The Aitken's $\triangle^2$ method is based on a new sequence defined as

$$\delta_n = \alpha_n - \frac{(\alpha_{n+1} - \alpha_n)^2}{(\alpha_{n+2} - 2\alpha_{n+1} + \alpha_n)} \tag{A.44}$$

which converges more rapidly than the original sequence $\{\alpha_n\}$. Additional details on Aitken's method can be found in Burden and Faires (1989).

By applying Aitken's method to a linearly convergent sequence obtained from fixed point (successive substitution) iteration, we can accelerate the convergence to quadratic order. This procedure is known as the Steffenson's method, which leads to Steffenson's algorithm as follows.

For a fixed point iteration, we generate the first three terms in the sequence α_n, that is, α_0, α_1, and α_2. Next, use the $\triangle^2$ Aitken method to generate δ_0. At this stage, we can assume that the newly generated δ_0 is a better approximation to α than α_2, and then apply the fixed point iteration to δ_0 to generate the next sequence of δ_0, δ_1 and δ_2. The Aitken method is now applied to the sequence $\{\delta_n; n = 0, 1, 2\}$ to generate γ_0, which is a better approximation to α, and the process continues.

A.7   REFERENCES

1. Burden, R. L., and J. D. Faires, "Numerical Analysis," PWS-Kent Publishing Company, Boston (1989).
2. Finlayson, B. A., "Nonlinear Analysis in Chemical Engineering," McGraw Hill, New York (1980).

Appendix **B**

Vectors and Matrices

Modeling of engineering systems gives rise to algebraic equations and/or differential equations. If the model equations are nonlinear algebraic equations, methods presented in Appendix A, such as the Newton–Raphson method, will reduce these nonlinear equations to a sequence of linear equations. Similarly, when the model equations are differential equations, methods such as finite-difference or orthogonal collocation will transform these equations to nonlinear algebraic equations if the starting equations are nonlinear. These resulting nonlinear algebraic equations are then converted to a sequence of linear algebraic equations. Thus, it is clear that systems of linear equations form the basic sets in the solution of any type of model equations, no matter whether they are nonlinear algebraic or nonlinear differential equations. This appendix will address linear algebraic equations, and the notation of vectors and matrices is introduced to allow the presentation of linear algebraic equations in a compact manner.

We will start with the definition of a matrix, with a vector being a special case of a matrix. Then we present a number of operations which may be used on matrices. Finally, we describe several methods for effecting the solution of linear equations.

B.1 MATRIX DEFINITION

A set of N linear algebraic equations with N unknowns, $x_1, x_2, \ldots, x_N$, may always be written in the form

$$a_{11}x_1 + a_{12}x_2 + a_{13}x_3 + \cdots + a_{1N}x_N = b_1$$

$$a_{21}x_1 + a_{22}x_2 + a_{23}x_3 + \cdots + a_{2N}x_N = b_2 \qquad \textbf{(B.1)}$$

$$\vdots$$

$$a_{N1}x_1 + a_{N2}x_2 + a_{N3}x_3 + \cdots + a_{NN}x_N = b_N$$

where x_i $(i = 1, 2, \ldots, N)$ are unknown variables, and b_i $(i = 1, 2, \ldots, N)$ are the constants representing the nonhomogeneous terms. The coefficients a_{ij}

$(i, j = 1, 2, \ldots, N)$ are constant coefficients, with the index i representing the ith equation and the index j to correspond to the variable x_j.

N is the number of equations, and it can be any integer number, ranging from 1 to infinity. If N is a large number, it is time-consuming to write those linear equations in the manner of Eq. B.1. To facilitate the handling of large numbers of equations, the notation of matrices and vectors will become extremely useful. This will allow us to write sets of linear equations in a very compact form. Matrix algebra is then introduced which allows manipulation of these matrices, such as addition, subtraction, multiplication, and taking the inverse (similar to division for scalar numbers).

The Matrix

A *matrix* is a rectangular array of elements arranged in an orderly fashion with rows and columns. Each element is distinct and separate. The element of a matrix is denoted as a_{ij}, with the index i to represent the ith row and the index j to represent the jth column. The size of a matrix is denoted as $N \times M$, where N is the number of rows and M is the number of columns. We usually represent a matrix with a boldface capital letter, for example $\mathbf{A}$, and the corresponding lowercase letter is used to represent its elements, for example, a_{ij}. The following equation shows the definition of a matrix $\mathbf{A}$ having N rows and M columns.

$$\mathbf{A} = \{a_{ij}; i = 1, 2, \ldots, N; j = 1, 2, \ldots, M\}$$

$$= \begin{bmatrix} a_{11} & a_{12} & a_{13} & \cdots & a_{1M} \\ a_{21} & a_{22} & a_{23} & \cdots & a_{2M} \\ & & & \cdots & \cdot \\ a_{N1} & a_{N2} & a_{N3} & \cdots & a_{NM} \end{bmatrix} \qquad \textbf{(B.2)}$$

where the bracket expression is the shorthand notation to describe the element as well as the size of the matrix.

The Vector

A *vector* is a special case of a matrix. A vector can be put as a column vector or it can be put as a row vector. A column vector is a matrix having a size of $N \times 1$. For example, the following vector $\mathbf{b}$ is a column vector with size $N \times 1$:

$$\mathbf{b} = \{b_i; i = 1, 2, \ldots, N\} = \begin{bmatrix} b_1 \\ b_2 \\ b_3 \\ \vdots \\ b_N \end{bmatrix} \qquad \textbf{(B.3)}$$

where b_i is the element associated with the row i.

The row vector is a matrix having a size of $1 \times N$. For example, a row vector **d** is represented as

$$\mathbf{d} = \{d_i; i = 1, 2, \ldots, N\} = \begin{bmatrix} d_1 & d_2 & d_3 \ldots d_N \end{bmatrix} \qquad \textbf{(B.4)}$$

B.2 TYPES OF MATRICES

Square Matrix

A *square matrix* is a matrix having the same number of rows and columns, that is, $\{a_{ij}; i, j = 1, 2, \ldots, N\}$. The elements a_{ii}, with $i = 1, 2, \ldots, N$, are called the major diagonal elements of the square matrix. The elements $a_{N1}, a_{N-1,2}$, to a_{1N} are called the minor diagonal elements.

Diagonal Matrix

A *diagonal matrix* is a square matrix having zero elements everywhere except on the major diagonal line. An identity matrix, denoted as **I**, is a diagonal matrix having unity major diagonal elements.

Triangular Matrix

A *triangular matrix* is a matrix having all elements on one side of the major diagonal line to be zero. An upper tridiagonal matrix **U** has all zero elements below the major diagonal line, and a lower tridiagonal matrix **L** has all zero elements above the diagonal line. The following equation shows upper and lower tridiagonal matrices having a size 3×3:

$$\mathbf{U} = \begin{bmatrix} a_{11} & a_{12} & a_{13} \\ 0 & a_{22} & a_{23} \\ 0 & 0 & a_{33} \end{bmatrix}; \qquad \mathbf{L} = \begin{bmatrix} a_{11} & 0 & 0 \\ a_{21} & a_{22} & 0 \\ a_{31} & a_{32} & a_{33} \end{bmatrix} \qquad \textbf{(B.5)}$$

Tridiagonal Matrix

A *tridiagonal matrix* is a matrix in which all elements that are not on the major diagonal line and two diagonals surrounding the major diagonal line are zero. The following equation shows a typical tridiagonal matrix of size 4×4

$$\mathbf{T} = \begin{bmatrix} a_{11} & a_{12} & 0 & 0 \\ a_{21} & a_{22} & a_{23} & 0 \\ 0 & a_{32} & a_{33} & a_{34} \\ 0 & 0 & a_{43} & a_{44} \end{bmatrix} \qquad \textbf{(B.6)}$$

The tridiagonal matrix is encountered quite regularly when solving differential equations using the finite-difference method (see Chapter 12).

Symmetric Matrix

The transpose of a $N \times M$ matrix $\mathbf{A}$ is a matrix $\mathbf{A}^T$ having a size of $M \times N$, with the element a_{ij}^T defined as

$$a_{ij}^T = a_{ji} \tag{B.7}$$

that is, the position of a row and a column is interchanged.

A symmetric square matrix has identical elements on either side of the major diagonal line, that is, $a_{ji} = a_{ij}$. This means $\mathbf{A}^T = \mathbf{A}$.

Sparse Matrix

A *sparse matrix* is a matrix in which most elements are zero. Many matrices encountered in solving engineering systems are sparse matrices.

Diagonally Dominant Matrix

A *diagonally dominant matrix* is a matrix such that the absolute value of the diagonal term is larger than the sum of the absolute values of other elements in the same row, with the diagonal term larger than the corresponding sum for at least one row; that is,

$$|a_{ii}| \geq \sum_{\substack{j=1 \\ j \neq i}}^{N} |a_{ij}| \qquad \text{for} \qquad i = 1, 2, \ldots, N \tag{B.8}$$

with

$$|a_{ii}| > \sum_{\substack{j=1 \\ j \neq i}}^{N} |a_{ij}| \tag{B.9}$$

for at least one row.

This condition of diagonal dominant matrix is required in the solution of a set of linear equations using iterative methods, details of which are seen in Section B.6.

B.3 MATRIX ALGEBRA

Just as in scalar operations, where we have addition, subtraction, multiplication and division, we also have addition, subtraction, multiplication and inverse (playing the role of division) on matrices, but there are a few restrictions in matrix algebra before these operations can be carried out.

Addition and Subtraction

These two operations can only be carried out when the sizes of the two matrices are the same. The operations are shown as follows.

$$\mathbf{A} + \mathbf{B} = \{a_{ij}\} + \{b_{ij}\} = \{c_{ij} = a_{ij} + b_{ij}\} = \mathbf{C} \tag{B.10}$$

$$\mathbf{A} - \mathbf{B} = \{a_{ij}\} - \{b_{ij}\} = \{c_{ij} = a_{ij} - b_{ij}\} = \mathbf{C} \tag{B.11}$$

Operations cannot be carried out on unequal size matrices.
    Addition of equal size matrices is associative and commutative; that is,

$$\mathbf{A} + (\mathbf{B} + \mathbf{C}) = (\mathbf{A} + \mathbf{B}) + \mathbf{C} \tag{B.12}$$

$$\mathbf{A} + \mathbf{B} = \mathbf{B} + \mathbf{A} \tag{B.13}$$

Multiplication

This operation involves the multiplication of the row elements of the first matrix to the column elements of the second matrix and the summation of the resulting products. Because of this procedure of multiplication, the number of columns of the first matrix, $\mathbf{A}$, must be the same as the number of rows of the second matrix, $\mathbf{B}$. Two matrices that satisfy this criterion are called conformable in the order of $\mathbf{A}\,\mathbf{B}$. If the matrix $\mathbf{A}$ has a size $N \times R$ and $\mathbf{B}$ has a size $R \times M$, the resulting product $\mathbf{C} = \mathbf{A} \cdot \mathbf{B}$ will have a size of $N \times M$, and the elements c_{ij} are defined as

$$c_{ij} = \sum_{r=1}^{R} a_{ir}b_{rj}; i = 1, 2, \ldots, N; j = 1, 2, \ldots, M \tag{B.14}$$

Matrices not conformable cannot be multiplied, and it is obvious that square matrices are conformable in any order.
    Conformable matrices are associative on multiplication; that is,

$$\mathbf{A}(\mathbf{BC}) = (\mathbf{AB})\mathbf{C} \tag{B.15}$$

but square matrices are generally not commutative on multiplication, i.e.

$$\mathbf{AB} \neq \mathbf{BA} \tag{B.16}$$

Matrices $\mathbf{A}$, $\mathbf{B}$, and $\mathbf{C}$ are distributive if $\mathbf{B}$ and $\mathbf{C}$ have the same size and if $\mathbf{A}$ is conformable to $\mathbf{B}$ and $\mathbf{C}$, then we have:

$$\mathbf{A}(\mathbf{B} + \mathbf{C}) = \mathbf{AB} + \mathbf{AC} \tag{B.17}$$

Multiplication of a matrix $\mathbf{A}$ with a scalar β is a new matrix $\mathbf{B}$ with the element $b_{ij} = \beta a_{ij}$.

Inverse

The inverse in matrix algebra plays a similar role to division in scalar division. The inverse is defined as follows:

$$\mathbf{A}\mathbf{A}^{-1} = \mathbf{I} \tag{B.18}$$

where $\mathbf{A}^{-1}$ is called the inverse of $\mathbf{A}$, and $\mathbf{I}$ is the identity matrix. Matrix inverses commute on multiplication, that is,

$$\mathbf{A}\mathbf{A}^{-1} = \mathbf{I} = \mathbf{A}^{-1}\mathbf{A} \tag{B.19}$$

If we have the equation,

$$\mathbf{A}\mathbf{B} = \mathbf{C} \tag{B.20}$$

where $\mathbf{A}$, $\mathbf{B}$, and $\mathbf{C}$ are square matrices, multiply the LHS and RHS of Eq. B.20 by $\mathbf{A}^{-1}$ and we will get

$$\mathbf{A}^{-1}(\mathbf{A}\mathbf{B}) = \mathbf{A}^{-1}\mathbf{C} \tag{B.21}$$

But since the multiplication is associative, the above equation will become

$$(\mathbf{A}^{-1}\mathbf{A})\mathbf{B} = \mathbf{B} = \mathbf{A}^{-1}\mathbf{C} \tag{B.22}$$

as $\mathbf{A}^{-1}\mathbf{A} = \mathbf{I}$ and $\mathbf{I}\mathbf{B} = \mathbf{B}$.

The analytical technique according to the Gauss–Jordan procedure for obtaining the inverse will be dealt with later.

Matrix Decomposition or Factorization

A given matrix $\mathbf{A}$ can be represented as a product of two conformable matrices $\mathbf{B}$ and $\mathbf{C}$. This representation is not unique, as there are infinite combinations of $\mathbf{B}$ and $\mathbf{C}$ which can yield the same matrix $\mathbf{A}$. Of particular usefulness is the decomposition of a square matrix $\mathbf{A}$ into lower and upper triangular matrices, shown as follows.

$$\mathbf{A} = \mathbf{L}\mathbf{U} \tag{B.23}$$

This is usually called the *LU* decomposition and is useful in solving a set of linear algebraic equations.

B.4   USEFUL ROW OPERATIONS

A set of linear algebraic equations of the type in Eq. B.1 can be readily put into vector-matrix format as

$$\mathbf{A}\mathbf{x} = \mathbf{b} \tag{B.24}$$

where

$$\mathbf{A} = \begin{bmatrix} a_{11} & a_{12} & a_{13} & \cdots & a_{1N} \\ a_{21} & a_{22} & a_{23} & \cdots & a_{2N} \\ & & \cdots & & \cdot \\ a_{N1} & a_{N2} & a_{N3} & \cdots & a_{NN} \end{bmatrix}; \quad \mathbf{x} = \begin{bmatrix} x_1 \\ x_2 \\ \vdots \\ x_N \end{bmatrix}; \mathbf{b} = \begin{bmatrix} b_1 \\ b_2 \\ \vdots \\ b_N \end{bmatrix} \tag{B.25}$$

Equation B.25 can also be written in the component form as

$$\sum_{j=1}^{N} a_{ij}x_j = b_i; \quad \text{for} \quad i = 1, 2, \ldots, N \qquad \text{(B.26)}$$

which is basically the equation of the row i.

There are a number of row operations which can be carried out and they don't affect the values of the final solutions **x**.

Scaling

Any row can be multiplied by a scalar, the process of which is called scaling. For example, the row i of Eq. B.26 can be multiplied by a constant α as

$$\sum_{j=1}^{N} \alpha a_{ij}x_j = \alpha b_i \qquad \text{(B.27)}$$

Pivoting

Any row can be interchanged with another row. This process is called pivoting. The main purpose of this operation is to create a new matrix that has dominant diagonal terms, which is important in solving linear equations.

Elimination

Any row can be replaced by a weighted linear combination of that row with any other row. This process is carried out on the row i with the purpose of eliminating one or more variables from that equation. For example, if we have the following two linear equations:

$$x_1 + x_2 = 2$$
$$3x_1 + 2x_2 = 5 \qquad \text{(B.28)}$$

Let us now modify the row 2; that is, equation number 2. We multiply the first row by (3) and then subtract the second row from this to create a new second row; hence we have:

$$x_1 + x_2 = 2$$
$$0x_1 + x_2 = 1 \qquad \text{(B.29)}$$

We see that x_1 has been eliminated from the new second row, from which it is seen that $x_2 = 1$ and hence from the first row $x_1 = 1$. This process is called elimination. This is exactly the process used in the Gauss elimination scheme to search for the solution of a given set of linear algebraic equations, which will be dealt with in the next section.

There are a number of methods available to solve for the solution of a given set of linear algebraic equations. One class is the direct method (i.e., requires no iteration) and the other is the iterative method, which requires iteration as the name indicates. For the second class of method, an initial guess must be provided. We will first discuss the direct methods in Section B.5 and the iterative methods will be dealt with in Section B.6. The iterative methods are preferable when the number of equations to be solved is large, the coefficient matrix is sparse and the matrix is diagonally dominant (Eqs. B.8 and B.9).

B.5   DIRECT ELIMINATION METHODS

B.5.1   Basic Procedure

The elimination method basically involves the elimination of variables in such a way that the final equation will involve only one variable. The procedure for a set of N equations is as follows. First, from one equation solve for x_1 as a function of other variables, $x_2, x_3, \ldots, x_N$. Substitute this x_1 into the remaining $N - 1$ equations to obtain a new set of $N - 1$ equations with $N - 1$ unknowns, $x_2, x_3, \ldots, x_N$. Next, using one of the equations in the new set, solve for x_2 as a function of other variables, $x_3, x_4, \ldots, x_N$, and then substitute this x_2 into the remaining $N - 2$ equations to obtain a new set of $N - 2$ equations in terms of $N - 2$ unknown variables. Repeat the procedure until you end up with only one equation with one unknown, x_N, from which we can readily solve for x_N. Knowing x_N, we can use it in the last equation in which x_{N-1} was written in terms of x_N. Repeat the same procedure to find x_1. The process of going backward to find solutions is called back substitution.

Let us demonstrate this elimination method with this set of three linear equations.

$$a_{11}x_1 + a_{12}x_2 + a_{13}x_3 = b_1 \tag{B.30}$$

$$a_{21}x_1 + a_{22}x_2 + a_{23}x_3 = b_2 \tag{B.31}$$

$$a_{31}x_1 + a_{32}x_2 + a_{33}x_3 = b_3 \tag{B.32}$$

Assuming a_{11} is not zero, we solve Eq. B.30 for x_1 in terms of x_2 and x_3 and we have

$$x_1 = \frac{b_1 - a_{12}x_2 - a_{13}x_3}{a_{11}} \tag{B.33}$$

Substitute this x_1 into Eqs. B.31 and B.32 to eliminate x_1 from the remaining two equations, and we have

$$a'_{22}x_2 + a'_{23}x_3 = b'_2 \tag{B.34}$$

$$a'_{32}x_2 + a'_{33}x_3 = b'_3 \tag{B.35}$$

where

$$a'_{ij} = a_{ij} - \frac{a_{i1}}{a_{11}} a_{1j}; \qquad b'_i = b_i - \frac{a_{i1}}{a_{11}} b_1 \qquad \text{for} \qquad i, j = 2, 3 \quad (\textbf{B.36})$$

Next, we solve Eq. B.34 for x_2 in terms of x_3 provided $a'_{22} \neq 0$; that is,

$$x_2 = \frac{b'_2 - a'_{23} x_3}{a'_{22}} \tag{B.37}$$

then substitute this x_2 into the last equation (Eq. B.35) to obtain

$$a''_{33} x_3 = b''_3 \tag{B.38}$$

where

$$a''_{33} = a'_{33} - \frac{a'_{32}}{a'_{22}} a'_{23}; \qquad b''_3 = b'_3 - \frac{a'_{32}}{a'_{22}} b'_2 \tag{B.39}$$

We see that the patterns of Eqs. B.36 and B.39 are exactly the same and this pattern is independent of the number of equations. This serial feature can be exploited in computer programming.

Thus, the elimination process finally yields one equation in terms of the variable x_3, from which it can be solved as

$$x_3 = \frac{b''_3}{a''_{33}} \tag{B.40}$$

Knowing x_3, x_2 can be obtained from Eq. B.37, and finally x_1 from Eq. B.33. This procedure is called back substitution.

B.5.2   Augmented Matrix

The elimination procedure described in the last section involves the manipulation of equations. No matter how we manipulate the equations, the final solution vector **x** is still the same. One way to simplify the elimination process is to set up an augmented matrix as

$$[\mathbf{A}|b] = \begin{bmatrix} a_{11} & a_{12} & a_{13} & b_1 \\ a_{21} & a_{22} & a_{23} & b_2 \\ a_{31} & a_{32} & a_{33} & b_3 \end{bmatrix} \tag{B.41}$$

and then perform the row operations described in Section B.4 to effect the elimination process.

Let us demonstrate this concept of an augmented matrix to the following example.

$$x_1 + 2x_2 + 3x_3 = 14$$
$$x_1 + x_2 - x_3 = 0 \qquad \textbf{(B.42)}$$
$$2x_1 + x_2 - x_3 = 1$$

For this set of three linear equations, we form an augmented matrix by putting the coefficient matrix first and then the RHS vector, shown as follows.

$$\left[\begin{array}{ccc|c} 1 & 2 & 3 & 14 \\ 1 & 1 & -1 & 0 \\ 2 & 1 & -1 & 1 \end{array}\right] \qquad \textbf{(B.43)}$$

Now we start carrying out row operations on the augmented matrix. First, we take the second row and subtract it from the first row to form a new second row, the result of which is shown as follows.

$$\left[\begin{array}{ccc|c} 1 & 2 & 3 & 14 \\ 0 & 1 & 4 & 14 \\ 2 & 1 & -1 & 1 \end{array}\right] \qquad \textbf{(B.44)}$$

The purpose of the last step is to eliminate x_1 from the second equation; that is, the new coefficient for x_1 in the new second equation is 0. This is the basic step of elimination. Now we do exactly the same to the third row. We multiply the first row by 2 and subtract the third row to form a new third row and get the result

$$\left[\begin{array}{ccc|c} 1 & 2 & 3 & 14 \\ 0 & 1 & 4 & 14 \\ 0 & 3 & 7 & 27 \end{array}\right] \qquad \textbf{(B.45)}$$

Thus, we have eliminated the variable x_1 from the second and third equations. Now, we move to the next step of the elimination procedure, that is to remove the variable x_2 from the third equation. This is done by multiplying the second row by 3 and subtracting the third row to form a new third row; that is,

$$\left[\begin{array}{ccc|c} 1 & 2 & 3 & 14 \\ 0 & 1 & 4 & 14 \\ 0 & 0 & 5 & 15 \end{array}\right] \qquad \textbf{(B.46)}$$

The last row will give a solution of $x_3 = 3$. Put this into the second equation to give $x_2 = 2$, and hence finally into the first equation to give $x_1 = 1$. This is the back substitution procedure. All the steps carried out are part of the Gauss elimination scheme. More details on this method will be presented in Section B.5.5.

Let us now come back to our present example and continue with the row operations, but this time we eliminate the variables above the major diagonal line. To do this, we multiply the third row by $(-4/5)$ and add the result to the second row to form a new second row, shown as follows.

$$\begin{bmatrix} 1 & 2 & 3 & | & 14 \\ 0 & 1 & 0 & | & 2 \\ 0 & 0 & 5 & | & 15 \end{bmatrix} \tag{B.47}$$

The last step is to remove the variable x_3 from the second equation. Finally, multiply the second row by (-2) and the third row by $(-3/5)$ and add the results to the first row to obtain a new first row as

$$\begin{bmatrix} 1 & 0 & 0 & | & 1 \\ 0 & 1 & 0 & | & 2 \\ 0 & 0 & 5 & | & 15 \end{bmatrix} \tag{B.48}$$

from which one can see immediately $x_1 = 1$, $x_2 = 2$ and $x_3 = 3$. The last few extra steps are part of the Gauss–Jordan elimination scheme, the main purpose of which is to obtain the inverse as we shall see in Section B.5.6.

This procedure of augmented matrix can handle more than one vector $\mathbf{b}$ at the same time, for example if we are to solve the following equations with the same coefficient matrix $\mathbf{A}$: $\mathbf{Ax}_1 = \mathbf{b}_1$, $\mathbf{Ax}_2 = \mathbf{b}_2$, we can set the augmented matrix as

$$\begin{bmatrix} \mathbf{A} & | & \mathbf{b}_1 & \mathbf{b}_2 \end{bmatrix} \tag{B.49}$$

and carry out the row operations as we did in the last example to obtain simultaneously the solution vectors $\mathbf{x}_1$ and $\mathbf{x}_2$.

B.5.3   Pivoting

The elimination procedure we described in Section B.5.1 requires that a_{11} is nonzero. Thus, if the diagonal coefficient a_{11} is zero, then we shall need to rearrange the equations, that is, we interchange the rows such that the new diagonal term a_{11} is nonzero. We also carry out this pivoting process in such a way that the element of largest magnitude is on the major diagonal line. If rows are interchanged only, the process is called partial pivoting, while if rows as well as columns are interchanged it is called full pivoting. Full pivoting is not normally carried out because it changes the order of the components of the vector $\mathbf{x}$. Therefore, only partial pivoting is dealt with here.

Partial pivoting not only eliminates the problem of zero on the diagonal line, it also reduces the round-off error since the pivot element (i.e., the diagonal element) is the divisor in the elimination process. To demonstrate the pivoting procedure, we use an example of three linear equations.

$$0x_1 + x_2 + x_3 = 5$$
$$4x_1 + x_2 - x_3 = 3 \tag{B.50}$$
$$x_1 - x_2 + x_3 = 2$$

Putting this set of equations into the augmented matrix form, we have

$$\begin{bmatrix} 0 & 1 & 1 & | & 5 \\ 4 & 1 & -1 & | & 3 \\ 1 & -1 & 1 & | & 2 \end{bmatrix} \tag{B.51}$$

We note that the coefficient a_{11} is zero; therefore, there is a need to carry out the pivoting procedure. The largest element of the first column is 4. Therefore, upon interchanging the first and the second rows, we will get

$$\begin{bmatrix} 4 & 1 & -1 & | & 3 \\ 0 & 1 & 1 & | & 5 \\ 1 & -1 & 1 & | & 2 \end{bmatrix} \tag{B.52}$$

Next, multiplying the third row by 4 and subtracting the first row to get the new third row will yield

$$\begin{bmatrix} 4 & 1 & -1 & | & 3 \\ 0 & 1 & 1 & | & 5 \\ 0 & -5 & 5 & | & 5 \end{bmatrix} \tag{B.53}$$

Although the pivot element in the second row is 1 ($\neq 0$), it is not the largest element in that column (second column). Hence, we carry out pivoting again, and this process is done with rows underneath the pivot element, not with rows above it. This is because the rows above the pivot element have already gone through the elimination process. Using them will destroy the elimination completed so far.

Interchange the second and the third row so that the pivot element will have the largest magnitude, we then have

$$\begin{bmatrix} 4 & 1 & -1 & | & 3 \\ 0 & -5 & 5 & | & 5 \\ 0 & 1 & 1 & | & 5 \end{bmatrix} \tag{B.54}$$

Next, multiply the third row by 5 and add with the second row to form a new third row, we get

$$\begin{bmatrix} 4 & 1 & -1 & | & 3 \\ 0 & -5 & 5 & | & 5 \\ 0 & 0 & 10 & | & 30 \end{bmatrix} \tag{B.55}$$

Finally, using the back substitution, we find that $x_3 = 3$, $x_2 = 2$, and $x_1 = 1$.

B.5.4   Scaling

When the magnitude of elements in one or more equations are greater than the elements of the other equations, it is essential to carry out scaling. This is done by dividing the elements of each row, including the **b** vector, by the largest element of that row (excluding the b element). After scaling, pivoting is then carried out to yield the largest pivot element.

B.5.5   Gauss Elimination

The elimination procedure described in the last sections forms a process, commonly called Gauss elimination. It is the backbone of the direct methods, and is the most useful in solving linear equations. Scaling and pivoting are essential in the Gauss elimination process.

The Gauss elimination algorithm is summarized as follows.

Step 1   Augment the matrix $\mathbf{A}(N \times N)$ and the vector $\mathbf{b}(N \times 1)$ to form an augmented matrix $\mathbf{A}$ of size $N \times (N + 1)$.

Step 2   Scale the rows of the augmented matrix.

Step 3   Search for the largest element in magnitude in the first column and pivot that coefficient into the a_{11} position.

Step 4   Apply the elimination procedure to rows 2 to N to create zeros in the first column below the pivot element. The modified elements in row 2 to row N and column 2 to column $N + 1$ of the augmented matrix must be computed and inserted in place of the original elements using the following formula:

$$a'_{ij} = a_{ij} - \frac{a_{i1}}{a_{11}} a_{1j}; \quad \text{for} \quad i = 2, 3, \ldots, N \quad \text{and} \quad j = 2, 3, \ldots, N + 1 \quad \textbf{(B.56)}$$

Step 5   Repeat steps 3 and 4 for rows 3 to N. After this is completely done, the resulting augmented matrix will be an upper triangular matrix.

Step 6   Solve for $\mathbf{x}$ using back substitution with the following equations:

$$x_N = \frac{a'_{N, N+1}}{a'_{N, N}} \quad \textbf{(B.57)}$$

$$x_i = \frac{a'_{i, N+1} - \sum_{j=i+1}^{N} a'_{ij} x_j}{a'_{ii}} \quad \text{for} \quad i = N - 1, N - 2, \ldots, 1 \quad \textbf{(B.58)}$$

B.5.6   Gauss–Jordan Elimination

Gauss–Jordan elimination is a variation of the Gauss elimination scheme. Instead of obtaining the triangular matrix at the end of the elimination, the Gauss–Jordan has one extra step to reduce the matrix $\mathbf{A}$ to an identity matrix. In this way the augmented vector $\mathbf{b}'$ is simply the solution vector $\mathbf{x}$.

The primary use of the Gauss–Jordan method is to obtain an inverse of a matrix. This is done by augmenting the matrix $\mathbf{A}$ with an identity matrix $\mathbf{I}$. After the elimination process in converting the matrix $\mathbf{A}$ to an identity matrix, the right-hand side identity matrix will become the inverse $\mathbf{A}^{-1}$. To show this, we use this example.

$$
\begin{bmatrix}
1 & 1 & 1 & | & 1 & 0 & 0 \\
2 & -1 & 1 & | & 0 & 1 & 0 \\
1 & -2 & 2 & | & 0 & 0 & 1
\end{bmatrix}
\qquad \textbf{(B.59)}
$$

Interchange the first and the second row to make the pivot element having the largest magnitude; hence, we have

$$
\begin{bmatrix}
2 & -1 & 1 & | & 0 & 1 & 0 \\
1 & 1 & 1 & | & 1 & 0 & 0 \\
1 & -2 & 2 & | & 0 & 0 & 1
\end{bmatrix}
\qquad \textbf{(B.60)}
$$

Now, scale the pivot element to unity (this step is not in the Gauss elimination scheme) to give

$$
\begin{bmatrix}
1 & -\frac{1}{2} & \frac{1}{2} & | & 0 & \frac{1}{2} & 0 \\
1 & 1 & 1 & | & 1 & 0 & 0 \\
1 & -2 & 2 & | & 0 & 0 & 1
\end{bmatrix}
\qquad \textbf{(B.61)}
$$

Following the same procedure of Gauss elimination with the extra step of normalizing the pivot element before each elimination, we finally obtain

$$
\begin{bmatrix}
1 & -\frac{1}{2} & \frac{1}{2} & | & 0 & \frac{1}{2} & 0 \\
0 & 1 & \frac{1}{3} & | & \frac{2}{3} & -\frac{1}{3} & 0 \\
0 & 0 & 1 & | & \frac{1}{2} & -\frac{1}{2} & \frac{1}{2}
\end{bmatrix}
\qquad \textbf{(B.62)}
$$

Now we perform the elimination for rows *above* the pivot elements, and after this step the original $\mathbf{A}$ matrix becomes an identity matrix, and the original identity matrix $\mathbf{I}$ in the RHS of the augmented matrix becomes the matrix inverse $\mathbf{A}^{-1}$; that is,

$$
\begin{bmatrix}
1 & 0 & 0 & | & 0 & \frac{2}{3} & -\frac{1}{3} \\
0 & 1 & 0 & | & \frac{1}{2} & -\frac{1}{6} & -\frac{1}{6} \\
0 & 0 & 1 & | & \frac{1}{2} & -\frac{1}{2} & \frac{1}{2}
\end{bmatrix}
\qquad \textbf{(B.63)}
$$

Obtaining the matrix inverse using the Gauss–Jordan method provides a compact way of solving linear equations. For a given problem,

$$
\mathbf{Ax = b} \qquad \textbf{(B.64)}
$$

we multiply the equation by A^{-1}, and obtain

$$A^{-1}(Ax) = A^{-1}b \qquad (B.65)$$

Noting that the multiplication is associative; hence, we have

$$(A^{-1}A)x = A^{-1}b, \quad \text{i.e.,} \quad x = A^{-1}b \qquad (B.66)$$

Thus, this inverse method provides a compact way of presenting the solution of the set of linear equations.

B.5.7 LU Decomposition

In the LU decomposition method, the idea is to decompose a given matrix A to a product LU. If we specify the diagonal elements of either the upper or lower triangular matrix, the decomposition will be unique. If the elements of the major diagonal of the L matrix are unity, the decomposition method is called the Doolittle method. It is called the Crout method if the elements of the major diagonal of the U matrix are unity.

In the Doolittle method, the upper triangular matrix U is determined by the Gauss elimination process, while the matrix L is the lower triangular matrix containing the multipliers employed in the Gauss process as the elements below the unity diagonal line. More details on the Doolittle and Crout methods can be found in Hoffman (1992).

The use of the LU decomposition method is to find solution to the linear equation $Ax = b$. Let the coefficient matrix A be decomposed to LU, i.e., $A = LU$. Hence, the linear equation will become

$$LUx = b \qquad (B.67)$$

Multiplying the above equation by L^{-1}, we have

$$L^{-1}(LU)x = L^{-1}b \qquad (B.68)$$

Since the multiplication is associative and $L^{-1}L = I$, the previous equation will become

$$Ux = b' \qquad (B.69)$$

where the vector b' is obtained from the equation

$$Lb' = b \qquad (B.70)$$

Equations B.69 and B.70 will form basic set of equations for solving for x. This is done as follows. For a given b vector, the vector b' is obtained from Eq. B.70 by forward substitution since L is the lower triangular matrix. Once b' is found, the desired vector x is found from Eq. B.69 by backward substitution because U is the upper triangular matrix.

B.6   ITERATIVE METHODS

When dealing with large sets of equations, especially if the coefficient matrix is sparse, the iterative methods provide an attractive option in getting the solution. In the iterative methods, an initial solution vector $\mathbf{x}^{(0)}$ is assumed, and the process is iterated to reduce the error between the iterated solution $\mathbf{x}^{(k)}$ and the exact solution $\mathbf{x}$, where k is the iteration number. Since the exact solution is not known, the iteration process is stopped by using the difference $\Delta x_i = x_i^{(k+1)} - x_i^{(k)}$ as the measure. The iteration is stopped when one of the following criteria has been achieved.

$$\left| \frac{(\Delta x_i)_{\max}}{x_i} \right| < \varepsilon; \qquad \sum_{i=1}^{N} \left| \frac{\Delta x_i}{x_i} \right| < \varepsilon; \qquad \left[\sum_{i=1}^{N} \left(\frac{\Delta x_i}{x_i} \right)^2 \right]^{1/2} < \varepsilon \quad \textbf{(B.71)}$$

The disadvantage of the iterative methods is that they may not provide a convergent solution. Diagonal dominance (Eqs. B.8 and B.9) is the sufficient condition for convergence. The stronger the diagonal dominance the fewer number of iterations required for the convergence.

There are three commonly used iterative methods which we will briefly present here. They are Jacobi, Gauss–Seidel and the successive overrelaxation methods.

B.6.1   Jacobi Method

The set of linear equations written in the component form is

$$b_i - \sum_{j=1}^{N} a_{ij} x_j = 0 \qquad \text{for} \qquad i = 1, 2, \ldots, N \qquad \textbf{(B.72)}$$

Divide the equation by a_{ii} and add x_i to the LHS and RHS (a similar procedure is used in Appendix A for the successive substitution method for solving nonlinear equations) to yield the equation

$$x_i = x_i + \frac{1}{a_{ii}} \left(b_i - \sum_{j=1}^{N} a_{ij} x_j = 0 \right) \qquad \text{for} \qquad i = 1, 2, \ldots, N \quad \textbf{(B.73)}$$

The iteration process starts with an initial guessing vector $\mathbf{x}^{(0)}$, and the iteration equation used to generate the next iterated vector is

$$x_i^{(k+1)} = x_i^{(k)} + \frac{1}{a_{ii}} \left(b_i - \sum_{j=1}^{N} a_{ij} x_j^{(k)} = 0 \right) \qquad \text{for} \qquad i = 1, 2, \ldots, N \quad \textbf{(B.74)}$$

The iteration process will proceed until one of the criteria in Eq. B.71 has been achieved.

The second term in the RHS of Eq. B.74 is called the residual, and the iteration process will converge when the residual is approaching zero for all value of i.

B.6.2 Gauss–Seidel Iteration Method

In the Jacobi method, the iterated vector of the $(k + 1)$th iteration is obtained based entirely on the vector of the previous iteration, that is, $\mathbf{x}^{(k)}$. The Gauss–Seidel iteration method is similar to the Jacobi method, except that the component $x_j^{(k+1)}$ for $j = 1, 2, \ldots, i - 1$ are used immediately in the calculation of the component $x_i^{(k+1)}$. The iteration equation for the Gauss–Seidel method is

$$x_i^{(k+1)} = x_i^{(k)} + \frac{1}{a_{ii}}\left(b_i - \sum_{j=1}^{i-1} a_{ij}x_j^{(k+1)} - \sum_{j=i}^{N} a_{ij}x_j^{(k)} = 0\right) \quad \text{(B.75)}$$

$$\text{for} \quad i = 1, 2, \ldots, N$$

Like the Jacobi method, the Gauss–Seidel method requires diagonal dominance for the convergence of iterated solutions.

B.6.3 Successive Overrelaxation Method

In many problems, the iterated solutions approach the exact solutions in a monotonic fashion. Therefore, it is useful in this case to speed up the convergence process by overrelaxing the iterated solutions. The equation for the overrelaxation scheme is modified from the Gauss–Seidel equation

$$x_i^{(k+1)} = x_i^{(k)} + \frac{w}{a_{ii}}\left(b_i - \sum_{j=1}^{i-1} a_{ij}x_j^{(k+1)} - \sum_{j=i}^{N} a_{ij}x_j^{(k)} = 0\right) \quad \text{(B.76)}$$

$$\text{for} \quad i = 1, 2, \ldots, N$$

where w is the overrelaxation factor. When $w = 1$, we recover the Gauss–Seidel method. When $1 < w < 2$, we have an overrelaxation situation. When $w < 1$, the system is underrelaxed. The latter is applicable when the iteration provides oscillatory behavior. When $w > 2$, the method diverges.

There is no fast rule on how to choose the optimum w for a given problem. It must be found from numerical experiments.

B.7 EIGENPROBLEMS

In this section, we will consider briefly the eigenproblems, that is, the study of the eigenvalues and eigenvectors. The study of coupled linear differential equations presented in the next section requires the analysis of the eigenproblems.

Let us consider this linear equation written in compact matrix notation

$$\mathbf{A}\mathbf{x} = \mathbf{b} \quad \text{(B.77)}$$

The homogeneous form of the equation is simply $\mathbf{A}\mathbf{x} = \mathbf{0}$, where $\mathbf{0}$ is the zero vector, that is, a vector with all zero elements. If all equations in Eq. B.77 are independent, then the trivial solution to the homogeneous equation is $\mathbf{x} = \mathbf{0}$. Now if we modify the matrix $\mathbf{A}$ to form $(\mathbf{A} - \lambda\mathbf{I})$, then the homogeneous equation will become

$$(\mathbf{A} - \lambda\mathbf{I})\mathbf{x} = \mathbf{0} \qquad (\mathbf{B.78})$$

Beside the obvious trivial solution to the above equation, there exists some value of λ such that the solution is nonzero. In such a case, the value of λ is called the *eigenvalue* and the solution vector $\mathbf{x}$ corresponding to that eigenvalue is called the *eigenvector*. The problem stated by Eq. B.78 is then called the eigenproblem.

This eigenproblem arises naturally during the analysis of coupled linear differential equations, as we shall see in the next section. The eigenvalues can be determined from the determinant $\det(\mathbf{A} - \lambda\mathbf{I}) = 0$, for which the equation is called the characteristic equation. Obtaining this equation for a large system is very difficult. Iteration procedures, such as the power method and its variations, provide a useful means to obtain eigenvalues and corresponding eigenvectors. Details of methods for solving for eigenvalues can be found in Hoffman (1992) and Fadeev and Fadeeva (1963).

B.8   COUPLED LINEAR DIFFERENTIAL EQUATIONS

The eigenproblem of Section B.7 is useful in the solution of coupled linear differential equations. Let these equations be represented by the following set written in compact matrix notation

$$\frac{d\mathbf{y}}{dt} = \mathbf{A}\mathbf{y} + \mathbf{f} \qquad (\mathbf{B.79})$$

subject to the condition

$$t = 0; \quad \mathbf{y} = \mathbf{y}_0 \qquad (\mathbf{B.80})$$

where $\mathbf{A}$ is the constant coefficient matrix of size $N \times N$ and $\mathbf{f}$ is the constant vector.

The general solution of linear differential equations is a linear combination of a homogeneous solution and a particular solution. For Eq. B.79, the particular solution is simply the steady-state solution; that is

$$\mathbf{A}\mathbf{y}_p = -\mathbf{f} \qquad (\mathbf{B.81})$$

which $\mathbf{y}_p$ can be readily obtained by the techniques mentioned in Sections B.5 and B.6.

The homogeneous solution must satisfy the equation

$$\frac{d\mathbf{y}}{dt} = \mathbf{A}\mathbf{y} \qquad (\mathbf{B.82})$$

which is the original equation with the forcing term removed. To solve the homogeneous equation, we assume the trial solution $\mathbf{x}\exp(\lambda t)$, where $\mathbf{x}$ and λ are yet to be determined vector and scalar quantities. If we substitute this trial solution into the homogeneous equation (Eq. B.82), we obtain the following algebraic equation

$$\mathbf{A}\mathbf{x} = \lambda\mathbf{x} \tag{B.83}$$

which is the eigenproblem considered in Section B.7. Assuming that there exists N eigenvalues and N corresponding eigenvectors, the homogeneous solution will be the linear combination of all equations; that is, the homogeneous solution is

$$\mathbf{y}_H = \sum_{i=1}^{N} c_i \mathbf{x}\exp(\lambda_i t) \tag{B.84}$$

The general solution is the sum of the homogeneous solution and the particular solution, that is,

$$\mathbf{y} = \mathbf{y}_p + \mathbf{y}_H \tag{B.85}$$

The constants c_i can be found by applying initial conditions.

B.9 REFERENCES

1. Fadeev, D. K., and V. N. Fadeeva, "Computational Methods of Linear Algebra," Freeman, San Francisco (1963).
2. Hoffman, J. D., "Numerical Methods for Engineers and Scientists," McGraw Hill, New York (1992).

Appendix C

Derivation of the Fourier–Mellin Inversion Theorem

We recall from Section 10.5.1 in Chapter 10, that solutions of the Sturm–Liouville equation, along with suitable Sturm–Liouville boundary conditions, always produced orthogonal functions. Thus, the functions

$$\varphi_n(x) = \sin\left(\frac{n\pi}{L}x\right) \qquad n = 1, 2, 3, \ldots \tag{C1}$$

form an orthogonal set in the range $0 \leq x \leq L$ if it satisfies

$$\frac{d^2 y}{dx^2} + \lambda y = 0; \quad y(0) = 0; \quad y(L) = 0 \tag{C2}$$

since this is a Sturm–Liouville equation if in Eq. 10.185 we stipulate $p = 1$, $q = 0$, $r = 1$, and $\beta = \lambda$. The eigenvalues are then seen to be $\lambda_n = n^2\pi^2/L^2$.

Thus, if we represent a function $f(x)$ in this interval by an expansion of such orthogonal functions

$$f(x) = A_1 \sin\frac{\pi x}{L} + A_2 \sin\frac{2\pi x}{L} + \cdots \tag{C3}$$

or, generally, as

$$f(x) = \sum_{n=1}^{\infty} A_n \sin\left(\frac{n\pi x}{L}\right) \tag{C4}$$

then the coefficients are obtained using the orthogonality condition, Eq. 10.174,

663

written for the problem at hand as

$$\int_0^L \varphi_n(x)\varphi_m(x)\, dx = 0 \qquad n \neq m \tag{C5}$$

hence, the coefficients A_n are obtained from

$$A_n \int_0^L \sin^2\left(\frac{n\pi x}{L}\right) dx = \int_0^L f(x) \sin\left(\frac{n\pi x}{L}\right) dx \tag{C6}$$

Performing the LHS integral, we obtain A_n

$$A_n = \frac{2}{L} \int_0^L f(x) \sin\left(\frac{n\pi x}{L}\right) dx \tag{C7}$$

The series representing $f(x)$ is called a Fourier-Sine series. Similarly, we can also express functions in terms of the Fourier-Cosine series

$$\varphi_n(x) = \cos\left(\frac{n\pi x}{L}\right) \qquad n = 0, 1, 2, 3, \ldots \tag{C8}$$

provided again the Sturm–Liouville equation is satisfied, with a different set of boundary conditions

$$\frac{d^2 y}{dx^2} + \lambda y = 0; \quad y'(0) = 0, \quad y'(L) = 0 \tag{C9}$$

It is important to note that $\varphi_0(x) = 1$ in this case, so to represent $f(x)$ write

$$f(x) = A_0 + A_1 \cos\left(\frac{\pi x}{L}\right) + A_2 \cos\left(\frac{2\pi x}{L}\right) + \cdots \tag{C10}$$

or generally,

$$f(x) = A_0 + \sum_{n=1}^{\infty} A_n \cos\left(\frac{n\pi x}{L}\right); \qquad 0 \leq x \leq L \tag{C11}$$

Using the orthogonality condition as before, noting

$$\int_0^L \cos^2\left(\frac{n\pi x}{L}\right) dx = \begin{cases} \dfrac{L}{2}; & n = 1, 2, 3 \ldots \\ L; & n = 0 \end{cases} \tag{C12}$$

so that the coefficients are computed to be

$$A_0 = \frac{1}{L} \int_0^L f(x)\, dx$$

$$A_n = \frac{2}{L} \int_0^L f(x) \cos\left(\frac{n\pi x}{L}\right) dx \tag{C13}$$

Thus, Eq. C10, along with C13, are known as the Fourier-Cosine representation of $f(x)$, in the interval $0 \leq x \leq L$ (Hildebrand 1962).

The question arises, how is it possible to obtain a similar representation in the complete semi-infinite interval with $x > 0$? It is clear if we replaced L with ∞ in

$$\varphi_n = \sin\left(\frac{n\pi x}{L}\right)$$

then φ_n would vanish. Suppose we consider the relation, as done by Hildebrand (1962):

$$\frac{2}{L}\int_0^L \sin(\lambda_1 x)\sin(\lambda_2 x)\, dx = \frac{\sin(\lambda_2 - \lambda_1)L}{(\lambda_2 - \lambda_1)L} - \frac{\sin(\lambda_2 + \lambda_1)L}{(\lambda_2 + \lambda_1)L} \quad \textbf{(C14)}$$

One can see the expression on the RHS vanishes if λ_1 and λ_2 are different integral multiples of π/L. Moreover, in the limit, the RHS becomes unity if λ_1 and λ_2 both take on the same integral multiple of π/L. We also see that as $L \to \infty$, the RHS tends to zero for any positive values of λ_1 and λ_2, so long as $\lambda_1 \neq \lambda_2$; but if $\lambda_1 = \lambda_2$, the RHS tends to unity.

Now, in general notation, we can write

$$\varphi_\lambda(x) = \sin(\lambda x) \quad \textbf{(C15)}$$

and obtain in the limit

$$\lim_{L \to \infty} \frac{2}{L}\int_0^L \varphi_{\lambda_1}(x) \cdot \varphi_{\lambda_2}(x)\, dx = \begin{cases} 0 & (\lambda_1 \neq \lambda_2) \\ 1 & (\lambda_1 = \lambda_2) \end{cases} \quad \textbf{(C16)}$$

This is rather like an orthogonality condition, for any positive values λ_1 and λ_2, in the interval $0 \leq x \leq \infty$. We are guided to the conclusion that a representation of a function $f(x)$ in such semi-infinite intervals must involve all possible functions of the type in Eq. C15, where λ is *not* restricted to discrete values, but can take on any positive number, as a continuum in $\lambda > 0$. Previously, we represented $f(x)$ by the infinite series in the region $0 \leq x \leq L$:

$$f(x) = \sum_{n=1}^{\infty} A_n \sin\frac{n\pi x}{L}$$

In the limit as $L \to \infty$, we can write the contribution $\sin(\lambda x)$ to the series representation of $f(x)$ as $A(\lambda)\sin(\lambda x)$. Now, since λ is regarded as a continuous variables we can rewrite the summation above as an infinite integral.

$$f(x) = \int_0^{\infty} A(\lambda)\sin(\lambda x)\, d\lambda; \quad 0 < x < \infty \quad \textbf{(C17)}$$

where x now resides in the semi-infinite plane.

By these arguments, we have developed a possible representation in the time domain, since representations in time must be semi-infinite in the sense $0 \leq t \leq \infty$.

Continuing, we must develop a procedure for finding $A(\lambda)$, much the same as in the discrete case to find A_n. In analogy with applications of the orthogonality condition, we multiply both sides of Eq C17 by $\sin(\lambda_0 x)$

$$f(x) \sin(\lambda_0 x) = \sin(\lambda_0 x) \int_0^\infty A(\lambda) \sin(\lambda x) \, d\lambda \qquad \text{(C18)}$$

Next, integrate both sides over the interval 0 to L (where eventually $L \to \infty$)

$$\int_0^L f(x) \sin(\lambda_0 x) \, dx = \int_0^L \sin(\lambda_0 x) \left[\int_0^\infty A(\lambda) \sin(\lambda x) \, d\lambda \right] dx \qquad \text{(C19)}$$

and if the order of integration can be interchanged, we get

$$\int_0^L f(x) \sin(\lambda_0 x) \, dx = \int_0^\infty A(\lambda) \left[\int_0^L \sin(\lambda x) \sin(\lambda_0 x) \, dx \right] d\lambda \qquad \text{(C20)}$$

We shall denote the RHS as F_L. Now, if we apply Eq. C14 to the RHS:

$$F_L = \frac{1}{2} \int_0^\infty \frac{A(\lambda)}{(\lambda - \lambda_0)} \sin L(\lambda - \lambda_0) \, d\lambda - \frac{1}{2} \int_0^\infty \frac{A(\lambda) \sin L(\lambda + \lambda_0)}{(\lambda + \lambda_0)} \, d\lambda \qquad \text{(C21)}$$

Now, since λ is continuous, we can replace it with another continuous variable; so, replace $L(\lambda - \lambda_0) = t$ in the first (noting that $dL(\lambda - \lambda_0) = L d\lambda = dt$ and replace $L(\lambda + \lambda_0) = t$ in the second integral, to get

$$F_L = \frac{1}{2} \int_{-L\lambda_0}^\infty A\left(\lambda_0 + \frac{t}{L}\right) \frac{\sin(t)}{t} \, dt - \frac{1}{2} \int_{L\lambda_0}^\infty A\left(-\lambda_0 + \frac{t}{L}\right) \frac{\sin(t)}{t} \, dt \qquad \text{(C22)}$$

Now, since $\lambda_0 > 0$, we note in the limit $L \to \infty$ that the second integral tends to zero, that is,

$$\int_\infty^\infty f(t) \, dt \to 0$$

and the first becomes

$$\lim_{L \to \infty} F_L = \frac{1}{2} \int_{-\infty}^\infty A(\lambda_0) \frac{\sin t}{t} \, dt \qquad \text{(C23)}$$

Carrying this further yields

$$\frac{1}{2} \int_{-\infty}^\infty A(\lambda_0) \frac{\sin t}{t} \, dt = \frac{A(\lambda_0)}{2} \int_{-\infty}^\infty \frac{\sin t}{t} \, dt = \frac{\pi}{2} A(\lambda_0) \qquad \text{(C24)}$$

To see this last result, take the imaginary part of the integral

$$\int_{-L}^{L} \frac{e^{it}}{t} \, dt = \left[\log t + t - \frac{t^3}{3 \cdot (3!)} + \cdots \right]_{-L}^{L}$$

in the limit as $L \to \infty$.

We now have formally shown for $\lambda_0 > 0$

$$\int_0^\infty f(x) \sin(\lambda_0 x) \, dx = \frac{\pi}{2} A(\lambda_0) \tag{C25}$$

Since λ is continuous, the above result is true for any λ, so replace λ_0 with λ to get the general result.

$$A(\lambda) = \frac{2}{\pi} \int_0^\infty f(x) \sin(\lambda x) \, dx \tag{C26}$$

This is formally equivalent to Eq. C7 for the case of discrete λ_n.

Now, to represent $f(x)$ for a continuum of λ, the integral representation is, formally,

$$f(x) = \int_0^\infty A(\lambda) \sin(\lambda x) \, d\lambda \tag{C27}$$

To prevent confusion with the dummy variable x in Eq. C26, we rewrite $A(\lambda)$ as

$$A(\lambda) = \frac{2}{\pi} \int_0^\infty f(t) \sin(\lambda t) \, dt \tag{C28}$$

hence, we now have

$$f(x) = \frac{2}{\pi} \int_0^\infty \sin \lambda x \left[\int_0^\infty f(t) \sin \lambda t \, dt \right] d\lambda \tag{C29}$$

This is called the *Fourier sine integral*. By similar arguments, we can derive the Fourier cosine integral as

$$f(x) = \frac{2}{\pi} \int_0^\infty \cos(\lambda x) \left[\int_0^\infty f(t) \cos \lambda t \, dt \right] d\lambda \tag{C30}$$

These are the two building blocks to prove the Fourier–Mellin inversion theorem for Laplace transforms.

The final stage in this proof is to extend the bounds of integration to $-\infty < x < \infty$. It is clear that the sine series represents $-f(-x)$ when $x < 0$, and the cosine series represents $f(-x)$ when $x < 0$. So, if $f(x)$ is an odd function[1] the representation in the sine integral can include all values of x ($-\infty$

---

[1] In general, a function $f(x)$ is called an odd function if $f(-x) = -f(x)$ (e.g., $f(x) = x^3$) and an even function if $f(-x) = f(x)$ (e.g., $f(x) = x^2$)

to $+\infty$). However, if $f(x)$ is an even function then the cosine integral can represent all values of x. Thus, it is possible to represent all values of x by using both sine and cosine components.

Thus, if we split a given function into even and odd parts

$$f(x) = f_e(x) + f_0(x)$$

we can easily see that

$$\frac{1}{\pi} \int_0^\infty \cos \lambda x \left[\int_{-\infty}^\infty f(t) \cos \lambda t \, dt \right] d\lambda$$

$$= \frac{1}{\pi} \int_0^\infty \cos \lambda x \left[\int_{-\infty}^\infty f_e(t) \cos \lambda t \, dt \right] d\lambda$$

$$= \frac{2}{\pi} \int_0^\infty \cos \lambda x \left[\int_0^\infty f_e(t) \cos \lambda t \, dt \right] d\lambda = f_e(x); \quad -\infty < x < \infty \quad \text{(C31)}$$

Similarly, for the odd component

$$\frac{1}{\pi} \int_0^\infty \sin \lambda x \left[\int_{-\infty}^\infty f(x) \sin \lambda t \, dt \right] d\lambda = f_0(x); \quad -\infty < x < \infty \quad \text{(C32)}$$

Adding the two, we have the representation

$$f(x) = \int_0^\infty [A(\lambda) \cos (\lambda x) + B(\lambda) \sin (\lambda x)] \, d\lambda \quad \text{(C33)}$$

where $A(\lambda)$ and $B(\lambda)$ are defined as

$$A(\lambda) = \frac{1}{\pi} \int_{-\infty}^\infty f(t) \cos (\lambda t) \, dt; \qquad B(\lambda) = \frac{1}{\pi} \int_{-\infty}^\infty f(t) \sin (\lambda t) \, dt \quad \text{(C34)}$$

Introducing these, and using trigonometric identities, we write the form

$$f(x) = \frac{1}{\pi} \int_0^\infty \left[\int_{-\infty}^\infty f(t) \cos[\lambda(t - x)] \, dt \right] d\lambda; \quad -\infty < x < \infty \quad \text{(C35)}$$

or equivalently,

$$f(x) = \frac{1}{2\pi} \int_{-\infty}^\infty \int_{-\infty}^\infty f(t) \cos[\lambda(t - x)] \, dt \, d\lambda \quad \text{(C36)}$$

This expression is called the complete Fourier integral representation and it can represent arbitrary $f(x)$ for all values of x in the usual sense; $f(x)$ is at least piecewise differentiable and the integral

$$\int_{-\infty}^\infty |f(x)| \, dx$$

exists. Next, by noting one half a function plus one half its complex conjugate recovers the function

$$\cos \lambda(t - x) = \frac{1}{2} \exp(i\lambda(t - x)) + \frac{1}{2} \exp(-i\lambda(t - x)) \qquad \text{(C37)}$$

then the complex form of Eq. C36 can be shown (Churchill 1963) to be

$$f(x) = \frac{1}{2\pi} \int_{-\infty}^{\infty} \int_{-\infty}^{\infty} f(t) e^{-i\lambda(t-x)} \, dt \, d\lambda \qquad \text{(C38)}$$

In this expression, t is simply a dummy variable and can be replaced with any convenient symbol. Later, we shall inspect arbitrary functions of time $f(t)$, instead of $f(x)$, so it is propitious to replace t with β to get

$$f(x) = \frac{1}{2\pi} \int_{-\infty}^{\infty} e^{i\lambda x} \left[\int_{-\infty}^{\infty} f(\beta) e^{-i\lambda\beta} \, d\beta \right] d\lambda \qquad \text{(C39)}$$

This form is now suitable for deducing the Fourier–Mellin inversion formula for Laplace transforms. In terms of real time as the independent variable, we can write the Fourier integral representation of any arbitrary function of time, with the provision that $f(t) = 0$ when $t < 0$, so Eq. C39 becomes

$$f(t) = \frac{1}{2\pi} \int_{-\infty}^{\infty} e^{i\lambda t} \left[\int_{0}^{\infty} f(\beta) e^{-i\lambda\beta} \, d\beta \right] d\lambda \qquad \text{(C40)}$$

Suppose we consider the product $\exp(-\sigma_0 t) f(t)$, then the integral in (C40) can be written

$$e^{-\sigma_0 t} f(t) = \frac{1}{2\pi} \lim_{\lambda \to \infty} \int_{-\lambda}^{\lambda} e^{i\lambda t} \left[\int_{0}^{\infty} f(\beta) e^{-\sigma_0 \beta} e^{-i\lambda\beta} \, d\beta \right] d\lambda \qquad \text{(C41)}$$

This representation is valid provided the integral for real values σ_0

$$J = \int_{0}^{\infty} e^{-\sigma_0 t} |f(t)| \, dt \qquad \text{(C42)}$$

exists. This simply reflects the requirement that all singularities of $F(s)$ are to the left of the line through σ_0, as illustrated in the first Bromwich path (Fig. 9.4). This guarantees that exponential behavior by $f(t)$ can always be dominated by suitable selection of positive, real values of σ_0. Suppose $f(t) = e^{\alpha t}$, then we have

$$J = \int_{0}^{\infty} e^{-\sigma_0 t} e^{\alpha t} \, dt = \int_{0}^{\infty} e^{-(\sigma_0 - \alpha)t} \, dt = \frac{1}{\sigma_0 - \alpha} \qquad \text{(C43)}$$

which is finite provided $\sigma_0 > \alpha$, hence the integral exists. Now that existence of the inner integral

$$\int_0^\infty f(\beta)e^{-\sigma_0 t}e^{-i\lambda\beta}\,d\beta$$

is guaranteed, we can write Eq. C41 as

$$f(t) = \frac{1}{2\pi}\lim_{\lambda\to\infty}\int_{-\lambda}^{\lambda}e^{(i\lambda+\sigma_0)t}\left[\int_0^\infty f(\beta)e^{-(i\lambda+\sigma_0)\beta}\,d\beta\right]d\lambda$$

We can now define the substitution of the complex variable $s = i\lambda + \sigma_0$, where σ_0 is a fixed real constant, hence, $ds = id\lambda$ so we get

$$f(t) = \frac{1}{2\pi i}\lim_{\lambda\to\infty}\int_{\sigma_0-i\lambda}^{\sigma_0+i\lambda}e^{st}\left[\int_0^\infty f(\beta)e^{-s\beta}\,d\beta\right]ds \qquad (\text{C44})$$

But the interior integral defines, in general, the Laplace transform

$$F(s) = \int_0^\infty f(\beta)e^{-s\beta}\,d\beta \qquad (\text{C45})$$

regardless of the symbol used for dummy variable; replacing the interior with $F(s)$, Eq. C44 becomes formally identical with Eq. 9.3, if we replace λ with ω

$$f(t) = \frac{1}{2\pi i}\lim_{\omega\to\infty}\int_{\sigma_0-i\omega}^{\sigma_0+i\omega}e^{st}F(s)\,ds \qquad (\text{C46})$$

The integral is taken along the infinite vertical line through $s = \sigma_0$ and parallel to the imaginary axis, as illustrated in Fig. 9.4. The existence of $F(s)$ is guaranteed as long as σ_0 is greater than the real part of any singularity arising from $F(s)$.

REFERENCES

1. Churchill, R. V., "Fourier Series and Boundary Value Problems," McGraw Hill, New York, 2nd Ed., p. 120 (1963).
2. Hildebrand, F. B., "Advanced Calculus for Applications," Prentice-Hall, Englewood Cliffs, New Jersey, pp. 236–239 (1962).

Appendix **D**

Table of Laplace Transforms[1]

$F(s)$	Number	$f(t)$
$\dfrac{1}{s}$	1	$u(t) = 1,\ t > 0$
$\dfrac{1}{s^2}$	2	$r(t) = t,\ t > 0$
$\dfrac{1}{s^n}\ (n = 1, 2, 3, \ldots)$	3	$\dfrac{t^{n-1}}{(n-1)!}$
$\dfrac{1}{\sqrt{s}}$	4	$\dfrac{1}{\sqrt{\pi t}}$
$\dfrac{1}{s^{3/2}}$	5	$2\sqrt{\dfrac{t}{\pi}}$
$\dfrac{1}{s^{\alpha}}\ (\alpha > 0)$	6	$\dfrac{t^{\alpha-1}}{\Gamma(\alpha)}$
$\dfrac{1}{s + a}$	7	e^{-at}
$\dfrac{1}{(s + a)^2}$	8	te^{-at}
$\dfrac{1}{(s + a)(s + b)}$	9	$\dfrac{1}{a - b}(e^{-bt} - e^{-at})$
$\dfrac{1}{(s + a)^n}\ (n = 1, 2, \ldots)$	10	$\dfrac{1}{(n - 1)!}t^{n-1}e^{-at}$
$\dfrac{1}{(s + a)^{\alpha}}\ (\alpha > 0)$	11	$\dfrac{t^{\alpha-1}e^{-at}}{\Gamma(\alpha)}$
$\dfrac{s}{(s + a)(s + b)}$	12	$\dfrac{1}{b - a}(be^{-bt} - ae^{-at})$
$\dfrac{1}{s^2 + a^2}$	13	$\dfrac{1}{a}\sin at$
$\dfrac{s}{s^2 + a^2}$	14	$\cos at$
$\dfrac{1}{s^2 - a^2}$	15	$\dfrac{1}{a}\sinh at$
$\dfrac{s}{s^2 - a^2}$	16	$\cosh at$
$\dfrac{1}{s(s^2 + a^2)}$	17	$\dfrac{1}{a^2}(1 - \cos at)$

(*continues*)

[1]From "Operational Mathematics," 2nd Ed., by R.V. Churchill, Copyright 1958, McGraw-Hill Book Co., Inc. Used by permission.

Table of Laplace Transforms (*Continued*)

$F(s)$	Number	$f(t)$
$\dfrac{1}{s^2(s^2 + a^2)}$	18	$\dfrac{1}{a^3}(at - \sin at)$
$\dfrac{1}{(s^2 + a^2)^2}$	19	$\dfrac{1}{2a^3}(\sin at - at\cos at)$
$\dfrac{s}{(s^2 + a^2)^2}$	20	$\dfrac{t}{2a}\sin at$
$\dfrac{s^2}{(s^2 + a^2)^2}$	21	$\dfrac{1}{2a}(\sin at + at\cos at)$
$\dfrac{s^2 - a^2}{(s^2 + a^2)^2}$	22	$t\cos at$
$\dfrac{s}{(s^2 + a^2)(s^2 + b^2)}$	23	$\dfrac{\cos at - \cos bt}{b^2 - a^2}(a^2 \neq b^2)$
$\dfrac{1}{(s - a)^2 + b^2}$	24	$\dfrac{1}{b}e^{at}\sin bt$
$\dfrac{(s - a)}{(s - a)^2 + b^2}$	25	$e^{at}\cos bt$
$\dfrac{3a^2}{s^3 + a^3}$	26	$e^{-at} - e^{(1/2)at}\left[\cos\left(\dfrac{at\sqrt{3}}{2}\right)\right.$ $\left. - \sqrt{3}\sin\left(\dfrac{at\sqrt{3}}{2}\right)\right]$
$\dfrac{4a^3}{s^4 + 4a^4}$	27	$\sin(at)\cosh(at) - \cos(at)\sinh(at)$
$\dfrac{s}{s^4 + 4a^4}$	28	$\dfrac{1}{2a^2}[\sin(at)\sinh(at)]$
$\dfrac{1}{s^4 - a^4}$	29	$\dfrac{1}{2a^3}[\sinh(at) - \sin(at)]$
$\dfrac{s}{s^4 - a^4}$	30	$\dfrac{1}{2a^2}[\cosh(at) - \cos(at)]$
$\dfrac{8a^3s^2}{(s^2 + a^2)^3}$	31	$(1 + a^2t^2)\sin(at) - at\cos(at)$
$\dfrac{1}{s}\left(\dfrac{s - 1}{s}\right)^n$	32	$\dfrac{e^t}{n!}\dfrac{d^n}{dt^n}(t^n e^{-t})$ $\left[n^{th}\text{ degree Laguerre polynomial}\right]$
$\dfrac{s}{(s - a)^{3/2}}$	33	$\dfrac{1}{\sqrt{\pi t}}e^{at}(1 + 2at)$
$\sqrt{s - a} - \sqrt{s - b}$	34	$\dfrac{1}{2\sqrt{\pi t^3}}(e^{bt} - e^{at})$
$\dfrac{1}{a + \sqrt{s}}$	35	$\dfrac{1}{\sqrt{\pi t}} - ae^{a^2t}\operatorname{erfc}(a\sqrt{t})$
$\dfrac{\sqrt{s}}{s - a^2}$	36	$\dfrac{1}{\sqrt{\pi t}} + ae^{a^2t}\operatorname{erf}(a\sqrt{t})$
$\dfrac{\sqrt{s}}{s + a^2}$	37	$\dfrac{1}{\sqrt{\pi t}} - \dfrac{2a}{\sqrt{\pi}}e^{-a^2t}\displaystyle\int_0^{a\sqrt{t}}e^{\beta^2}\,d\beta$

(*continues*)

Table of Laplace Transforms (*Continued*)

$F(s)$	Number	$f(t)$
$\dfrac{1}{\sqrt{s}\,(s-a^2)}$	38	$\dfrac{1}{a}e^{a^2t}\,\mathrm{erf}(a\sqrt{t})$
$\dfrac{1}{\sqrt{s}\,(s+a^2)}$	39	$\dfrac{2}{a\sqrt{\pi}}e^{-a^2t}\displaystyle\int_0^{a\sqrt{t}}e^{\beta^2}\,d\beta$
$\dfrac{1}{\sqrt{s}\,(a+\sqrt{s})}$	40	$e^{a^2t}\,\mathrm{erfc}(a\sqrt{t})$
$\dfrac{1}{(s+a)\sqrt{s+b}}$	41	$\dfrac{1}{\sqrt{b-a}}e^{-at}\,\mathrm{erf}\left[\sqrt{t(b-a)}\right]$
$\dfrac{\sqrt{s+2a}}{\sqrt{s}}-1$	42	$ae^{-at}[\mathrm{I}_1(at)+\mathrm{I}_0(at)]$
$\dfrac{1}{\sqrt{(s+a)(s+b)}}$	43	$e^{-\frac{1}{2}(a+b)t}\,\mathrm{I}_0\!\left(\dfrac{a-b}{2}t\right)$
$(\alpha>0)\,\dfrac{\Gamma(\alpha)}{(s+a)^\alpha(s+b)^\alpha}$	44	$\sqrt{\pi}\left(\dfrac{t}{a-b}\right)^{\alpha-1/2}e^{-\frac{1}{2}(a+b)t}$ $\cdot\,\mathrm{I}_{\alpha-1/2}\!\left(\dfrac{a-b}{2}t\right)$
$\dfrac{1}{(s+a)^{1/2}(s+b)^{3/2}}$	45	$te^{-\frac{1}{2}(a+b)t}\left[\mathrm{I}_0\!\left(\dfrac{a-b}{2}t\right)\right.$ $\left.+\mathrm{I}_1\!\left(\dfrac{a-b}{2}t\right)\right]$
$\dfrac{\sqrt{s+2a}-\sqrt{s}}{\sqrt{s+2a}+\sqrt{s}}$	46	$\dfrac{1}{t}e^{-at}\mathrm{I}_1(at)$
$\dfrac{1}{\sqrt{s^2+a^2}}$	47	$J_0(at)$
$\dfrac{\left[\sqrt{s^2+a^2}-s\right]^\alpha}{\sqrt{s^2+a}}\,(\alpha>-1)$	48	$a^\alpha J_\alpha(at)$
$\dfrac{1}{(s^2+a^2)^\alpha}\,(\alpha>0)$	49	$\dfrac{\sqrt{\pi}}{\Gamma(\alpha)}\left(\dfrac{t}{2a}\right)^{\alpha-1/2}J_{\alpha-1/2}(at)$
$\left[\sqrt{s^2+a^2}-s\right]^\alpha\,(\alpha>0)$	50	$\dfrac{\alpha a^\alpha}{t}J_\alpha(at)$
$\dfrac{\left[s-\sqrt{s^2-a^2}\right]^\alpha}{\sqrt{s^2-a^2}}\,(\alpha>-1)$	51	$a^\alpha\mathrm{I}_\alpha(at)$
$\dfrac{1}{(s^2-a^2)^\alpha}\,(\alpha>0)$	52	$\dfrac{\sqrt{\pi}}{\Gamma(\alpha)}\left(\dfrac{t}{2a}\right)^{\alpha-1/2}\mathrm{I}_{\alpha-1/2}(at)$
$\dfrac{1}{s}e^{-\alpha/s}$	53	$J_0(2\sqrt{\alpha t})$
$\dfrac{1}{\sqrt{s}}e^{-\alpha/s}$	54	$\dfrac{1}{\sqrt{\pi t}}\cos(2\sqrt{\alpha t})$
$\dfrac{1}{\sqrt{s}}e^{\alpha/s}$	55	$\dfrac{1}{\sqrt{\pi t}}\cosh(2\sqrt{\alpha t})$
$\dfrac{1}{s^{3/2}}e^{-\alpha/s}$	56	$\dfrac{1}{\sqrt{\pi\alpha}}\sin(2\sqrt{\alpha t})$

(*continues*)

Table of Laplace Transforms (*Continued*)

$F(s)$	Number	$f(t)$
$\dfrac{1}{s^{3/2}}e^{\alpha/s}$	57	$\dfrac{1}{\sqrt{\pi\alpha}}\sinh(2\sqrt{\alpha t})$
$\dfrac{1}{s^{\mu}}e^{-\alpha/s}(\alpha>0)$	58	$\left(\dfrac{t}{\alpha}\right)^{(\mu-1)/2}J_{\mu-1}(2\sqrt{\alpha t})$
$\dfrac{1}{s^{\mu}}e^{\alpha/s}(\alpha>0)$	59	$\left(\dfrac{t}{\alpha}\right)^{(\mu-1)/2}I_{\mu-1}(2\sqrt{\alpha t})$
$e^{-\alpha\sqrt{s}}(\alpha>0)$	60	$\dfrac{\alpha}{2\sqrt{\pi t^3}}\exp\left(-\dfrac{\alpha^2}{4t}\right)$
$\dfrac{1}{s}e^{-\alpha\sqrt{s}}(\alpha\geq0)$	61	$\operatorname{erfc}\left(\dfrac{\alpha}{2\sqrt{t}}\right)$
$\dfrac{1}{\sqrt{s}}e^{-\alpha\sqrt{s}}(\alpha\geq0)$	62	$\dfrac{1}{\sqrt{\pi t}}\exp\left(-\dfrac{\alpha^2}{4t}\right)$
$s^{-3/2}e^{-\alpha\sqrt{s}}(\alpha\geq0)$	63	$2\sqrt{\dfrac{t}{\pi}}\exp\left(-\dfrac{\alpha^2}{4t}\right)-\alpha\operatorname{erfc}\left(\dfrac{\alpha}{2\sqrt{t}}\right)$
$\dfrac{1}{s}\log s$	64	$\Gamma'(1)-\log t\,[\Gamma'(1)=-0.5772]$
$\dfrac{1}{s^k}\log s(k\geq0)$	65	$t^{k-1}\left\{\dfrac{\Gamma'(k)}{[\Gamma(k)]^2}-\dfrac{\log t}{\Gamma(k)}\right\}$
$\dfrac{\log s}{s-a}(a\geq0)$	66	$e^{at}[\log a-Ei(-at)]$
$\dfrac{\log s}{s^2+1}$	67	$Si(t)\cdot\cos(t)-Ci(t)\cdot\sin(t)$
$\dfrac{s\log s}{s^2+1}$	68	$-Si(t)\cdot\sin(t)-Ci(t)\cdot\cos(t)$
$\dfrac{1}{s}\log(1+ks)(k>0)$	69	$-Ei(-t/k)$
$\log\left(\dfrac{s-a}{s-b}\right)$	70	$\dfrac{1}{t}(e^{bt}-e^{at})$
$\dfrac{1}{s}\log(1+k^2s^2)$	71	$-2Ci(t/k)$
$\dfrac{1}{s}\log(s^2+a^2)(a>0)$	72	$2\log(a)-2Ci(at)$
$\dfrac{1}{s^2}\log(s^2+a^2)(a>0)$	73	$\dfrac{2}{a}[at\cdot\log(a)+\sin(at)-at\cdot Ci(t)]$
$\log\left(\dfrac{s^2+a^2}{s^2}\right)$	74	$\dfrac{2}{t}[1-\cos(at)]$
$\log\left(\dfrac{s^2-a^2}{s^2}\right)$	75	$\dfrac{2}{t}[1-\cosh(at)]$
$\arctan\left(\dfrac{k}{s}\right)$	76	$\dfrac{1}{t}\sin(kt)$
$\dfrac{1}{s}\arctan\left(\dfrac{k}{s}\right)$	77	$Si(kt)$
$e^{k^2s^2}\operatorname{erfc}(ks)(k>0)$	78	$\dfrac{1}{k\sqrt{\pi}}\exp\left(-\dfrac{t^2}{4k^2}\right)$

(*continues*)

Table of Laplace Transforms (*Continued*)

$F(s)$	Number	$f(t)$
$\frac{1}{s}e^{k^2s^2}\text{erfc}(ks)(k>0)$	79	$\text{erf}\left(\frac{t}{2k}\right)$
$e^{ks}\text{erfc}(\sqrt{ks})(k>0)$	80	$\frac{\sqrt{k}}{\pi\sqrt{t}\,(t+k)}$
$\frac{1}{\sqrt{s}}\text{erfc}(\sqrt{ks})$	81	$\begin{cases}0 & \text{when } 0<t<k \\ (\pi t)^{-1/2} & \text{when } t>k\end{cases}$
$\frac{1}{\sqrt{s}}e^{ks}\text{erfc}(\sqrt{ks})(k>0)$	82	$\frac{1}{\sqrt{\pi(t+k)}}$
$\text{erf}\left(\frac{k}{\sqrt{s}}\right)$	83	$\frac{1}{\pi t}\sin(2k\sqrt{t})$
$\frac{1}{\sqrt{s}}e^{k^2/s}\text{erfc}\left(\frac{k}{\sqrt{s}}\right)$	84	$\frac{1}{\sqrt{\pi t}}e^{-2k\sqrt{t}}$
$K_0(ks)$	85	$\begin{cases}0 & \text{when } 0<t<k \\ (t^2-k^2)^{-1/2} & \text{when } t>k\end{cases}$
$K_0(k\sqrt{s})$	86	$\frac{1}{2t}\exp\left(-\frac{k^2}{4t}\right)$
$\frac{1}{s}e^{ks}K_1(k\sqrt{s})$	87	$\frac{1}{k}\sqrt{t(t+2k)}$
$\frac{1}{\sqrt{s}}K_1(k\sqrt{s})$	88	$\frac{1}{k}\exp\left(-\frac{k^2}{4t}\right)$
$\frac{1}{\sqrt{s}}e^{k/s}K_0\left(\frac{k}{s}\right)$	89	$\frac{2}{\sqrt{\pi t}}K_0(2\sqrt{2kt})$
$\pi e^{-ks}I_0(ks)$	90	$\begin{cases}[t(2k-1)]^{-1/2} & \text{when } 0<t<2k \\ 0 & \text{when } t>2k\end{cases}$
$e^{-ks}I_1(ks)$	91	$\begin{cases}\frac{(k-t)}{\pi k\sqrt{t(2k-1)}} & \text{when } 0<t<2k \\ 0 & \text{when } t>2k\end{cases}$
$-e^{as}Ei(-as)$	92	$\frac{1}{t+a}(a>0)$
$\frac{1}{a}+se^{as}Ei(-as)$	93	$\frac{1}{(t+a)^2}(a>0)$

The exponential, cosine, and sine integral functions are defined as

$$Ei(t)=\int_{-\infty}^t\frac{e^\beta}{\beta}\,d\beta$$

$$Ci(t)=\int_\infty^t\frac{\cos\beta}{\beta}\,d\beta$$

$$Si(t)=\int_0^t\frac{\sin\beta}{\beta}\,d\beta$$

These are tabulated in most mathematical handbooks and are also discussed in Chapter 4.

Appendix E

Numerical Integration

This appendix provides procedures and formulas for numerical integration. This is the only recourse when the integral in question cannot be integrated analytically. Many engineering problems give rise to integrals wherein analytical solutions are not always possible. For example, the solution of boundary value problems using the method of Galerkin gives rise to integrals which in general cannot be integrated if the defining equations are nonlinear. In such cases the integrands are known functions. However, there are cases where values of the function y_j at discrete points x_j are known and the integral of y with respect to x is required. The evaluation of the integral in this case must be dealt with numerically. We will deal with these two distinct cases in this appendix.

The integration of a function $y(x)$ with respect to x from a to b is defined as:

$$I = \int_a^b y(x)\, dx \qquad (E.1)$$

where $y(x)$ can be a known function or only its values at some discrete points. The process of replacing an integral with the sum of its parts is often referred to as "quadrature."

E.1 BASIC IDEA OF NUMERICAL INTEGRATION

The basic idea of the numerical integration is to approximate the function $y(x)$ by a polynomial of degree N, $P_N(x)$, and then to perform the integration of this approximating polynomial exactly since each term x^j in the polynomial can be analytically integrated. The accuracy of the numerical integration depends on how well we choose our approximating polynomial.

If the function $y(x)$ is known, we simply choose discrete positions x (for example, 2, 3, or 4 points) within the domain of integration ($a < x < b$) and then cause an approximating polynomial to pass through these points. We can then perform the integration of the approximating polynomial. If the discrete points are unequally spaced, the Lagrange polynomial developed in Chapter 8 can be used to fit the data, while if the points are equally spaced the Newton forward difference presented in the next section will be particularly useful. The Lagrange polynomial also works for equally spaced points. If the $N + 1$ points are chosen, the fitting polynomial will be an Nth degree polynomial.

When the function $y(x)$ is known, we have the flexibility of choosing the discrete points. With this flexibility, we can choose points such that the accuracy of the numerical integration can be enhanced.

If the function $y(x)$ is described by a collection of discrete points, then it is probably best to fit a polynomial to these points. The Lagrangian polynomial can be used, which can be fitted exactly. Alternately, the method of least square could be used. For the latter method, the polynomial may not pass through the discrete points. The Lagrangian polynomial, described in Chapter 8, can be used to exactly fit the discrete data for unequally spaced data points. For equally spaced data, the Newton forward difference polynomial will be very useful for the integration procedure. The following section will deal with equally spaced data.

E.2   NEWTON FORWARD DIFFERENCE POLYNOMIAL

Assume that we have a set of equally spaced data at equally spaced points, x_0, $x_1, ..., x_n, x_{n+1}, ...$ and let y_j be the values of y at the point x_j. The forward difference is defined as follows:

$$\Delta y_n = y_{n+1} - y_n \qquad \text{(E.2)}$$

One can apply this forward difference operator to Δy_n to obtain $\Delta^2 y_n$; that is,

$$\Delta^2 y_n = \Delta(\Delta y_n) = \Delta y_{n+1} - \Delta y_n = (y_{n+2} - y_{n+1}) - (y_{n+1} - y_n)$$

$$\Delta^2 y_n = y_{n+2} - 2y_{n+1} + y_n \qquad \text{(E.3)}$$

Similarly, we can proceed to obtain $\Delta^3 y_n$:

$$\Delta^3 y_n = \Delta(\Delta^2 y_n) = \Delta^2 y_{n+1} - \Delta^2 y_n$$

$$= (y_{n+3} - 2y_{n+2} + y_{n+1}) - (y_{n+2} - 2y_{n+1} + y_n)$$

$$\Delta^3 y_n = y_{n+3} - 3y_{n+2} + 3y_{n+1} - y_n$$

The same procedure can be applied to higher order differences.

The Newton forward difference polynomial (Finlayson 1980) is defined as

$$P_n(x) = y_0 + \alpha \Delta y_0 + \frac{\alpha(\alpha - 1)}{2!} \Delta^2 y_0$$

$$+ \cdots + \frac{\alpha(\alpha - 1)(\alpha - 2)\ldots(\alpha - n + 1)}{n!} \Delta^n y_0 + \varepsilon \qquad \text{(E.4)}$$

where

$$\varepsilon = \binom{\alpha}{n+1} h^{n+1} y^{(n+1)}(\xi) \qquad \text{where} \qquad x_0 < \xi < x_n \qquad \textbf{(E.5)}$$

$$\binom{\alpha}{n+1} = \frac{\alpha!}{(n+1)!(\alpha-n-1)!}; \qquad \alpha = \frac{x-x_0}{h}; \qquad h = \Delta x = x_1 - x_0$$

$$\textbf{(E.6)}$$

The Newton forward difference is derived by allowing the nth degree polynomial to pass through the points, y_0 to y_n. Note that when $x = x_1$, the value of $\alpha = 1$; $x = x_2$, $\alpha = 2$; and $x = x_n$, $\alpha = n$.

It is useful at this point to note that the Newton forward difference formula is utilized here for the development of the numerical integration formula, while the Newton backward difference formula was previously used (in Chapter 7) for the integration of ordinary differential equations of the initial value type.

E.3   BASIC INTEGRATION PROCEDURE

Having found the Nth degree Newton forward difference polynomial to approximate the integrand $y(x)$ ($y \approx P_n(x)$), the integral of Eq. E.1 can be readily integrated analytically after the integrand $y(x)$ has been replaced by $P_n(x)$ as

$$I = \int_a^b y(x)\, dx \approx \int_a^b P_n(x)\, dx = h \int_{\alpha_a}^{\alpha_b} P_n(\alpha)\, d\alpha \qquad \textbf{(E.7)}$$

where α_a and α_b are values of α at $x = a$ and $x = b$, respectively. If we substitute the explicit form of the Newton forward difference formula from Eq. E.4 into Eq. E.7 and integrate analytically, we then have the approximate numerical integration of the integral I.

If $x_0 = a$ and $x_1 = b$, the integration of Eq. E.7 will become

$$I = \int_{x_0}^{x_1} y(x)\, dx \approx h \int_0^1 P_n(\alpha)\, d\alpha \qquad \textbf{(E.8)}$$

that is,

$$I \approx h \int_0^1 \left[y_0 + \alpha\, \Delta y_0 + \frac{\alpha(\alpha-1)}{2!} \Delta^2 y_0 + \cdots \right] d\alpha \qquad \textbf{(E.9)}$$

E.3.1   Trapezoid Rule

If two terms are retained in the approximating polynomial of Eq. E.9, the integration will become

$$I \approx h \left[y_0 \alpha + \Delta y_0 \frac{\alpha^2}{2} \right]_0^1 = \frac{h}{2}(y_0 + y_1) \qquad \textbf{(E.10)}$$

where we have defined $\Delta y_0 = y_1 - y_0$.

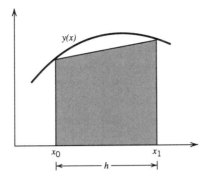

Figure E.1 Graphical representation of the Trapezoid rule integration.

This is commonly called the Trapezoid formula. Figure E.1 shows graphically the Trapezoid rule of integration, where the shaded area is the integration value obtained by the Trapezoid formula.

To find the error arising from the Trapezoid formula, we start with the error of the approximating polynomial (see Eqs. E.4 and E.5) when only two terms are retained as

$$\frac{\alpha(\alpha - 1)}{2} h^2 y''(\xi) \qquad \text{where} \qquad x_0 < \xi < x_1 \qquad \text{(E.11)}$$

Integration of this error with respect to x will give the error for the Trapezoid formula, which is simply

$$\text{Error} = \int_{x_0}^{x_1} \frac{\alpha(\alpha - 1)}{2} h^2 y''(\xi) \, dx = h^3 y''(\xi) \int_0^1 \frac{\alpha(\alpha - 1)}{2} \, d\alpha = -\frac{1}{12} h^3 y''(\xi)$$

$$\text{(E.12)}$$

Thus, the error of integrating the integral from x_0 to x_1 is of order of h^3, denoted as $O(h^3)$. The order notation is discussed in Chapter 6.

We can apply this useful Trapezoid rule over the range from x_0 to x_n by carrying out the Trapezoid formula over n subintervals (x_0, x_1), $(x_1, x_2), \ldots, (x_{n-1}, x_n)$ as shown in Fig. E.2; that is,

$$I = \int_{x_0}^{x_n} y(x) \, dx = \int_{x_0}^{x_1} y(x) \, dx + \int_{x_1}^{x_2} y(x) \, dx + \cdots + \int_{x_{n-1}}^{x_n} y(x) \, dx \quad \text{(E.13)}$$

The integral of each subinterval is then evaluated using the Trapezoid formula, and the result is

$$I \approx \left[\frac{h}{2}(y_0 + y_1) \right] + \left[\frac{h}{2}(y_1 + y_2) \right] + \cdots + \left[\frac{h}{2}(y_{n-1} + y_n) \right]$$

$$\therefore I \approx \frac{h}{2}(y_0 + 2y_1 + 2y_2 + \cdots + 2y_{n-1} + y_n) \qquad \text{(E.14)}$$

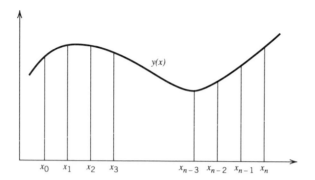

Figure E.2

The error of this composite formula is the sum of all errors contributed by each subintervals given in Eq. E.12

$$\text{Error} = \sum_{j=1}^{n} \left(-\frac{1}{12} \right) h^3 y''(\xi) = -\frac{n}{12} h^3 y''(\xi) \qquad \text{where} \qquad x_0 < \xi < x_n$$

$$(\textbf{E.15})$$

Here, n is the number of intervals and is related to the spacing h as $n = (x_n - x_0)/h$. The error can be written explicitly in terms of the spacing h as

$$\text{Error} = -\frac{1}{12}(x_n - x_0)h^2 y''(\xi) \qquad\qquad (\textbf{E.16})$$

Thus, the global error of the Trapezoid rule formula carried out over n intervals is of order of h^2, while the local error (which is the error of integration of one interval) is of order of h^3.

E.3.2 Simpson's Rule

Let us now integrate the integral I from x_0 to x_2, we then have

$$I = \int_{x_0}^{x_2} y(x)\, dx \qquad\qquad (\textbf{E.17})$$

$$\approx h \int_0^2 \left[y_0 + \alpha\,\Delta y_0 + \frac{\alpha(\alpha - 1)}{2!}\Delta^2 y_0 + \frac{\alpha(\alpha - 1)(\alpha - 2)}{3!}\Delta^3 y_0 + \cdots \right] d\alpha$$

Keeping three terms in the approximating polynomial expansion, the value of this approximating polynomial can then be integrated analytically to give

$$I \approx \frac{h}{3}(y_0 + 4y_1 + y_2) \qquad\qquad (\textbf{E.18})$$

This is Simpson's formula, one of the most widely used integration routines. The error for this formula is the integration of the error of the approximating polynomial, that is,

$$\text{Error} = \int_{x_0}^{x_2} \frac{\alpha(\alpha - 1)(\alpha - 2)}{3!} h^3 y'''(\xi)\, dx$$

$$= h \int_0^2 \frac{\alpha(\alpha - 1)(\alpha - 2)}{3!} h^3 y'''(\xi)\, d\alpha = 0 \qquad \textbf{(E.19)}$$

This null result does not simply mean that the error of the Simpson rule formula is zero, but rather the integration of the third term is zero. We need to go to the fourth term of the approximate polynomial, which is

$$\frac{\alpha(\alpha - 1)(\alpha - 2)(\alpha - 3)}{4!} h^4 y^{(4)}(\xi) \qquad \text{where} \qquad x_0 < \xi < x_2 \quad \textbf{(E.20)}$$

to determine this error, which is simply the integration of Eq. E.20

$$\text{Error} = h \int_0^2 \frac{\alpha(\alpha - 1)(\alpha - 2)(\alpha - 3)}{4!} h^4 y^{(4)}(\xi)\, d\alpha = -\frac{h^5}{90} y^{(4)}(\xi) \quad \textbf{(E.21)}$$

The error for the one step calculation of the Simpson's rule is of order of h^5, a significant advantage over the Trapezoid rule, which has one step error of order of h^3.

Now, we generalize the integration from x_n to x_{n+2} and get

$$I = \int_{x_n}^{x_{n+2}} y(x)\, dx \qquad \textbf{(E.22)}$$

$$\approx h \int_0^2 \left[y_n + \alpha\, \Delta y_n + \frac{\alpha(\alpha - 1)}{2!}\, \Delta^2 y_n + \frac{\alpha(\alpha - 1)(\alpha - 2)}{3!}\, \Delta^3 y_n + \cdots \right] d\alpha$$

where the variable α is now defined as

$$\alpha = \frac{x - x_n}{h} \qquad \textbf{(E.23)}$$

Integration of the RHS of Eq. E.22 term by term would give

$$I \approx \frac{h}{3} (y_n + 4y_{n+1} + y_{n+2}) + O(h^5) \qquad \textbf{(E.24)}$$

Thus, when $n = 0$, we recover the formula in Eq. E.18.

The formula (E.24) is sometimes called the Simpson's 1/3 rule. This was obtained by using the second order polynomial to fit $y(x)$ in Eq. E.17. If a third order polynomial is used (i.e., by keeping four terms in the approximating polynomial expansion, Eq. E.17), we will obtain the Simpson's 3/8 formula (see Hoffman 1992 for more details on this formula).

Simpson's rule can now be extended to the range (x_0, x_n), which is divided into sets of two subintervals. This means that the total number of intervals must be a multiple of two (i.e., even),

$$I = \int_{x_0}^{x_2} y(x)\, dx + \int_{x_2}^{x_4} y(x)\, dx + \cdots + \int_{x_{n-2}}^{x_n} y(x)\, dx \qquad \text{(E.25)}$$

Substituting Eq. E.24 into E.25 (i.e., $n = 0$ for the first integral, $n = 2$ for the second integral, and so on until $n = n - 2$ for the last integral of Eq. E.25) yields

$$I = \frac{h}{3}(y_0 + 4y_1 + 2y_2 + 4y_3 + \cdots + 2y_{n-2} + 4y_{n-1} + y_n) \qquad \text{(E.26)}$$

The global error of Eq. E.26 will be of order of $O(h^4)$.

The Trapezoid and Simpson's rule belong to a family of integration formulas called the *Newton–Cotes family*. Abramowitz and Stegun (1964) provide a family of 10 Newton–Cotes formulas. They also present six additional Newton–Cotes formulas of the open type, that is, the functional values at end points (y_0 and y_n) are not included in the integration formula. These latter formulas are particularly useful when the function values at the end points are unbounded. The first two Newton–Cotes formula of the open type are

$$\int_{x_0}^{x_3} y(x)\, dx = \frac{3h}{2}(y_1 + y_2) + \frac{h^3}{4} y''(\xi)$$

$$\int_{x_0}^{x_4} y(x)\, dx = \frac{4h}{3}(2y_1 - y_2 + 2y_3) + \frac{28h^5}{90} y^{(4)}(\xi)$$

E.4 ERROR CONTROL AND EXTRAPOLATION

Knowledge of the magnitude of the error in the integration formulas presented in Section E.3 is very useful in the estimation of the error as well as in the improvement of the calculated integral.

The integral I numerically calculated using the spacing h is denoted as $I(h)$, and if we denote the exact integral as I_{exact} we then have

$$I_{exact} = I(h) + Ah^n \qquad \text{(E.27)}$$

where h^n is the order of magnitude of the integration formula used.

Applying the same integration formula but this time use the spacing of h/P, where P ($P > 1$) is some arbitrary number (usually 2), then we have the formula

$$I_{exact} = I\left(\frac{h}{P}\right) + A\left(\frac{h}{P}\right)^n \qquad \text{(E.28)}$$

If we equate Eqs. E.27 and E.28 and solve for Ah^n, which is the error of the integration result using the spacing h, we obtain

$$\text{Error} = Ah^n = \frac{P^n}{P^{n-1}}\left[I\left(\frac{h}{P}\right) - I(h)\right] \qquad \text{(E.29)}$$

This formula provides the estimate of error incurred by the approximation using the spacing of h. If this error is larger than the prespecified error, the spacing has to be reduced until the prespecified error is satisfied. When this is the case, the better estimate of the approximate integral is simply Eq. E.27 with Ah^n given by Eq. E.29. This process is called extrapolation.

E.5   GAUSSIAN QUADRATURE

Gaussian quadrature is particularly useful in the case where the integrand $y(x)$ is known; that is, we have the flexibility of choosing the discrete points, x_j (also called the quadrature points). The Gaussian quadrature formula for N quadrature points is of the form

$$I = \int_a^b y(x)\, dx \approx \sum_{j=1}^N w_j y(x_j) \qquad \text{(E.30)}$$

where w_j are called the quadrature weights and x_j are called the quadrature points.

The definition of the Gaussian quadrature formula in Eq. E.30 implies that the determination of this formula is determined by the selection of N quadrature points and N quadrature weights; that is, we have $2N$ parameters to be found. With these degrees of freedom ($2N$ parameters), it is possible to fit a polynomial of degree $2N - 1$. This means that if x_j and w_j are properly chosen, the Gausssian quadrature formula can exactly integrate a polynomial of degree up to $2N - 1$.

To demonstrate this point, consider this example of integration with respect to x from 0 to 1 using two quadrature points

$$I = \int_0^1 y(x)\, dx = \sum_{j=1}^2 w_j y(x_j) \qquad \text{(E.31)}$$

There are four parameters to be found in the above integral, namely two quadrature points x_1, x_2, and two quadrature weights w_1 and w_2. We shall choose these values such that the integrals of the integrands 1, x, x^2, and x^3 are satisfied *exactly*. For the integrand of 1, we have

$$\int_0^1 1\, dx = 1 = w_1 + w_2 \qquad \text{(E.32)}$$

where $y(x_1) = y(x_2) = 1$ have been used in the above equation.

Similarly for the other integrands, x, x^2, and x^3, we have

$$\int_0^1 x \, dx = \frac{1}{2} = w_1 x_1 + w_2 x_2 \tag{E.33}$$

$$\int_0^1 x^2 \, dx = \frac{1}{3} = w_1 x_1^2 + w_2 x_2^2 \tag{E.34}$$

$$\int_0^1 x^3 \, dx = \frac{1}{4} = w_1 x_1^3 + w_2 x_2^3 \tag{E.35}$$

Solving the above set of four equations (Eqs. E.32 to E.35), we obtain these values for the required four parameters

$$x_1 = \frac{1}{2} - \frac{1}{2\sqrt{3}}; \qquad x_2 = \frac{1}{2} + \frac{1}{2\sqrt{3}}; \qquad w_1 = w_2 = \frac{1}{2} \tag{E.36}$$

Note that these quadrature points, x_1 and x_2, lie within the domain of integration.

This set of four parameters for the Gaussian quadrature formula can integrate any polynomial up to degree 3 exactly, for only two quadrature points.

Equation E.31 considers the range of integration as $(0, 1)$. Any range, say (a, b) can be easily transformed to the range $(0, 1)$ by using the simple formula,

$$z = \frac{x - a}{b - a}$$

where x lying in the domain (a, b) is linearly mapped to z, which lies in the domain $(0, 1)$.

Now if we use 3 quadrature points in the Gaussian formula, we then have six parameters in the quadrature equation

$$I = \int_0^1 y(x) \, dx = \sum_{j=1}^{3} w_j y(x_j) \tag{E.37}$$

These six parameters $(x_1, x_2, x_3, w_1, w_2, w_3)$ are found by forcing the above Gaussian quadrature equation to integrate the integrands exactly, 1, x, x^2, x^3, x^4, and x^5; that is

$$\int_0^1 x^j \, dx = \frac{1}{j + 1} = w_1 x_1^j + w_2 x_2^j + w_3 x_3^j \qquad \text{for} \qquad j = 0, 1, 2, 3, 4, 5 \tag{E.38}$$

Solving this set of six equations, we obtain these values for the six parameters

$$x_1 = \frac{1}{2} - \frac{\sqrt{0.6}}{2}; \qquad x_2 = \frac{1}{2}; \qquad x_3 = \frac{1}{2} + \frac{\sqrt{0.6}}{2} \tag{E.39}$$

$$w_1 = \frac{5}{18}; \qquad w_2 = \frac{4}{9}; \qquad w_3 = \frac{5}{18}$$

Again, it is noted that these quadrature points lie within the domain of integration.

Thus, we see that the Gaussian quadrature formula using N quadrature points

$$I = \int_0^1 y(x)\, dx = \sum_{j=1}^N w_j y(x_j) \qquad (\text{E.40})$$

can integrate a polynomial of degree up to $2N - 1$ exactly. Abramowitz and Stegun (1964) tabulate values for x_j and w_j for a number of quadrature points.

We can also apply the Gaussian quadrature formula to the case when we have a weighting function in the integrand as

$$I = \int_0^1 \left[x^\beta (1 - x)^\alpha \right] y(x)\, dx = \sum_{j=1}^N w_j y(x_j) \qquad (\text{E.41})$$

where the weighting function is $W(x) = x^\beta (1 - x)^\alpha$. Using the same procedure

Table E.1   Quadrature Points and Weights for Gaussian Quadrature

α	β	$N = 1$	$N = 2$	$N = 3$
0	0	$x_1 = 0.5$ $w_1 = 1.0$	$x_1 = 0.211325$ $x_2 = 0.788675$ $w_1 = 0.500000$ $w_2 = 0.500000$	$x_1 = 0.112702$ $w_2 = 0.500000$ $x_3 = 0.887298$ $w_1 = 0.277778$ $w_2 = 0.444444$ $w_3 = 0.277778$
0	1	$x_1 = 0.666667$ $w_1 = 0.500000$	$x_1 = 0.355051$ $x_2 = 0.844949$ $w_1 = 0.181959$ $w_2 = 0.318041$	$x_1 = 0.212340$ $x_2 = 0.590533$ $x_3 = 0.911412$ $w_1 = 0.069827$ $w_2 = 0.229241$ $w_3 = 0.200932$
1	0	$x_1 = 0.333333$ $w_1 = 0.500000$	$x_1 = 0.155051$ $x_2 = 0.644949$ $w_1 = 0.318042$ $w_2 = 0.181958$	$x_1 = 0.088588$ $x_2 = 0.409467$ $x_3 = 0.787660$ $w_1 = 0.200932$ $w_2 = 0.229241$ $w_3 = 0.069827$
1	1	$x_1 = 0.500000$ $w_1 = 0.166667$	$x_1 = 0.276393$ $x_2 = 0.723607$ $w_1 = 0.083333$ $w_2 = 0.083333$	$x_1 = 0.172673$ $x_2 = 0.500000$ $x_3 = 0.827327$ $w_1 = 0.038889$ $w_2 = 0.088889$ $w_3 = 0.038889$

Table E.2   Quadrature Points and Weights for Laguerre Quadrature for $N = 3$ and 5

N	x_j	w_j
3	0.41577456	0.71109301
	2.29428036	0.27851773
	6.28994051	0.01038926
5	0.26356032	0.52175561
	1.41340306	0.39866681
	3.59642577	0.07594245
	7.08581001	$0.36117587 \times 10^{-2}$
	12.64080084	$0.23369972 \times 10^{-4}$

we presented above to determine the quadrature points and quadrature weights, we can list in Table E.1 below those values for a number of combinations of α and β.

Integrals of the type

$$\int_0^\infty e^{-x} y(x)\, dx \quad \text{and} \quad \int_{-\infty}^\infty e^{-x^2} y(x)\, dx \qquad (\text{E.42})$$

can also be numerically obtained using the Laguerre and Hermite formulas

$$\int_0^\infty e^{-x} y(x)\, dx = \sum_{j=1}^N w_j y(x_j) \qquad (\text{E.43})$$

and

$$\int_{-\infty}^\infty e^{-x^2} y(x)\, dx = \sum_{j=1}^N w_j y(x_j) \qquad (\text{E.44})$$

respectively. Tables E.2 and E.3 list values of quadrature points and weights for the Laguerre and Hermite formulas (Abramowitz and Stegun 1964).

Table E.3   Quadrature Points and Weights for Hermite Quadrature for $N = 5$ and 9

N	$\pm x_j$	w_j
5	0.00000000	0.94530872
	0.95857246	0.39361932
	2.02018287	0.01995324
9	0.00000000	0.72023521
	0.72355102	0.43265156
	1.46855329	$0.88474527 \times 10^{-1}$
	2.26658058	$0.49436243 \times 10^{-2}$
	3.19099320	$0.39606977 \times 10^{-4}$

E.6 RADAU QUADRATURE

The Gaussian quadrature presented in the previous section involves the quadrature points which are within the domain of integration $(0, 1)$. If one point at the boundary, either at $x = 0$ or $x = 1$, is included in the quadrature formula, the resulting formula is called the Radau quadrature.

Let us start with the inclusion of the boundary value at $x = 1$ in the quadrature formula, and use the following example of one interior quadrature point, i.e., the total number of quadrature points is two, one interior point and the point at $x = 1$.

$$I = \int_0^1 y(x)\, dx = w_1 y(x_1) + w_2 y(x = 1) \tag{E.45}$$

where w_1 and w_2 are quadrature weights at x_1 and $x = 1$, respectively. Thus, there are three unknowns (w_1, w_2, and x_1) in this quadrature equation. To find them, we enforce the previous requirement by integrating three polynomials 1, x, and x^2 *exactly*. This means

$$\int_0^1 1\, dx = 1 = w_1 + w_2 \tag{E.46}$$

$$\int_0^1 x\, dx = \frac{1}{2} = w_1 x_1 + w_2 \tag{E.47}$$

$$\int_0^1 x^2\, dx = \frac{1}{3} = w_1 x_1^2 + w_2 \tag{E.48}$$

Solving the above three equations, we obtain

$$x_1 = \frac{1}{3}; \quad x_2 = 1; \quad w_1 = \frac{3}{4}; \quad w_2 = \frac{1}{4} \tag{E.49}$$

Thus, we see that the Radau quadrature formula using 1 interior quadrature point can integrate a polynomial of degree up to 2 exactly. In general, a Radau quadrature formula using N interior quadrature points and one point at the boundary $x = 1$ can integrate a polynomial of degree up to $2N$ exactly. We list in Table E.4 the values for the quadrature weights and points for the Radau quadrature formula

$$I = \int_0^1 y(x)\, dx = \sum_{j=1}^{N} w_j y(x_j) + w_{N+1} y(x = 1) \tag{E.50}$$

Similarly, when the Radau quadrature formula is used with the boundary point at $x = 0$ included instead of $x = 1$, the Radau quadrature formula is

$$I = \int_0^1 y(x)\, dx = w_0 y(x = 0) + \sum_{j=1}^{N} w_j y(x_j) \tag{E.51}$$

Table E.5 lists the quadrature points and weights for this case.

Table E.4 Quadrature Points and Weights for Radau Quadrature with the Last End Point Included: Weighting Function $W(x) = 1$

N	Quadrature Points	Quadrature Weights
1	$x_1 = 0.333333$	$w_1 = 0.750000$
	$x_2 = 1.000000$	$w_2 = 0.250000$
2	$x_1 = 0.155051$	$w_1 = 0.376403$
	$x_2 = 0.644949$	$w_2 = 0.512486$
	$x_3 = 1.000000$	$w_3 = 0.111111$
3	$x_1 = 0.088588$	$w_1 = 0.220462$
	$x_2 = 0.409467$	$w_2 = 0.388194$
	$x_3 = 0.787660$	$w_3 = 0.328844$
	$x_4 = 1.000000$	$w_4 = 0.062500$

Similar to inclusion of the boundary point at $x = 1$, the quadrature formula for including the boundary point at $x = 0$ (in addition to the N interior quadrature points) can integrate a polynomial of degree up to $2N$ exactly.

Heretofore, we have considered the quadrature formula for the integral of a function $y(x)$ with a weighting function $W(x) = 1$. The Radau quadrature formula for the following integral with the weighting function $W(x) = x^\beta(1-x)^\alpha$

$$I = \int_0^1 \left[x^\beta(1-x)^\alpha \right] y(x)\, dx \tag{E.52}$$

is given by the following two formulas

$$I = \int_0^1 \left[x^\beta(1-x)^\alpha \right] y(x)\, dx = \sum_{j=1}^N w_j y(x_j) + w_{N+1} y(x=1) \tag{E.53}$$

Table E.5 Quadrature Points and Weights for Radau Quadrature with the First End Point Included: Weighting Function $W(x) = 1$

N	Quadrature Points	Quadrature Weights
1	$x_0 = 0.000000$	$w_0 = 0.250000$
	$x_1 = 0.666667$	$w_1 = 0.750000$
2	$x_0 = 0.000000$	$w_0 = 0.111111$
	$x_1 = 0.355051$	$w_1 = 0.512486$
	$x_2 = 0.844949$	$w_2 = 0.376403$
3	$x_0 = 0.000000$	$w_0 = 0.062500$
	$x_1 = 0.212341$	$w_1 = 0.328844$
	$x_2 = 0.590533$	$w_2 = 0.388194$
	$x_3 = 0.911412$	$w_3 = 0.220462$

and

$$I = \int_0^1 \left[x^\beta (1-x)^\alpha \right] y(x)\, dx = w_0 y(x=0) + \sum_{j=1}^N w_j y(x_j) \quad \textbf{(E.54)}$$

When $\alpha = \beta = 0$, we recover the Radau formula obtained earlier.

Using the same procedure of determining quadrature weights and quadrature points described earlier (that is we force the quadrature formula to fit polynomials up to degree $2N$), we will obtain the N interior quadrature points and the $N + 1$ quadrature weights. The other quadrature point will be either at the point $x = 0$ or $x = 1$. Tables E.6 and E.7 list these values for a number of combinations of α and β.

Table E.6   Quadrature Points and Weights for Radau Quadrature with the Last End Point Included: Weighting Function $W(x) = x^\beta (1-x)^\alpha$

α	β	$N = 1$	$N = 2$	$N = 3$
1	0	$x_1 = 0.250000$	$x_1 = 0.122515$	$x_1 = 0.072994$
		$x_2 = 1.000000$	$x_2 = 0.544152$	$x_2 = 0.347004$
		$w_1 = 0.444444$	$x_3 = 1.000000$	$x_3 = 0.705002$
		$w_2 = 0.055556$	$w_1 = 0.265016$	$x_4 = 1.000000$
			$w_2 = 0.221096$	$w_1 = 0.169509$
			$w_3 = 0.013889$	$w_2 = 0.223962$
				$w_3 = 0.101529$
				$w_4 = 0.005000$
0	1	$x_1 = 0.500000$	$x_1 = 0.276393$	$x_1 = 0.172673$
		$x_2 = 1.000000$	$x_2 = 0.723607$	$x_2 = 0.500000$
		$w_1 = 0.333333$	$x_3 = 1.000000$	$x_3 = 0.827327$
		$w_2 = 0.166667$	$w_1 = 0.115164$	$x_4 = 1.000000$
			$w_2 = 0.301503$	$w_1 = 0.047006$
			$w_3 = 0.083333$	$w_2 = 0.177778$
				$w_3 = 0.225217$
				$w_4 = 0.050000$
1	1	$x_1 = 0.400000$	$x_1 = 0.226541$	$x_1 = 0.145590$
		$x_2 = 1.000000$	$x_2 = 0.630602$	$x_2 = 0.433850$
		$w_1 = 0.138889$	$x_3 = 1.000000$	$x_3 = 0.753894$
		$w_2 = 0.027778$	$w_1 = 0.061489$	$x_4 = 1.000000$
			$w_2 = 0.096844$	$w_1 = 0.028912$
			$w_3 = 0.008333$	$w_2 = 0.079829$
				$w_3 = 0.054926$
				$w_4 = 0.003333$

Table E.7   Quadrature Points and Weights for Radau Quadrature with the First End Point Included: Weighting Function $W(x) = x^\beta(1-x)^\alpha$

α	α	$N = 1$	$N = 2$	$N = 3$
1	0	$x_1 = 0.000000$	$x_1 = 0.000000$	$x_1 = 0.000000$
		$x_2 = 0.500000$	$x_2 = 0.276393$	$x_2 = 0.172673$
		$w_1 = 0.166667$	$x_3 = 0.723607$	$x_3 = 0.500000$
		$w_2 = 0.333333$	$w_1 = 0.083333$	$x_4 = 0.827327$
			$w_2 = 0.301503$	$w_1 = 0.050000$
			$w_3 = 0.115164$	$w_2 = 0.225217$
				$w_3 = 0.177778$
				$w_4 = 0.047006$
0	1	$x_1 = 0.000000$	$x_1 = 0.000000$	$x_1 = 0.000000$
		$x_2 = 0.750000$	$x_2 = 0.455848$	$x_2 = 0.294998$
		$w_1 = 0.055556$	$x_3 = 0.877485$	$x_3 = 0.652996$
		$w_2 = 0.444444$	$w_1 = 0.013889$	$x_4 = 0.927006$
			$w_2 = 0.221096$	$w_1 = 0.005000$
			$w_3 = 0.265016$	$w_2 = 0.101528$
				$w_3 = 0.223962$
				$w_4 = 0.169509$
1	1	$x_1 = 0.000000$	$x_1 = 0.000000$	$x_1 = 0.000000$
		$x_2 = 0.600000$	$x_2 = 0.369398$	$x_2 = 0.246106$
		$w_1 = 0.027778$	$x_3 = 0.773459$	$x_3 = 0.566150$
		$w_2 = 0.138889$	$w_1 = 0.008333$	$x_4 = 0.854410$
			$w_2 = 0.096844$	$w_1 = 0.003333$
			$w_3 = 0.061489$	$w_2 = 0.054593$
				$w_3 = 0.079829$
				$w_4 = 0.028912$

E.7   LOBATTO QUADRATURE

The last section illustrated the Radau quadrature formula when one of the quadrature points is on the boundary either at $x = 0$ or $x = 1$. In this section, we present the Lobatto quadrature formula, which includes both boundary points in addition to the N interior quadrature points. The general formula for the Lobatto quadrature is

$$I = \int_0^1 y(x)\, dx = w_0 y(x = 0) + \sum_{j=1}^{N} w_j y(x_j) + w_{N+1} y(x = 1) \quad \textbf{(E.55)}$$

To demonstrate the Lobatto formula, we use the case of one interior point; that is, $N = 1$, and the quadrature formula is

$$I = \int_0^1 y(x)\, dx = w_0 y(x = 0) + w_1 y(x_1) + w_2 y(x = 1) \qquad (\textbf{E.56})$$

where x_1 is the interior quadrature point.

There are four parameters to be found in Eq. E.56, namely w_0, w_1, w_2, and x_1. To find them we stipulate that the formula would integrate four polynomials 1, x, x^2, x^3 exactly; that is,

$$\int_0^1 1\, dx = 1 = w_0 + w_1 + w_2 \qquad (\textbf{E.57}a)$$

$$\int_0^1 x\, dx = \frac{1}{2} = w_0(0) + w_1 x_1 + w_2 \qquad (\textbf{E.57}b)$$

$$\int_0^1 x^2\, dx = \frac{1}{3} = w_0(0) + w_1 x_1^2 + w_2 \qquad (\textbf{E.57}c)$$

$$\int_0^1 x^3\, dx = \frac{1}{4} = w_0(0) + w_1 x_1^3 + w_2 \qquad (\textbf{E.57}d)$$

**Table E.8   Quadrature Points and Weights for Lobatto Quadrature:
Weighting Function $W(x) = 1$**

N	Quadrature points	Quadrature weights
1	$x_0 = 0.000000$	$w_0 = 0.166667$
	$x_1 = 0.500000$	$w_1 = 0.666667$
	$x_2 = 1.000000$	$w_2 = 0.166667$
2	$x_0 = 0.000000$	$w_0 = 0.083333$
	$x_1 = 0.276393$	$w_1 = 0.416667$
	$x_2 = 0.723607$	$w_2 = 0.416667$
	$x_3 = 1.000000$	$w_3 = 0.083333$
3	$x_0 = 0.000000$	$w_0 = 0.050000$
	$x_1 = 0.172673$	$w_1 = 0.272222$
	$x_2 = 0.500000$	$w_2 = 0.355556$
	$x_3 = 0.827327$	$w_3 = 0.272222$
	$x_4 = 1.000000$	$w_4 = 0.050000$
4	$x_0 = 0.000000$	$w_0 = 0.033333$
	$x_1 = 0.117472$	$w_1 = 0.189238$
	$x_2 = 0.357384$	$w_2 = 0.277429$
	$x_3 = 0.642616$	$w_3 = 0.277429$
	$x_4 = 0.882528$	$w_4 = 0.189238$
	$x_5 = 1.000000$	$w_5 = 0.033333$

Table E.9 **Quadrature Points and Weights for Lobatto Quadrature:**
Weighting Function $W(x) = x^\beta(1 - x)^\alpha$

α	β	$N = 1$	$N = 2$	$N = 3$	$N = 4$
1	0	$x_0 = 0.000000$	$x_0 = 0.000000$	$x_0 = 0.000000$	$x_0 = 0.000000$
		$x_1 = 0.400000$	$x_1 = 0.226541$	$x_1 = 0.145590$	$x_1 = 0.101352$
		$x_2 = 1.000000$	$x_2 = 0.630602$	$x_2 = 0.433850$	$x_2 = 0.313255$
		$w_0 = 0.125000$	$x_3 = 1.000000$	$x_3 = 0.753894$	$x_3 = 0.578185$
		$w_1 = 0.347222$	$w_0 = 0.066667$	$x_4 = 1.000000$	$x_4 = 0.825389$
		$w_2 = 0.027778$	$w_1 = 0.271425$	$w_0 = 0.041667$	$x_5 = 1.000000$
			$w_2 = 0.153574$	$w_1 = 0.198585$	$w_0 = 0.028582$
			$w_3 = 0.008333$	$w_2 = 0.184001$	$w_1 = 0.147730$
				$w_3 = 0.072414$	$w_2 = 0.171455$
				$w_4 = 0.003333$	$w_3 = 0.113073$
					$w_4 = 0.037583$
					$w_5 = 0.001587$
0	1	$x_0 = 0.000000$	$x_0 = 0.000000$	$x_0 = 0.000000$	$x_0 = 0.000000$
		$x_1 = 0.600000$	$x_1 = 0.369398$	$x_1 = 0.246106$	$x_1 = 0.174611$
		$x_2 = 1.000000$	$x_2 = 0.773459$	$x_2 = 0.566150$	$x_2 = 0.421815$
		$w_0 = 0.027778$	$x_3 = 1.000000$	$x_3 = 0.854410$	$x_3 = 0.686745$
		$w_1 = 0.347222$	$w_0 = 0.008333$	$x_4 = 1.000000$	$x_4 = 0.898648$
		$w_2 = 0.125000$	$w_1 = 0.153574$	$w_0 = 0.003333$	$x_5 = 1.000000$
			$w_2 = 0.271425$	$w_1 = 0.072414$	$w_0 = 0.001587$
			$w_3 = 0.066667$	$w_2 = 0.184001$	$w_1 = 0.037582$
				$w_3 = 0.198585$	$w_2 = 0.113073$
				$w_4 = 0.041667$	$w_3 = 0.171455$
					$w_4 = 0.147730$
					$w_5 = 0.028571$
1	1	$x_0 = 0.000000$	$x_0 = 0.000000$	$x_0 = 0.000000$	$x_0 = 0.000000$
		$x_1 = 0.500000$	$x_1 = 0.311018$	$x_1 = 0.211325$	$x_1 = 0.152627$
		$x_2 = 1.000000$	$x_2 = 0.688982$	$x_2 = 0.500000$	$x_2 = 0.374719$
		$w_0 = 0.016667$	$x_3 = 1.000000$	$x_3 = 0.788675$	$x_3 = 0.625281$
		$w_1 = 0.133333$	$w_0 = 0.005556$	$x_4 = 1.000000$	$x_4 = 0.847373$
		$w_2 = 0.016667$	$w_1 = 0.077778$	$w_0 = 0.002380$	$x_5 = 1.000000$
			$w_2 = 0.077778$	$w_1 = 0.042857$	$w_0 = 0.001191$
			$w_3 = 0.005556$	$w_2 = 0.076190$	$w_1 = 0.024576$
				$w_3 = 0.042857$	$w_2 = 0.057567$
				$w_4 = 0.002381$	$w_3 = 0.057567$
					$w_4 = 0.024576$
					$w_5 = 0.001190$

Solving these four equations, we obtain the four parameters as

$$x_1 = 0.5; \quad w_0 = \frac{1}{6}; \quad w_1 = \frac{4}{6}; \quad w_2 = \frac{1}{6} \qquad \textbf{(E.58)}$$

Table E.8 lists values of quadrature points and weights for the Lobatto quadrature formula.

For the general integral of the type

$$I = \int_0^1 \left[x^\beta (1 - x)^\alpha \right] y(x)\, dx \qquad \textbf{(E.59)}$$

with the weighting function $W(x) = x^\beta (1 - x)^\alpha$, the Lobatto quadrature equation is written as

$$I = \int_0^1 \left[x^\beta (1 - x)^\alpha \right] y(x)\, dx = w_0 y(x = 0) + \sum_{j=1}^{N} w_j y(x_j) + w_{N+1} y(x = 1)$$

$$\textbf{(E.60)}$$

Table E.9 lists values for the quadrature points and weights for a number of combinations of α and β.

E.8 CONCLUDING REMARKS

This appendix has illustrated the most valuable integration formulas for engineering applications. The most popular formulas are the Trapezoid and Simpson rules, and they are recommended for initial application of the integration formula since they are so simple to use. Error control and extrapolation could then be used to improve the estimation of the integral. When the integrand has unbounded values at the boundaries, the Newton–Cotes formula of the open type will prove to be a suitable choice. The Gaussian, Laguerre, Hermite, Radau, and Lobatto quadrature formulas are particularly useful when the function $y(x)$ is known, since the summation process is so simple and easily programmed. They are also attractive in cases requiring the utilization of the collocation method, explained in Chapter 8.

E.9 REFERENCES

1. Abramowitz, M., and I.A. Stegun, "Handbook of Mathematical Functions," National Bureau of Standards Applied Mathematics Series (1964).
2. Finlayson, B., "Nonlinear Analysis in Chemical Engineering," McGraw Hill, New York (1980).
3. Hoffman, J.D., "Numerical Methods for Engineers and Scientists," McGraw Hill, New York (1992).

Appendix **F**

Nomenclature

a_n	coefficient used in the Frobenius expansion, Eq. 3.32
a_{ij}	component of the matrix $\mathbf{A}$
A	cross sectional area (Chapter 1)
$\mathbf{A}$	first derivative matrix, Eq. 8.102
$Ai(x)$	Airy function, Eq. 3.75
b_n	coefficient used in the Frobenius expansion
b_{ij}	component of the matrix $\mathbf{B}$
B	parameter, Eq. 7.36
$\mathbf{B}$	second derivative matrix, Eq. 8.102
B_1	residue, Eq. 9.76
B_N	coefficient in the Laurent expansion, Eq. 9.75
Bi	Biot number for heat transfer, Eq. 1.81b
$Bi(x)$	Airy function, Eq. 3.76
c	variable index, used in Frobenius expansion, Eq. 3.32
$c_N^{(\alpha,\beta)}$	coefficient, defined in Eq. 8.110
C	fluid concentration
C^*	fluid concentration in equilibrium with the adsorbed phase
$C_{A,B}$	fluid concentration
Ci	cosine integral, Eq. 4.52
C_p	specific heat
d	inner pipe diameter, used in Example 2.25
D	diffusion coefficient
D_e	effective diffusivity in porous particle, Eq. 8.21
D_E	diffusion coefficient along the bed length, Eq. 2.215
E	activation energy
Ei	exponential integral, Eq. 4.50
erf	error function
f	functional
$f(x)$	forcing function in the RHS of ODEs
F	volumetric flow rate, Eq. 7.36
F	functional, for example, Eq. 7.46
$F(s)$	Laplace transform of $f(t)$, Eq. 9.1
g_j	coefficient, defined in Eq. 8.88

G_{N-k}	Function defined in Eq. 8.86a
h	step size used in integration methods of Chapter 7
h	heat transfer coefficient
h_j	coefficient, defined in Eq. 8.88
h_G	heat transfer coefficient of gas phase, Eq. 1.52
h_L	heat transfer coefficient of liquid phase, Eq. 1.52
Ha	Hatta number, Eq. 2.242
i	counting index
i	imaginary number $= \sqrt{-1}$
$I(x)$	integrating factor, used in Chapter 2
I_n	Bessel function
J	flux $=$ mass transfer per unit area
J	Jacobian matrix, for example, Eq. 8.160d
J_n	Bessel function
$J_N^{(\alpha,\beta)}$	Jacobi polynomial, Eq. 8.78, Eq. 8.79
k	thermal conductivity
$k_{1,2}$	reaction rate constant, used in Example 2.5
k_c	mass transfer coefficient per unit interfacial area, Eq. 1.33
$k_c a$	mass transfer coefficient per unit bed volume, Eq. 1.33
K	affinity constant of adsorption equilibrium, defined in Eq. 1.32
K	equilibrium constant between two phases, Eq. 5.3
K_n	Bessel function, Eq. 3.162
$K_n(x)$	kernel of the integral transform, used in Chapter 11
$l_j(x)$	Lagrangian interpolation polynomial, Eq. 8.90
L	length
L	heavy phase flow rate, Eq. 5.4
L	differential operator, Eq. 8.1
m	parameter, Eq. 1.61
n	parameter, Eq. 1.73
Nu	Nusselt number for heat transfer, defined in Problem 1.1
p	dy/dx, defined in Eq. 2.87
$p_N(x)$	rescaled polynomial of the Jacobi polynomial of degree N, Eq. 8.86b
$p_{N+1}(x)$	node polynomial, Eq. 8.91
$P(x)$	function of x, used in Chapter 2
P_j	coefficient, Eq. 3.30
Pe	Peclet number, Eq. 2.242
Pr	Prandt number, defined in Problem 1.1
q	concentration in the adsorbed phase
q	heat flux $=$ heat transfer per unit area
q_r	heat flux in the r direction
q_z	heat flux in the z direction
Q	heat transfer (Chapter 1)
$Q(x)$	function of x, used in Chapter 2
Q_j	coefficient, Eq. 3.31
r	radial coordinate
R	residual, Eq. 8.3
R_b	boundary residual, Eq. 8.4

R	tube radius or particle radius
R	chemical reaction rate
$R(x)$	function of x, used in Chapter 2
Re	Reynolds number
s	complex variable, Eq. 9.2
S	cross-sectional area of a tube, used in Example 7.3
Si	sine integral, Eq. 4.51
t	time
t^*	inner time variable, Eq. 6.98
T	temperature
T_w	wall temperature (Chapter 1)
T_0	inlet temperature (Chapter 1)
x	independent variable
x^*	inner variable, Eq. 6.62
x_j	jth interpolation point used in Chapter 8
X	mass solute/mass carrier solvent, Eq. 5.2
X	reaction conversion, Eq. 7.38
$X(\tau)$	penetration front, used in Chapter 12
y	dependent variable
y_a	approximate dependent variable, Eq. 8.5
y_j	value of y at the interpolation point x_j, used in Chapter 8
$y_j(x)$	coefficient in the asymptotic expansion, Eq. 6.5
$y_j^{(i)}(x)$	coefficient in the inner asymptotic expansion, Eq. 6.64
$y_j^{(0)}(x)$	coefficient in the outer asymptotic expansion, Eq. 6.59
Y	mass solute/mass extracting solvent, Eq. 5.1
u	independent variable, $= x^2$, used in Chapter 8 for symmetry, for example, Eq. 8.117
u	nondimensional temperature, Eq. 1.81a
u	superficial velocity, for example, used in Example 7.3
U	overall heat transfer coefficient in heat exchanger, used in Example 2.25
v	y/x, used in Chapter 2
v_0	mean velocity (Chapter 1)
v_z	parabolic velocity, defined in Chapter 1, Eq. 1.21
V	nondimensional variable, defined in Eq. 2.73
V	light solvent mass flow rate, Eq. 5.4
V	reservoir volume used in Example 7.1
V_p	particle volume, Eq. 8.28
w_j	quadrature weight, Eq. 8.107
$w_k(x)$	test function, Eq. 8.7
W	Wronskian determinant, defined in Eq. 2.359
W_A	mass transfer rate of the species A, Eq. 2.120
x	coordinate
z	axial coordinate
α	independent variable, Eq. 7.79, Eq. 7.81
α	parameter, defined in Eq. 2.225
α	nondimensional parameter, Eq. 1.54

α	exponent in the weighting function of the Jacobi polynomial, Eq. 8.83
β	exponent in the weighting function of the Jacobi polynomial, Eq. 8.83
β	parameter, defined in Eq. 2.225
β	parameter, defined in Eq. 5.6
β_n	eigenvalue, Eq. 1.86b, used in Chapter 10, for example, Eq. 10.185
δ	nondimensional parameter, Eq. 1.81b
$\delta(x)$	Dirac delta function
δ_n	asymptotic sequence, Eq. 6.37
ε	bed porosity, Eq. 1.34
ε	particle porosity
ε	small parameter used in Chapter 6 for perturbation analysis
ε	error, Eq. 7.21
$\gamma_{N,i}$	coefficient of the Jacobi polynomial of degree N
γ	Euler constant, Eq. 3.153
ρ	density
λ	parameter, defined in Chapter 1, Eq. 1.9
λ	decay constant used in Chapter 7 for ODEs, Eq. 7.20a
$\lambda_{i,0}$	parameter, Eq. 2.387 and Eq. 2.388
μ	viscosity
θ	temperature, defined in Eq. 1.11
ψ	nondimensional parameter, defined in Eq. 1.18
ζ	nondimensional parameter, defined in Equation 1.19
λ	eigenvalue, used in coated wall reactor, Eq. 10.182, or in Example 10.3, Example 10.4, Example 10.5
ζ	nondimensional length, Eq. 1.81a
η	effectiveness factor
η	combined variable used in the combination of variables method, Eq. 10.15
η	parameter, defined in Eq. 1.63
ξ	nondimensional coordinate, Eq. 1.81a
ξ_n	eigenvalue used in Chapter 11
ϕ	Thiele modulus, Eq. 8.24
$\phi_i(x)$	trial functions, Eq. 8.5
φ	exact function, Eq. 2.23
φ	polar angle of a complex number
$\varphi(k)$	function, Eq. 3.112
$\varphi_n(x)$	orthogonal polynomial used in Chapter 10, for example, Eq. 10.184
σ	real part of a complex number
ω	imaginary part of a complex number
Ω	vector defined in Eq. 11.187

Postface

After finishing this text, the reader has become aware of the importance of a solid foundation in algebra. Such is the final state in solving problems, since both ODE and PDE solutions finally end in algebraic terms. Knowledge of this basic body is the platform to build on higher levels. New developments will also have their underpinnings in classical mathematics. Of these, the reader can see the need for constructing methods of coping with nonlinearities. Toward this end, there is evolving a body of applied mathematics which constructs nonlinear operators to solve differential equations as handily as Heaviside and Laplace operators.

Foremost among linear problems is the analysis of coupled sets of PDE, with obvious applications in systems and control engineering. Often the Laplace transform can be used for resolving such complexities, through Laurent's expansion and residue theory. Residue theory is, as the reader can verify, more a problem of pedagogy than practice; once learned, it is easy to apply. What is the proper way to teach such abstract ideas as the "complex domain"? How can communication barriers be overcome? On what philosophical basis should higher mathematics be taught? Many teach by the deductive mode, that is, moving from the general case to a specific one. The engineer more often operates in the inductive mode, by studying many individual cases from which a general principle evolves. Here, the dichotomy into two basic teaching styles daily confronts students. Each has attractive features, but it would appear in practice that the inductive style is best for undergraduates, whereas the deductive mode appeals to graduate level students. By trial and error, teachers learn the optimum for time and place. Elementary feedback control is the operative mechanism. Linear methods work at the introductory level. Learning to cope well with scalar problems must necessarily precede vector arrays.

Toward this end, we have attempted to reduce exposition using vector terminology. Occasionally, this was not possible, so an appendix was developed to provide a review of elementary principles for vector-matrices operation. Linear operators in Hilbert space was briefly mentioned, mainly as a means of coping with the case of infinite eigenfunctions. Last, but not least, a careful

698

treatment of modern methods for numerical solutions for PDE was incorporated.

Several generations have been nurtured on the classic text *Transport Phenomena*. By and large, the original version by Bird, Stewart, and Lightfoot has survived the test of time. You, the reader, will ultimately determine the usefulness of the present exposition of applicable mathematics. Every effort was made to eliminate errors in the text, but some may survive, so please inform the authors if errors are discovered.

Index

A

Absorption:
  gas–liquid mass transfer, 32, 471
  with chemical reaction, 34, 53, 181, 471
Activity, catalyst decay, 220
Adams–Bashforth method, 251
Adams–Moulton method, 253
Addition of matrices, 648
Addition of vectors, 648
Adsorption, packed bed, 10, 444
  batch, 521
Airy equation, 113
  function, 115
Aitkens Δ^2 method, 642
Algebra, complex, 332
Algebraic equations:
  linear, matrix method solution, 658
  numerical solution methods, 630
Ammonia–water distillation, 182
Amplitude of complex number, 332
Analytic functions (of a complex variable), 337
Argument of complex number, 332
Associated eigenproblem, 499
Associative law of algebra, 78
Asymptotic expansions, 191, 192

Asymptotic sequence, 191, 192
Augmented matrix, 652
Axial dispersion, packed bed, 68

B

Batch reactor, 46
Benzoic acid, dissolution of, 29
Bernoulli equation, 45
Bessel functions:
  first kind, 128
  modified, 130
  properties, 135
  tables of values, 136, 137
Bessel's equation, 127
  generalized, 131
  modified, 130
Beta function, 152
  relation to Gamma function, 152
Binomial power series, 80, 108
Binomial theorem, 80
Biot number, 26
Bisection algorithm, 630
Blood dialysis, modeling, 460
Boundary conditions:
  homogeneous type, 14
  in Laplace transforms, 376
  in numerical solutions, 229
  in Sturm–Liouville equation, 499
  time dependent, 513
Boundary value problem:
  collocation methods for, 268
  nonlinear, 304

Branch cut, 380
Branch points, 379
Brinkman correction, 147
Bromwich path, 350, 378
Bubble coalesence, 56
Bubble column, mass transfer models, 32, 101
Bubble rise, 102

C

Capillary tube diffusometer, 482, 483
Catalyst pellet, 139, 145, 212, 220
Catalytic reactor, modeling, 34
Cauchy:
  integral formula (second integral theorem), 391
  theorem, 341, 343
Cauchy–Riemann conditions, 337
Chain rule, total differential, 38
Characteristic equation, 64, 90
Chebyshev polynomial, 128
Classification, of ODE, 39
  of PDE, 402
Coefficients, linear, first order ODE, 49
Collocation methods, 271, 277
  orthogonal collocation methods, 284, 290
Column matrix, 645
Combination of variables, 400, 405, 409

Commutative law of algebra, 78
Complementary solution, 61
Complex conjugate, 333
Complex number:
  amplitude or modulus of, 332
  argument of, 332
  imaginary part, 332
  real part, 332
  trigonometrical, exponential identities, 334
Complex variables:
  analytic functions of, 337
  Cauchy's integral formula, 391
  Cauchy's theorem, 341, 343
  derivatives of, 337
  evaluation of residues, 347, 349
  integration of, 341
  Laurent's expansion, 346
  multivalued functions of, 335, 340
  singularities of, 338
    branch points, 379
    essential, 340
    poles, 345
  theory of residues, 345
Composite solutions, 198
Conformable matrices, 648
Consecutive reactions, 98
Conservation law, general, 5
Contour integration, inversion of Laplace by 331
Contour plots, 342, 344
Convergence:
  acceleration of, 589, 642
  radius of, 107
  rate of, 639
Convergent series, 107
Convolution for Laplace, 366
Cosine integral, 156
Countercurrent extraction, 165
Crank–Nicolson equation, 582
Crank–Nicolson method, 584
CSTR, definition, 165
Cubic equation, asymptotic expansion, 212
CVD reactor, 415

D
Danckwerts conditions, 70
Darcy's law, 147
Demoivre's theorem, 335
Derivative:
  partial, 397
  substitution method, 52
Difference(s):
  backward, 247, 581
  forward, 247, 677
Difference equations:
  characteristic equation, 167
  differential, solution by numerical methods, 247
  finite, 166
    degree and order, 166
  linear finite,
    complementary solution, 167
    particular solution, 172, 175
  nonlinear finite
    analytical, 177
    Riccati, 176
Difference formula, 247
Difference operator, 247
Differential equations, ordinary:
  Airy's equation, 113
  Bessel's equation, 127
  Chebychev's equation, 128
  complementary solution of, 61
  Duffing equation, 52
  Frobenius solution method, 108
  Kummer's equation, 122, 127
  Lane–Emden equation, 52
  matrix representation of, 89, 661
  nonlinear, 38, 43, 45, 46, 50, 51
  order and degree, 50
  particular solution of, 50, 62–63, 72
  solution by Laplace, 368
  solution by matrix methods, 661
  Van der Pol equation, 52

Differential equations, partial:
  derivation of, 9
  particular solutions by combination of variables, 399, 409
  solution by Laplace transforms, 443
  solution by separation of variables, 406, 420
  solution by numerical methods, 546
  solution by Sturm–Liouville transforms, 487
  superposition of solutions, 425
Differential equations, simultaneous, 89
Differential operator, 77
Differential properties of Bessel functions, 187
Dirac Delta function, 362, 383
Dirichlet boundary conditions, 296, 304
Distillation column, 169, 177
Distributive law of algebra, 77, 648
Draining tank, 97
Duffing equation, 52

E
Effectiveness factor, 21, 139, 141
Efficiency, Murphree plate, 181
Eigenfunctions, 433, 490
Eigenproblem, 490, 660
Eigenvalues, 425, 490
Elimination of independent variable, 90–91
Elimination of dependent variable, 89–90
Elliptic integrals, 152
  first kind, 154
  second kind, 154
Error function, 148
  complementary, 150
  properties of, 149
  table of values, 150
Errors:
  global, 251
  local, 250

Euler formula, complex variables, 67, 335
Exactness, 39, 41
Exponential function:
  identities for, 334
  series representation, 104
  transform of, 335
Exponential integral, 156
Exponential power series, 104
Extraction, liquid–solid, 408, 477
Extraction cascade, 165

F
Factor, integrating, 39
Falling film, for gas absorption, 471
Fin, temperature profile, 142
Finite differences:
  operators, 247
  solving differential equations, 231, 572
  staged equations, 165
First Bromwich path, complex variables, 350
Fixed bed catalytic reactor, 34, 68
Fluid flow:
  packed bed, 147
  transient start–up for tube flow, 465
Fourier–Mellon integral, 331, 663
Fourier series approximation, numerical inversion, Laplace, 388
Fourier's law, 10
Frobenius method, 108
Functions, orthogonal, 426, 496
Functions and definite integrals:
  Beta, 152
  Elliptic, 154
  error, 148
  Gamma, 150
Fundamental lemma of calculus, 6

G
Gamma function, 150
  in Laplace transforms, 335
  relation to Beta function, 152

Gamma function, incomplete, 151
Gas absorption:
  in bubble columns, 32, 72
  in falling film, 471
  in packed columns, 10
  with reaction, 34, 53, 181
Gaussian quadrature, 683
Gauss–Jacobi quadrature formula, 293
Gauss–Jordan elimination, 656
Generalized Bessel equation, 131
Generalized Sturm–Liouville transform, 521
Graetz equation (heat transfer), 143, 432

H
Hankel transform, 495
Heat conduction equation:
  boundary conditions, 25
  cylindrical coordinates, 25, 501
  spherical coodinates, 455
  steady state, 440
  unsteady state, 440, 440
Heat conduction solutions:
  Graetz problem, 432
  Nusselt problem, 450
  unsteady linear, 435, 455
Heat exchanger, double–pipe, 91, 472, 473
Heat loss:
  from fins, 142, 144, 146
  from packed beds, 480
  from pin promoters, 132
  from rod bundles, 18
  from rod promoters, 99
Heat transfer coefficient, 19
Heaviside operator, 77
Hermite formulas, 686
Hierarchy of models, 17
Hilbert space, 492
Homogeneous equations, 13, 61, 403
Homogeneous functions, 43, 58
Hyperbolic function, 69

I
Image (Sturm–Liouville transform), 491

Indicial equation, 109
Infinite series, 106
Inhomogeneous boundary conditions, 440, 504
Inhomogeneous equations (PDE), 434, 511
Initial value, problems, numerical solution of, 225, 227
Inner product, 270, 493
Inner variable, 567
Integral properties of Bessel function, 138
Integrals, line, 350
Integration, numerical procedure, 676
Integrating factor, 39
Interpolation formula, 246, 247
  Lagrange, 289
Inverse operator, 72, 77, 79
Inverse transform:
  by contour integration for Laplace, 331, 350
  by convolution integral for Laplace, 366
  for Heaviside operators, 77
  by partial fractions for Laplace, 363
  by Sturm–Liouville transform, 498, 534
Inversion theorem, Laplace, 331, 350

J
Jacobian matrix, 245, 637
Jacobi polynomials, 128, 143, 285, 286
Jacobi's equation, 128, 143

K
Kernel, Sturm–Liouville transform, 495
Kronecker delta function, 582
Kummer's equation, 122, 127

L
Lagrange interpolation formula, 289
Lagrange polynomials, 314
Laguerre polynomial, 127
Lane–Emden equation, 52
Laplace operator, 354

Laplace transform, 350, 354
  convolution theorem for, 366
  differentiation of, 358
  inverse transforms,
      solution of ODE, 368
      using partial fractions, 363
Laplace transforms:
  of derivatives, 357, 360
  of integral functions, 358
  method of solving ODE, 368
  method of solving PDE, 443
  properties of, 356
  shifting theorem for, 360
  step functions, 361
      unit impulse, 361
      ramp function, 361
  table of, 671
Laurent's expansion, 346
Least square method, 189
Legendre's equation, 128, 144
Legendre's polynomial, 128
Leibnitz, rule for
      differentiating an integral, 148
Lerch's theorem, Laplace
      transforms, 356
Leveque solution, 417, 459
L'Hopital's rule, 348, 363
Linear coefficients first order
      ODE, 49
Linearly independent,
      definition for ODE, 62
Lobatto quadrature, 690
Log–mean ΔT, 96
Loschmidt diffusion cell, 459
Lumped parameters, 6

M
Mass transfer:
  condition, 408
  diffusion equation, 477
  integro-differential
      boundary
  by molecular diffusion
      (Fick's law), 16
  unsteady, to spheres, 455
Matched asymptotic
      expansion, 195, 202
Matching, 197, 202
Matrices:
  addition, 648

application to sets of ODE, 661
  augmented, 652
  characteristic equations, 661
  commutable, 648
  conformable, 648
  decomposition, 649
  diagonally dominate, 647
  diagonals, 646
  elimination methods, 650
  inverse, 649
  Jacobian, 637
  matrix algebra, 647
  multiplication, 648
  operator, 527
  pivoting, 654
  solution of linear algebraic
      equations, 658
  sparse, 647
  square, 646
  subtraction, 648
  symmetric, 647
  transpose, 647
  triangular, 646
  tridiagonal, 646
Mellin–Fourier theorem, 331,
      350, 663
Membrane transport, 216, 327,
      468
Moment method, 358
Multivalued function, 380

N
Newton, method of solving
      algebraic equations, 635
Newton backward
      interpolation formula, 247
Newton forward difference
      polynomial, 677
Numbers:
  absolute value of, 337
  Biot, 26
  complex, 332
  conjugate, 333
  imaginary, 332
  Nusselt, 29
Numerical methods:
  algebraic equations, 630
      bisection method, 630
      Newton–Raphson
      method, 635
      secant method, 637

successive substitution, 632
  Crank–Nicolson method, 253
  derivative boundary
      conditions, 229
  Euler method:
      explicit, 231, 250
      implicit, 232, 252
  first order, ODE, 226, 249
  higher order ODE:
      boundary value problems, 268
      initial value problems, 228
  Newton–Raphson method, 635
  partial differential equations
      collocation method, 592
      finite difference, 572
      finite element, 603
      polynomial
      approximation, 546
Nusselt problem, heat transfer
      in tube flow, 450

O
One point collocation method,
      277, 309
Operator:
  boundary, 499
  differential, 263
  finite difference, 247
  general, 61, 499
  Laplace, 331
Ordering concept, gauge
      function, 189
Order symbols, 190
Orthogonal functions, 426,
      493, 498
Orthogonality property, 426,
      493, 498

P
Parameter estimation in
      Laplace domain, 358
Partial differential equations:
  boundary conditions, 406
  combination of variables,
      399, 409
  formulation, 8
  Fourier transform
      (Sturm–Liouville
      transform), 495
  inhomogeneous, 434

Partial differential equations:
(*Cont.*)
  initial value type:
    Laplace transform, 444,
      450
    Sturm–Liouville
      transform, 487
  numerical methods, 546
  orthogonal function, 426
    Sturm–Liouville
      equation, 426
  particular solution, 405
  separation of variables, 420
    coated wall reactor, 421
    steady-state heat transfer,
      440, 450
    superposition of solutions,
      425
    unsteady heat transfer,
      409, 413, 440, 455
Partial differentiation, 397
  changing independent
    variables, 399
Partial fractions for Laplace
  transforms, 363
Particular solution, 62–63, 72,
  405
Pendulum equation, 52
Perturbation method, 184, 562
Piston, dynamic of movement,
  31, 394
Poles:
  in complex variables, 347
  first order, 347
  inversion of Laplace
    transforms, 352
  second order, 349
Polynomial approximation, 546
Polynomials:
  Graetz, 143, 432
  Jacobi, 128, 143, 285, 287
  Lagrange, 314
  Laguerre, 127
  Legendre, 128
Power Series, 106
Plug flow model, 4
Predictor-corrector method,
  253
Propagation of errors, 679

Q
Quadrature, definition, 293
  formulas, Gauss–Jacobi, 293

R
Radau–Lobatto quadrature
  formula, 295
Radau quadrature, 687
Radius of convergence, 107
Range of convergence, 107
Ratio test, 107
Reactor batch, 46
Reactor coated wall, 421
Reactors, tanks in series, 179
  transient response, 395
Reactors, tubular:
  coated wall, 421
  packed bed, 34, 68
Recurrence relation, 109, 111
  for Bessel functions, 138
Regular behavior, 142, 338
Regular perturbation, 188
Relative volatility, 177
Residual, weighted residuals,
  268
Residues:
  evaluation of, 347
  evaluation for multiple
    poles, 349
  theory of, 345
Reynolds number, 8
Riccati difference equation,
  176
Riccati ODE, 45
Robin boundary condition, 301
Rodrigues formula, 144, 285
Row matrix, 645
Runge–Kutta formula, 255,
  256
Runge–Kutta–Gill formula,
  256
Runge–Kutta method, 253

S
Self-adjoint operator, 500, 517,
  532
Separation constant, 423
Series:
  Bessel's equation, 127
    modified, 130
    properties, 135
  convergent, 106
  indicial equation, 109
  infinite,
    properties of, 106

power, 104, 106
  solution by Frobenius
    method, 108
Shifting theorem, 360
Shooting method, 267
Sign conventions, 13
Simpson's rule for integration,
  680
Simultaneous ODE, 89
Sine integral, 156
Singularities:
  branch points, 379
  of complex variables, 338
  essential, 340
  pole type, 340
Singular perturbation solution,
  188, 562
Singular solutions, ODE, 51,
  60
Solvent extraction cascade, 165
Spheres:
  adsorption onto, 462
  dissolution of gas within
    bubble, 102
Stability of numerical
  methods, 232
Step function:
  unit impulse, 361
  unit step, 361
Stiffness, numerical methods,
  243
Stirling formula, 162
Strained coordinates, 218
Sturm–Liouville equation,
  426, 495
Sturm–Liouville transforms,
  487
Subdomain method, 188
Successive substitution
  method, 632

T
Tank draining, 97
Tanks in Series, 179
Taylor series expansion, 142
Temperature, surface
  variation, 415
Temperature distribution:
  in semi-infinite slab, 409,
    413
  in circular fins, 142
  in flowing fluid, 48, 431, 457

Test function, 270
Thermocouple, model of, 30, 31
Total differential, 38
Transformation of matrices, 647
Transforms:
  Fourier, 495
  Hankel, 495
  Laplace, 331
  Laplace, table of, 671
  Sturm–Liouville, 486
Transpose of a matrix, 647
Trapezoid rule for numerical integration, 232, 253, 678
Trial functions, 270

Trigonometric functions:
  inverse operator on, 78
  Laplace transforms of, 356

U
Undetermined coefficients, method for ODE, 85
Unsteady state operations, 11, 368, 408, 435, 444

V
Variables:
  complex, 332
  dependent, 90
  dummy, 148

independent:
  changing, 90
  combination of, 399, 409
  method of separation, 420
Variation of parameters, 85
Vectors, 644
Velocity profile, tube flow, 8, 465
Vibration of drumhead, 463
Volterra integral, 392

W
Weighted residuals, method of, 268
Wronskian, definition, 87

Z
Zakian method, inverting Laplace transforms, 383